STUDIES IN GEOLOGY NO. 6

Contributions to

The Geologic Time Scale

*Papers given at the Geological Time Scale Symposium 106.6
25th International Geological Congress
Sydney, Australia, August 1976*

Edited by
George V. Cohee, Martin F. Glaessner,
and Hollis D. Hedberg

Published by
The American Association of Petroleum Geologists
Tulsa, Oklahoma, U.S.A., 1978

Published July 1978
Library of Congress Catalog Card No. 78-60977
ISBN: 0-89181-010-2

The AAPG staff responsible:

Ronald Hart, Project Coordinator
Connie Teer, Production
Nancy Wise, Production

Printed by Edward Brothers, Inc.
Ann Arbor, Michigan, U.S.A.

Contributions to

The Geologic Time Scale

Contents

Introduction

During the Montreal meetings of the International Subcommission on Stratigraphic Classification (ISSC) held in August, 1972, Dr. George V. Cohee presented a proposal, on behalf of the Geologic Names Committee of the U.S. Geological Survey, that ISSC sponsor a symposium on the international geochronologic scale.

After discussion in Montreal, this proposal was circulated for comments and discussion by the chairman of the ISSC to the entire membership of the Subcommission through the medium of ISSC circulars. A vote was called for in ISSC Circular-45 of October 1973, with the result that the members almost unanimously favored going ahead with plans for such a symposium to be held in conjunction with the 25th International Geological Congress in Australia in 1976. Furthermore, it was determined that Dr. Cohee and Professor Martin Glaessner of the University of Adelaide should be appointed as co-chairmen of a committee to plan, organize, and conduct the symposium. These men selected, as the other members of the committee: R. E. Folinsbee, W. B. Harland, H. D. Hedberg, N. deB. Hornibrook, E. Jäger, I. McDougall, S. S. Oriel, Z. E. Peterman, H. H. Renz, and A. I. Tougarinov.

It was decided that the symposium should consist of four sessions. The first would include general papers on methods of dating, geochronologic scales, stratotypes, biochronology, and the magnetic-polarity time scale. Session two would be organized by the Subcommission on Geochronology, directed by its chairman Emily Jäger, and would deal with the physical time scale and related geochronological problems. The third and fourth sessions would consist of reports on dating and problems needing resolution within each individual geologic system, presented largely by members of the International Subcommissions on these systems within the IUGS Commission on Stratigraphy.

Through the efforts of the co-chairmen (Dr. Cohee and Prof. Glaessner), and of the organizer of the 2nd session (Professor Jäger), an outstanding program of 36 pertinent papers was assembled and successfully presented at the 25th International Geological Congress at Sydney in four half-day sessions (August 18, 19, 20, and 23, 1976). Unfortunately, the man who initiated the proposal to ISSC for the symposium, worked on its preparation, and who, more than anyone, is responsible for its success— Dr. George Cohee—was unable to be present for reasons of health. However, the participants were most appreciative of Dr. Cohee's major role in this project and all who may benefit from the Symposium and its published papers must be grateful for his initiative and dedication. Dr. Cohee's place at Sydney was ably taken over by his colleague from the U.S. Geological Survey, Dr. Douglas Kinney, who shared with Prof. Glaessner the management of the meetings.

The four sessions were chaired as follows: first session, M. F. Glaessner and D. M. Kinney; second session, R. E. Folinsbee and A. Salvador; third session, E. Jäger and R. H. Steiger; fourth session, I. McDougall and D. M. Kinney.

This 1976 Symposium on the Geochronologic Scale constitutes another milestone of progress toward improvement of our record of the relation of rocks and geologic events to the passage of geologic time, following such prior organized collaborative efforts as the 1961 Conference on Geochronology of Rock Systems (published by the New York Academy of Sciences), and the 1964 Symposium on The Phanerozoic

Time Scale published by the Geological Society of London. In addition to many concrete accomplishments in putting new age data on record, in reviewing and evaluating methods of age determination, and in considering various philosophies of geochronology, the symposium has been of immense value in clarifying the present state of progress toward a better time scale and in defining problems which must be solved. Whereas it truly may be called a milestone of progress, the symposium also brought out clearly that there are many miles yet to go.

All who have been associated with the meeting are appreciative of the vision of the 25th International Geological Congress (Australia) in holding this Symposium, and of the American Association of Petroleum Geologists in publishing the papers. The editorial board for the papers here published consisted of H. D. Hedberg, M. F. Glaessner, G. V. Cohee, Douglas M. Kinney, J. B. Obradovich, S. S. Oriel, Z. E. Peterman, and T. W. Stern.

Hollis D. Hedberg
past-Chairman,
International Subcommission
on Stratigraphic Classification

George V. Cohee   **Martin F. Glaessner**   **Hollis D. Hedberg**

Dating and Correlation, A Review[1]

D. J. MCLAREN[2]

Abstract Geologic understanding depends on interpreting earth history using relative time scales derived from the study of positional relationships of rock units and mineral bodies. Age determinations of specific rocks and events are used to calibrate a relative time scale. The fundamental problem in geochronology is correlation. Scientists date fossils, rocks, minerals, and remanent magnetism by inferring their position in a geologic succession and correlating them to a relative time scale. The only real evidence for an interval of time is relative position which is extended by hypotheses of correlation. Fossils and radiometric specimens, by their succession, may indicate the direction of time because of evolutionary changes or relative decay of elements. In magnetostratigraphy, apparent polar wandering may similarly indicate time direction but successive paleomagnetic reversals show no such direction and must therefore be "dated" by some other method before accurate magnetic correlation can be used.

The biochronological time scale is an empirical artifact that has grown haphazardly over the last 200 years in areas determined by extraneous factors in regard to development of geologic concepts. The pitfalls of biochronology are known, but the vast and increasing body of empirical knowledge continues to reduce major possibilities of error. Although paleontology provides the "empirical basis for evolutionary theory" it must not be forgotten that the concept of evolution is derived from superposition and correlation. Within the Phanerozoic, accuracy of biochronology does not decrease with distance in time, although application may be strongly facies controlled. In radiometric dating, the percentage error may remain constant, therefore decreasing resolution with increasing age. The difficulty in applying radiometric systems is in knowing what event has been dated; the materials may have undergone successive changes subsequent to their formation. Similar considerations apply to paleomagnetic evidence.

Since the first relative time scales were constructed, new information and insights have emerged from both continental areas and the ocean floor. New methods of correlation now allow hope for an accurate time scale unifying biochronology, radiometry, and magnetostratigraphy. Multiple systems must be used within each major discipline. Thus correlation of many animal or plant groups may achieve greater precision in biochronology than single zonal indices. Similarly, several independent radiometric dating systems may increase confidence or help distinguish successive events that may be detectable by such methods. The most important single task of geochronology is to construct a unified time scale using the strengths of all methods and to test the time scale repeatedly by stratigraphic reproducibility. The potential accuracy of a unified time scale has scarcely been realized.

Introduction

The International Geochronologic Scale symposium was organized by the committee[3] acting on behalf of the Subcommission on Stratigraphic Classification of the Commission on Stratigraphy, which is a body of the International Union of Geological Sciences (IUGS). The first session of this symposium was devoted to examining differ-

[1] Manuscript received, October 12, 1976.

[2] Geological Survey of Canada, Ottawa, Canada.

[3] M. F. Glaessner, G. V. Cohee, R. E. Folinsbee, W. B. Harland, H. D. Hedberg, N. deB. Hornibrook, E. Jäger, I. McDougall, S. S. Oriel, Z. E. Peterman, H. H. Renz, and A. I. Tugarinov.

Article Identification Number: 0149-1377/78/SG06-0001/$03.00/0

1

ent methods of time correlation and of constructing time scales. The other three sessions were concerned with physical time, commonly referred to as radiometric or isotopic. It is interesting to see the degree of interdependence of the various systems currently used to construct geologic time scales and to correlate them. I have decided therefore, to comment on this interdependence. This is both dangerous and difficult, because, although it has been done many times before, points of view tend to be polarized by the specialization of the commentator. I hope that my special prejudices are minimized in this introduction, and if I satisfy no one, I shall assume that I have proved my impartiality. This, then, is an essay rather than a reasoned paper. I have attempted to cover a very large field, and in doing so have tried to be brief and simple. There is a minimum of references cited and I cannot acknowledge all those with whom I have discussed such matters over many years. The Commission on Stratigraphy is central and implicit in all I say. The advances made in international understanding of stratigraphic classification during the last decade or so and in the understanding of the need for mutual interaction between different systems of constructing geologic time scales have been remarkable. The international representation presided over by IUGS, of which the Commission on Stratigraphy and the ISSC are part, must take much of the credit for encouraging the development of such mutual cooperation. However, I should like to single out Hollis Hedberg for acknowledgement as being one of the central figures in stimulating international interest and understanding, whether those so stimulated agreed with him or not.

Unidirectional History

Throughout the 19th century there was a continuing but largely unconscious controversy between protagonists of two rival concepts. One was of a stable, Huttonian earth in which cycle followed cycle, and neither beginning nor end could be visualized or was required. Opposed to this theory was Thompson's mechanical model of the earth as a physical body obeying the newly enunciated laws of thermodynamics, with a strongly directional but relatively short history, and a rapidly approaching cold death. Burchfield (1975) retold the alarming story. Geologists could not accept Thompson's short time span, but were nevertheless forced to consider the implications of unidirectional history imposed by the physicist. As a result of the discovery of radioactivity, the physicists subsequently returned an almost limitless amount of time to geologists, but the lesson had been learned and although the entropy death had been considerably extended, the concept of directional history remained. . .The Lord gave, and the Lord hath taken away.

Uniformitarianism versus evolutionism is still a rich source of argument for the hobbyist in the history of geology. When will geologists learn to trust conclusions drawn from their own observations and not be misled by pronouncements by physicists, based on premises that cannot be examined?

If earth history is unidirectional (Hubbert, 1967) then there is a possibility of erecting a time scale by the study of past events. Because of the enormous duration of earth history, some 4.5×10^9 years, physical changes may have been slow. What directional events have been recorded? What can be used to indicate the direction of "Time's Arrow"? There are not very many.

Time Scale

When William Thompson (Lord Kelvin) gave geologists 100 million years in which to fit all of earth history, they began trying to find ways to question such an apparent absolute truth by quantifying their own time scale. They were remarkably unsuccessful. Sediment thickness, rates of erosion, salt in the oceans, and certain astronomical arguments all produced results which have subsequently proved to be grossly wrong. Paradoxically the much maligned example suggested by Darwin (in the first edition of the *Origin of Species*) for the rate of erosion of the Kentish Weald was closer than most of the others; we now know that he erred only by a factor of three or four.

If we forsake the problems of duration for a moment we can turn to the main contribution of geology to our understanding of time: the relative ordering of events of earth history. The tool used for putting events in order is the principle of positional relationships of rock and mineral bodies. All time observations in geology depend on this principle, and the law of superposition is a special case. The application of the principle of positional relationships is central to constructing time scales, and provides the only objective test that can be applied to all methods. Therefore a system of feedbacks allows continual testing and refinement, so that in any time scale the relative ordering of all observable events must be satisfied. From this generalization let us now briefly examine some of the systems currently used for defining time scales.

Life

The most visible directional phenomenon in earth history has been the evolution of life. It is likely that some form of life existed soon after the oldest sedimentary rocks were formed (Cloud, 1974). Effects of life in the Archean are hard to assess, but examples have been suggested. By the Proterozoic, oxygen in the atmosphere must be ascribed almost entirely to photosynthesis, and the effects of life on the geologic cycle became all pervading. Simpson (1970) recognized that biohistory has been directional throughout much of geologic time, particularly in the Phanerozoic, but that there have been long spans of time when definite directional changes in geohistory have not been conspicuous. Although the Phanerozoic occupies only about one-ninth of the total age of the earth and perhaps one-sixth of the total geologic history, nevertheless it is hard not to believe that some directional changes are detectable in the Phanerozoic. It is probably wrong to contrast geohistoric with biohistoric changes, because of their necessarily intimate association. Taken together, arguments seem convincing that many geologic systems of the Phanerozoic have their own unmistakable style and are unique in an evolutionary sense. Plate tectonics might support this postulate, as well as the concept of irreversible geochemical differentiation, and the profound effect of life on the surface of the lithosphere and the hydrosphere (see Ager, 1973, for discussion on the "uniqueness" of different systems).

It must not be forgotten that the sole means of establishing the directional aspects of life development is by collecting fossil remains from successions that allow their relative ages to be established. Positional relationships then, allow the succession of life to be ordered in a relative time scale, thereby allowing study of life development. Such development provides a valuable and sensitive means of setting up hypotheses of time correlation from place to place.

Isotopes

Another method of establishing time scales in earth history is radiometric or isotope dating. Unlike fossils, this dating system or family of dating systems has a built-in directional component and is dependent on the principle that radioactive decay takes place at an unchanging rate. That is to say, that half-lives measured today have been constant throughout geologic history. Consequently, unidirectional, secular change can be translated into years and the hope of a numerical time scale may become a reality. As with fossils, a mineral or whole rock containing isotopes that can be analyzed by the dating systems must be set into the relative time framework by the principle of positional relationships. The mineral or rock must be fixed geologically by its position or any determination of its age in years would be irrelevant. However, proportions of isotopes may also be affected by subsequent events, thermal or tectonic, and consequently the scientist must attempt to make a judgement on the most likely event that "set the clock" in the mineral or rock to be used. The difficulty therefore, is in deciding what is being dated. As with fossils, radiometric data derived from rocks may be used to suggest correlation. Unlike paleontology however, radiometric dating at its best allows an age in years to be assigned to the particular mineral or rock being analyzed.

Remanent Magnetism

Implicit in the discovery of remanent magnetism is the potential for extrapolating accurate time lines over the whole earth, in rocks of the right facies, both igneous and sedimentary. Such a system depends on the discoveries that the relative positions of the continental masses to the earth's dipole appear to have changed with time, and that at various times in earth history the dipole has reversed. The former discovery has allowed polar wandering curves to be constructed for different continental plates, and this may allow an approximate dating system to evolve, dependent on knowing the position of the pole at different times in the history of a particular continent. Polarity reversal, on the other hand, gives hope of worldwide time correlation in which the term "instantaneous" is not an exaggeration. However, horizons must be recognized by biostratigraphic or radiometric methods just as a bentonite bed must be accurately identified in each locality before being used for regional correlation. There is no directional aspect to polarity reversal history and neither relative nor absolute dating can be derived directly from the evidence—at least on land. In this sense, the method differs markedly from biostratigraphy or radiometric geochronology.

Correlation

In the foregoing remarks on paleontology, radiometric time scales, and remanent magnetism, the term correlation has been used. To the geologist, without modification, it normally means equivalence in time or a hypothesis of equivalence in time (Rodgers, 1959). Without correlation, successions of time derived in one area are unique and contribute nothing to understanding earth history elsewhere; the importance of accurate time correlation from place to place is paramount. In both biostratigraphic and radiometric methods, there is an irreducible margin of uncertainty in all correlation, although both systems have not reached their full potential accuracy. Polarity stratigraphy may prove the most accurate system of correlation available, at least for certain specific horizons in the stratigraphic column.

Resolution

In this brief essay it is difficult to go into detail on how accurate these systems may be; however, one or two principles are worth examining. Paleontologic correlation depends on comparing similar fossil remains between localities more or less widely separated. All fossils are dependent to some degree on facies, and in addition, there are dangers of misidentification, mimicry, reworking, and so on. Nevertheless, the relatively crude systems of biostratigraphic correlation developed in the past have resulted in faunal zones that may be accurate to within one or two million years. An advantage of the system lies in the fact that accuracy does not diminish with age as far back as the base of the Phanerozoic. At that point there is a sudden decrease in precision and Proterozoic acritarch or stromatolite zones are by two or three orders of magnitude less precise even according to the most optimistic claims of their proponents. Nevertheless, such zonations have a high potential of usefulness.

Multiple correlation with many animal and plant groups used simultaneously may afford an accuracy perhaps as much as 10 times greater than current methods (Kauffman, 1970). It seems probable that recognition of individual horizons rather than zones is the way that paleontologic correlation may make the greatest contribution to constructing worldwide time scales.

Lambert (1964) discussed precision and accuracy of radiometric ages in the Phanerozoic and suggested that an individual determination will not be better than 3% (except for some Rb-Sr determinations). Such a figure refers to experimental error and both Lambert and others in the same volume pointed out other factors that must be considered. A second source of error or uncertainty lies in the time of the "setting of the clock" which may depend on the rate of cooling of an intrusion containing a mineral to be dated. A third source of error relates to undetected subsequent alteration of the mineral under investigation; and a fourth source to the accuracy of the constants employed. Such ambiguities are widely recognized and can be dealt with in regard to the statistical problem of setting up a time scale in years for the Phanerozoic using the methods of correlation to calibrate the scale. Furthermore, several independent radiometric dating systems may increase confidence or help in distinguishing successive events.

All time correlation can be improved by the use of multiple systems and by ongoing testing through stratigraphic reproducibility. Such testing can lead to statistical improvement, although some methods are not particularly susceptible to statistical generalizations. For instance, the evidence from a few selected fossils may be far more significant than analyses of very large faunas. Witness the relatively few radiometric dates that have contributed significantly to calibrating the earlier part of the Phanerozoic time scale.

Correlation in the Phanerozoic by radiometric methods is rarely as effective as biochronology or magnetochronology. However, in the Precambrian the radiometric system is the primary method for both dating and correlation.

There appears to be some uncertainty at present in discussion of correlation by magnetic polarity, whether the polarity interval or the boundary of the interval is of greatest stratigraphic importance. Irving and Pullaiah (1976) stress that the boundaries of the quiet intervals are global time markers potentially observable in both oceanic crust and terrestrial beds. Therefore, precision in regard to magnetostratigraphy is largely a question of correctly identifying the magnetic event.

Application

The three methods discussed previously seem to be complementary rather than overlapping. They reinforce each other and their relative importance changes with position in the time scale. Thus, in Phanerozoic rocks, paleontology is by far the most powerful method for relative dating, and is likely to become increasingly accurate for correlation. It offers no direct assistance in assigning ages in years, although it provides the necessary correlation and directional succession. Radiometric age dating is relatively weak in its capacity to assign accurate relative ages to Phanerozoic rocks or as a means of correlation, although it is the only widespread system currently used in the Precambrian. Any attempt to produce a coherent time scale in years for the whole geologic column must be based on this method. Magnetostratigraphy is a powerful tool for correlation, but its potential has scarcely been reached or its importance realized.

Why Time Scales?

Geology has come a long way without an accurate time scale. Stratigraphy and biochronology have built up a hierarchy of divisions expressing relative age which have achieved a remarkable degree of sophistication and accuracy. Nevertheless, there are far too many obvious geologic problems whose solutions wait for positive answers requiring accurate information about duration and correlation on a worldwide scale.

Some of these geological problems and questions include: (1) rates of tectonic processes; (2) rates of sedimentation and accurate basin history; (3) correlation of geophysical and geological events; (4) correlation of tectonic and eustatic events; (5) are epeirogenic movements worldwide (They are currently being claimed as a cause of widespread extinctions; see Sloss, 1976); (6) have there been simultaneous extinctions of unrelated animal and plant groups; (7) what happened at era boundaries; (8) have there been catastrophes in earth history which have left a simultaneous record over a wide region or worldwide; and (9) are there different kinds of boundaries in the geologic succession (That is, "natural" boundaries marked by a worldwide simultaneous event versus "quiet" boundaries, man-made by definition).

The list could be extended indefinitely according to the special interests of the author.

Definition

Time scales must have marker points. This commonly implies some kind of terminology. If and when the whole of earth history can be arranged in sequence against a numerical time scale, names will become unnecessary, although they might continue to be convenient. However, we are far from that and the relative geologic time scale requires its days, weeks, and months to be distinguished. The last 200 years has seen the haphazard development of a hierarchy of various terms which has served well, but astonishingly we discover that many of the terms lack precise definition. It is a source of wonder to read such authoritative works as the "Phanerozoic Time Scale" and its Supplement (Harland, Smith, and Wilcock, 1964; Harland and Francis, 1971) and realize that the suggested physical ages are plotted against a relative time scale with many ambiguous terms. This is not a trivial comment because system boundaries in the

Phanerozoic may differ in concept from one area of the world to another by as much as two stages, or several million years, and the same is true for many lesser boundaries. Ill-defined terms are a major barrier to communication.

The *International Stratigraphic Guide* (1976), steered to completion by Hollis Hedberg, sets out reasonable rules and suggestions for defining stratigraphic units. Controversy will continue as to nomenclature but the world is rapidly moving to general acceptance of the idea that stratigraphic names should be defined physically at a single point if meaningful correlation is to be made elsewhere. I believe that we shall continue to move towards the concept of defining only the base of stratigraphic units, or even time horizons, which may in themselves be named. In any case, "stratotypes" will be required so that all the systems of correlation may be applied equally.

Integration

The problem that faced the symposium and faces geologists in general is the integration of all systems of dating rocks and their correlation. Two developments in the recent past have changed the probability of achieving this quickly: (1) the discovery of remanent magnetism and, combined with age determination systems, its potential for accurate time correlation; and (2) the growing knowledge about the geologic history of ocean sediments through deep-sea drilling. Advances are rapid, not only in knowledge mostly confined to the late Mesozoic and Cenozoic, but also in methodology. Results are numerous and are rapidly increasing, but I should like to select the recent papers by van Hinte (1976a, b) as an illustration of method and as a paradigm for the future. In his paper "A Jurassic Time Scale," he summarized the method as an ". . .intricate approach using a paleomagnetic time scale for linear extrapolation between paleontologically correlated, radiometrically dated calibration points. . ." (Van Hinte, 1976a, p. 489).

Van Hinte would be the first to claim that his syntheses were trials or progress reports on a continuing process. The importance of such methods lies in the fact that all evidence is presented and all assumptions are exposed. Such syntheses should lead to rapid advances in the process of constructing an accurate time scale for the history of the earth. In older rocks, without the benefit of deep-sea information, syntheses must be attempted. Paleozoic quiet-interval boundaries may be recognizable and could give accurate time-marker points on a worldwide scale which would be valuable in constructing overall time scales.

There is a long way to go and large errors are easily found in any of the syntheses so far attempted. Geology is worldwide and the time is long past when a scientist could generalize his observations only from a local or regional point of view. It is hoped that the symposium will prove to be an important milestone in advancing the full integration of stratigraphy and geochronology.

References Cited

Ager, D. V., 1973, The nature of the stratigraphical record: New York, John Wiley & Sons, 114 p.

Burchfield, J. D., 1975, Lord Kelvin and the age of the earth: New York, Science History Publications, 260 p.

Cloud, P., 1974, Evolution of ecosystems: Am. Scientist, v. 62, p. 54-66.

Harland, W. B., A. G. Smith, and B. Wilcock, eds., 1964, The Phanerozoic time-scale; a symposium: Geol. Soc. London Quart. Jour., v. 120, supp., 458 p.

—— and E. H. Francis, eds., 1971, The Phanerozoic time-scale; a supplement: Geol. Soc. London Spec. Pub. 5, 356 p.

Hubbert, M. K., 1967, Critique of the principle of uniformity, *in* C. C. Albritton, ed., Uniformity and simplicity: Geol. Soc. America Spec. Paper 89, p. 3-33.

International Subcommission on Stratigraphic Classification (ISSC), 1976, H. D. Hedberg, ed., International stratigraphic guide: New York, John Wiley & Sons, 200 p.

Irving, E. and G. Pullaiah, 1976, Reversals of the geomagnetic field, magnetostratigraphy, and relative magnitude of paleosecular variation in the Phanerozoic: Earth-Sci. Rev., v. 12, p. 35-64.

Kauffman, E. G., 1970, Population systematics, radiometrics and zonation—a new biostratigraphy: North Am. Paleont. Conv. Proc., Pt. F, p. 612-666.

Lambert, R. St. J., 1964, The relationship between radiometric ages obtained from plutonic complexes and stratigraphical time *in* W. B. Harland et al, eds., The Phanerozoic time-scale; a symposium: Geol. Soc. of London Quart. Jour., v. 120, supp., p. 43-52.

Rodgers, J., 1959, The meaning of correlation: Am. Jour. Sci., v. 257, p. 684-691.

Simpson, G. G., 1970, Uniformitarianism. An inquiry into principle, theory, and method in geo-history and biohistory, *in* M. K. Hecht and W. C. Steere, eds., Essays in evolution and genetics in honor of Theodosius Dobzhansky: New York, Appleton-Century-Crofts, p. 43-96.

Sloss, L. L., 1976, Areas and volumes of cratonic sediments, western North America and eastern Europe: Geology, v. 4, p. 272-276.

Van Hinte, J. E., 1976a, A Jurassic time scale: AAPG Bull., v. 60, p. 489-497.

—— 1976b, A Cretaceous time scale: AAPG Bull., v. 60, p. 498-516.

Geochronologic Scales[1]

W. B. HARLAND[2]

Abstract The word *geochronology* is derived from H. S. Williams' (1893), use of geochrone which was a unit of time (equivalent to the duration of the Eocene period) whereby estimated ages were quantified. His estimate of ages back to Cambrian time, done before the discovery of radioactivity, were remarkably accurate. Since the use of radiometric methods, the term geochronology, for most scientists, has referred to the measurement of time in units of a year, generally measured radiometrically but interpolated by thickness and fossil content of strata.

However, another scale was conceived for the natural chapters of earth history, and was calibrated by successive creations of life with intervening revolutions. That was refined into a biostratigraphic scale until, in recent years, it became clear that standardization for international use was impossible without stratotype boundaries. So a new scale is in the process of definition. The International Subcommission of Stratigraphic Classification has referred to the time element of these global chronostratigraphic divisions as geochronologic.

Geochronology might in any case be thought to signify geologic time. Several scales, different in their construction, that apply to geologic time have been proposed. Consequently, geochronology is now a general term referring to the whole science of geologic time, and it is an essential discipline in stratigraphy.

Time correlation is the primary activity to improve the time-space framework. To achieve international correlation a degree of standardization is necessary so that divisions shall have agreed names and definitions. Two kinds of scales are used to provide a time frame to support, and be supported by geological events.

One is the obvious *chronometric scale* of the physicist in caesium seconds or of the astronomer (and seemingly geologist) in earth-years. The other will be the *chronostratic scale*. Each provides a common language to express the time relationships of geological phenomena. The symposium was our effort to calibrate the chronostratic scale in boundary stratotypes against the chronometric scale.

Introduction

The purpose of the symposium was to improve knowledge and understanding of the time scale of geohistory. This paper attempts to clarify the essential concepts that are necessary. I distinguish between matters of *convention* (e.g. what name or term to use for an object or concept) and matters of *scientific principle* (e.g. what questions are capable of scientific demonstration). It is difficult to separate convention and principle because when we use language for scientific communication it can be translated and deliberately constructed for the purpose; but at the same time, it influences our understanding of the matters of scientific substance.

In an effort to focus on matters of principle, I shall choose whatever word seems best to me at the moment and resist the temptation to justify or comment on it. Because we cannot ignore different usages, I will give a brief indication of different meanings in a glossary. By doing this I hope to avoid the contentious discussion in

[1] Manuscript received, January, 10, 1977.
[2] Sedgwick Museum, Cambridge, CB2 3EQ, England.

Article Identification Number: 0149-1377/78/SG06-0002/$03.00/0

the text that is often associated with terminology and nomenclature. If we have slightly different concepts it is often sterile to argue about the terms for them. On the other hand differences of concept become clearer when applied to particular examples.

The distinction between *artificial* and *natural* time scales is analagous to the distinction between *conventional* and *scientific* principle. Just as we can invent a word and define it, so we can construct a time scale that will be determinate; whereas natural or phenomenon scales are essentially indeterminate. The element of convention, artifact, axiom, or determination is a *normative* element. Scientific, natural, phenomenic, and indeterminate aspects of science are *positive*. These are useful terms. A normative question can be settled by vote—a positive one cannot.

A further, somewhat analogous contrast is that between *informal* and *formal* usage (ISSC, 1976, p. 13). This is not trivial. Formal nomenclature is established or accepted according to some distinct procedure. Its advantage is that the names are unique, unambiguous, and not subject to frequent change. Informal usage has no such constraint and is more suited to positive scientific development because it is more flexible and experimental. The difference is important for us in so far as any development of formal language needs some international discussion.

Geochronology: An Historical Note

I make one exception to my rule not to discuss words, because *geochronologic scales* was the subject I was invited to introduce, and the word *geochronology* has a complex history that is worth noting as it is bound up with the history of geology.

Geologic time expressed as earth history was the singular scientific achievement of the first part of the nineteenth century. It began with the appreciation of superposition, characterization of strata by fossils, and the gradual elaboration of what has become the chronostratigraphic scale. At first it was assumed that time itself presented no problem and that the real question was the rate, intensity, and even the naturalness of earth processes. In due course when evolution as the supreme natural process, appeared to be understood, the scale or magnitude of time became the dominant question in the second part of the nineteenth century.

H. S. Williams was one of the more successful and far-sighted among his contemporaries, and finding that the real duration of earth history (since the beginning of Cambrian time) was then impossible to assess in years, he invented the *geochrone*, which was the span of time represented by the duration of the Eocene period. By weighing the relative thicknesses of rocks and amount of history to be accommodated in each period, he devised one of the first time scales. Because it is little-known, I have shown it as Figure 1. It was a remarkable achievement based on thicknesses of strata, amount of evolutionary progress made, and an intuitive feel for the quantity of history that had to be accommodated in each division.

Geochronology has been widely used for the measurement of geologic time in units of a specified duration. At first it was geochrones, then with the application of radioactivity it became possible to use seconds or years as the units of time. Geochronology in this sense has perhaps been used more in Europe. The word chronometry also has been used (or geochronometry; Harland et al, 1972) and it is now used by the International Subcommission on Stratigraphic Classification (ISSC, 1976, p. 15).

Geochronology has been used in a different sense by the American Commission on Stratigraphic Nomenclature (1961, p. 659) and also in the circulars and publications of the Subcommission on Stratigraphic Classification (Hedberg, 1972, p. 29) for the time aspect of divisions that constitute the chronostratigraphic scale. Whether or not this is a correct historical application of the geochrone concept is not now my point. The fact is that geochronology also has been used widely in what I refer to as the chronostratic (or geochronostratic) sense.

In the spirit of the symposium I accepted the word geochronology (*sensu lato*) as containing both the chronometric and chronostratic concepts. Once the meaning of a word has been broadened in any direction it is difficult to restrict it, and we now have good terms to refer to the two narrow senses that might be claimed for it. The two

Recent	} 1	
Quaternary		
Pliocene	} 1	} 3
Miocene		
Eocene	1	
Cretaceous	4	
Jurassic	3	} 9
Triassic	2	
Carboniferous	6	
Devonian	5	
Upper Silurian	4	} 45
Lower Silurian or Ordovician	15	
Cambrian	15	

FIG. 1—Time scale according to H. S. Williams (1893) "...standard time scale of geochronology, on the basis of the Eocene period for a time unit or geochrone..." (p. 295). Note the Eocene index equals 1 geochrone.

aspects *metric* and *stratic* are the two essential kinds of time scales with which this symposium will be concerned. The two scales *chronometric* and *chronostratic* may be used in many ways but when standardized for global use they may assume a *geo* prefix: *geochronometric* and *geochronostratic*. It could also be argued that geochronology also contains within itself another and essentially different time scale, the natural or phenomenon scale, which will be distinguished as the *chronologic* scale.

Elements of the Time Concept

Time as a Dimension

The familiar time concept is to view experience and nature as unfolding in a four-dimensional framework in which time is the unique dimension—a linear progression and one in which the three dimensions of space give the possibility for a different solid configuration of nature corresponding to each instant of time.

In this analogy time and space are easily interchanged. We imagine time as linear and refer to a "point" or "position" in time rather than an "instant" or "moment," and also to a span of time. Conversely, temporal terms are commonly used for space—as when we say "large pebbles occur frequently in a section," we mean abundantly.

The essential difficulty of geologic time is not its immensity but our construction of it. In discussing stratigraphy and in expressing stratigraphic operations, we interpret time from rock in space in order to express sequences of events.

Geological Time and Geoscientist Time

When we say "sandstone beds occur frequently in a section," three aspects of time may be in use. On the one hand, there is the ready interchange of space and time terms in the dimensional concept. In this case it means that there are many sandstone beds. In a superpositional sense, we imply that many beds in sequence represent a sequence in time of many sandstone depositional events. Perhaps even more fundamentally is the thought that something in nature is commonly or frequently observed. In this sense the frequency in time is a frequency of observation. And so we can distinguish, and must distinguish, what I refer to as *geologic time* and *geoscientist time*. Moreover, geoscientist time is what we ourselves experience and what we infer about the written (historic) record.

So I distinguish history and geohistory. There is no suggestion that these are two different time processes but in so far as time is conceived, there are different ways of realizing it—by geohistorical, archaeological, historical, bibliographical, and autobiographical activity.

The Concept of Sequence

In elucidating a sequence from a rock complex there are commonly two operations: (1) to establish an order; and (2) to establish the direction of aging. These two elements in a sequence are distinguishable as follows.

Order

A sequence without polarity is referred to here as an *order*. This means that the order ABC is the same as the order CBA. To establish an order is often the first step in elucidating a stratigraphic complex when many observed fragments of order may or may not be pieced together.

Polarity of Time Sequence

A sequence with polarity is a directional sequence. It is the "arrow of time" by which ABC cannot be the same sequence as CBA. It is the ordinal quality of numbering. The argument for there being a time sequence in nature has been reasoned from at least two points of view. On the one hand (other things being equal) *entropy* or disorder increases with time. On the other hand the complexity of history gives increasing information with time (Layzer, 1975).

This is an observable effect of evolution (perhaps even its essence). By evolution in this sense I refer not only to the development of life but to development of the world and its parts. Figure 2 shows how a complex situation contains information of the past, and of direction of time, thus enabling ABC to be distinguished from CBA. Only when an apparently simple system is examined does it seem that there is no evidence for direction. This applies to the classical concepts of atomic motions. From a distance, the rise and fall of the tide might seem reversible, but not when the detail is taken into account.

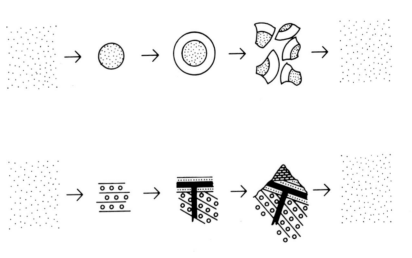

FIG. 2—Illustration of the opposing tendencies of evolution (which gives information of an increasing historical past) and entropy (which reduces structure to an atomic level of disorganization). For some 10^{10} years in our planetary environment, sufficient information will be available to plot a significant history.

The principle of superposition is both the most obvious, unnecessary, uninteresting, and intellectually insulting of considerations, and also the most subtle as it relates time and space and essentially derives the time sequence from the space order (see Fig. 3).

Time-Relation

Time Sequence

The essential relationship derived from the above is that any two instants in time may be placed in a time sequence, so that in the time sequence, *A* is pre-*B* and *C* is post-*B* etc. The implication is that events anywhere (if close to, then also remote from each other in space) have such a possible relationship.

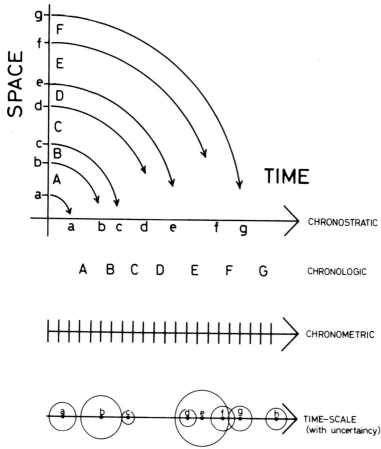

FIG. 3—Illustration of how one dimension in space, visualized as superposition can be conceived as a dimension of time in a chronostratic scale with chronologic events. This may be compared with an independently generated chronometric scale. To attempt the calibration of one against the other is the time scale exercise of which this symposium is a current expression.

Space-Time Concept

Indeed it cannot even be said that two events now interpreted at the same place represented the same point in space because with respect to space and time all is relative. Nevertheless, the Euclidian frame that is easy to conceive serves us well within most limits of geohistory. In this context the limits relate to planetary environments.

Time Correlation and Synchroneity

The concept of synchroneity (i.e. of isochronous events) is familiar, as is that of diachronous events. For most purposes the expression of such relationships would be expressed in the pre-*C*–post-*A* form. Logically as well as practically we can do without the concept of anything observable being precisely synchronous. One event *A* may be both pre-*B* and post-*B*, spanning event *B*.

Recognizable Time-Instant and Interval

A theoretical instant in time that is not in any way identifiable is of less interest than the unique instants that are potentially recognizable. To identify an instant in time it is necessary to relate it to a sufficiently complex history. This statement generally relates to a point in rock with a sufficiently complex character to identify it uniquely in space and by interpretation in time. The concept that a point in rock may correspond to a point in time at an instant when that rock was formed is essential to the concept of time correlation.

If a point in rock is sufficiently useful for correlation or if it represents a convenient event in local history it may usefully be labeled. This is done typically in measured sections where, for example, particular samples were collected or where a position is identified for relocation. Such a point in rock (if diagenetic and other changes are excluded) refers very precisely to the point in time of initial formation which is thereby defined. This is the chronostratic way of defining an instant in time.

It is possible to express an interval of time precisely by defining the two limiting points precisely. In this way the interval is an historical interval and if interpreted from strata it is a chronostratic interval.

Duration

The time concept of *duration* is familiar. It is the distance in time between two points measurable in units.

A duration may be expressed numerically as a duration of, say, X years at some time (e.g. the duration of a typical ocean-spreading cycle). More generally it is expressed as a duration measured backwards in time from the "present." Both the time in history and its duration can be expressed by the duration from the present to the beginning and end of the required interval. These numerical expressions are chronometric. It is the cardinal or scalar quality of numbers.

The Concept of Age

The age of a rock is one of its most sought-after characters. An age may be given chronostratically and/or chronometrically, each as a point in or an interval of time. The terms true and apparent are also used in discussion of chronometric ages, although strictly speaking, all age estimates are in some degree apparent ages. The word "age" (*sensu lato*) is more convenient than the words "time of formation."

Phenomenon Sequences—Geohistory

Time Scales

The concept of time would be impossible without knowledge of natural processes. We interpret each experience in a space-time framework. The framework depends on phenomena for identification while phenomena depend on the framework for comprehension. While comprehension of nature depends on a space-time frame to display it, the frame has no separate existence. Moreover, our only interest in the frame is the history it displays.

Phenomenon and Framework Time Scales

I consider the phenomena of geology as generating or potentially generating an indefinite number and variety of natural time scales. Any geologic process can be conceived as being a sequence of *events*, each corresponding to a moment of time. If greater precision is required, these events may be conceived as separated by instants (or points) in time, each of no duration.

The names *phenomenon* and *framework* for two contrasting kinds of scales are preferred to the alternative *natural* and *artificial*, insofar as they are all artifically constructed for scientific purposes and all are natural in that, from a geologic point of view, natural processes or events are selected from a natural continuum. The distinction here is fundamental to international scientific work. I have referred to the two kinds of scale as phenomenon and framework scales (Harland, 1973). This distinction is intended to refer to the different purposes of the two kinds of scales.

Framework scales are constructed by selecting certain standards or definitions to provide a reference independent of historical interpretation in order that it might be used to calibrate history. These chronometric and chronostratic scales will be referred to later. They are, I believe, the only time scales we can usefully standardize.

Phenomenon scales use no fixed standard but any phenomena from the infinite variety of nature. Observations on relative ages of phenomena may be useful for a particular investigation. It is not necessary to conceive the sequence of phenomena as providing scales. Their interest is intrinsic, and they have also been used as scales. It is therefore worth examining their nature, and their particular advantages and disadvantages.

Essentially, a phenomenon scale may be constructed from any sequence of instants or events, each of which can be identified in some degree. We may refer to this more generally as the *chronology* of the phenomenon. If the events are interpreted from a rock sequence, the chronology is essentially derived by the principle of superposition and by correlation (Fig. 3).

Some Common Chronologic Scales of Phenomena

Biostratigraphic

The traditional stratigraphic scale is calibrated and certainly correlated by the fossil content of strata. The reasons for departing from it are: (1) the difficulty of selecting and sufficiently defining a biologic change that is time dependent and ideally isochronous; (2) the difficulty that, in any sequence, new data may arise and the ranges and composition of assemblages are always likely to vary with new discoveries and/or with new interpretations; (3) there is always a subjective element of definition of the phenomena in establishing the scale; and (4) the difficulty of describing evolution against a scale that is itself based on an interpretation of evolution.

The international move to establish standard boundaries arises from these considerations. This development in no way limits the activity or importance of biostratigraphy, for precisely the same activities serve the purpose of correlation. Indeed, until the chronostratic divisions come to be defined by boundary stratotypes, we are still using a largely biostratigraphic scale. In this connection a whole range of terms has been proposed (George et al, 1969, p. 149-150; ISSC, 1976, p. 50-61).

Radiometric

Because radiometric age determinations are by far the most powerful method of placing rocks in a scale it is commonly assumed that the chronometric scale is a radiometric scale; but this is not so. The chronometric scale depends only on the definition of the unit of duration used and its multiplication. Many methods may be used to place events. Before the discovery of radioactivity, some reasonable estimates were made from the rates of evolution and of deposition, and from seasonal varves or tree rings which are probably more precise within their range than radiometric methods. Nevertheless, radiometric methods do provide the basic data for the chronology of rocks that can be so determined. In this sense a *radiometric scale* would be a sequence of well-determined rocks that would slip a bit if a decay constant were changed or if a better determination were made.

With biostratigraphy in the chronostratic scale and with radiometric methods in the chronometric scale, the very progress of science challenges the scale of age determinations of rocks. Yet it is precisely this scientific activity that requires the use of a geochronometric scale.

Paleomagnetic

The chronology of reversals (more crudely, polar wandering) is providing another chronology that is becoming widely accepted.

The relatively isochronous and ubiquitous nature of geomagnetic change makes the magnetic method almost ideal, however there is ambiguity of the two alternative states (positive and negative), and of the problem of their identity. If, however, magnetic changes are to provide a standard scale for earth history, its value will depend on the degree of calibration and correlation with other scales. This is not meant to minimize the importance of the method but rather to release it fully to serve as a method of correlation rather than of definition.

Tectogenetic

The concept of revolutions punctuating earth history provided the first notion of a time scale at the beginning of the last century, and was somewhat refined in this century by Stille (1924) and by others. The use of tectonic phases and episodes, including the effects of metamorphism, is a powerful tool in the interpretation of regional chronology. It has suffered very conspicuously from the circular argument, whereby tectonic phases were recognized as part of the global scale and then used to correlate discordances with the global scale. Rate of ocean spreading is another tectonic phenomenon with considerable potential for chronometry, but it may apply only to the last 200 Ma or so of earth history.

Lithogenetic

It is convenient to mention some of the very wide range of phenomena that may be interpreted from rock sequences whether sedimentary or igneous. Rate of subsidence, especially with shallow water sedimentation (or of rate deposition, especially of oceanic sediments), were among the earliest phenomena to be used and are still used to interpolate and extrapolate from points where other evidence of age is available. Changes of sea level have a eustatic as well as a diastrophic cause and are almost instantaneous. Their potential is not realized because identity is difficult to establish.

Climatic chronology also has the possibility of world-wide correlation provided the identity of climatic fluctuations can be established. Volcanic dust is also said to be a potential factor for correlation. Geochemical changes, although at a very slow rate, are of importance partly because of the possibility of time-dependent compositions.

Conclusions

Each kind of phenomenon has its own chronology. The understanding of that chronology is best served by relating it to other natural chronologies, and to the standard scales in order to use phenomena effectively for correlation. A framework with objectivity, stability, and independence of opinions will be useful to describe the complex evolution of the planets.

Time Correlation

Time correlation or chronocorrelation is discussed because the value of chronologic scales is entirely dependent on correlation.

Chronotaxis and Homotaxis

Correlation in a geochronologic context is understood as time correlation. However, it is worth making a clear distinction between correlation which is intended to be chronologic (i.e. *chronotaxis* of Weller, 1960) and correlation of other characters (for example the *homotaxis* of Huxley, 1862).

Weller (1960) admitted that chronotaxis is strictly unattainable. But all positive science has an element of uncertainty even though that uncertainty can be reduced to virtual certainty in some cases and to a clear statement of probability or likelihood in many others. Since the beginning of stratigraphy a major preoccupation has been to improve methods of correlation and to improve an understanding of their limits. To increase precision in correlation will continue as a difficult yet important objective.

Correlation Activity and Objectives

Chronocorrelation

The word *correlation* refers to the activity of correlation with the objective of making some positive chronotaxial statement. It does not imply any particular degree of success. The activity of time correlation involves propositions of the kind—*A* : (1) is pre-*B*; (2) is post-*B*; (3) spans *B*; (4) is spanned by *B*; (5) is syn-*B*; or (6) is coeval, penecontemporaneous, etc. Generally the search is for a combination of different time-dependent characters that are otherwise unrelated.

Lithocorrelation

We must distinguish the quite different activity (see Fig. 4) of correlation when equating two successions within a named formation. This activity may be referred to as *lithocorrelation*. It is the basis of mapping and of describing and classifying local rock units. It is partly positive insofar as there is uncertainty as to whether rocks belong to the same unit; but it is mostly normative in that the author, as authority, is determining the classification to be used till a change is required.

The author could be wrong, so it is a positive assertion. To include one rock in this formation rather than another may well be a matter in which the author, as authority, sets up his preferred classification. In this sense, if he says that something is the case then it remains so until a later authority makes a new classification. Changing the rank of a formation up to a group or down to a member may be a mere matter of convenience in the growing complexity of an historical stratigraphical scheme and requires only the authority of a competent scientist.

FORMATIONS

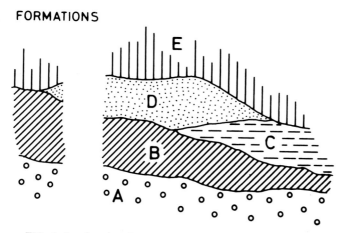

FIG. 4—Local rock units are used to organize rocks for description into conveniently defined packages. The validity of formations A to E rests on the procedure of the author in defining and describing them including his choices of convenient boundaries. Lithocorrelation is only a problem when postulating that isolated outcrops (as in the left of the figure) were once continuous with the established formations on the right. Original continuity is a criterion—time is not. The operation is a survey expressed in maps, stratigraphic sections, cross sections, and descriptions containing all observed characters of each unit (lithology, fossils, petrophysics, etc.).

Correlation Method

The method of correlation is to consider characters that are common to the two rocks or events being correlated. If there are many characters of a somewhat similar nature, a numerical comparison may be attempted. Roughly, more characters in common give greater confidence of chronotaxis or homotaxis.

Critical however, is an understanding of the nature of the characters and their individual correlation value. Generally the objects being correlated will each contain a sequence, so that the strength of correlation characters will be modified by the detail and the number of the sequences being matched.

Correlation Properties

Geologic phenomena each have their peculiar properties which are useful for correlation:

(1) A recognizable stage in evolutionary sequence (for example organic evolution and geochemical evolution) distinguishes certain phases in earth history; (2) pulsatory or reversible events (climatic, seasonal, and tidal oscillation, and magnetic reversals); (3) similarity of signature or pattern of events; (4) speed of propagation (biological migration and dispersal), volcanic ejecta, sea level, and magnetic change; (5) distance of propagation; (6) continuity in space of phenomena (as with a magnetic field, connected water levels, evolution with interbreeding, or of magmas) within a province; (7) constancy of rate of change of processes disconnected in space, identical rates of nuclear decay in normal planetary environments as compared with evolutionary rates in separate provinces and environments; and (8) replication of phenomena, especially of biologic individuals which is the basis for paleontological methods.

Precision and Uncertainty

Most refinement in correlation derives from a fuller understanding of the properties of the characters employed. There is always room for improvement, however there are several reasons why an apparent age will differ more or less from the true age.

Most obvious is nonavailability of appropriate characters and gaps in the record which lead to indeterminate results. In other cases human skill or error affect the uncertainty as in observational and experimental errors. These can in some degree be measured as standard errors, or as systematic errors due to mistaken assumptions (e.g. decay constant) or due to inadequate understanding of the characters used, taxonomic lumping, splitting, or mistakes.

Measure of Correlation

Elaborate systems have been proposed to introduce some numerical measure of correlation of characters for time, province, or any other affinity. The advantage is that this will make it possible to compare results more objectively. These systems are not necessarily more objective or more reliable than good personal judgement or intuition; the advantage is that differences between estimates are capable of evaluation and comparison.

Correlation Statement

Finally the correlation attempt comes out in a statement of chronotaxis or homotaxis. It will be in the form indicated previously, qualified by an assessment of the characters used and assessment of uncertainty, etc. In whatever form it is given it will be a proposition or opinion that is not immutable. Therefore author and date need to be added for completeness, as well as the evidence available to the author and other qualifications.

The value of this correlation will then depend partly on the supporting evidence presented and partly on the reputation of the authority. Nowadays the authority is often a committee of experts.

It is a short but doubtful step to say that such an authority determines the age of a rock, or assigns it to some division. Here it is important to distinguish the correlation applied in surveying local rock units where successive authorities determine (without uncertainty) the classification of local rock units. The uncertainty of age in a correlation statement may be reduced by increasing the span of the division with which the rock is correlated. A hierarchy of divisions serves this purpose well for ages in the middle of a division of higher rank but not near its margin.

Standard Scales

The Two Scales

Chronometric and *chronostratic* scales are both framework scales rather than phenomenon scales. Both artificial in the sense that they can be determined and constructed for use, they need to be agreed to as international conventions. We need to have as few scales as possible to express ages in a common language. I have argued elsewhere that not more than two scales are necessary and that two are needed (Harland, 1975). If these two scales can be agreed on by international action, they will provide an unambiguous way of expressing ages in any language or community.

Statement of Age

Some scientists misconceive that the individuals who have devised these scales imagine that correlation can thereby be done more precisely. The limits of correlation have been discussed in the last section and precision of correlation is not affected by establishing the standard scales. On the other hand the statement of age is improved if the standard is clear and unambiguous.

There are two elements in a statement of age: (1) the names and terms in which it is expressed; and (2) the limit of chronotaxis or correlation. The error or uncertainty of (2) is in some degree unavoidable but the uncertainty in (1) is entirely avoidable. If ambiguity, imprecision, or boundary "fuzz" is allowed to persist in definitions of the chronostratic scale, then the uncertainty in a statement of age is both (1) and (2) and not just (2).

Requirements of a Standard Scale

A standard scale should:

(1) express any age in any place; (2) express broad and general, also detailed and particular ages; (3) be understood, clear and unambiguous; (4) be independent of opinion and therefore have some objective reference that is accessible; (5) be stable, that is not subject to frequent change; (6) be agreed and used internationally in all languages; and (7) be in use as soon as possible.

These requirements can be met by agreement through the Commission of Stratigraphy of the IUGS.

The setting up of a standard is a sufficiently formidable procedure so that it should be changed only with good reason. On the other hand there is a serious danger that so much discussion will take place as to delay for several years the completion of the scales. In my view this would be a mistake because the scales are essentially artifacts; there is no truth or correctness in them—they are for convenience only. It will be most convenient to approve the new standards soon.

Decisions Needed

For each of the two scales (chronometric and chronostratic) at least four kinds of decisions have to be made: (1) the classification scheme of proposed divisions including size of divisions, ranks of hierarchy if any, and how the divisions relate to traditional usage; (2) the nomenclature for the divisions; (3) the standardization of the divisions; and (4) the name for the standardized global scale to distinguish it from any other scales with which it might be confused, and the terms for its parts.

The Classification Scheme

Chronometric Scale

The first point is not in question. Already a hierarchy at intervals of 10^3 exists. For most geologic purposes, 10^6 years is a common unit (m.y. or Ma). For late Quaternary studies, 10^3 years is convenient (ka), and for the scale of earth history and Precambrian and planetary geology, 10^9 (Ga) is in use. (Billion should not be used because of its dual meaning throughout the world.)

Then there is the debatable question whether any other groupings are desirable, such as dividing earth history at 1 Ga, 1.5 Ga, 2.5 Ga, etc.

Chronostratic Scale

A hierarchy is well established with the successive ranks chron (chronozone), age (stage), epoch (series), period (system), era (erathem), and eon (eonothem). One question arises as to whether in all cases the boundaries of successively higher ranks should coincide with some lower ranks as in Figure 5—P and Q with T but not X. It is convenient that it should.

Another question was raised (Hughes et al, 1967; 1968) as to whether an alternative point system of describing spans of time is not better. Whether or not better, each has its use and will in due course be used. The larger ranks of the hierarchy can be chosen for expressing a general time, for example late Mesozoic (not quite the same as Cretaceous). The point system is simpler and more flexible but will take time to establish.

The Nomenclature of Divisions

Chronometric Scale

There is no need for names when all ages on this scale can be expressed in figures. A terminology is needed for a duration such as 1 Ga (Gigennium). Eon has been suggested, but it conflicts with usage in the chronostratic hierarchy. It has also been suggested that the older part of the chronometric scale be divided into Early (Lower), Middle and Late (Upper) Precambrian time at defined divisions (e.g. 1.5 and 2.5 Ga according to the Precambrian Subcommission of IUGS).

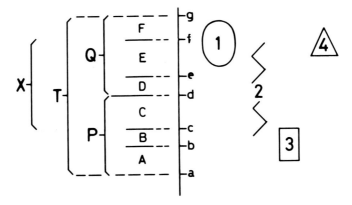

FIG. 5—A scale is a linear set of points in space or time—not necessarily equally spread. The chronostratic scale of points **a** to **g** may also be labelled by intervals **A** to **F**. For convenience these may be combined into two units **P** and **Q** of higher rank. The time of phenomena 1 to 4 can be expressed as ages on either the point or interval scale.

I have suggested that because the chronometric scale is in fact used for all geohistory, then the major divisions on it should be made at (round) numbers to be decided. I suggested a series of Latin root names for these phases of earth and planetary history (Harland, 1975), and I now modify the divisions to conform more closely to the opinions of the Precambrian Subcommission and redefine the bounding times (Table 1).

Table 1. Chronometric Scale

Time-division	Years (Ga)	Subdivision (at 0.5 Ga)
Novo-time	1.5	(Early, Middle, Late)
Medio-time	2.5	(Early, Late)
Antiquo-time	4.0	(Early, Middle, Late)
Prisco-time	—	(with scope for subdivisions for early history of solar system)

I suggested this because it seemed unnecessary to restrict the nomenclature of the chronometric scale to Precambrian time when a nomenclature for the chronostratic scale is being extended into Precambrian time. Indeed a time scale for planetary evolution would be ridiculous if it stopped with the advent of Phanerozoic history on the earth.

Chronostratic Scale

There is a long history of nomenclature, so that for most proposed divisions there is at least one available name that is suitable. It is generally recognized that rather than introduce new names for each division, an old name in common use can be adopted when the division has been more precisely defined. These names are for spans of time whether for chrons or eras.

It will also be desirable for at least three reasons, for the formal Upper, Middle, and Lower divisions that recur in the Geochronostratic Scale to be named and defined: there is often ambiguity in the use of Upper, Middle, and Lower because of different usages; names will often avoid the need for transposing Early with Lower, and Late with Upper. The terms early/lower, middle, and late/upper will be released for informal use with any divisions.

The alternative point system will almost certainly acquire a new set of names from the locality where the boundary stratotype is selected (Klonk for the Siluro-Devonian boundary; McLaren, 1973). If care is taken on later boundaries, the exercise is likely to impress key names on memory and they will be available for use when preferred. At this point I would urge committees setting up such boundary reference points to adopt the most conveniently memorable and usable name.

Standardization of Divisions

Chronometric Scale

There are two decisions to be made—practically they are trivial, but logically they are fundamental. The first is the datum. For the scale to be measured backwards in time, its zero point might be the calendar BC-AD divide Natus Christus (NC), giving negative values to BC and positive values to AD (Needham, 1954). Alternatively the year 1950 was selected by radiocarbon workers as "Present" so values are given as BP (before 1950).

The second decision is the standard unit of duration. Astronomers use a scale with *Ephemeris Time* in which the ultimate standard is the *ephemeris second*, adopted in 1957 as the fundamental unit of time, and defined as $1/31,556,925.9747$ of the tropical year at 1900, January, 0 days, 12 hours E.T. The standard second is not the same as that defined by the International Conference of Weights and Measures in terms of atomic transition (of caesium) as "atomic clock time obtained by continuous additions of multiples of this unit" (see George et al, 1969).

The various definitions of the second and derivative years cannot be defined one in terms of the other since they are based on different observations. They are approximately the same so it makes no difference to the values we use but it would be satisfactory to know what standard we intend to use.

I once argued gently for the second, rather than the year, as the geochronometric unit to be consistent with the Systeme Internationale, and to make geophysical rates of processes easily convertible to physical processes expressed in SI. On this basis, 1 Ma = 31.557 Ts, and 1 Ga = 31.557 Ps (10^{12} = tera; 10^{15} = peta; 10^{18} = exa). But I admitted defeat and accepted the year—but only one of the two years—as the standard.

Given a datum and a standard unit, the periodic scale is then defined with equal intervals. This is not the same as calendar or earth years (of our orbiting planet) which vary in duration. The geochronometric scale is a periodic scale of units—it tells us nothing at all of earth history and there is no suggestion that it took any great achievement to set it up. It already is used. The point is to understand the distinction between the scale itself and the values it is used to express.

Chronostratic Scale

Standardization of the scale by reference points (in stratotypes) that refine the mutual boundaries of adjacent divisions is now generally accepted, (Hedberg, 1972; ISSC, 1976). The divisions are defined by boundaries so it is not only redundant to define them by body or unit stratotype but positively misleading since there may be overlap and conflict with the boundary point (ISSC, 1976, p. 84-86).

This is the philosophy of the golden spike. It was familiar in discussion of boundary questions in my undergraduate days at Cambridge, possibly because of the aftermath of the Sedgwick-Murchison controversy. It is seen in embryo in the IGC 1888 (1891). It was applied in principle in the London IGC (1948) proposed by W. B. R. King for the Pliocene-Pleistocene boundary at the base of the Calabrian sequence in Italy, though a point was not specified. It appears for example in J. Challinor's dictionary under "Systems, Stratigraphic" (1961, and in later editions under Stratigraphic Systems). It was elaborated by the Stratigraphy Committee of the Geological Society of London, by George et al (1967; 1969).

Name for Standardized Global Scales

So far I have used the contracted form chronometric and chronostratic to distinguish the type of scale. These are useful general terms. Names are also needed to refer to the newly standardized scales.

Chronometric

There is little competition. For the scale in years numbered backwards (from 1950) and to distinguish it from any other chronometric scale, I merely recommend the already established uncontracted form, *Geochronometric Scale*.

Chronostratic

There is more difficulty here because not only is there a choice of names but we are in a transition from a scale which is largely biostratigraphically defined, or rather vaguely defined in rock, to a scale with increasing numbers of boundaries standardized in stratotype sections. I suggest that the present hybrid still be referred to in general terms as the stratigraphic scale. The scale that is being superseded could be named the Traditional Stratigraphic Scale (TSS).

The scale as it becomes newly defined has already been referred to both as the *Standard Stratigraphic(al) Scale* (George et al, 1967; Lafitte et al, 1972; Harland et al, 1972) and the *Global Chronostratigraphic (Geochronologic) Scale* of ISSC (Hedberg, 1972; ISSC, 1976). With due respect for the priority and clarity of these names I tentatively suggest that they might better be abbreviated to the name: *Geochronostratic Scale*. The Geo- prefix to the general term chronostratic would signify that it is the global standard.

Regional stratigraphic scales are also chronostratic. The normal procedure in stratigraphy is to describe the rocks in local units and select good sections in any area for regional correlation. These sections and reference points may compete for selection as boundary stratotypes in the geochronostratic scale. But until they are selected they must retain their local nomenclature in order to avoid ambiguity in the use of geochronostratic names. Such points not selected (as golden spikes) may yet serve a useful purpose (as silver spikes) for local correlation. But they cannot be regarded as extensions of the global scale; nor does this procedure remove the need to estimate ages also against the global scale.

Conclusions for Scale Names

On this basis, geochronologic scales would comprise: (1) chronologic scales—of events in sequences of phenomena; (2) chronometric and chronostratic—as general terms for the two kinds of framework scale; and (3) geochronometric and geochronostratic—as names for the two global standard framework scales.

To save these names from misuse by forgetfulness, confusion, reference to other possibilities, hybrid scales, etc, it may be useful to reserve for any occasion the general, comprehensive term *stratigraphic scale* with appropriate qualifiers.

I believe these suggestions to be in harmony with the efforts of the ISSC and other associated bodies.

The Chronomere-Stratomere Problem

The terms *chronomere* and *stratomere* were introduced by the Geological Society of London Stratigraphy Committee (George et al, 1967) only for theoretical discussion of a question that still persists. They have little other use but the question is perhaps an important one.

Examples of chronomeres are eras, periods, epochs, ages, and chrons. Examples of equivalent stratomeres are erathems, systems, series, stages, and chronozones. There is no longer any question about these hierarchies of terms and their equivalence but there is a question about the derivation of one kind of division from the other.

The way many scientists express it (American Commission, 1961; some ISSC statements; and many geologists) is that, to take period-system as an example, the system is primary, being rock, and the period, being time—a derivative concept is therefore secondary. On this view it is better to refer to the rock (i.e. system) when in doubt. I shall refer to this opinion as the *rock-time model*. For example in the ISSC guide (1976) the couplet is represented "system (period)"; in the ISSC publication (ISSC, 1964) the title was *Definition of Geologic Systems*.

An alternative view (labeled here for the purpose of this argument as the *time-rock model*) is that the period is first defined. The system in turn becomes that rock formed during the defined period. This was expressed as: "A chronostratigraphic unit is a body of rock strata which is unified by representing the rocks formed during a specific interval of geologic time." (ISSC, 1961, p. 23). On this model the period is primary. The question is then how to define the period. This is done by specifying two points in rock that are interpreted as two instants in time that define the period.

There is an advantage with the rock-time model with respect to: (1) *local units* (e.g. formations) because the rock is first defined (with diachronous boundaries) and the equivalent time is interpreted directly from the events that formed the rock, and (2) in the *Traditional Stratigraphic Scale* which was largely defined in rock units and so generally had diachronous boundaries.

In my view, for the global or geochronostratic scale, the rock-time model is untenable. I am conscious of going against the popular view which includes the American Code, the ISSC Guides based on it, and many stratigraphers in most countries. I put this challenge rather deliberately because I have already put it more delicately and it has not been taken up (Harland, 1973, p. 570-571). It is a question that cannot be settled by a decision of any authority or majority because it is a matter of scientific principle and not of convention.

I suspect the rock-time model is reinforced by and reflected in the tendency to use period or system names as nouns or substantives. In the German language this is the case as with *Der Jura*; in English, Jurassic is more safely adjectival but many talk with conviction of the *The* Jurassic. Similarly *The* Precambrian is referred to by the ISSC (1976, p. 81).

One difficulty is that this rock-time model is implicit in the original statement of the American Commission (1961), especially the distinction made between "time-stratigraphic (chronostratigraphic) units" and "geologic-time (geochronologic) units" in Articles 26 and 36 respectively. The difficulty of making good sense of these two articles has led many to conclude that this is a somewhat mysterious or esoteric field best left to theoreticians, when indeed the conceptual model itself may be misleading. The alternative time-rock model fits normal stratigraphic operation better (see Fig. 6).

The argument for the time-rock model is that, in the general case the boundary reference points are first defined. These are points in rock but they gain their significance by conversion to the concept of events in time. Except at each point the boundary of the system cannot be seen or known with any certainty, and at some distance or in different facies there may be considerable uncertainty. This is inevitable and there is no logic to overcome it. Therefore the "geologic system" exists in theory by extension of a conceptual time surface from the boundary points. It exists in practice too but as an alleged system which as soon as it is delineated is almost certainly not the true system. We may then say "possibly," "probably," or "certainly of Jurassic age." In consequence of this difficulty, many would like to avoid the use of the system,

SYSTEMS

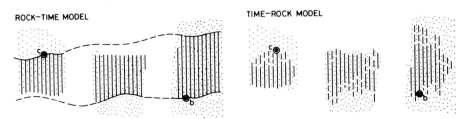

ROCK—TIME MODEL   TIME—ROCK MODEL

FIG. 6–Chronostratic divisions cannot be established or used to organize strata with the kind of authority or certainty employed in lithocorrelation. Two alternative models of the stratomere (e.g. a system) are illustrated. (Left) The *rock-time model* supposes that, as time is interpreted from rock, the system is primary and so defines the period (which some regard as only a secondary linguistic convenience). On this model the boundaries of the system are conceived as surfaces, isochronous with the two defining boundary points. Because the surfaces cannot be identified or extended away from the points, for some authors the system is a kind of ideal and for others it is conceived as the best possible delineation—in some parts older or younger than the period. (Right) The *time-rock model* supposes that the time interpreted from the rock is at the boundary points which determine the span of the period. The system is that rock is conceived as forming during the period, and it can be estimated only with uncertainty near the boundaries.

series, stage hierarchy as being both unnecessary and confusing.

For those who do not wish to face or resolve this conflict, it is possible to manage with a compromise wording which accepts that both chronomere and stratomere are derivative from the reference points (Harland et al, 1972).

Application of Standard Geochronologic Scales

In conclusion, the standard scales are for stating the ages of rocks and events for unambiguous international communication. Whatever other chronologic scales of phenomena are in use it will be good to express them in terms of one or both of these two standard scales and aim at fitting all geohistorical data to a common framework.

The two standard scales have their independent uses and one cannot be expressed in terms of the other except with uncertainty.

The Time Scale Calibration
Relationship Between the Two Standard Scales

Given the two standard scales, Geochronostratic and Geochronometric, there is possible use for them every time a geologic age is stated. The two scales have their peculiar properties that make it convenient on occasion to give ages in chronostratic or alternately in chronometric terms.

It would be simpler if one of these scales could be replaced by the other. In general, for ages that are not precisely stated this is already done but for precise use the age on one scale may be known with an uncertainty factor 10 or more times less than on the other. I have used the spatial analogy of position by geographical coordinates or position with reference to other points (Harland, in press); each has its uses and misuses (it would be inappropriate to lay a table on board ship by astrofix).

Definition of the Time Scale

At any one time there are working calibrations of the chronostratic scale by the chronometric scale. Such a calibration is what has come to be known as the *time scale*. If a more technical term were needed, it could be referred to as the *Geochronologic Scale* because it combines the Geochronometric and the Geochronostratic scales as defined above; indeed it is the title for this paper. But time scale is generally understood well enough.

The principal object of the symposium was to improve this time scale. As the uncertainties in calibration are reduced, more and more purposes will be adequately served by one or other of the standard scales alone because the conversion from one to the other will be more exact.

But for the most precise use some parts of the two scales can never be interchangeable because calibration depends on what evidence is available, and there is no guarantee that rocks will be found that allow satisfactory calibration in every case.

The attempt to give numerical values for the ages of the Traditional Stratigraphic Scale was begun by Holmes (1911), and continued by him in his scales of 1947 and 1961. In honor of this work the Geological Society of London published *The Phanerozoic Time Scale* in 1964. In the meantime many valuable contributions had been made. Part of the history was summarized in a supplement to the 1964 volume (Harland and Francis, 1971). At that time I recounted some other scales and expressed some disappointment that little further progress had been made when so much was needed. The situation has developed rapidly since then. New contributions to the Geochronological Scale have been proposed by Berggren (1972), and J. E. Van Hinte (1976a, b) and so the prospect of a new and greatly improved scale is before us.

It is not for me to forecast what will result from the symposium, so I will conclude by listing only some directions where advances evidently can be made.

Calibration of the Two Scales

The essential method involves an irreducible element of luck. That is the occurrence of rock that contains either directly, or by good evidence for correlation, age data both chronometric and chronostratic. The critical data then known were listed as numbered "items" in the Geological Society's *Phanerozoic Time Scale* volumes of 1964 and 1971.

An ideal example is where interdigitated volcanics and sediments give both radiometric ages and biostratigraphic and magnetic correlation with the geochronostratic scale. More commonly one kind of evidence brackets the other more widely, and the luck then, is how narrow is the bracket.

Generally good calibration points are so scarce on the scale that the best possible use must be made of them. This involves estimates by extrapolation and interpolation. These estimates depend on the rate of processes. Traditional rates are still used as the best available, e.g. rate of sedimentation and rate of evolution. Both of these have been refined somewhat as a result of deep-sea investigations, and also added to by estimates of the seafloor spreading rate. Early seafloor spreading rates accepted some time scales as given; now it is time to use spreading data as part of the evidence.

Another improvement in calibration is by better correlation using the added precision of, for example, marine planktonic studies, magnetic reversals, and climatic and eustatic change.

Nevertheless, the primary method still depends on the luck of finding critical rocks in which those age estimates can be compared. And new rocks as well as new methods are still being discovered—for example, the Ordovician-Silurian boundary (Lanphere, et al, 1977).

As the Precambrian geochronostratic scale is extended to older rocks each new point will need calibrating in years. Possibly the attempt at calibration will accompany the new definition (see Fig. 7).

Further improvements will be incorporated in the new scale, namely better apparent radiometric ages from improved technique and better agreed decay constants, on the one hand, and better definitions of the chronostratic scale.

Glossary of Usage

This glossary is an extension of the text. It is intended to isolate, for convenient reference, matters of definition and difference of opinion, and to simplify the text. It is not intended as a glossary in its own right, being selected for the purpose of this paper.

Where alternative uses are given that are in some conflict, they are indicated by the letters (a) and (b). In this case (a) is the meaning preferred by the author and the alternative given under (b) is only where there is substantial opinion in its favor and so the matter is controversial. Sense (a) has the property of an attempted consistent usage within this paper.

Geochronostratic Scale					Geochronometric Scale
EON	ERA	PERIOD	EPOCH	AGE	
	Cenozoic				65 – 66 Ma
	Mesozoic				235 – 247 Ma
Phanerozoic		Cambrian	(Late) e.g. Tremadoc (Middle) e.g. Merioneth (Early) e.g. St. Davids e.g. Comley		
				e.g. Tommotian	565 – 575 Ma
		e.g. Vendian	e.g. Ediacaran e.g. Varangian		0.68 – 0.7 Ga
e.g. Proterozoic	e.g. Riphean	e.g. Karatavian	e.g. Sturtian		0.95 – 1.0 Ga
		e.g. Yurmatinian			ca1.35 Ga
		e.g. Burzyanian			1.5 – 1.65 Ga
	e.g. Aphebian				ca2.5 Ga
e.g. Archaean					

FIG. 7–Extension of time scale to Precambrian time (compare with Table 1).

Alternative uses that seem to coexist and are therefore not in conflict, are distinguished by the numbers (1), (2), (3), etc.

absolute age—commonly used for geochronometric ages which are numerical but in no sense absolute. Holmes and the IUGS Precambrian Subcommission discouraged this term as misleading. When used, it is often contrasted with *relative age:* a contrast in this paper between chronometric and chronostratic ages.

age—(1) *sensu stricto* is the chronostratic division of rank between epoch and chron, and would be capitalized for a particular division. (2) *sensu lato* is a convenient geochronologic term for the chronometric or chronostratic time of any event.

apparent age—the actual value for the geochronometric age obtained in a radiometric laboratory after standard corrections have been made. The numerical standard error generally associated is a measure only of the difference of individual determinations, and leaves out most other reasons for uncertainty.

assignment—(a) of rock to a local rock unit (see *lithostratigraphic unit*).

(b) of age (see *estimation*).

auxiliary type section—see *stratotype*.

biochron—the time interval represented by a *biozone:* (a) this is a useful informal concept in discussion of correlation.

(b) because of the indefiniteness of biozones, the formal biochron as the smallest division of the traditional stratigraphic scale is giving way to a *chron*, defined by reference points in type sections.

biohorizon—(a) preferred as a positively identified occurrence of distinctive fossils and facies.

(b) rather than a theoretical surface (of no thickness) delimiting biostratigraphic units.

biostratigraphic unit—see *biozone.*

biozone—is a (biostratigraphic) body or unit of rock based on, and identified by, the occurrence of recognizable fossils. For effective use the following matters need to be clear: (1) a biozone is a rock with appropriate fossils, and not a time equivalent (see *biochron*), (2) a biozone depends on the definition and identification of appropriate taxa, and so depends on the authority and date of determination, (3) a biozone is based on fossils actually found. A degree of judgment is applied when interpolating and extrapolating from recorded occurrences of fossils—a biozone is thus subject to indefinite revision with further collection, (4) derived fossils, contamination, etc. shall be assessed, (5) the taxa selected as diagnostic may be chosen as follows: if one taxon is used the biozone may be identified wherever it occurs (*range-zone*), or by abundant occurrence (*acme-zone*). Alternatively, an occurrence of two taxa may be diagnostic (*concurrent range zone*) or more (*assemblage zone*), (6) a very distinctive occurrence of taxa in a widely distributed facies of small thickness may be referred to as a *bio-horizon*, and (7) an *interval zone* has been suggested for the rock between two distinctive biohorizons or biozones (for fuller details and more terms see ISSC, 1976, p. 45-65).

It is preferred to include biostratigraphic data in descriptions of rock sequences and to use the previously listed concepts for discussion of facies, correlation etc. (see *litho-stratigraphic unit*), and not to set up formal biostratigraphic units because they are subject to opinion and revision.

body stratotype—*unit stratotype* (see *stratotype*).

boundary reference point— see *stratotype*.

boundary stratotype—see *stratotype*.

categories of classification—a category is a class of classes, and so reflects a way of thinking. Categories of stratigraphic concepts are often, therefore, controversial. (a) In this paper *phenomenon* and *framework* categories of stratigraphic division, or *norma-tive* and *positive* propositions, are separated as essentially different.

(b) To be discouraged is the treatment of rock and time (framework categories) as of equal rank with indefinite numbers of categories based on individual phenomena. For example, the American Commission and ISSC list categories of units: lithostrati-graphic, biostratigraphic, chronostratigraphic, and suggest many others.

chron—(a) is the lowest rank of *chronomere* in the geochronostratic scale, to be defined, as with any other division in it, by reference points in type sections. Many of the reference points bounding chrons will also serve divisions of higher rank. When established chrons replace the divisions that are currently based on biozones in the Traditional Stratigraphic Scale, they may take the traditional zonal names.

(b) therefore chron should not be used for time divisions otherwise based.

chronocorrelation—time correlation (1) a scientific activity and (2) a proposition, or correlation statement, (in this sense, equal to *chronotaxis*).

chronohorizon—not necessary; but if used, then (a) prefer for time corresponding to a horizon which has finite thickness.

(b) and not as *isochronous surface* or *time surface* (ISSC, 1976, p. 14 and 67-68).

chronomere—a chronostratic interval of time (George et al, 1967) used as equivalent to *stratomere*.

chronometry—measurement of time.

chronostratic—abbreviated from chronostratigraphic (Hedberg, 1972, p. 28) and referring to the time divisions and related rock divisions defined by stratotypic refer-ence points.

chronostratigraphy—concerned with *chronostratic* operations.

chronotaxis—time equivalence or chronocorrelation.

chronozone—(a) lowest rank of stratomere in geochronostratic scale and so corre-sponds to *chron*.

(b) not to be used as time equivalent of any other kind of zone.

climatochron—if a term is needed for the time equivalent of the climatozone which is rock.

climatozone—(a) as with biozone and magnetozone an informally described body of rock characterized by paleoclimatic indicators.

(b) prefer not to establish a formal hierarchy of rock and time equivalents, except as local rock units (e.g. tillites).

coeval—approximately, but not alleged to be exactly, chronotaxial.

contemporaneous—allegedly chronotaxial, but of disparate events.

correlation—(1) an activity, and (2) a proposition (geologic correlation by any character or criteria, not only time correlation).

determination—(a) properly used where the matter can be settled by authority (e.g. establishing a taxon or identifying a fossil); also the definition of a rock unit.

(b) better avoided for matters of *estimation*.

division—(1) *sensu lato*—a general term for any part of any stratigraphic sequence (George et al, 1967), and (2) *sensu stricto*—a term for any part of a geochronologic scale, excepting a geochronometric *unit* (e.g. Jurassic, Tremadocian, middle Precambrian—2.5 to 1.5 Ga).

Eon—(a) a chronomere of the highest rank in the geochronostratic scale, compounded of eras. The equivalent stratomere is eonothem.

(b) therefore avoid use for chronometric unit of 1 Ga equals 1×10^9 years—for which gigennium (or billion) is correct.

(c) and for a long time.

episode—the time interval of a named event.

epoch—(a) a chronomere in geochronostratic hierarchy.

(b) therefore avoid use for time equivalent of magnetozone.

error—standard error commonly quoted in radiometric age estimates measures only the scatter of actual observations, and in no sense measures any other source of uncertainty.

estimation—an age estimation based either on laboratory work or on the opinion of a geoscientist. For most uses this is a better word than *determination* or *assignment*.

event—(a) a happening, generally conceived from observation and interpretation of rock, and of relatively simple structure. Usage is best kept entirely informal.

(b) avoid use of event in formal nomenclature (e.g. in magnetostratigraphy).

formal stratigraphic usage—contrasts with *informal*—requires agreement as to terms and procedures, definitions, and nomenclature. Formal names are capitalized (e.g. local rock units, geochronologic divisions).

Geochronologic Scale—(a) the time scale calibrating geochronostratic divisions by geochronometric estimates, by any and every means.

(b) without capitalization, may apply to any time scale for use in geohistory, but for general use, *time scale* is preferred.

geochronology—(a) *sensu lato*—". . .the science of dating and determining the time-sequence of events in the history of the earth. . ." (ISSC, 1976, p. 15.)

(b) *sensu stricto*—*geochronometry* (e.g. of Subcommission of Geochronology), or *chronostratigraphy* (e.g. American Commission, 1961, and earlier ISSC publications).

Geochronometric Scale—the scale in standard units of duration, or years, for geohistory, with or without named divisions, compounded of units.

geochronometry—"that branch of geochronology that deals with the quantitative measurement of geologic time (normally in years)" (ISSC, 1976, p. 15; Harland et al, 1972).

Geochronostratic Scale—(a) proposed here for scale of chronostratic divisions when standardized.

(b) equivalent to, but of shorter form than, *Standard Stratigraphic Scale* (SSS) of George et al, 1967; Harland et al, 1972; Laffitte et al, 1972; and *Standard Global Chronostratigraphic (Geochronologic) Scale* of ISSC (Hedberg, 1972; ISSC, 1976).

geohistory—a contraction for geologic history or earth history.

golden spike—(or *peg*)—a colloquial expression for the agreed reference point in a boundary stratotype, for the Geochronostratic Scale (see also silver spike).

global standard—(ISSC, 1976, p. 76-81) see *geochronostratic*.

geologic time—to do with geohistory.

geoscientist time—to do with scientific history, so that a sequence of dated interpretations in *geoscientist time* may relate to a distinct sequence of events in geologic time. Contrasts with geologic time.

geoscientist time—to do with geohistory, contrasts with geologic time.

hierarchy—a convenience in classification whereby larger entities comprise smaller entities without overlap or gap. Examples: (1) for local rock units: complex, supergroup, group, formation, member, bed, (2) for geochronostratic divisions: eon—eonothem, era—erathem, period—system, epoch—series, age—stage, and chron—chronozone, (3) for compounds of geochronometric unit—Ga, Ma, ka, a, (4) for divisions of Geochronometric Scale, (e.g. middle Precambrian, 2.5 to 1.5 Ga), or else *aevum* (1.5 Ga) and *aetos* (0.5 Ga) of Harland (1975), and (5) for biomagneto, etc.: -superzone, -zone, -subzone.

historical geology—stratigraphy *sensu lato*, but includes geohistorical interpretation of evidence other than strata (e.g. origin of solar system).

homotaxis—"Similarity of serial arrangement for groups of strata in different localities showing similar vertical sequences of faunas." (Huxley, 1862).

horizon—(a) prefer a thin unit of rock recognizable over a wide area, typically homotaxial and not necessarily chronotaxial.

(b) avoid usage as isochronous surface.

informal stratigraphic usage—(compared to *formal*) has the positive advantage, that by not being fixed by definition it is capable of indefinite revision and improvement. It is therefore appropriate to positive scientific work as with biostratigraphy, magnetostratigraphy, tectonic, climatic, correlation, etc.

interpretative stratigraphy—(compared to *descriptive stratigraphy*) includes geochronology, paleogeology, paleoenvironmental studies, etc.

instant—(a) a point in time of no duration.

(b) do not favor equivalence to *chronohorizon*.

interval—(a) a *time interval*, for example: (1) a division of time between stated points or events, and (2) informally used as time equivalent of any zone, (e.g. magnetozone).

(b) also used for "the body of strata between two stratigraphic markers" as referenced, *stratigraphic interval* of ISSC (1976, p. 14).

lithologic—to do with petrologic character, easily recognizable in the field and therefore suitable as one element in a rock description.

lithostratigraphic unit—is a local rock unit defined and named from a type section according to a well-established routine that uses the hierarchical scheme formation, etc. Unfortunately the term conceals different usages that may confuse:

(a) the unit as based on any characters that are convenient for description and distinction from adjacent units—lithologic, paleontologic, petrophysical, etc. (George et al, 1967).

(b) the unit as based on lithologic (petrologic) characters only (American Commission, 1961; Hedberg, 1960). This usage is supplemented by a parallel category of biostratigraphic units, etc.

Some who prefer practice (a) but acknowledge the strength of tradition (b) avoid this term for sense (a) and use instead rock stratigraphic or local units (Harland et al, 1972; Laffitte et al, 1972).

lithostratigraphy—(a) descriptive or rock stratigraphy which gives the recorded data of stratigraphy, (e.g. in measured stratigraphic sections with fossil lists and other information, and in large scale maps and so divides the earth's crust into convenient, named units to which all data and all interpretations may be referred.

(b) the same definition, only restricted to purely lithologic characters. Therefore, this becomes a subdiscipline of stratigraphy.

local rock unit—local unit or rock unit (see *lithostratigraphic unit*).

magnetochron—is the time equivalent of *magnetozone*. An alternative use for this chronomere is X-time, or X-interval, for X-magnetozone (i.e. equivalent stratomere), when X is a proper name. Magnetochron is preferred to magnetic epoch.

magnetostratigraphy—is one of the many subdisciplines of stratigraphy (see *magnetozone*).

magnetozone—for magnetic characters (compare to *biozone*), generally to be named from the locality to which it may be referred in type section. If a geomagnetic polarity zone is in use, it will either be clear from the use of the proper name as defined and from the qualifying words, *normal* or *reversed*. This leaves it open to use zones for unclear or other magnetic properties if convenient. Because of their global nature and great value in correlation, a distinct nomenclature is useful. However, the main value is to relate the data to the Geochronologic Scale, and additional terminology should be minimized. A magnetozone is rock; for the equivalent time see *magnetochron*.

moment—(a) a short and specified finite-time interval

(b) do not favor the "geochronologic equivalent of a chronohorizon" (ISSC, 1976, p. 68) because a chronohorizon is itself the time corresponding to a horizon.

original section—(a) the stratal section originally described and of historical rather than contemporary reference value.

(b) holostratotype (see *stratotype*).

paleoclimatic zone—see *climatozone*.

penecontemporaneous—nearly but not contemporaneous, compared to *coeval*, which is approximately contemporaneous.

petrophysical stratigraphic units—(e.g. magnetozone) seismic, density, porosity zones

(a) the characters are appropriate for definition of local rock units, subsurface units especially.

(b) otherwise to be treated informally for petrophysical correlation.

phenomonic—of phenomenon (i.e. natural or geohistorical).

prostratigraphy—from German usage of Schindewolf (1957), for descriptive stratigraphy (see *lithostratigraphy*).

radiometric age—based on analysis of radioactive and radiogenic elements and isotopes. Usage is preferred over isotopic age which excludes the early classic estimates.

rank—see *hierarchy*.

reference point—see *stratotype*.

reference section—(1) any section described well enough and useful for reference; and (2) a *stratotype*.

regional stratigraphic scale—(a) preferred is the procedure to describe good sections anywhere and establish reference points in them so as to focus correlation estimates. Such reference points may be candidates for silver or golden spikes. They should be related to local names until (exceptionally) a global chronostratic reference point is established.

(b) the view could be misleading that in some way the Geochronostratic Scale can be extended approximately through local stratotypes (ISSC, 1976, p. 81). No more than one point may be used to define any geochronostratic boundary.

relative age—(a) age of one event relative to another, or

(b) avoid relative age (and *absolute age*) prefer chronostratic age (and chronometric age).

rock unit—see *lithostratigraphic unit*.

scale—conceived as a device in one dimension with divisions to calibrate phenomena. A scale may be of equal units (e.g. chronometric scale), or of many kinds with unequal divisions (e.g. chronostratic divisions).

series—(a) agreed as stratomere for epoch (e.g. Lower Jurassic Series or rock of Early Jurassic Epoch). Epoch/series names associated with the terms Early (Lower), Middle, and Upper (Late), need to be replaced by proper names when standardized (e.g. Lias).

(b) correct earlier usage for a body of strata now to be avoided.

strata—rock layers almost any shape and any lithology if formed in sequence.

stratal—a useful adjective to replace stratigraphic when concerned only with strata, (e.g. stratal thickness, facies, sequence, etc.).

stratigraphic scale—suggested as a completely general term for use when a more restricted usage is not intended.

stratigraphy—(a) *sensu lato*—". . .is the study of rocks and their distribution in space and time with the object of reconstructing the history of the earth, and eventually of extra-terrestrial bodies. . ." (Laffitte et al, 1972). It may be divided into descriptive stratigraphy (see lithostratigraphy, definition a), and interpretative stratigraphy (including geochronology, geohistory, environmental paleogeography etc.)

(b) *sensu stricto*—do not favor either: (1) descriptive stratigraphy; or (2) geochronology and geohistory (or historical geology).

stratomere—a division of rock formed in a distinct chronomere. It is therefore bounded by unknown isochronous surfaces (George et al, 1967) as in the geochronostratic scale, system (stratomere) corresponds to period (chronomere).

stratotype—a stratigraphic section, selected, named, and published for reference. As originally set up it is a *holostratotype*. If necessary when inadequate it may be replaced, even at a distance, by a *lectostratotype*, or because destroyed or nullified by a *neostratotype*, (ISSC, 1976, p. 26).

(1) a *boundary stratotype* is a section in which the *reference point* that defines the boundary of the division in a chronostratic scale is agreed (golden spikes for the Geochronostratic Scale, silver spikes for any regional stratigraphic scale)—there can be only one reference point (i.e. only one stratotype) for one chronostratic boundary; and

(2) a *body* or *unit stratotype* is the typical, or type section, referred to when a local rock or lithostratigraphic unit is defined. An *auxiliary* type, or reference, section may also be used to typify the unit with other characters or in other places (and may be termed a *parastratotype* if used in the original definition by the original author, or a *hypostratotype* if subsequently designated: (ISSC, 1976, p. 26). Auxiliary type sections cannot be used to extend chronostratic boundaries.

The stratotype analogy with taxonomic procedure in biology is not a good one and may be avoided by using only: *boundary reference points* (for chronostratic scale), and *type sections* and *auxiliary type sections* (for local rock units) with author and date.

synchroneity—is possible for events of appreciable duration; only instants can be isochronous.

tectonic nomenclature—(a) tectonic phenomena on a large scale, (e.g. orogens, tectogenesis) or on a small scale (e.g. tectonic phases, unconformities) are generally defined and named in some region on the basis of rock structure.

(b) tectonic names so related (e.g. Caledonian, Cimmerian) are misleading when applied to the time of allegedly synchronous events outside the eponymous area except as an explicit time-correlation hypothesis.

time correlation—estimation of the relative ages of events interpreted from rock by placing them in sequence with other events—chronostratigraphic correlation (ISSC, 1976, p. 14); but do not favor so narrow a definition as "demonstrates correspondence in age" as the logic of the demonstration is uncertain (see *chronotaxis*).

time-rock model—(a) for a conceptual model as explained here in contrast to a *rock-time model* (see text).

(b) usage not recommended for any kind of stratigraphic unit or division.

time-scale—see *geochronologic scale*.

time surface—(a) an *isochronous surface*—a useful concept, but can hardly be demonstrated except locally.

(b) avoid use of *chronohorizon* which has duration, and avoid *time plane* except in a correlation diagram.

true age—the true age (geochronometric or geochronostratic) is unknown except at the points of definition of the scales, but the concept is useful in discussion and motivation.

type—the term needs further qualification to be useful.

type-section—see *stratotype*.

unit—see *division*—(a) used in two senses (Harland et al, 1972) (1) for unit in chronometry in mathematical sense, and (2) for part of a rock sequence which is distinguished by characters that are said to give it a unity.

(b) is used by American Commission, and ISSC generally, for parts of a chronostratic scale, although division is preferred in this sense.

zone—the term needs further qualification to be useful.

References Cited

American Commission on Stratigraphic Nomenclature, 1961, Code of stratigraphic nomenclature: AAPG Bull., v. 45, no. 5, p. 645-665.

Berggren, W. A., 1972, A Cenozoic time scale—some implications for regional geology and paleobiogeography: Lethaia, v. 5, p. 195-215.

Challinor, J., 1974 (1961, 1964, 1967), A dictionary of geology: New York, Univ. Wales Press-Univ. Oxford Press, 350 p.

George, T. N., et al, 1967, The stratigraphical code—report of the stratigraphical code sub-committee: Geol. Soc. London Proc., no. 1638, p. 75-87.

—— et al, 1969, Recommendation on stratigraphical usage: Geol. Soc. London, Proc., no. 1638, p. 139-166.

Harland, W. B., 1970, Time, space and rock (an essay on some fundamentals of stratigraphy): West Commemoration Volume (Paridabad, India), p. 17-42.

—— 1973, Stratigraphic classification, terminology and usage–essay review of: An international guide to stratigraphic classification, terminology and usage (H. D. Hedberg, ed.): Geol. Mag., v. 110, no. 6, p. 567-574.

—— 1975, The two geological time scales: Nature, v. 253, p. 505-507.

—— (in press), The main concepts of the time-stratigraphic scale for the Precambrian (Moscow Symposium Sept. 1975).

—— and W. H. Francis, eds., 1971, The Phanerozoic time scale–a supplement: Geol. Soc. London, Spec. Pub. 5, 356 p.

—— A. G. Smith, and B. Wilcock, eds., 1964, The Phanerozoic time scale–a symposium: Geol. Soc. London Quart. Jour. v. 120, Supp., 458 p.

—— et al, 1972, A concise guide to stratigraphical procedure: Geol. Soc. London Quart. Jour., v. 128, p. 295-305.

Hedberg, H. D., ed., 1972, ISSC report 7: Lethaia, v. 5, no. 3, p. 283-323.

Holmes, A., 1911, The association of lead with uranium in rock minerals and its application to the measurement of geological time: Royal Soc. London Proc. Ser. A, v. 85, p. 248-256.

—— 1947, The construction of a geological time scale: Glasgow Geol. Soc. Trans., v. 21, p. 117-152.

—— 1959, A revised geological time scale: Edinburgh Geol. Soc. Trans., v. 17, p. 183-216.

Hughes, N. F., and J. C. Moody-Stuart, 1969, A method of stratigraphic correlation using Early Cretaceous miospores: Palaeontology, v. 12, pt. 1, p. 84-111.

—— et al, 1967, A use of reference points in stratigraphy: Geol. Mag., v. 104, no. 6, p. 634-635.

—— et al, 1968, Hierarchy in stratigraphical nomenclature (reply to discussion by P. C. Sylvester Bradley): Geol. Mag. v. 195, no. 1, p. 79.

Huxley, T., 1862, The anniversary address: Geol. Soc. London Quart. Jour., no. 18, p. xl-liv.

International Geological Congress, London 1888 (1891), Appendices A, B, and C.

International Subcommission on Stratigraphic Classification (ISSC), 1961, Principles of stratigraphic classification and terminology: 21st Int. Geol. Cong. (Norden) Proc., Rept. 1, Part 25, p. 38.

—— 1964, Definition of geologic systems: AAPG Bull., v. 49, no. 10, p. 1694-1703.

—— 1976, H. D. Hedberg, ed., International stratigraphic guide: New York, John Wiley & Sons, 200 p.

Kulp, J. L., 1961, Geologic time scale: Science, v. 133, no. 3459, p. 1105-1114.

—— 1961, (ed.), Geochronology of rock systems: New York Acad. Sci. Annal., v. 91, p. 159-594.

Laffitte, R., et al, 1972, Some international agreement on essentials of stratigraphy: Geol. Mag., v. 109, p. 1-15.

Lanphere, M. A., M. Churkin, Jr., and G. D. Eberlein, 1977, Radiometric age of the *Monograptus cyphus* graptolite zone in southeastern Alaska–an estimate of the age of the Ordovician-Silurian boundary: Geol. Mag., v. 114, p. 15-24.

Layzer, D., 1975, The arrow of time: Sci. American, v. 233, p. 56-69.

McLaren, D. J., 1973, The Silurian-Devonian boundary: Geol. Mag., v. 110, p. 302-303.

Needham, J., 1954 (*et seq*), Science and civilisation in China: London, Cambridge University Press.

Schindewolf, O. H., 1957, Comments on some stratigraphic terms: Am. Jour. Sci., v. 255, no. 6, p. 394-399.

Stille, H., 1924, Grundfragen der vergleichenden tektonik: Berlin, Borntraeger, 443 p.

Tarling, D. H., and J. G. Mitchell, 1976, Revised Cenozoic polarity time scale: Geology, v. 4, no. 3, p. 133-136.

Van Hinte, J. E., 1976a, A Jurassic time scale: AAPG Bull., v. 60, no. 4, p. 489-497.

—— 1976b, A Cretaceous time scale: AAPG Bull., v. 60, no. 4, p. 498-516.

Weller, J. M., 1960, Stratigraphic principles and practice: New York, Harper and Brothers, 725 p.

Williams, H. S., 1893, Elements of the geological time scale: Jour. Geology v. 1, p. 283-295.

Stratotypes and an International Geochronologic Scale[1]

HOLLIS D. HEDBERG[2]

Abstract An international geochronologic scale is needed in order to provide a single universal standard of reference for dating rock strata, or events recorded in rock strata, anywhere in the world.

There are many possible means of relating rock strata and the geologic events they record to the passage of time—physical relations of strata (law of superposition), degree of isotopic decay, stage of organic evolution indicated by fossils, and other methods. Each is useful, but each is fallible under certain circumstances. The ideal standard, therefore, is one which does not depend on any one method but allows and encourages the utilization of all methods of age determination and time correlation.

All methods of geologic age determination and time correlation must be based fundamentally on features of the rock strata; and thus the rock strata constitute the best register in which to inscribe the standard reference points—*stratotypes*—for whatever units or other horizons we wish to recognize in a global geochronologic scale. Only thus, by the designation of standard unit stratotypes, boundary stratotypes, and other horizon stratotypes, can we establish unequivocal definitions of points on this scale in a manner which lends itself to the utilization of all methods of age determination and time correlation, reinforcing the evidence from each by that from all others, and limited for each only by its capacity to usefully contribute.

The practical value and utility of points on a global geochronologic scale is dependent on the extent and accuracy with which chronohorizons coincident with the stratotypes of these points can be traced or identified elsewhere in the world. Therefore, effort should be made to designate these stratotypes at places in the stratigraphic sequence which, through their coincidence with, or relation to time-significant features (isotopic dates, fossils, magnetic reversals, etc), particularly lend themselves to reliable widespread time correlation. Individual chronohorizons are no less important than chronostratigraphic units and their boundaries in the reference base for a geochronologic scale.

Some geologists have questioned a basic tenet of chronostratigraphy—theoretical boundaries of a chronostratigraphic unit should be everywhere isochronous—on the grounds that erosion or nondeposition has locally altered the position of these boundaries from those seen at the stratotypes. However in such cases, the boundary is still theoretically present at the same time level, although included within the time value of the hiatus or unconformity.

The stratotype concept should be applied to the global geochronologic scale, including both Quaternary and Precambrian.

Introduction

What is a geochronologic scale? Literally, it is a ladder with rungs indicating positions in geologic time. It is a sequential arrangement of identifiable time horizons (chronohorizons) and units of geologic time. This time scale provides a standard reference system for expressing the age of a rock or the age of a geologic event and its position with respect to earth history.

An international geochronologic scale is one which has worldwide application and is

[1] Manuscript received, October 4, 1976.
[2] Princeton University, Princeton, NJ 08540.

Article Identification Number: 0149-1377/78/SG06-0003/$03.00/0

globally accepted. It provides a single universal standard of reference for dating rock strata or events of earth history with respect to the passage of geologic time.

What is geologic time? Geologic time is simply time used in the context of earth history, determined by geologic methods. Our only record of the passage of geologic time lies in the Earth's rock strata. This record is imperfect, but it is our only one.

What methods can be used to determine position in geologic time? There are many different methods which have been used more or less successfully:

(1) In a normal depositional rock sequence, the uppermost strata are youngest, and age increases with depth. Thus, physical position of the strata provides the basis for our oldest kind of geochronologic scale, which is still valid with respect to relative age for any locally observed sedimentary sequence.

(2) Fossils in a rock may often indicate its position with respect to the irreversible course of organic evolution, and thus relative age. Most of the divisions of our current international scale are based on fossils.

(3) The isotopic decay of certain elements in certain minerals can often provide information which can be translated into age in years or millions of years. Isotopic dating is the major hope for chronostratigraphic organization of the vast, poorly fossiliferous, older part of the rock sequence, and it is contributing greatly to the dating of all parts of the column.

(4) Numerous other aids in estimating age and the passage of geologic time have been developed. These include methods based on rates of sedimentation, growth increments of invertebrates, tree-rings, varves, and many others.

All of these methods are useful under certain circumstances, but all are also fallible under certain circumstances. For example, the normal depositional sequence of beds may be broken or even reversed by structural disturbances. Fossils may be absent or may be long-ranging forms that do not allow sharp positioning in the evolutionary scale. Minerals suitable for isotopic age determination may be lacking, or isotopic dates obtained may be those of the last thermal metamorphism of a rock rather than of its true age, and so on.

Consequently, the ideal basis for a geochronologic scale will be one which need not depend exclusively on any one of these methods but will allow *all* to be fully utilized. Fortunately, such a basis is available to us. Our record of the passage of geologic time lies in the rock strata and only in the rock strata; all methods of age determination mentioned previously depend on analysis of rock strata features. Therefore, rock strata constitute the most natural and the most practical register in which to inscribe the positional standards of whatever markers or units of an international geochronologic scale we wish to recognize. Thus we can combine on one scale the contributions of every means of age determination, with disadvantage to the capabilities of none, and with immense strengthening of the end result by reinforcing one line of evidence with another.

Because such a scale is based on physical *marker points* and *divisions* of the rock column, it is fundamentally a chronostratigraphic scale; however, whereas the purpose of the horizons and units recognized on the scale is to provide a geochronologic reference framework for earth history, it may be referred to appropriately as either a chronostratigraphic scale or a geochronologic scale. In the International Stratigraphic Guide (ISSC, 1976, p. 10-11, 66, and 76-81), we have chosen to refer to this scale as the *Global Chronostratigraphic (Geochronologic) Scale*.

Is A New Scale Needed?

In graphical representations of the scale, vertical distance is now generally made proportional to geologic time in millions of years, thanks to the indispensable quantitative contributions of isotopic dating. Useful points inscribed on the scale may be of several types: (1) positions of reliable isotopic age measurements, (2) chronohorizons of important paleontologic or other geologic events, and (3) limits (boundary stratotypes) of classic chronostratigraphic units, etc. Although the scale may be called a ladder, it differs from a carpenter's ladder in that the distance between the steps or rungs may be unequal and quite irregular. Its function is simply to provide the best possible reference framework for recording the timing of important events in earth history, regardless of their spacing.

During the last century, a hierarchy of conventional units of rock strata became established for the scale to express intervals of geologic time of several different orders of magnitude. These terms as presently used are shown in Figure 1. Globally recognized formal names have now been given to many of these—at least two eonothems (eons), three erathems (eras), twelve systems (periods), and numerous series (epochs) and stages (ages). These named units of our current international geochronologic scale historically have served a useful purpose in the development of global geology by providing a universally accepted reference scheme for equating ages of rocks throughout the world.

Chronostratigraphic   Geochronologic

Eonothem . . . . *Eon*
Erathem . . . . *Era*
System . . . . *Period*
Series . . . . *Epoch*
Stage. . . . . *Age*

Chronozone . . . . *Chron*

FIG. 1—Conventional hierarchy of chronostratigraphic and geochronologic terms. (From the International Stratigraphic Guide, ISSC, 1976, p. 68).

Many geologists have criticized this existing scale and the basis on which it rests. They have said that its units are too artificial, that they are not artificial enough, that it is too strongly based on paleontology, that it rests on outmoded concepts from the early history of geology, that its units are poorly defined, that they are too unequal in rank, that the scale overstresses time intervals (units) as contrasted with individual events (horizons), that it over emphasizes the last few hundred million years of earth history and neglects the preceding thousands of millions of years, that it is not sufficiently quantitative, and so on.

Much of this criticism is justified, but the fact remains, imperfect as it may be, that this scheme has been the heart of our remarkable progress in stratigraphy and earth history during the last 200 years. Moreover, even if a better scheme could be devised, so much of our data on world stratigraphy has already been entered in the vast geologic literature in terms of the units of this scheme that to abandon it completely would be almost suicidal. The answer, it seems to me, is not to change to a different scheme but rather to seek to remedy the defects of the present scheme and to improve it for the future.

The Task at Hand

In my opinion, the *first* and most urgent task in connection with our present international geochronologic scale is to achieve a better definition of its units and horizons so that each will have a standard fixed-time significance, and the same time significance for all geologists everywhere. Most of the named international chronostratigraphic (geochronologic) units still lack precise globally accepted definitions and consequently their limits are controversial and variably interpreted by different workers. This is a serious and wholly unnecessary impediment to progress in global stratigraphy. What we need is simply a single permanently fixed and globally accepted standard definition for each named unit or horizon, and this is where the concept of stratotype standards (particularly *boundary* stratotypes and other *horizon* stratotypes) provides a satisfactory answer.[3]

[3] A stratotype is the original or subsequently designated type of a named stratigraphic unit or boundary or other horizon identified as a specific interval or a specific point in a specific sequence of rock strata, and constituting the standard for the definition and recognition of that stratigraphic unit, boundary or other horizon. See International Stratigraphic Guide (ISSC, 1976, p. 24-29).

The most fundamental, and the only permanently stable, base for recording a position in geologic time is a designated point or interval in a specific sequence of continuously deposited rock strata to serve as its stratotype. The stratotype standard is more primary and fundamental than an isotopic age in years, or a relative age based on fossils, because both of these latter are abstractions, subject to variation in judgment or in techniques. The standards are derived secondarily by means of analysis and interpretation of evidence in the rock strata, whereas a stratotype is marked directly on the time calendar of the earth's sequence of rock strata. My definition of a named chronostratigraphic unit as ". . .the rocks formed during the interval between two instants in time each represented by clearly designated points in sequences of continuously deposited strata (boundary stratotypes). . ." fixes its time scope in that sequence incontrovertibly. Similarly, the identification of a named chronostratigraphic horizon with a certain designated point in a continuously deposited sequence of strata (horizon stratotype) incontrovertibly fixes its position in time in that sequence. Thus, through these stratotypes we can at least have single and definite standards for reference in attempts to identify these units or horizons elsewhere.

Some geologists have objected to stratotype standards for an international geochronologic scale on the grounds that they are too exact, too rigid, too restrictive, and too difficult to correlate, among other reasons. Some geologists have proposed that the standards be paleontologic concepts, or tectonic concepts, or age in years as determined by isotopic methods, or that they be based on a combination of isotopic and magnetostratigraphic data. There can be no objection to attempts by others to make scales based on these features for appropriate purposes, but these should not be allowed to impair or nullify the fundamental stratotype-based scale which alone can utilize the combined contributions of all these features.

The *second* major task with respect to an international geochronologic scale is that of the widespread extension by time correlation of the stratotype points of the scale to other areas, worldwide. No matter how well a point in time may be identified in a stratotype, it is of little use unless the same horizon-in-time can be, at least roughly, identified elsewhere. And here another advantage of stratotype standards becomes evident. The concept of the stratotype standard lends itself not only to the use of *all* methods of age determination in the setting up of the stratotype, but also to *all* methods of time correlation in extending the horizon or interval of the stratotype away from its type locality to other parts of the world—lithologic, paleontologic, isotopic, magnetostratigraphic, and other methods—because these all utilize and depend on rock strata features.

Time correlation is often very difficult and rarely very exact but the objective is simply to extend horizons away from the stratotype at as *nearly an isochronous position as possible*. Certainly, the more methods and cross checks we can apply, the better our results and the closer our approach to a never-perfectly-attainable goal. Care should be taken in the first place to designate stratotype points at horizons which, through their relation to time-significant features of the rock strata, lend themselves to reliable worldwide or regional time correlation by as many and as effective means as possible. It is particularly desirable to choose stratotype points at or near horizons which can be accurately dated isotopically.

Some geologists have questioned a basic tenet of chronostratigraphy—that a chronohorizon should be of the same age everywhere and that the theoretical boundaries of a chronostratigraphic unit should be isochronous. They say that the boundaries of a chronostratigraphic unit cannot be everywhere isochronous because erosional unconformities and hiatuses in deposition, in some places, have reduced the section to only a small part of its time scope as seen at the stratotype, so that the uppermost and lowermost limits of the unit in these incomplete sections are at quite different horizons than in the stratotype and hence cannot possibly be considered isochronous. However, this supposed anomaly is merely the failure to realize that in an area where the boundary sequence of strata known at the stratotype is incomplete or missing due to erosion or nondeposition, the boundary is still present in theory as an isochronous horizon merged with the hiatus or surface of unconformity (Fig. 2). The theoretical time scope of the unit between its two isochronous boundary horizons remains the same even

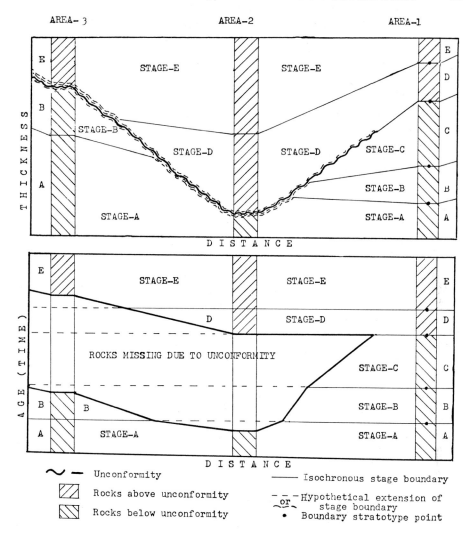

FIG. 2–These two diagrams are the same section–the upper one has a vertical scale indicating thickness and the lower one has a vertical scale indicating time. In both, the horizontal scale is a geographic distance. The right-hand column on both diagrams (Area 1) shows a hypothetical stratotype section through five consecutive stages. The central and left-hand columns (Areas 2 and 3) show the boundaries of these stages traced from the stratotype section into areas where a major unconformity has cut out several of the stages.

In the lower diagram, where the vertical scale is time, the isochronous boundaries of the stages are horizontal lines. The time value of the unconformity at any point is shown by the length of the vertical line through the trapezoidal area in the middle of the diagram through which the isochronous stage boundaries pass as horizontal dashed lines. In the upper diagram, where the vertical scale is actual thickness of rock strata, the boundaries of the stages are still isochronous, but are not horizontal lines because of lateral variation in rate of deposition and lateral variation in amount of erosion. The isochronous stage boundaries merge laterally with the surface of unconformity in which they are comprised.

though part of the unit is physically missing. One could as well object to recognition of a unit as isochronous on the grounds that due to internal diastems some parts of the time scope of the unit were not everywhere represented by rock strata.

The *third* task, with respect to the international geochronologic scale is for more emphasis on individual *chronohorizons* regardless of their relation to conventional named chronostratigraphic units of the scale. These chronohorizons may represent important geologic events or they may simply be based on points of particularly reliable and significant isotopic age determinations. They may mark important fossil extinctions or important magnetic polarity reversals. They are no less important features of the international geochronologic scale and no less important to earth history just because they may not happen to coincide with classic chronostratigraphic unit boundaries.

In conclusion, I believe that the stratotype concept is the best means of assuring definite universally acceptable standards for important points of the international geochronologic scale, that it should be applied to the definition both of *units* and of *horizons* throughout the whole range of the scale (including both Quaternary and Precambrian), and that particular care should be taken to select boundary stratotypes or other horizon stratotypes at appropriate positions that are highly amenable to extensive or worldwide time correlation.

References Cited

International Subcommission on Stratigraphic Classification (ISSC), 1976, H. D. Hedberg, ed., International stratigraphic guide: New York, John Wiley & Sons, 200 p.

Biochronology[1]

W. A. BERGGREN[2] and J. A. VAN COUVERING[3]

Abstract Biochronology is the organization of geologic time according to the irreversible process of evolution in the organic continuum. It is an ordinal framework which measures all but youngest Phanerozoic time with greater resolution (ca 1 m.y., the average age range of species in rapidly evolving lineages), if with less accuracy, than radiochronology. Both are aspects of geochronology.

In contrast, biostratigraphy (strictly speaking) is iterative, consisting of observed (not predicted) superpositional sequences of fossils without inherent chronologic significance. (This is to say that an upside-down biostratigraphy or one with huge time gaps is perfectly useful as long as it is consistent.) It is the arrangement and correlation in time of biostratigraphies that constitute the often unappreciated role of biochronology. The basis of biochronologic correlation is any notable singular occurrence, or "datum event," in the fossil record which has a geographic range overlapping the spatial limits of coeval but disjunct biostratigraphical zones. However, biostratigraphic sequences are the milieu in which datum events in the biochronology of different fossil lineages are compared and radiometrically calibrated. The concept of biochronology is illustrated by reference to the essentially isochronous First Appearance Datum (FAD) of the planktonic foraminiferal species, *Globigerina nepenthes*, and the three-toed horse, *Hipparion*, in the marine and continental stratigraphic record, respectively, during the late middle Miocene, ca 12.5 m.y.

Introduction

Interpretations of earth history depend on two different systems of logic, both of which arrange geologic observations into sequences of events. The first and most widely used is the logic of superposition: the ordering of events iteratively in a system of invariant properties simply by determining the physical relationship of features in the rocks. This is what is meant by the word *stratigraphy*. The second logical system depends on the recognition of an ordinal progression which links a series of events in a system of irreversibly varying properties. This provides a theoretical basis outside of

[1] Manuscript received, January 17, 1977.

[2] Woods Hole Oceanographic Institution, Woods Hole, Massachusetts 02543, and Dept. of Geology, Brown University, Providence, Rhode Island 02912.

[3] University of Colorado Museum, Boulder, Colorado 80302.

Paper prepared at the request of Dr. George V. Cohee, Co-Chairman of Symposium 106.6: The International Geological Time Scale, and presented orally at the 25th International Geological Congress at Sydney, Australia. The ideas expressed are the result of recent attempts to achieve a greater degree of biostratigraphic resolution between marine and continental stratigraphies. Discussions with numerous colleagues aided in the formulation of the ideas expressed here but we would like to single out in particular M. B. Cita, R. Z. Poore, I. Premoli-Silva, W. B. F. Ryan, and H. D. Hedberg for their important roles.

Much of the discussion of the *Hipparion* Datum is taken, almost verbatim, from parts of the paper in press by Van Couvering, P. Robinson, and C. C. Black. We thank our colleagues for the use of these sections.

Material is based on research supported by the National Science Foundation under grants DES-74-21983 and OCE-76-21274.

This is Woods Hole Oceanographic Institution Contribution Number 3910.

Article Identification Number: 0149-1377/78/SG06-0004/$03.00/0

the preserved geologic record by which the nature and relation of the events in the progression can be recognized or predicted, and according to which missing parts of the record can be identified.

Geology is an historical philosophy, so the ordinal progressions to which we refer are progressions in time, just as geologic time is perceived by the progress in one or another ordinal series of events. This is what is meant by the word *geochronology*.

In the following pages, we discuss the relationship of geochronologic systems based on biologic evolution (biochronology) and on decay rates of unstable isotopes (radio-chronology) to the stratigraphic record of fossils (biostratigraphy) and geomagnetic polarity reversals (magnetostratigraphy) during the Neogene. We have chosen an example from the continental and marine stratigraphic record, namely the essentially simultaneous First Appearance Datum (FAD) of the three-toed horse, *Hipparion*, and the planktonic Foraminifera, *Globigerina nepenthes* Todd, during the late middle Miocene. (It should be noted that there is, as yet, no accepted theory which specifies the existence and duration of each individual geomagnetic polarity interval, so that a true magnetochronology does not exist.)

Biostratigraphy and Biochronology

Evidence from paleontology is relevant both to stratigraphy and to geochronology. By noting the fossils that marked different sedimentary units and the way that the units succeeded one another, William Smith was able to make a good living as a bio-stratigrapher. Whether or not he speculated about the cause of morphologic changes in the fossil shells and tests from one stratum to the next, from a practical point of view he did not need any explanation. On the other hand, many theories about the process leading to variations in the fossil record were proposed before Darwin and Wallace could gather enough information to come up with a useful description of evolution. It is now common practice for biostratigraphic arguments to include some appeal to adaptive selection or other evolutionary principles, but it remains immaterial whether *Bolivina* "B" is descended from *Bolivina* "A," for a local biostratigraphy based on key taxa such as these to be valid, and this is probably very fortunate.

Nevertheless, litho-, bio-, and magneto-facies in the rocks are physically discontinuous and are subject to iterative confusion. This means that most long-distance correlations are geochronologic in substance. Because fossils are more abundant than datable horizons in Phanerozoic sediments, and biologic events can be correlated in time more precisely than radiometric dates in all but the youngest levels, the long-distance correlations are primarily biochronologic in this milieu. Biochronologic correlations are based, in effect on three procedures: (1) recognizing the most widespread and distinctive events in biologic history, using for the most part the FAD or LAD (Last Appearance Datum) of key taxa (Van Couvering et al, in press); (2) locating such events in local biostratigraphies and evaluating their age with respect to as many reinforcing criteria as possible; and (3) stratigraphically relating such events to evidence for other biochronologic datum events and to radiometrically dated or calibrated levels, such as a tuff bed or a paleomagnetic boundary.

As observations such as these are synthesized, a time scale of dated events can be built up. But no matter how plausible and internally self-consistent it may be, a geo-chronologic history is still subject to stratigraphic cross-examination. Thus, the Neogene, as a biochron, is framed in the biostratigraphy of the Mediterranean region; and it is to *this* biostratigraphy that information from stratigraphic sequences of Neogene age, in the deep sea or in other lands, must be conveyed if such information is to have more than provincial significance. The means of conveyance, as we have pointed out, is geochronology in the form of the Neogene time scale.

To illustrate the way in which biostratigraphy and biochronology are combined, imagine an isolated, otherwise unremarkable outcrop of sandstone which is correlated to a level in a thick series of sandstones on the basis of its paleontology. Is this correlation due to biostratigraphy (the recognition of the fossils as features which uniquely characterize a known stratigraphic level in a reference section) or to biochronology (the recognition of the fossils as having an evolutionary grade or "age" which falls at a known point in the span of evolutionary time measured by fossils of a reference section)? Obviously there are many correlation decisions which are based on a combination of the two, because the choice is rarely obvious.

The biochronologic system most familiar to vertebrate paleontologists is the organization of mammalian history into regional land mammal "ages" (Wood et al, 1941). As Tedford (1970) and the ISSC (1976) have pointed out, these are not ages or zones in the time-stratigraphic sense but *biochrons*, and are subdivisions of time (not strata). The present North American mammal biochronology followed an earlier attempt to set up a regional biostratigraphic zonation modeled after ammonite zones in Europe, which foundered on the practical difficulties of observing stratigraphic ranges (vertical and lateral) of many fewer fossils and a much more compartmentalized and patchy stratigraphic milieu.

Advocates of a return to biostratigraphic regional "zones" in continental paleontology point to new methods of recovering small-mammal remains in abundance, but they fail to clearly recognize that mammalian biostratigraphy is still extremely localized and that extending any biostratigraphic unit vertically or laterally beyond the stratigraphic range of any one of the defining taxa of an assemblage-zone is an exercise in biochronology. The literature is studded with "closet biochrons" which correspond in name to such time-stratigraphic units as zones or stage/ages—but which are correlated biochronologically far beyond the limits defined by stratotype lithology or faunal assemblage. Because these closet biochrons are based on biostratigraphic units there has been much heated and unnecessary argument over regional correlations where the "type" assemblage, or one of its important members, displays notable diachronism or where the biochron interval is recognized in biostratigraphic series in which the original type-assemblage cannot be identified at all. Because of this, those who use biochrons in trans-facies correlations (including those who work with worldwide "zones" of marine planktonic microfossils) tend (almost unconsciously) to replace original zone titles, which commonly include the name of a fossil taxon, with numbers, letters, or geographic names. Some additional effort to distinguish between a stage/age (e.g. Chattian, Maastrichtian) and the biochron representing the same span of time is the next step forward.

FADs and LADs: Are Datum Events Instantaneous and Synchronous?

In order to correlate and synchronize the paleontologic record beyond the "faunule" or paleo-ecofacies limits, we must use biochronology assisted where possible by radiochronology. And, in order to use biochronology most effectively, we use the features in the paleontologic record which mark the most widespread, easily identified, and rapidly propagated events. These are biochronologic "datum events," and are commonly the extinction or the immigration of a particular taxon. The first appearance of a newly evolved taxon must logically be caused by immigration everywhere it is observed, except in the evolving population itself, but the dispersal of sequential members in an evolving population may be so extensive and rapid (as in planktonic microfossils) that the distinction is without significance. Groups of taxa may also seem to appear or to disappear in the record more or less jointly, but biochronologic precision is improved by selecting the best-suited individual taxon from such a group upon which to base the datum—unlike the group concept of "assemblage zones" and other biostratigraphic entities. Because the datum's genesis must be inferred from the geologic record, we will use the conservative approach and refer to biochronologic datum events only as FAD (First Appearance Datum) or LAD (Last Appearance Datum).

The spreading out or *prochoresis* (cf Gabunia and Rubinstein, 1968, p. 14) of an immigrant continental taxon in the fossil record can be very rapid (Elton, 1958). In climatically hospitable regions the most successful immigrants (e.g. rabbits in Australia, English sparrows and starlings in North America, muskrats and coypus in Europe) are not restrained by food supply, speed, or predation, so much as by basic reproduction rate and ethologic factors, such as intraspecific territorial inhibitions. Under such conditions, not only is the rate of prochoresis relatively rapid, reaching continental distances at rates between 10 and 100 km/yr, but the immigrant taxa quickly become very numerous in the colonized regions. The fossil record (discussed below) shows that

Hipparion appears in notable abundance in its earliest levels in Eurasia and Africa, and on this basis alone appears to have been a similarly successful immigrant. Even at a relatively slow rate of prochoresis of 10 km/yr the genus could have extended its range from the Bering Straits to France or Kenya in only 1,000 years (or 0.001 m.y.).

Speciation in planktonic Foraminifera and other planktonic microorganisms is a topic beset by many problems, because basic knowledge is still incomplete about their life history, morphologic variability, and population dynamics, as well as the long-term dynamics in the water masses they inhabit. Nevertheless it is possible to imagine that new forms (if not certainly new species) of planktonic microfossils such as radiolaria, diatoms, coccoliths, tintinnids, and forams could have been distributed across oceanic areas to the limits of their adaptive range by the normal mixing and meandering of current gyres in a few tens or hundreds of years, whatever their mode of origin. Known dispersal rates of sessile organisms with passively transported planktonic larvae clearly support this possibility (Scheltema, 1968, 1971a, b), and it would seem that a problem facing students of evolution in fully planktonic groups is to identify a readily available way in which populations can achieve genetic isolation for a period long enough to speciate.

Furthermore, the sequence of FADs and LADs observed in biostratigraphic studies of planktonic microfossils in different parts of the world shows an astonishingly regular order. Micropaleontologists take this almost for granted, but it suggests (despite rare, and therefore infamous, anomalies) rates of dispersal or extinction that do not show appreciable mutual diachroneity on a worldwide scale.

What is the time frame of this apparent synchroneity? Even on the stratigraphic level, bioturbation, accidents of preservation, collection methods, and analytical bias combine to make a given "pick" uncertain by several centimeters *at least* in any given stratigraphy. This represents uncertainty on the order of thousands of years at oceanic depositional rates under the best of conditions; furthermore, in the examples we have chosen from Deep Sea Drilling Project (DSDP) records, variation in the stratigraphic position of various datum events from core to core represents tens or hundreds of thousands of years of deposition. These variations are not obviously systematic and experience has shown that essentially random variations on this magnitude are normal in such studies. This amounts to stratigraphic and methodologic "noise" masking the relative chronology within and between datum events at this scale which could only possibly be reduced by comparing many time-equivalent biostratigraphies.

Pending such an analysis, we rely on the consensus of observations (Hays and Shackleton, 1976, on the level of global synchroneity in the LAD of a cosmopolitan radiolarian species) and on the reasonable probability that dispersal and extinction rates are direct or indirect functions of the rate of oceanic mixing to estimate that the FAD of *Globigerina nepenthes*, like dozens of others in the microfossil record, had a duration of less than ten thousand years.

As Figure 1 suggests, some geologic events can be considered both "instantaneous" and "synchronous," when looked at through a geochronologic perspective. These are phenomena such as volcanic tephra falls, tsunamis, submarine gravity flows, etc., which in the pre-Pleistocene Neogene (23 to 2 m.y.) have no measurable difference in age from local beginning to end ("instantaneous") or from place to place ("synchronous"). Examples of Neogene events which are "noninstantaneous" and/or "nonsynchronous," are orogenic pulses, climate changes, glacial oscillations, and the replacement of one fossil assemblage by another. Because they are exclusive propositions, FADs and LADs are necessarily instantaneous events, even in modern situations. Furthermore, since Neogene events which fall within 0.01 m.y. of one another cannot be geochronologically ordered (Fig. 1), so the FAD of a highly successful immigrant mammal taxon would be geochronologically synchronous throughout its range.

Of course, simple logic or stratigraphic evidence such as comparison with a paleomagnetic event can demonstrate that a FAD such as the spread of *Hipparion* or *Globigerina nepenthes* required a finite amount of elapsed time and was thus actually diachronous, but it can still be synchronous within the power of science to resolve its age.

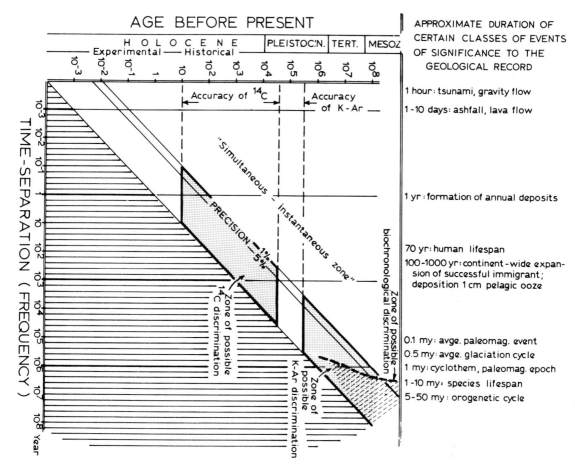

FIG. 1—Resolving power of geochronologic systems in the Cenozoic. The values for accuracy and duration are generalized. The effect of geochronologic perspective in the quantitative (radiometric) system is intuitively obvious, but note that the evolutionary (biochronologic) precision is based on the lifespan of species. This system is thus capable of discrimination with almost undiminished precision between events of (ca) 1 m.y. frequency to the limit of Phanerozoic time. The sway in the lower part of the nearly vertical limiting curve reflects the fact that shorter-lived species are necessarily selected in Pleistocene biochronology.

Biochronology of the *Globigerina nepenthes* FAD

The first observed occurrence of *Globigerina nepenthes* Todd in DSDP cores from sites in the Pacific (62.1, 77, 289) and Indian Ocean basins (214, 238) is consistently related with other microfossil datum events within a relatively narrow stratigraphic range (Fig. 2). Data from the Atlantic Ocean basin are inadequate, but the datum event under discussion appears in a similar context in the Cassinasco-Mazzapiedi section of the Serravallian Stage in northern Italy (Ryan et al, 1974). The biostratigraphy in the DSDP cores varies from site to site (Fig. 3), but by comparing them with one another and with information from land exposures the biochronology of the *G. nepenthes* FAD and its radiochronologic calibration can be estimated.

Calcareous Nannoplankton

The *Discoaster hamatus* FAD, which marks the base of Zone NN9, is closely associated with the *G. nepenthes* FAD in all of the studied cores and in the Serravallian Stage. With the exception of Site 62.1, this datum event is consistently located just above the *G. nepenthes* FAD. Furthermore, in the two cores where *Catinaster coalitus*, is recorded (the FAD of which marks the base of Zone NN8), it is coincident with or below the *G. nepenthes* FAD (Site 289: Andrews et al, 1975, p. 249; Site 214: Gartner, 1974, p. 585). The first occurrence of *G. nepenthes* in the Cassinasco-Mazzapiedi section is also shown by Ryan et al, (1974, Fig. 6) to correspond with a level within Zone NN8.

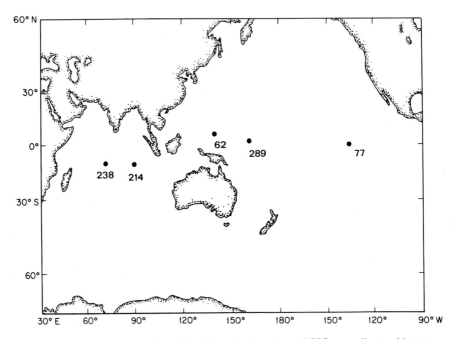

FIG. 2—Map showing location of Pacific and Indian Ocean DSDP cores discussed in text.

According to the biostratigraphy of the other cores, the *G. nepenthes* FAD is correlated respectively to lower NN9 (Site 62.1: Martini and Worseley, 1971, p. 1482-1483), to upper Zone NN6-NN7 (Site 77: Bukry, 1972, p. 824) and even to NN5 (Site 238: Roth, 1974, p. 975; Vincent, 1974, p. 1130). In the latter two cores, *C. coalitus* is not recorded so that the correlation to pre-NN8 levels in the nannoplankton zonation is based largely on negative evidence. It should be noted that the upper limits of the ranges of *Discoaster exilis* and *Discoaster kugleri* are virtually coincident with the base of Zone NN9, marked by *D. hamatus* FAD (Sites 62.1 , 289, 238), and therefore overlap nearly all of Zone NN8 rather than being restricted within the uppermost part of Zone NN7.

Without *C. coalitus*, therefore, the correlation of the *G. nepenthes* FAD to Zones NN6-NN7 in Site 77 is uncertain, and the assignment of the datum to a level corresponding with Zone NN5 in Site 238 is even more questionable since it is based mainly on the persistence of *Sphenolithus heteromorphus* into what are otherwise middle Miocene assemblages.

Radiolaria

The LAD of *Dorcadospyris alata*, the zonal marker for RN4, is consistently below the *G. Nepenthes* FAD in all the studied cores. However, the observed FAD of *Cannartus pettersoni*, which marks the base of Zone RN5, is below both of the other two datum levels in two cores (Site 62.1: Bronniman and Resig, 1971, p. 1246; Riedel and Sanfilippo, 1971, p. 1537, 1540, 1564; Site 289: Holdsworth, 1975, p. 525) but is above both of them in two other cores (Site 77: Hays et al, 1972, p. 66, 73-74; Site 214: Johnson, 1974, p. 528).

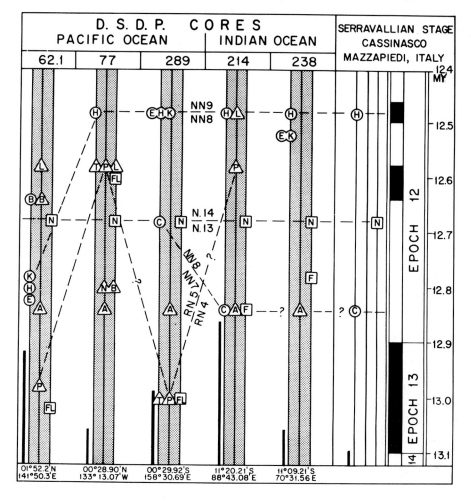

FIG. 3–Biochronology of *Globigerina nepenthes* Datum. In each core, and in the Italian reference section, the bar indicates 6 m of actual thickness; the vertical scales are individually adjusted for varying depositional rates to give approximately time-equivalent biostratigraphy. Magnetostratigraphy according to Blakely (1974), recalibrated and referred to Serravallian Stage (cf Ryan et al, 1974; Van Couvering and Berggren, this volume). Radiometric year-ages are approximate at the scale shown.

Other radiolarian datum events are also inconsistent: the LAD of *Lithopera baueri* varies in the opposite sense from the *C. pettersoni* FAD in Sites 62.1 and 77, but the LAD of *Cyrtocapsella tetrapera* is apparently coincident with the *C. pettersoni* FAD in Sites 77 and 289. The LAD of *Cannartus laticonus*, finally, is consistently above the

G. nepenthes FAD in Sites 62.1, 77, and 214, in contradiction to the variation in the relationship of the *C. pettersoni* FAD to this datum in these three sites. It should be noted that the *D. hamatus* FAD is everywhere higher in the sections studied than the *C. pettersoni* FAD.

Planktonic Foraminifera

The only planktonic foraminiferal datum events of significance in this narrow interval, except for the one that forms the subject of the investigation, are the LADs of *Globorotalia fohsi lobata* and *Globorotalia fohsi s.l.* In the Pacific Ocean sites, the *G. fohsi lobata* LAD is covariant almost simultaneously with the *C. pettersoni* FAD, below the *G. nepenthes* FAD in Site 62.1 (Bronnimann et al, 1971, p. 1732, 1734) and Site 289 (Andrews et al, 1975, p. 252) and above it in Site 77 (Hays et al, 1972, p. 66).

In the Indian Ocean sites, the *G. fohsi s.l.* LAD is recorded below the *G. nepenthes* FAD in both Site 214 (Berggren, Lohmann, and Poore, 1974, p. 642) and Site 238 (Fisher et al, 1974, p. 475; Vincent, 1974, p. 1130). However, Srinivasan (1975) noted an overlap in the stratigraphic ranges of *Globigerina nepenthes* and *Globorotalia fohsi lobata* in other Indian Ocean cores and in the Hut Bay Formation of the Andaman Islands. Unfortunately, the *G. fohsi* morphoseries is not known from sites within the Mediterranean basin, although it has been observed by Berggren in subsurface cores in North Africa to terminate near the *G. nepenthes* FAD.

Two explanations are available for the discrepancies between the biostratigraphy of FADs and LADs in different cores. First, it is possible that the apparent first and last appearances of given taxa are truly diachronous. This is probably more likely to be true of LADs than FADs, but no systematic pattern to inter-core variations can be discerned that would support an argument for diachroneity. All the sites lie within 11° of the present Equator, and all sample bottom-sediments that accumulated well above the calcium carbonate compensation depth (CCD), so that bias due to paleo-environment or selective dissolution apparently did not enter into the observed varia-tions. Second, it is possible that operator error (the bias introduced by different specialists making taxonomic determinations on the same groups, in some cases under difficult conditions) is responsible. In particular, recovery of radiolarian specimens from essentially calcareous samples is generally poor, as noted by the respective investigators of Sites 214 (Johnson, 1974, p. 528), Site 238 (Sanfilippo and Riedel, 1974, p. 999) and especially Site 289 (Holdsworth, 1975, p. 525).

Reworking of calcareous nannoplankton into higher levels is also a notorious problem even to the most careful workers, but with the sole exception of the reported *Sphenolithus heteromorphus* in middle Miocene levels of Site 238 (which we have discounted) the variations in nannoplankton biostratigraphy with respect to the radio-larian and planktonic foraminiferal datum events do not appear to have been in-fluenced by this particular bias.

Despite inconsistencies in the positions of the LADs and FADs discussed here, a pattern is nevertheless apparent. Discounting zonal assignments based on incomplete evidence, it appears that the *G. nepenthes* FAD is bracketed by the *Catinaster coalitus* FAD (below) and the *Discoaster hamatus* FAD (above), and thus correlates to a level within nannoplankton Zone NN8. The datum is closely associated with and (judging from the better evidence from Sites 77 and 214, and considering that the datum is displaced upward in Site 62.1) probably slightly older than the *Cannartus pettersoni* FAD, which would place it in the uppermost part of the *Dorcadospyris alata* Zone, RN4.

Finally, the consensus of the evidence is that the FAD of *G. nepenthes*, at the base of planktonic foraminiferal Zone N14, is concurrent with (or only slightly older than) the LAD of *Globorotalia fohsi lobata*, although the latter datum event has generally been placed within Zone N13 (Blow, 1969). The type locality of Zone N13 contains a calcareous nannofossil assemblage which Martini (1971) assigned to the *Discoaster exilis-Discoaster kugleri* concurrent-range zone (NN7), but as this index-combination is seen to extend through most, if not all, of Zone NN8 in the cores examined here it is possible that part of Zone N13 overlaps Zone NN8. According to the rule, "Base defines boundary," however, it cannot overlap the next succeeding planktonic foraminiferal zone; thus, it is possible that the upper part of Zone N13 in the type area is actually assignable to Zone N14 according to the observed overlap of the ranges of *Globorotalia fohsi lobata* and *Globigerina nepenthes*. The short interval of overlap suggests that Zone NN8 is a very short zone correlating to uppermost Zone N13 and lowermost Zone N14 (Berggren and Van Couvering, 1974).

Radiochronology of the *Globigerina nepenthes* FAD

Second-order correlations that suggest a correlation of the *G. nepenthes* FAD to a level within Magnetic Epoch 12 (see Berggren and Van Couvering, 1974) have been confirmed by direct observations of the magnetostratigraphy and biostratigraphy of the Cassinasco-Mazzapiedi section (Nakagawa et al, 1974; Ryan et al, 1974), and are further supported by the cross-match between the (slightly delayed) *G. nepenthes* FAD in DSDP Equatorial Pacific Site 62.1, and a mid-Epoch 12 level in the overlapping Equatorial Pacific core sequences of RC12-62 and M70-17 (Ryan et al, 1974). In the much more extended, and therefore more precisely measured magnetostratigraphic sequence of the Italian Serravallian section, Ryan et al (1974) placed the *G. nepenthes* FAD below the lower of the two normal events in Epoch 12 and thus in the lower third of that interval; the base of nannoplankton Zone NN8 is near the base of Epoch 12, and the base of nannoplankton Zone NN9 is in the upper of the two normal events about midway in Epoch 12. This confirms the zonal correlations that can be deduced from comparing the biostratigraphy of the DSDP sites referenced in the previous section, and further indicates that the *G. nepenthes* FAD in Site 62.1, above the base of Zone NN9, is probably displaced upward in this core.

Correlation of the planktonic microfossil biochronology, represented by the datum events that correspond to the zonal boundaries, with the magnetostratigraphy allows estimation of both the duration and the age in radiometric years of the biochronologic intervals. Using the duration of Epoch 12 and the component events recorded in sea-floor magnetic lineations (Blakely, 1974) rather than in core magnetostratigraphy (Theyer and Hammond, 1974; Opdyke et al, 1974) because we assume spreading rates to be more constant over time intervals of this length than sedimentation rates, it appears that Zone NN8 (between mid-Epoch 12 and lowermost Epoch 12, located in the Cassinasco-Mazzapiedi section) has a duration of approximately 0.3 to 0.4 m.y.

In the radiometric-year calibration of the magnetostratigraphy in this section, Ryan et al (1974, p. 655, 667) estimated paleomagnetic reversal boundaries according to a linear-regression "fit," extrapolated between selected radiometrically dated points with the (approximately located) calibration points backfitted to give the straightest possible time-magnetostratigraphy curve (Ryan et al, 1974, Fig. 7). In this curve dates given to ash layers in the experimental Mohole, correlated with levels within nannoplankton Zones NN6 and NN8 (Martini and Bramlette, 1963; Martini, 1971) with ages of 12.3 ± 0.4 m.y. and 11.4 ± 0.6 m.y. respectively (Dymond, 1966), would produce sharply anomalous inflections and are therefore rejected by these writers. The extrapolated calibration curve of Ryan et al (1974) places the *G. nepenthes* FAD (base Zone N14), in lower Epoch 12, at approximately 12.7 m.y. Although we earlier recommended an age of about 12.0 m.y. for this datum (Berggren and Van Couvering, 1974), the review of marine-nonmarine correlations in this paper agrees with the conclusions drawn by Ryan et al (1974), from the correlations used in our previous work[4] and strongly supports the revision of the time scale of marine microfossil datum events in the middle Miocene.

Biochronology of the *Hipparion* FAD

The nature of the *Hipparion* FAD has been obscured by naive statistical treatments of the exceptionally abundant and variable fossil material. As a result, an awesome

---

[4] Ryan et al (1974, Table 6 and p. 672-673) referred to a number of mid-Miocene continental calibration points, used by Berggren and Van Couvering (1974), which have been abandoned or revised since: Arroudjaoud granite (poorly documented), Baccinello (Late Turolian V3 fauna, not Vallesian, = N. 16), and the Beglia Formation of Tunisia. See also Van Couvering et al, "Geochronology of the *Hipparion* Datum," (in press).

number of *Hipparion* species have been named from upper Miocene deposits. Nevertheless, because of the distinctiveness and the initial abundance of its representatives, the genus itself provides a FAD which can be correlated to a very narrow biochronologic interval in the Old World. No other features of the Miocene mammalian fossil record have as wide a geographic extent as the *Hipparion* datum itself (which supports its usefulness), but in western Eurasia (and China?) and in North Africa the appearance of *Hipparion* is preceded by an interval most notably characterized by the dispersal of primitive giraffids of the *Paleotragus-Zarafa* type. The localities which show this influence range from Fort Ternan, Kenya (dated at 14 m.y., cf Hamilton, 1973) to Beni Mellal, Algeria (Jaeger et al, 1973; Gentry, 1970), Bled Douarah, Tunisia (Robinson and Black, 1974), Can Mata and Can Ponsic, Barcelona (Crusafont, 1972) and the upper Chinji beds of the Siwaliks (Gentry, 1970).

Paleotragus is also abundant in Chinese *Hipparion* faunas if not earlier (Kurtén, 1952). "Sarmatian" localities in central Europe are also included in this pre-*Hipparion* interval, notably Opole, Poland, and St.-Gaudens, France, but these lack giraffids. Berggren and Van Couvering (1974) recommended the general adoption of the term "Oeningian" for this biochron, based on the stage type section containing pre-Vallesian mammals (Tobien, 1971) and bracketed by K-Ar-dated tuffs ranging from 14.0 to 12.7 m.y. (Lippolt et al, 1963). It succeeds mammalian levels generally attributed to "Vindobonian" age and is overlain stratigraphically by tuffs with the earliest local *Hipparion* fauna at Höwenegg. However, V. Fahlbusch (written communication, 1975) indicated that European vertebrate paleontologists have agreed to use "Astaracian" in place of "Vindobonian" and Oeningian intervals in the continental biochronology.

In classical terms the early *Hipparion* faunas of Eurasia are assigned to "Pontian" age, but this is seriously compromised by incompatible usages from brackish and marine sequences (Van Couvering and Miller, 1971). The usage currently adopted in Europe for this biochron is "Vallesian," in which the FAD of murine rodents (*Progonomys* sp.) closely follows the arrival of *Hipparion* in Europe (Crusafont-Pairo, 1950; Crusafont and Villalta, 1954; Thaler, 1966; Bruijn, 1973) and also in North Africa (Jaeger et al, 1973). Earliest Vallesian levels are also marked, at least locally, by the persistence of the Eurasian early Miocene equid *Anchitherium*. In central Europe, Fahlbusch (1975) pointed out the FAD of *Leptodontomys* (Eomyidae) just prior to the *Hipparion* datum.

Höwenegg and Öhningen

Because abundant Miocene terrestrial fossils are associated with rift-valley type vulcanism in the region of the upper Rhine graben, that part of the region centered on the western end of the Bodensee (Lac Constance) is the main source of K-Ar ages bracketing the *Hipparion* FAD in Europe. The mittened "hand" of the Bodensee grasps the end of the Hegau ridge, where numerous exhumed volcanic centers, including that of Höwenegg, are prominent features of the landscape. Tephra falls from the Hegau spread into adjacent parts of southern Germany and northern Switzerland where they are prominent constituents of the continental Obere Süsswassermolasse (OSM). At quarries in this formation near Öhningen, some 15 km south of Höwenegg where the Bodensee empties into the Rhine, abundant plant and animal fossil remains have been the object of attention for over 300 years and are the basis for the Oeningian "stage," outlined by O. Heer in 1858 (cf Tobien, 1971).

Figure 4 presents a much simplified outline of the relationships of lithology, fossils, and interpretations in the middle Miocene of the Bodensee region, based on studies of the Oeningian type area by Rutte (1956), the Obere Süsswassermolasse in parts of northern Switzerland (e.g., Hoffmann, 1956, 1958a, b; Buchi, 1958; Pavoni, 1958), and equivalent strata in the Hegau (Schreiner, 1963; Lippolt et al, 1963). These studies show that the typical Oeningian fossil beds are correlated over a wide area as a bentonitic and paludal sedimentary interval in the upper part of the OSM. In addition, numerous fossils have come from other parts of the formation.

The beginnings of Hegau vulcanism, within the middle OSM, is marked by a "lower bentonite" with distinct, heavy-mineral assemblages succeeded by bedded hornblende-rich volcaniclastics of the Deckentuffe. Mammalian remains in the OSM (and its cor-

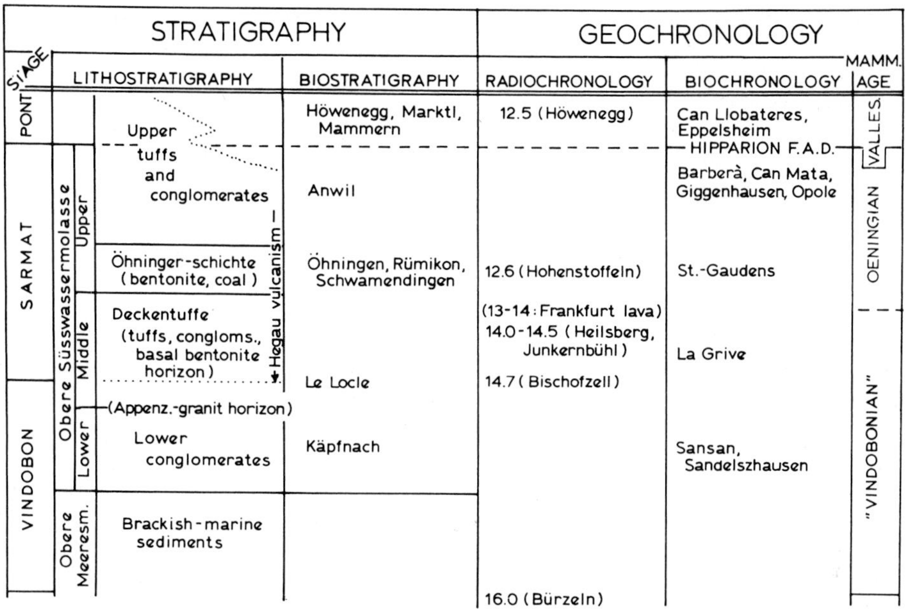

	STRATIGRAPHY		GEOCHRONOLOGY		
STAGE	LITHOSTRATIGRAPHY	BIOSTRATIGRAPHY	RADIOCHRONOLOGY	BIOCHRONOLOGY	MAMM. AGE
PONT.	Upper tuffs and conglomerates	Höwenegg, Marktl, Mammern	12.5 (Höwenegg)	Can Llobateres, Eppelsheim — HIPPARION F.A.D.	VALLES.
SARMAT		Anwil		Barberà, Can Mata, Giggenhausen, Opole	OENINGIAN
SARMAT	Öhninger-schichte (bentonite, coal)	Öhningen, Rümikon, Schwamendingen	12.6 (Hohenstoffeln)	St.-Gaudens	
SARMAT	Deckentuffe (tuffs, congloms., basal bentonite horizon)		(13-14: Frankfurt lava) 14.0-14.5 (Heilsberg, Junkernbühl)	La Grive	"VINDOBONIAN"
VINDOBON	(Appenz.-granit horizon) Lower conglomerates	Le Locle	14.7 (Bischofzell)		"VINDOBONIAN"
VINDOBON		Käpfnach		Sansan, Sandelszhausen	"VINDOBONIAN"
VINDOBON	Brackish-marine sediments				
VINDOBON			16.0 (Bürzeln)		

FIG. 4—Relations of the Obere Susswassermolasse in the upper Rhine drainage.

relative strata in Bavaria) have been grouped into "Vindobonian" and "Sarmatian" levels, on the basis of large mammals (Thenius, 1959, p. 78-79, 81-82), or into three levels (Dehm, 1955) distinguished by small-mammal evolutionary stages (Fahlbusch, 1964, 1970, 1975; Engesser, 1972). *Paleotragus*-like giraffids have not been reported from this region, perhaps because of the latitude, but *Hipparion primigenium* is abundant in local faunas just younger than the OSM sequence, and locally in the uppermost conformable strata of the OSM itself (e.g. Höwenegg, Marktl, and Hammerschmiede).

Radiochronology of the *Hipparion* Datum

The cited evidence converges on an age of 12.5 m.y. for the FAD of *Hipparion (H. primigenium)* in central and western Eurasia and in northern Africa (see Fig. 5). The evolutionary origin of hipparions, if in North America, is not appreciably older than this, but the first reports of the *Hipparion* FAD in sub-saharan Africa place it at least 1 m.y. younger.

The K-Ar radiochronology of the datum is based principally on the recently rechecked dating of the faunal sequence in the Obere Süsswassermolasse (H. Tobien, written commun., 1976). Contradictory dating reported from the Pannonian-Pontian strata of the central Paratethys and the Neogene of Turkey (suggesting a younger age for the datum), directly conflicts with other late Miocene dates (see also Van Couvering and Miller, 1971) and can be shown to be suspect because *a priori* assumptions of regional biochronology are incorporated (cf review in Van Couvering et al, in press).

The argument that the *Hipparion* FAD in Eurasia and North Africa is effectively synchronous and instantaneous is valid in the context of biochronology. It is, so far, impossible to demonstrate any appreciable difference in the evolutionary levels of the oldest local faunas in which *Hipparion* is recorded in the area from India (if not China) to Spain. The real time difference in the observed FAD in this vast region is probably less than 0.5 m.y., although there are at present no data to support this estimate apart from the inconclusive radiochronology of earliest North American hipparions; on theoretical grounds the prochoresis from Beringia (or China) to the Mediterranean region might have taken as little as 0.1 m.y.

LOCUS	BIOSTRATIGRAPHIC POSITION	BIOCHRONOLOGIC POSITION	SIGNIFICANT LOCAL CORRELATION	RADIOMETRIC CALIBRATION, Ma
California	Lower faunal levels of Rosamond, Santa Margarita, Mint Canyon Fms.	Middle Clarendonian after evolution of hipparions	Equivalent to Lower Mohnian Stage, = N14-15 (NN8-NN9)	ca. 11-12 nonmarine; ca. 11-12 marine
Siwaliks	Middle Nagri Stage	Early (?Earliest) Vallesian	Faunal sequence	(In preparation)
Anatolia	Pisidic Fm., Esme-Akcakoy faunal level	Vallesian	Faunal sequence (indirect, in regional correl.)	Pre-Vallesian faunas overlain by ca. 11.5 tuffs; Turolian fauna at ca. 9.0
Bavaria	Uppermost faunal levels of Obere Susswassermolasse	Earliest Vallesian	Faunal sequence	12.5 (Höwenegg); 14 to 12.6 in underlying sequence
Vienna Basin	"Unter Pannon" continental	Early Vallesian	Equivalent to mid-Sarmat marine, overlies N10-12 (NN6-NN7)	(Basal Sarmat 13.5, mid-Sarmat ca. 12)
France	"Helvetien marin" sands, Vaucluse	Early Vallesian	Overlain by marine Tortonian, ca. N.15/16	None
Barcelona	"Vallesian tipico" Valles-Penedes graben	Earliest Vallesian	Faunal sequence; pre-Vallesian underlain N.8	None
Tunisia	Middle Beglia Fm.	Earliest Vallesian	Faunal sequence; = N.14; pre-Vallesian underlain N.12?	None
Algeria	Bou Hanifia Fm., lower faunal level	Early Vallesian	Faunal sequence; late Vallesian overlain N.16/17	12.2 (Oued el Hammam); late Vallesian = ca. 9.5
Kenya	Middle Ngorora Fm.	Undefined	Faunal sequence	9.8 (bracketed by 12-8.5)

FIG. 5—Synoptic review of the *Hipparion* datum. "Faunal sequence" in the "significant local correlations" column refers to the observation of the datum as a first occurrence in a stratigraphic sequence of mammalian faunas. The correlation of lower Mohnian to Zones N.14-15 (and to Zones NN8-NN9), rather than to earlier zones (Warren, 1972; Lipps and Kalisky, 1972) in accordance with current work of R. Z. Poore (personal commun.), agrees with unpublished dating of basal tuffs in the type Mohnian (J. D. Obradovich, personal communication, 1975) and the calibration presented by Ryan et al (1974). For other details see Van Couvering et al (in press).

Correlation of the *G. nepenthes* and *Hipparion* FAD

The marine correlation is based primarily on the lateral correlation of the Beglia Formation of central Tunisia to its marine equivalent on the Isle of Zembra (Robinson and Black, 1974; Wiman, written commun., 1976), and is entirely consistent with a number of observations which place Zone N.12 microfaunas below pre-*Hipparion* mamal faunas and Zone N.16 microfaunas as equivalent to Turolian mammal faunas (see Fig. 5; Van Couvering et al, in press). The paleomagnetic calibration of the marine zonation, in part because it has been influenced by previous reports of correlation of the *Hipparion* datum with the marine sequence, also places the midpart of Zone N.14 at 12.5 m.y. (Ryan et al, 1974).

The comparison of the marine micropaleontologic record to magnetostratigraphy and to high-frequency variations in the deep sea, oxygen-isotope record indicates that FADs and LADs can approach isochroneity even more closely than had been assumed. Certain worldwide Pleistocene extinction datums apparently occurred in less than a few thousand years (Hays and Shackleton, 1976), and many Neogene datum events are consistently located with respect to one another and to detailed magnetostratigraphy in cores from different ocean basins (Ryan et al, 1974), indicating to us a synchroneity on the order of 0.01 m.y., if not less, as discussed above.

The *Hipparion* FAD offers the best known opportunity for measuring a mammalian datum event against magnetostratigraphy because of its wide extent and distinctiveness. If the correlation of the *Hipparion* datum to Zone N.14 is correct, magnetostratigraphic calibration of marine microplanktonic zones (Ryan et al, 1974) indicates that the datum should fall within the mainly reversed-polarity interval of Magnetozone 12, just earlier than the distinctive polarity signature of the 9 to 12 m.y. period which corresponds to Seafloor Anomaly 5-5A (Fig. 6).

Studies also place the *G. nepenthes* datum in the lower part of Magnetic Epoch 12, with an inferred (extrapolated) age of approximately 12.7 m.y. The radiometric age of the *Hipparion* datum, 12.5 m.y., therefore closely agrees with the correlation of the two datum events on biostratigraphic grounds.

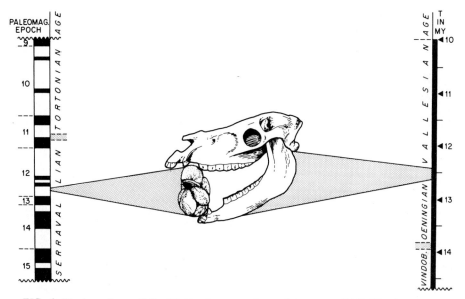

FIG. 6—Biochronology of the *Globigerina nepenthes* and *Hipparion* FAD. The former datum has been linked to paleomagnetic Epoch 12 (with an estimated age of 12.7 m.y.), the latter has been radiometrically dated at 12.5 m.y.

Conclusions and Summary

Most species do not range over global distances and the limits of their ranges can shift with time more or less independently of limits for other species. Because of this it is unwise to attempt isochronous correlations over very large distances solely by comparing assemblages of fossil mammals. On the other hand, because the fossil record is so fragmentary and accidental—defects which the assemblage zone is designed to cope with—it is also unwise to attempt to subdivide local biostratigraphic sequences solely on the presence or absence of some key species, no matter how remarkable or widespread the species may be.

Nevertheless, the regional datum event is essential to transcontinental and transoceanic biochronology, and should not be scorned simply because it has been wrongly emphasized in unsophisticated biostratigraphic studies. We have tried to show that there can be good FADs and LADs, using the prochoresis of *Hipparion primigenium* and the FAD of *G. nepenthes* as two of the most important examples available. With critical study, datum events such as these can tie together distant and diverse biostratigraphic sequences with essentially "instantaneous" geochronologic linkages. In fact, a competent biostratigrapher can project the time of a significant datum event throughout local sequences once it has been established in a few local faunas—for instance, the way in which the *Hipparion primigenium* datum has been related to Vallesian small-mammal sites, in which no equids have been found.

To go further, we believe that in extensive correlation networks the time boundaries must be established on the basis of biochronologic datum events and not on biostratigraphic assemblage zones, which have an inherently "fuzzy" chronologic validity outside of the original faunule. The essential first step, worth taking in itself, is to clarify terminology so that biostratigraphic and biochronologic units are distinguished from one another, and fruitless and needless "wars" of definition can end.

The essential desire to extend correlations in time, which is mistakenly put forward as an activity in stratigraphy and has resulted in the completely overinflated concept of chronostratigraphy embodied in the "stage" described by the International Stratigraphic Guide (ISSC, 1976), is properly an exercise in geochronology itself. In other words it is the comparison of local stratigraphies with one another in chronologic context, and it is the establishment of radiochronologic or, more often biochronologic, calibration within each stratigraphy that allows the correlation. The strata, as such, are not correlated but the time units represented in the strata are.

The planktonic microfossil zones—calcareous nannoplankton, radiolaria, and foraminifera for the best examples—can be thought of as true biostratigraphic zones, the limits of which are three-dimensional in stratigraphic space, and which are defined by the presence of characteristic fossils in the rocks. They are also used, with very little confusion due to their geochronologically "instantaneous" development, as biochrons. This means that the "zones" are recognized in strata barren of fossils (that is, unfossiliferous or poorly fossiliferous intervals are attributed to time units called zones) and are correlated with (if not extended into) continental biochronology, as at Kastellion, Greece (Bruijn et al, 1971).

In most correlation charts the zones are used, like magnetostratigraphic zones, as closely calibrated stratigraphic units which, due to the chronologic significance generated by the essentially isochronous boundaries of the calibrated sequence, are understood as having a secondary time value. In other words, while the geochronologic "clocks" are radiometric decay and biologic evolution, there are closely calibrated stratigraphic sequences like the planktonic microfossil zones or magnetostratigraphic epochs that can be thought of as geochronologic yardsticks. They have no intrinsic time value but they can be used to measure time according to their own calibration.

References Cited

Andrews, J. E., et al, 1975, Site 289, *in* J. E. Andrews et al, eds., Initial reports of the Deep Sea Drilling Project: Washington, D.C., U.S. Govt., v. 30, p. 231-398.

Berggren, W. A., and J. A. Van Couvering, 1974, The Late Neogene: Palaeogeography, Palaeoclimatology, Palaeoecology, v. 16, nos. 1-2, p. 1-216.

—— G. P. Lohmann, and R. Z. Poore, 1974, Shore laboratory report on Cenozoic planktonic Foraminifera, Leg 22, *in* C. C. von der Borch et al, Initial reports of the Deep Sea Drilling Project, v. 22, Washington, D.C., U.S. Govt. Printing Office, p. 635-655.

Blakely, R. J., 1974, Geomagnetic reversals and crustal spreading rates during the Miocene: Jour. Geophys. Research, v. 79, p. 2979-2985.

Blow, W. H., 1969, Late middle Eocene to Recent planktonic foraminiferal biostratigraphy, *in* E. J. Brill, 1967, First Int. Conf. Planktonic Microfossils (Geneva) Proc.: v. 1, p. 199-421.

Bronnimann, P., and J. Resig, 1971, A Neogene Globigerinaceous biochronologic time scale of the southwestern Pacific, Leg 7, *in* E. L. Winterer et al, Initial reports of the Deep Sea Drilling Project, v. 7, pt. 2, Washington, D.C., U.S. Govt. Printing Office, p. 1235-1469.

—— et al, 1971, Biostratigraphic synthesis: late Oligocene and Neogene of the western tropical Pacific, Leg 7, *in* E. L. Winterer et al, Initial reports of the Deep Sea Drilling Project, v. 7, pt. 2, Washington, D.C., U.S. Govt. Printing Office, p. 1723-1745.

Bruijn, H. de, 1973, Analysis of the data bearing on the correlation of the Messinian with the succession of land mammals, *in* C. W. Drooger, ed., Messinian events in the Mediterranean: Amsterdam, North-Holland, p. 260-262.

—— P. Y. Sondaar, and W. J. Zachariasse, 1971, Mammalia and Foraminifera from the Neogene of Kastellios Hill (Crete): a correlation of continental and marine biozones, pt. I: Nederlandse Akad. Wet., Proc., Ser. B, v. 74, no. 5, p. 1-22.

Buchi, Ulrich P., 1958, Zur Geologie der Oberen Susswassermolasse (OSM) zwischen Toss- und Glattal: Ecologae Geol. Helvetiae, v. 51, no. 1, p. 73-106.

Bukry, D., 1972, Coccolith stratigraphy, leg 9, Deep Sea Drilling Project, *in* J. I. Tracey, Jr., et al, eds., Initial reports of the Deep Sea Drilling Project: Washington, D.C., U.S. Govt., v. 9, p. 817-832.

Crusafont, M., 1972, Les Ischyrictis de la transition Vindobonien-Vallesien: Palaeovert., v. 5, p. 253-260.

—— and J. F. Villalta, 1954, Caracteristicas bioticas del Pontiense espanol: 19th Int. Geol. Cong., (Algiers), Sec. 12, p. 119-126.

Crusafont-Pairo, M., 1950, La cuestion del llamado Meotico espanol: Arrahona, Sabadell, v. 1, no. 1.

Dehm, R., 1955, Die Saugertierfaunen in der Oberen Susswassermolasse und ihre Bedeutung fur die Gliederung: Erlaut. Geol. Ubersichstk. Suddeutsch. Molasse, Bayerisch Geol. Landsamt.

Dymond, J., 1966, Potassium-argon geochronology of deep-sea sediments: Science, v. 152, p. 1239-1241.

Elton, C. S., 1958, The ecology of invasions by animals and plants: New York, John Wiley and Sons, 181 p.

Engesser, B., 1972, Die obermiozaene Saeugertierfauna von Anwil (Baselland): Basel Univ., diss. [327 p.] Naturforsch. Ges. Bassenland Taetigkeitsber., v. 28, p. 37-363.

Fahlbusch, V., 1964, Die Criceteden (Mamm.) der Oberen Susswassermolasse Bayerns: Bayerische Akad. Wiss. Abh., Math.-Naturw. Kl. (N.F.), v. 118, p. 1-136.

—— 1970, Phylogenie und stratigraphische Bedeutung der Miozanen Criceteden (Mammalia, Rodentia) Suebayerns: Proc. Committee on Mediterranean Neogene Stratigraphy, 4th Session, Pt. I, (Bologna) Giorn. Geol., v. 35, no. 1, p. 153-160.

—— 1975, Die Eomyiden (Rodentia, Mammalia) der Oberen Susswassermolasse Bayerns: Bayerische Staatssamml. Palaontol. Hist. Geol., Mitt., v. 15, p. 63-90.

Fisher, R. L., et al, 1974, Initial reports of the Deep Sea Drilling Project: v. 24, Washington, D.C., U.S. Govt. Printing Office, 1183 p.

Gabunia, L., and M. Rubenstein, 1968, On the correlation of the Cenozoic deposits of Eurasia and North America based on the fossil mammals and the absolute age data: 23d Internat. Geol. cong., Proc., Sec. 10, p. 9-17.

Gartner, S., Jr., 1974, Nannofossil biostratigraphy, Leg 22, *in* C. C. von der Borch et al, Initial reports of the Deep Sea Drilling Project, v. 22, Washington, D.C., U.S. Govt. Printing Office, p. 577-599.

Gentry, A. W., 1970, The bovidae (Mammalia) of the Fort Ternan fossil fauna, *in* L. S. B. Leakey and R. J. G. Savage, Fossil vertebrates of Africa, v. 2: New York, Academic Press, p. 243-324.

Hamilton, W. R., 1973, Lower Miocene ruminants of Gebel Zelten, Libya: British Mus. (Nat. History) Bull. Geology, v. 21, no. 3, p. 75-150.

Hays, J. D., and N. J. Shackleton, 1976, Synchronous extinction of the radiolarian *Stylatractus universus*: Geology, v. 4, no. 11, p. 649-652.

—— et al, 1972, Initial reports of the Deep Sea Drilling Project, v. 9, Washington, D.C., U.S. Govt. Printing Office, 1205 p.

Heer, O., 1858, Flora Tertiana Helvetiae, Vol. 2: Winterthur, 110 p.

Hofmann, F., 1956, Die obere Susswassermolasse in der Ostschweiz und im Hegau: Ver. Schweiz. Petrol. Geol. Ing. Bull., v. 23, no. 64, p. 23-34.

—— 1958a, Vulkanische tuffhorizonte in der Oberen Susswassermolasse des Randen und Reiat, Kanton Schaffhausen: Ecologae Geol. Helvetiae, v. 51, p. 371-377.

—— 1958b, Das bentonitvorkommen von Le Locle (Kanton Neuerberg): Ecologae Geol. Helvetiae, v. 51, p. 65-71.

Holdsworth, B. K., 1975, Cenozoic radiolaria biostratigraphy, tropical and equatorial Pacific, Leg 30, *in* J. E. Andrews et al, Initial reports of the Deep Sea Drilling Project: v. 30, Washington, D.C., U.S. Govt. Printing Office, p. 499-537.

International Subcommission on Stratigraphic Classification (ISSC), 1976, H.D. Hedberg, ed., International stratigraphic guide: New York, John Wiley & Sons, 200 p.

Jaeger, J.-J., J. Michaux, and D. David, 1973, Biochronologie du Miocene moyen et superieur continental du Maghreb: Acad. Sci. Comptes Rendus, Ser. D., v. 277, no. 22, p. 2477-2480.

Johnson, D., 1974, Radiolaria from the eastern Indian Ocean, Leg 22, *in* C. C. von der Borch et al, Initial reports of the Deep Sea Drilling Project, v. 22, Washington, D.C., U.S. Govt. Printing Office, p. 521-575.

Kurten, B., 1952 (1954), The Chinese *Hipparion* fauna: a quantitative study with comments on the ecology of the machairodonts and hyaenids and the taxonomy of the gazelles: Soc. Sci. Fennica, Commentat Biol., v. 13, no. 4, p. 1-82.

Lippolt, H. J., W. Gentner, and W. Wimmenauer, 1963, Altersbestimmungen nach der Kalium-Argon-Methode an tertiaren Eruptivgesteinen Sudwestdeutschlands: Baden-Wurttemberg Geol. Landesamt Jahreshefte, Jh. 6, p. 507-538.

Lipps, J., and M. Kalisky, 1972, California Oligo-Miocene calcareous nannoplankton biostratigraphy and paleoecology, *in* F. H. Stinemeyer, ed., Pacific Coast Miocene Biostratigraphy

Symposium, Pacific Sect. SEPM (Bakersfield, Calif.) Proc.: p. 239-254.

Martini, E., 1971, Standard Tertiary and Quaternary calcareous nannoplankton zonation, *in* A.
—— and M. N. Bramlette, 1963, Calcareous nannoplankton from the experimental Mohole drilling: Jour. Paleontology, v. 37, p. 845-856.

Farinacci, ed., 2d Planktonic Conference, Rome, Proc.: v. 2, p. 739-785.

—— and T. Worsley, 1971, Tertiary calcareous nannoplankton from the western Equatorial Pacific, Leg 7, *in* E. L. Winterer et al, Initial reports of the Deep Sea Drilling Project: v. 7, pt. 2, Washington, D.C., U.S. Govt. Printing Office, p. 1471-1511.

McGowran, B., 1974, Foraminifera, Leg 22, *in* C. C. von der Borch et al, Initial reports of the Deep Sea Drilling Project: v. 22, Washington D.C., U.S. Govt. Printing Office, p. 609-627.

Nakagawa, H., et al, 1974, Preliminary results on magnetostratigraphy of Neogene stage stratotype sections in Italy: Riv. Italiana Paleontologia e Stratigrafia, v. 80, no. 4, p. 615-630.

Opdyke, N. D., L. H. Burckle, and A. Todd, 1974, The extension of the magnetic time scale in sediments of the central Pacific Ocean: Earth and Planetary Sci. Letters, v. 22, p. 300-306.

Pavoni, N., 1958, Neue bentonitvorkommen in der Zurcher Molasse: Eclogae Geol. Helvetiae, v. 51, p. 299-304.

Riedel, W. R., and A. Sanfilippo, 1971, Cenozoic radiolaria from the western tropical Pacific, Leg 7, *in* E. L. Winterer et al, Initial reports of the Deep Sea Drilling Project, v. 7, pt. 2, Washington, D.C., U.S. Govt. Printing Office, p. 1529-1672.

Robinson, P., and C. C. Black, 1959, Note preliminaire sur les vertebres fossiles du Vindobonien (Formation Belgia) du Bled Bouarah, Governeurat de Gafsa, Tunisie: Travaux de Geologie tunisienne, no. 2, Tunis, Serv. Geol., Notes 31, p. 67-70.

—— —— 1974, Vertebrate faunas from the Neogene of Tunisia: Egypt, Geol. Surv. Ann. 4, p. 319-332.

Roth, P., 1974, Calcareous nannofossils from the northwestern Indian Ocean, Leg 24, *in* R. L. Fisher et al, Initial reports of the Deep Sea Drilling Project, v. 24, Washington, D.C., U.S. Govt. Printing Office, p. 969-994.

Rutte, E., 1956, Die geologie des Schienerberges (Bodensee) und der Ohninger Funstatten: Neues Jb. Geol. Palaont., Abh. v. 102, no. 2, p. 143-282.

Ryan, W. B. F., et al, 1974, A paleomagnetic assignment of Neogene stage boundaries and the development of isochronous datum planes between the Mediterranean, the Pacific, and Indian oceans in order to investigate the response of the world ocean to the Mediterranean "salinity crisis": Riv. Italiana Paleontologia e Stratigrafia, v. 80, p. 631-688.

Sanfilippo, A., and W. R. Riedel, 1974, Radiolaria from the west-central Indian Ocean and Gulf of Aden, Leg 24, *in* R. L. Fisher et al, Initial reports of the Deep Sea Drilling Project, v. 24, Washington, D.C., U.S. Govt. Printing Office, p. 997-1035.

Scheltema, R. S., 1968, Dispersal of larvae by equatorial ocean currents and its importance to the zoogeography of shoal-water tropical species: Nature, v. 217, p. 1159-1162.

—— 1971a, Larval dispersal as a means of genetic exchange between geographically separated populations of shallow-water benthic marine gastropods: Biol. Bull., v. 140, no. 2, p. 284-322.

—— 1971b, The dispersal of the larvae of shoal-water benthic invertebrate species over long distances by currents, *in* D. J. Crisp, ed., Fourth European marine biology symposium: Cambridge, Cambridge Univ. Press: p. 7-28.

Schreiner, A., 1963, Geologische Untersuchungen am Howenegg/Hegau: Baden-Wurttemberg Geol. Landesemt Jahreshefte, 6, p. 395-420.

Shafik, S., 1975, Nannofossil biostratigraphy of the southwest Pacific: Leg 30, *in* J. E. Andrews et al, Initial reports of the Deep Sea Drilling Project, v. 30, Washington, D.C., U.S. Govt. Printing Office, p. 549-598.

Srinivasan, M. S., 1975, Middle Miocene planktonic foraminifera from the Hut Bay Formation, Little Andaman, Bay of Bengal: Micropaleontology, v. 21, p. 133-150.

Tedford, R. H., 1970, Principles and practices of mammalian geochronology in North America: North Am. Paleont. Conv., Proc., Pt. F., p. 666-703.

Thaler, L., 1966, Les Tongeurs fossiles du Bas-Languedoc dans leurs rapports avec l'histoire des faunes et al stratigraphie du Tertiare d'Europe: Mus. Natl. Hist. Nat., (Paris) Mem. Ser. C., v. 217, p. 1-295.

Theyer, F., and S. R. Hammond, 1974, Paleomagnetic polarity sequence and radiolarian zones, Brunhes to polarity epoch 20: Earth and Planetary Sci. Letters, v. 22, p. 307-319.

Thenius, E., 1959, Handbuch der Stratigraphischen Geologie, v. III, Tertiar.; Pt. 2, Wirbeltifer-faunen: Stuttgart, Enke Verlag, 328 p.

Tobien, H., 1971, Oeningian, *in* G. C. Carloni et al, eds., Stratotypes of Mediterranean Neogene Stages: Geologia Gior., v. 37, no. 2, p. 135-143.

Van Couvering, J. A., and J. A. Miller, 1971, Late Miocene marine and non-marine time scale in Europe: Nature, v. 230, p. 559-563.

—— P. Robinson, and C. Black (in press), Geochronology of the *Hipparion* Datum, *in* M. O. Woodburne, ed., Mammalian paleontology as an exercise in geochronology: California Univ.

Pub. Geol. Sci.

Vincent, E., 1974, Cenozoic planktonic biostratigraphy and paleooceanography of the tropical western Indian Ocean, *in* R. L. Fisher et al, eds., Initial reports of the Deep Sea Drilling Project: Washington, D.C., U.S. Govt., v. 24, p. 111-1150.

Von der Borch, C. C., et al, 1974, Initial reports of the Deep Sea Drilling Project: v. 22, Washington, D.C., U.S. Govt. Printing Office, 890 p.

Warren, A. D., 1972, Luisian and Mohnian biostratigraphy of the Monterey Shale at Newport Lagoon, Orange County, California, *in* E. H. Sinemeyer, ed., Pacific Coast Miocene Biostratigraphy Symposium: Pacific Sect. SEPM (Bakersfield, Calif.) Proc.: p. 27-36.

Winterer, E. L., et al, 1971, Initial reports of the Deep Sea Drilling Project: v. 7, Washington, D.C., U.S. Govt. Printing Office, 1757 p.

Wood, H. E., 2nd Chairman, et al, 1941, Nomenclature and correlation of the North American continental Tertiary: Geol. Soc. America Bull., v. 52, p. 1-48.

The Magnetic Polarity Time Scale: Prospects and Possibilities in Magnetostratigraphy[1]

M. W. MCELHINNY[2]

Abstract The pattern of geomagnetic reversals for the past 150 m.y. has been well established from potassium-argon and polarity measurements on lava sequences, polarity measurements on continuous sequences in deep-sea sediment cores, and the interpretation of marine magnetic anomalies. Interest in the wider use of magnetostratigraphic methods has increased through the development of the extremely sensitive SQUID magnetometer, capable of fast measurement of virtually all rock types. The frequency of reversals appears generally to have been lower in pre-Cenozoic times when the magnetic field exhibited a cyclic behavior in polarity bias. Analysis of land-based paleomagnetic data suggests the existence of quiet (no reversals) and disturbed (many reversals) intervals throughout the Phanerozoic. The delineation of the quiet intervals offers the most encouraging prospect for their use in intercontinental correlation. The discovery of a quiet interval just below the Precambrian/Cambrian boundary suggests a possible method for arriving at a definition of this boundary.

Introduction

Our knowledge that the earth's magnetic field has reversed its polarity frequently in the past has provided geologists with a unique tool for use in stratigraphic correlation. Polarity changes are globally synchronous and are recorded by all rock types providing horizons of potentially unparalleled resolution for use by the stratigrapher. The most spectacular use of the geomagnetic polarity time scale has followed its recognition in the form of linear magnetic anomalies resulting from seafloor spreading.

For the past 5 m.y., the polarity time scale is well defined through potassium-argon studies of lava flows from all parts of the world (Cox, 1969). Studies of deep-sea sediments have provided continuous records well back into the Miocene. On land however, difficulties have arisen because of lack of sensitivity of currently available magnetometers, especially because of their inability to measure carbonate sequences in detail. Indeed the types of sediments most frequently studied by paleomagnetists in the past have been mainly red beds and these are, commonly, those of least interest to the stratigrapher. However breakthroughs in superconducting physics have led recently to the development of the SQUID (Superconducting Quantum Interference Device) magnetometer (Goree and Fuller, 1976). These magnetometers, with a sensitivity at least 100 times that of previous types and measurement speeds nearly 10 times faster, promise to produce a revolution in magnetostratigraphy over the next few years. Essentially the prospect is that all rock types now become accessible to magnetostratigraphic studies. As such not only does the instrument become the ultimate tool for magnetostratigraphy, but it shows promise that the whole process will develop rapidly

[1] Manuscript received, January 17, 1977.
[2] Australian National University, Canberra, A.C.T., Australia.
Article Identification Number: 0149-1377/78/SG06-0005/$03.00/0

as the ultimate tool for the stratigrapher. I shall outline some of the most obvious regions of the stratigraphic column where immediate applications can be worthwhile.

Subcommission on Magnetic Polarity Time Scale

A Subcommission of the International Commission on Stratigraphy was established to facilitate an unambiguous development of the polarity time scale and its associated nomenclature. The first meetings were held at the Montreal International Geological Congress in August 1972. Reports of these and subsequent meetings were published in *Geotimes* (Watkins, 1973, 1975, 1976). The recommended unit terms and hierarchies in paleomagnetic stratigraphy are outlined in Table 1. The basic local magnetostratigraphic unit is the *magnetozone* but the term "zone" may be used without prefix when the usage is clear. A magnetozone has its upper and lower limits defined by the positions of the boundaries, which are termed *transitions*, marking a change of polarity between opposite senses.

TABLE 1. Summary of recommended unit-terms and hierachies in paleomagnetic stratigraphy.

Magnetostratigraphic Units	Approx. duration (years)	Chronostratigraphic units	Chronologic Units
Polarity subzone	$10^4 - 10^5$	Polarity subinterval	Polarity event
Polarity zone	$10^5 - 10^6$	Polarity interval	Polarity epoch
Polarity superzone	$10^6 - 10^7$	Polarity superinterval	Polarity period
Polarity hyperzone	$10^7 - 10^8$	Polarity hyperinterval	Polarity era

The natural geomagnetic polarity reversal spectrum is believed to extend from the order of 10^4 to 10^7 years per reversal (McMahon and Strangway, 1967). Short-period reversals may have occurred, but may represent large rapid swings of the geomagnetic field from its usual configuration. Such behavior often is characterized by migration of the virtual geomagnetic pole to the opposite hemisphere for a period of perhaps 10^2 to 10^3 years. The recommended name for these occurrences is the term *polarity excursion*, defined as "...a sequence of virtual geomagnetic poles which may reach intermediate latitudes and which may extend beyond $135°$ of latitude from the pole, for a short interval of time, before returning to the original polarity" (Watkins, 1976).

Three systems of terminology have arisen in magnetic polarity stratigraphy. Originally the magnetic polarity time scale was derived from paleomagnetic and potassium-argon age analyses of volcanic rocks. From these were derived the four named polarity epochs (Brunhes, Matuyama, Gauss, Gilbert) for the Pleistocene and Pliocene. Extension of this scale has been derived largely from measurements of deep-sea sedimentary cores, and a system of numbering epochs has developed. Likewise in magnetic-polarity stratigraphy derived from analyses of marine magnetic anomalies, geologists have established a system of identifying features in these anomalies by a *magnetic anomaly number*. The three different systems permit the user to tie interpretations to a polarity time scale related as closely as possible to the particular type of data being analyzed. The Subcommission felt that, until the ultimate correlation of the different time scales can be achieved, there was merit in this loose pluralistic approach. To encourage the use of magnetostratigraphic methods and terms, three members of the Subcommission have drawn up a relevant correlation chart which is reproduced in three parts: Neogene, Paleogene, and Late Cretaceous. It is emphasized that these correlation diagrams (Figs. 1-3) are not endorsed as the ultimate system. The system will change as refinements occur.

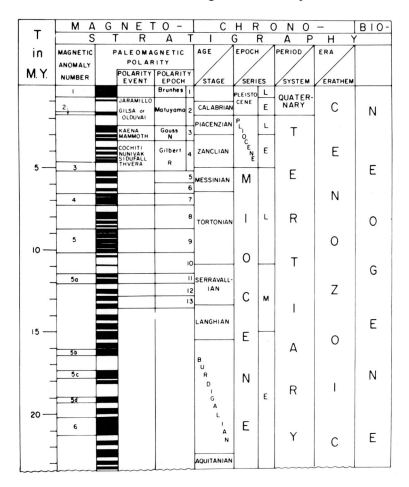

FIG. 1—Correlations between established stratigraphic zonations and magnetic polarities for Neogene (after Watkins, 1976). Chart is to encourage use of magnetostratigraphic methods and terms, and is not intended as final unambiguous correlation scheme. (Chart compiled by William A. Berggren, Neil D. Opdyke, and Norman D. Watkins; Printed with permission of *Geotimes*.)

Unlike other fossil identifications in rocks throughout the world, the identification of a particular fossil magnetization does not necessarily guarantee that it was derived at the time the rock was originally deposited. Ancient field directions may be imperfectly recorded in sediments because of dynamic distortion during original deposition. Furthermore, postdepositional processes, especially those of a chemical nature, may lead to overprinting of the original detrital remanent magnetism. The uninitiated need to beware of defining magnetostratigraphic sequences unless multiple sampling of horizons has been done and laboratory experiments have included methods appropriate to the removal of unstable components or overprinting. Formal magnetostratigraphy based on sediments should be derived from sequences involving different sedimentation rates, lithologies, and depositional and postdepositional environments.

Frequency of Reversals

A glance at Figures 1 through 3 shows that, as time increases, the number of polarity changes decreases. From the early studies of marine magnetic anomalies and their

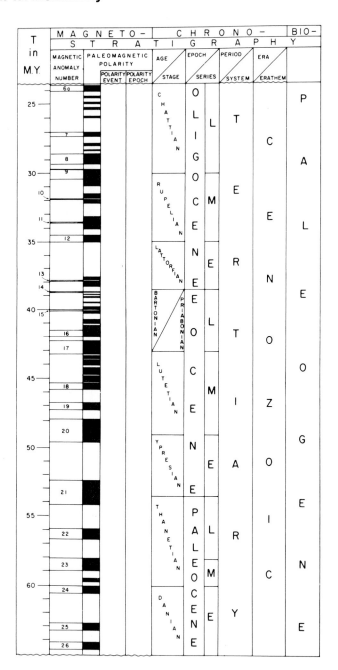

FIG. 2—As for Figure 1, correlation chart, correlations between established stratigraphic zonations and magnetic polarities for the Paleogene. (Printed with permission of *Geotimes*.)

extension into the Late Cretaceous, Heirtzler et al (1968) first noted that there was a significant drop in reversal frequency prior to 45 m.y. ago. The time scale from marine magnetic anomalies prior to 76 m.y. ago and as long ago as 153 m.y., has been deduced by Larson and Pitman (1972) and updated by Larson and Hilde (1975). Cox (1975)

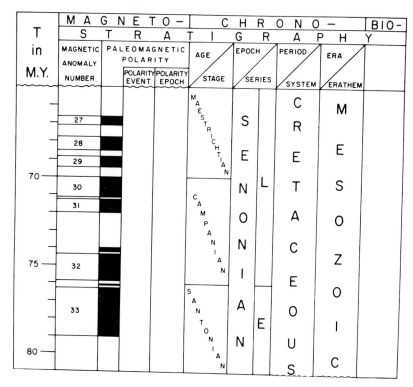

FIG. 3—As for Figure 1, correlation chart, correlations between established stratigraphic zonations and magnetic polarities for part of Cretaceous. (Printed with permission of *Geotimes*.)

used the results of Heirtzler et al (1968) and Larson and Pitman (1972) to analyze the variations in the average frequency of reversals over the past 150 m.y. as seen through a sliding window 10 m.y. long. The result of this analysis is illustrated in Figure 4, which shows that the average frequency of reversals was reduced prior to 45 m.y. ago and was zero during part of the Cretaceous. The interval between 85 and 107 m.y. corresponds to the Cretaceous quiet normal interval during which there were no reversals (Helsley and Steiner, 1969).

A long quiet reversed interval in the late Paleozoic is known (Irving and Parry, 1963). In magnetostratigraphy, and its applications in intercontinental correlation, these quiet intervals hold the key. If intervals of no polarity changes, of length greater than 10 m.y., exist in the stratigraphic column, they should be identified readily for all continents and should then be correlated. Therefore it is of interest to determine whether or not such quiet intervals were present during the Phanerozoic.

In general, the reversal pattern of the earth's magnetic field over the long time scale can be divided into quiet and disturbed intervals reflecting the absence of reversals or the occurrence of many reversals. For such quiet intervals the magnetic field must have a polarity bias so that over a period of time one of the two possible states (normal or reversed) is preferred. Polarity bias can be determined by examining the polarity ratio of paleomagnetic results as observed on land epoch by epoch.

The first attempt to delineate the polarity bias of the magnetic field during the Phanerozoic was carried out by McElhinny (1971). The analysis showed a clear trend of bias toward reversed polarity during the late Paleozoic and toward normal polarity during the Mesozoic. A more recent determination of polarity ratios, using 50-m.y. overlapping averages, is shown in Figure 5. The previously established features are clear and there is some evidence that the early Paleozoic may have significant reversed bias.

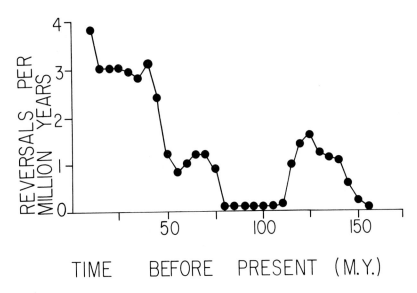

FIG. 4–Variations in average frequency of reversals, seen through sliding window 10 m.y. long. From 76 m.y. to present, rates are from Heirtzler et al (1968), and prior to 76 m.y., rates are from Larson and Pitman (1972). After Cox (1975).

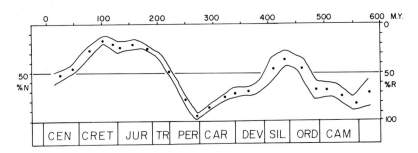

FIG. 5–Overlapping 50-m.y. averages of polarity ratios from land-based paleo-magnetic data, through Phanerozoic (after Irving and Pullaiah, 1976). Means and limits of standard errors are shown.

A careful and detailed analysis of these polarity ratios led Irving and Pullaiah (1976) to conclude that the following were the best estimates of the quiet intervals in the late Paleozoic and Mesozoic: Cretaceous Normal (KN), 81-110 m.y.; Jurassic Normal (JN), 145-164 m.y.; Triassic Normal (TRN), 205-220 m.y.; and Permo-Carboniferous Reversed (PCR), 227-313 m.y. Other, less certain, quiet intervals that may have existed during the early Paleozoic include an Ordovician Reversed Interval (Early to Middle Ordovician) and a Cambrian Reversed Interval (Middle to Late Cambrian). The only attempt so far to undertake any detailed magnetostratigraphy for the early Paleozoic has been by Soviet workers using the extensive sequences along the Lena River in Siberia (Khramov, 1974).

Phanerozoic Reversal Pattern

Figure 6 provides a summary diagram showing the overall pattern that has emerged from paleomagnetic studies on land and from marine magnetic anomalies. Because of the balance between the normal and reversed polarities in the Cenozoic it is doubtful

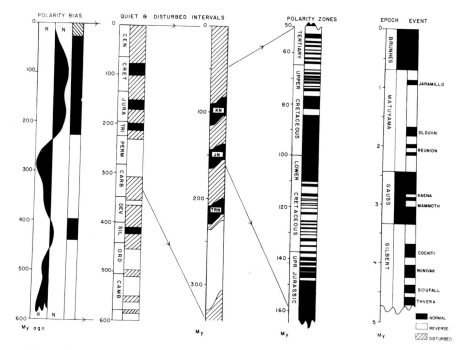

FIG. 6—Patterns of geomagnetic field through Phanerozoic. Left to right: Polarity bias, quiet and disturbed intervals, Cretaceous and Late Jurassic magnetic anomaly sequence, polarity time scale for past 5 m.y.

that the sequence would have been established in such detail without the aid of the marine anomaly sequences. The balance also is present in Early Triassic and in Late Silurian and Early Devonian. These times, therefore, probably will be the least useful in correlation because the tendency for more frequent reversals will make the matching of reversal sequences more difficult.

The pattern of quiet and disturbed intervals holds the greatest promise for magneto-stratigraphy. The upper and lower boundaries of quiet intervals are potentially the most useful tool in stratigraphic correlation especially on an international and inter-continental scale. The use of the most spectacular of the quiet intervals, the late Paleozoic reversed interval, has been discussed in this respect for over a decade (e.g., Irving and Parry, 1963; McElhinny, 1969; Irving and Pullaiah, 1976).

Finally, it may be appropriate, and indeed simple, to use magnetostratigraphy to resolve one of the most important problems in stratigraphy—the definition of the Precambrian-Cambrian boundary. Kirschvink (1976) has been examining the polarity sequences across the Precambrian-Cambrian boundary in the Amadeus basin of central Australia. The pattern suggests that a quiet reversed zone is present just below the possible boundary and this is followed by a zone of mixed polarities (disturbed zone). The top of the quiet interval that appears to precede the Precambrian-Cambrian boundary ultimately could be used as the definition of this boundary. Kirschvink (1976) already has noted that there are similarities in the polarity sequences in central Australia and the Desert Range, Nevada (Fig. 7). During the next few years this sequence could be established with some certainty.

References Cited

Cox, A., 1969, Geomagnetic reversals: Science, v. 163, p. 237-245.
—— 1975, The frequency of geomagnetic reversals and the symmetry of the nondipole field: Rev. Geophys. Space Phys., v. 13, p. 35-51.

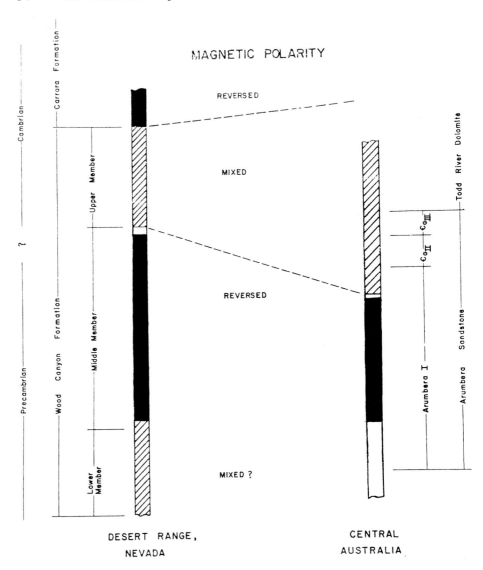

FIG. 7–Correlation of polarity sequences across Precambrian-Cambrian boundary between central Australia and Desert Range, Nevada (Kirschvink, 1976).

Goree, W. S., and M. D. Fuller, 1976, Magnetometers using RF-driven Squids and their application in rock magnetism and paleomagnetism: Rev. Geophys. Space Phys., v. 14, p. 591-608.

Heirtzler, J. R., et al, 1968, Marine magnetic anomalies, geomagnetic field reversals, and motions of the ocean floors and continents: Jour. Geophys. Research, v. 74, p. 2119-2136.

Helsley, C. E., and M. C. Steiner, 1969, Evidence for long intervals of normal polarity during the Cretaceous Period: Earth and Planetary Sci. Letters, v. 5, p. 325-332.

Irving, E., and L. G. Parry, 1963, The magnetism of some Permian rocks from New South Wales: Royal Astron. Soc. Geophys. Jour., v. 7, p. 395-411.

—— and G. Pullaiah, 1976, Reversals of the geomagnetic field, magnetostratigraphy and relative magnitude of paleosecular variation in the Phanerozoic: Earth-Sci. Rev., v. 12, p. 35-64.

Khramov, A. N., 1974, Paleozoic paleomagnetism: Leningrad, Nedra, 236 p.

Kirschvink, J. L., 1976, The magnetic stratigraphy of late Proterozoic to Early Cambrian sediments of the Amadeus basin, central Australia: a palaeomagnetic approach to the Precambrian/Cambrian boundary problem: Paper presented at 25th Internat. Geol. Cong., Sydney.

Larson, R. L., and T. W. C. Hilde, 1975, A revised time scale of magnetic reversals for the Early Cretaceous and Late Jurassic: Jour. Geophys. Research, v. 80, p. 2586-2594.

—— and W. C. Pitman, 1972, World-wide correlation of Mesozoic magnetic anomalies and its implications: Geol. Soc. America Bull., v. 83, p. 3645-3662.

McElhinny, M. W., 1969, The paleomagnetism of the Permian of southeast Australia and its significance regarding the problem of intercontinental correlation: Geol. Soc. Australia Spec. Pub., v. 2, p. 61-67.

—— 1971, Geomagnetic reversals during the Phanerozoic: Science, v. 172, p. 157-159.

McMahon, B. E., and D. W. Strangway, 1967, Kiaman magnetic interval in the western United States: Science, v. 155, p. 1012-1013.

Watkins, N. D., 1973, Magnetic polarity time scale: Geotimes, v. 18, no. 5, p. 21-22.

—— 1975, Correlating stratigraphic zones and magnetic polarities: Geotimes, v. 20, no. 6, p. 26-27.

—— 1976, Polarity Subcommission sets up some guidelines: Geotimes, v. 21, no. 4, p. 18-20.

Subcommission on Geochronology: Convention on the Use of Decay Constants in Geochronology and Cosmochronology[1]

R. H. STEIGER[2] and E. JÄGER[3]

Abstract On August 24, 1976 the IUGS Subcommission on Geochronology[4] met in Sydney, Australia, during the 25th International Geological Congress. They unanimously agreed to recommend the adoption of a standard set of decay constants and isotopic abundances in isotope geology. Values have been selected, based on current information and usage, to provide for uniform international use in published communications. The Subcommission urges that all isotopic data be reported using the recommended values (see appendix).

The recommendation represents a *convention* for the sole purpose of achieving interlaboratory standardization. The Subcommission does not intend to endorse specific methods of investigation or to specifically select the works of individual authors, institutions, or publications. All selected values are open to and should be the subjects of continuing critical scrutinizing and laboratory investigation. Recommendations will be reviewed by the Subcommission from time to time to bring the adopted conventional values in line with significant new research data.

Introduction

At the 1972 International Geological Congress in Montreal, the Subcommission on Geochronology decided to work towards the acceptance of standard decay constants. To achieve this aim the Subcommission organized the Paris 1974 symposium on the status of the decay constants and a concomitant inquiry. The mimeographed report of this meeting (McDougall, Steiger, and Jäger, 1974) was sent to all known geochronology laboratories together with a questionnaire, the results of which were presented at the 1976 International Geological Congress and are briefly summarized in this paper.

In addition, a session on the physical time scale and its related problems was held at the *Symposium on an International Geochronologic Scale*, 25th International Geological Congress.

Questionnaire Response

Two hundred and forty questionnaires were sent to laboratories or individuals; 160 of these individuals were thought to be active geo- or cosmo-chronologists. Eighty answers were received representing 121 individuals or 75 percent of the scientists active in the field. The following is a summation of the responses to the questions:

1. Do you think the Subcommission on Geochronology should assist with the decay constants problem? *Yes: 90%; uncertain* or *no: 10%*.

[1] Manuscript received, February 18, 1977.

[2] Institute for Cristallography and Petrography, ETH, Zurich, Switzerland.

[3] Mineralogisch-Petrographisches Institute, University of Berne, Berne, Switzerland.

[4] The Subcommission on Geochronology is a body of the Commission on Stratigraphy of the International Union of Geological Sciences (IUGS). The Subcommission for the 1972-1976 term included: E. Jäger, R. H. Steiger, Switzerland; G. D. Afanass'yev/A. I. Tugarinov, USSR; U. G. Cordani, Brazil; R. E. Folinsbee, Canada; J. R. Lancelot, France; I. McDougall, Australia; L. O. Nicolaysen, South Africa; K. Shibata, Japan; L. T. Silver, USA; N. J. Snelling, United Kingdom.

Article Identification Number: 0149-1377/78/SG06-0006/$03.00/0

2. Do you wish the Subcommission to make a recommendation for the adoption of a standard set of decay constants? *Yes: 90%; yes* (conditional): *5%; no: 5%.*

3. When should such a recommendation be made? *As soon as possible: 4%; before IGC, Sydney 1976: 11%; at the IGC, Sydney 1976: 60%; after testing of results available up to 1976: 25%.*

4. (Uranium) What constants do you currently use? *Those obtained by Jaffey et al (1971): 68%; Fleming et al (1952): 32%.*

5a. If you are presently using other Uranium decay constants do you plan to switch to the values by Jaffey et al (1971)? *Yes: 100%.*

5b. How will you do that? *On your own: 17%; if agreement is reached: 83%.*

6a. Are the Uranium decay constants by Jaffey et al (1971) superior to the previously used values? *Yes: 97%; not sure: 3%.*

6b. What are your reasons for preference? *Higher precision; more consistent U/Pb and Pb/Pb ages; most careful and complete experiment, details considered seem exhaustive, computations described in every detail;* and *high reputation of laboratory and scientists involved.*

7. Which $^{238}U/^{235}U$ ratio do you currently use? *137.88: 65%; 137.8: 28%; 137.7: 7%.*

8. (Thorium) Which ^{232}Th decay constant do you currently use? *4.9475: 63%; 4.99: 31%; 4.88: 4%; 4.98: 2%* (all constants multiplied by 10^{-11}/y).

9. (Rubidium) Which ^{87}Rb decay constant do you currently use? *1.39 x 10^{-11}/y: 60%; 1.47 x 10^{-11}/y: 36%; other values: 4%.* (The 1.47 x 10^{-11}/y value is mainly used in Europe.)

10a. Do you plan to switch to another ^{87}Rb decay constant? *Yes: 39%; yes (conditional): 40%; no: 21%.*

10b. If yes, which value will you use in the future? *1.41 - 1.43 x 10^{-11}/y: 100%.*

10c. Will you wait for an international agreement before changing? *Yes: 81%; no: 11%; not sure: 8%.*

11. Comments on the validity of the Neumann and Huster (1974) counting experiment: *Precision not sufficient: 8%; should be checked by other methods: 42%; looks good: 11%; it agrees with other methods: 39%.*

12. Which $^{85}Rb/^{87}Rb$ ratio do you currently use? *2.591 or similar: 55%; 2.59265 or similar: 28%; others: 17%.*

13a. (Strontium) Which $^{86}Sr/^{88}Sr$ ratio do you currently use? *0.1194: 98%; 0.1186: 2%.*

13b. Which $^{84}Sr/^{86}Sr$ ratio do you currently use? *0.05655 or similar: 36%; 0.0568 to 0.057 or similar: 54%; others: 10%.*

14. (Potassium) Which ^{40}K decay constants do you currently use? λ_β = 4.72, λ_e = 0.585, *or 0.584 (Wetherill, 1966): 92%;* $\lambda_{\beta-}$ = 4.905, $\lambda_e + \lambda'_e$ = 0.575 *(Beckinsale and Gale, 1969): 8%* (all constants multiplied by 10^{-10}/y).

15a. Do you plan to switch to other ^{40}K constants? *Yes: 18%; yes (conditional): 41%; no: 41%.*

15b. If yes, which values would you prefer? *Beckinsale and Gale, 1969: 100%.*

16. Which $^{40}K/K$ mole to mole ratio are you using? *1.19: 79%; 1.18: 14%; others: 7%* (all constants multiplied by 10^{-4}).

17. (Argon) Which $^{40}Ar/^{36}Ar$ atomic ratio are you using for atmospheric argon? *295.5: 65%; 296.0: 19%; 295.0: 3%; others: 13%.*

Questionnaire Summary

The following conclusions were drawn from the 1975 inquiry: (1) the poll reflects the opinion of a large majority of the active geo- and cosmo-chronologists; (2) the Subcommission on Geochronology has been entrusted to deal with the decay constants problem; and (3) the Subcommission is urged to adopt and recommend a standard set of decay constants before the close of 1976.

The following conditions are necessary to assure a wide acceptance of such a recommendation: (1) the uranium decay constants by Jaffey et al (1971) must be the basis for any standard set of decay constants; (2) the new value for the ^{87}Rb decay constant should lie within 1.41 and 1.43 x 10^{-11}/y, but there is some reluctance to change to such a new value before current critical experiments are completed; and (3) the decay constants for ^{40}K should be discussed and the values by Beckinsale and Gale (1969) taken into consideration.

Working Session on Decay Constants

During the session, "The Physical Time Scale and Related Problems" all scientists interested were invited by the Subcommission to attend a working session devoted to the discussion and evaluation of the decay constants. This meeting was held on August 20, 1976, with 23 geochronologists participating[5].

The great majority of the attending scientists supported the idea of proposing a standard set of decay constants. It was suggested that such a set should be issued in the form of a recommendation and that the numbers should be presented as given in the original publication, not rounded off.

Based on the results of the 1975 inquiry by questionnaire, it was moved that the uranium decay constants by Jaffey et al (1971) be the mainstay for a standard set. This was accepted together with the ^{238}U/^{235}U ratio of 137.88 (Shields, 1960) used by the majority of workers. Later it was suggested that the thorium decay constant be included in a recommendation. The value by Le Roux and Glendenin (1963), now adopted by most laboratories, was approved without further discussion.

The questionnaire results revealed that a two-thirds majority of the laboratories were using the 1.39 x 10^{-11}/y ^{87}Rb decay constant (Wetherill, 1966) which is founded on the uranium decay constants by Fleming et al (1952). While most participants of the working session agreed that a change in the ^{87}Rb decay constant was mandatory because of the new uranium values, there was disagreement as to how far the adjustment should go. Mere adaption to the new uranium constants requiring a change to about 1.41 x 10^{-11}/y was supported by some. Others argued that a value obtained by independent physical measurement such as the counting experiment by Neumann and Huster (1974) and the new value by the University of Alberta group, Edmonton (Davis et al, 1976) should be preferred. The Edmonton result of 1.417 ± 0.011 (95% C. L.) obtained by measurement of ^{87}Sr accumulated in strontium-free rubidium salt had been announced at the symposium just two days earlier. Further arguments for a ^{87}Rb decay constant of 1.42 x 10^{-11}/y included the results of intercomparison of coexisting minerals in terrestrial rocks (Afanass'yev et al, 1974), and the comparison of Rb-Sr and U-Pb ages in meteorites (Wetherill, 1975). An interesting new comparison of K-Ar and Rb-Sr ages in rapidly cooled igneous rocks was presented at the meeting by Tetley et al (1976). Their ^{87}Rb constant estimated on the basis of Beckinsale and Gale (1969) decay constants for ^{40}K gave 1.42 ± 0.01 x 10^{-11}/y.

It was then moved that in view of the uncertainties still involved, a value for the ^{87}Rb decay constant be accepted by convention and recommended for provisional use. Although the need for additional experiments was recognized, some participants were concerned as to whether a provisional recommendation would have much impact. While the participants subsequently acceded to the idea of a convention recommending a value of 1.42 x 10^{-11}/y, it was agreed that the problem of the ^{87}Rb decay constant was not definitely solved. Further efforts to obtain high quality determinations were generally advocated.

A proposal recommending the use of the ^{87}Rb isotopic abundance (^{85}Rb/^{87}Rb = 2.59265; Catanzaro et al, 1969) and the strontium isotopic abundance (^{84}Sr/^{86}Sr = 0.05655 or an updated value; Moore et al, 1977) as measured at the U.S. National Bureau of Standards was accepted[6]. On the other hand it was agreed that the

[5]Participants at working session, 25th International Geological Congress, Sydney, Australia, August 20, 1976: Australia—W. Compston, J. A. Cooper, C. M. Gray, I. McDougall, D. A. Nieuwland, R. Pidgeon, J. Richards, G. Riley, N. Tetley, J. S. Williams; Brazil—U. Cordani; Canada—R. L. Armstrong, R. E. Folinsbee; France—G. S. Odin, M. Roques, M. Vachette; Japan—K. Shibata; Switzerland—E. Jäger, R. H. Steiger; United States—P. Damon, M. Lanphere, M. Shafiqullah, L. T. Silver.

[6]The new values measured at the U.S. National Bureau of Standards are 0.119353 ± 3 for ^{86}Sr/^{88}Sr and 0.056562 for ^{84}Sr/^{86}Sr (personal commun., I. L. Barnes, 1977). To obtain a ^{84}Sr/^{86}Sr ratio compatible with the conventionally used ^{86}Sr/^{88}Sr ratio of 0.1194, the absolute ^{84}Sr/^{86}Sr ratio of 0.056562 was normalized to a value of 0.056584 using a factor of 0.1194/0.119353.

$^{86}Sr/^{88}Sr$ ratio of 0.1194 (Nier, 1938), universally applied in the normalization of strontium isotopic data, should be further used by convention, regardless of a possible minor change suggested by measurements currently under way at the U.S. National Bureau of Standards.

The paper read by Tetley et al (1976) had quite an impact on the discussion regarding the ^{40}K constant. It was argued that the Beckinsale and Gale (1969) values for the ^{40}K decay were based on a number of newer and independent counting experiments, which yielded fairly consistent results. The audience was informed about some criticism of the Beckinsale and Gale (1969) paper, which had been raised in response to the questionnaire. Objections were directed at the methods used to derive the new ^{40}K decay constants and not at the values themselves. A convention combining the Beckinsale and Gale (1969) values with the new potassium abundances (^{40}K = 0.01167 atom percent) determined by Garner et al (1976) was proposed and agreed on.

Appendix: Recommended Values

Uranium: $\lambda_{238_U} = 1.55125 \times 10^{-10}/y$ (Jaffey et al, 1971)

$\lambda_{235_U} = 9.8485 \times 10^{-10}/y$

atomic ratio $^{238}U/^{235}U = 137.88$ (Shields, 1960; Cowan and Adler, 1976)

Thorium: $\lambda_{232_{Th}} = 4.9475 \times 10^{-11}/y$ (Le Roux and Glendenin, 1963)

Rubidium: $\lambda_{87_{Rb}} = 1.42 \times 10^{-11}/y$ (Afanass'yev et al, 1974; Davis et al, 1977; Neumann and Huster, 1974, 1976; Tetley et al, 1976)

atomic ratio $^{85}Rb/^{87}Rb = 2.59265$ (Catanzaro et al, 1969)

Strontium: atomic ratios $^{86}Sr/^{88}Sr = 0.1194$ (Nier, 1938)

atomic ratio $^{84}Sr/^{86}Sr = 0.056584$ (Moore et al, 1977)

Potassium: $\lambda_{40_{K_{\beta^-}}} = 4.962 \times 10^{-10}/y$

$\lambda_{40_{Ke}} + \lambda'_{40_{Ke}} = 0.581 \times 10^{-10}/y$ (Beckinsale and Gale, 1969; Garner et al, 1976)

Isotopic abundance (atomic per cent): $^{39}K = 93.2581\%$; $^{40}K = 0.01167\%$; $^{41}K = 6.7302\%$. (Garner et al, 1976)

Argon: atomic ratio $^{40}Ar/^{36}Ar$ atmospheric = 295.5   (Nier, 1950)

References Cited

Afanass'yev, G. D., S. I. Zykov, and I. M. Gorokhov, 1974, Correlativity of geochronometric ages yielded by co-existing (cogenetic) minerals with different ^{40}K, ^{87}Rb, and some other radioactive elements decay constants: unpublished manuscript, 19 p.

Beckinsale, R. D., and N. H. Gale, 1969, A reappraisal of the decay constants and branching ratio of ^{40}K: Earth and Planetary Sci. letters, v. 6, p. 289-294.

Catanzaro, E. J., et al, 1969, Absolute isotopic abundance ratio and atomic weight of terrestrial rubidium: Nat. Bur. Standards (USA), Research Jour., A. Physics and Chemistry, v. 73A, p. 511-516.

Cowan, G. A., and H. H. Adler, 1976, The variability of the natural abundance of ^{235}U: Geochim. et Cosmochim Acta, v. 40, p. 1487-1490.

Davis, D. W., et al, 1977, Determination of the ^{87}Rb decay constant: Geochim. et Cosmochim. Acta, (in press).

Fleming, E. H., Jr., A. Ghiorso, and B. B. Cunningham, 1952, The specific alpha-activities and half-lives of ^{234}U, ^{235}U, and ^{238}U: Phys. Rev., v. 88, p. 642-652.

Garner, E. L., et al, 1976, Absolute isotopic abundance ratios and the atomic weight of a reference sample of potassium: Nat. Bur. Standards (USA), Research Jour., A. Physics and Chemistry, v. 79A, p. 713-725.

Jaffey, A. H., et al, 1971, Precision measurements of half-lives and specific activities of ^{235}U and ^{238}U: Phys. Rev. Ser. C., v. 4, p. 1889-1906.

Le Roux, L. J., and L. E. Glendenin, 1963, Half-life of ^{232}Th: Nat. Mtg. on Nucl. Energy (Pretoria, South Africa), Proc., p. 83-94.

McDougall, I., R. H. Steiger, and E. Jäger, 1974, Report on the activities of the IUGS Subcommission on Geochronology at the International Meeting for Geochronology, Cosmochronology, and Isotope Geology: unpublished, 5 p. (Paris, August, 1974).

Moore, L. J., et al, 1977, The absolute abundance ratios and the atomic weight of a reference sample of strontium: Nat. Bur. Standards (USA), Research Jour., A. Physics and Chemistry, v. 79a, p. 713-725.

Neumann, W., and E. Huster, 1974, The half-life of ^{87}Rb measured as difference between the isotopes ^{87}Rb and ^{85}Rb: Physik, v. 270, p. 121-127.

—— —— 1976, Discussion of the ^{87}Rb half-life determined by absolute counting: Earth and Planetary Sci. letters, v. 33, p. 277-288.

Nier, A. O., 1938, Isotopic constitution of Sr, Ba, Bi, Tl, and Hg: Phys. Rev., v. 54, p. 275-278.

—— 1950, A redetermination of the relative abundances of carbon, nitrogen, oxygen, argon, and potassium: Phys. Rev., v. 77, p. 789-793.

Shields, W. R., 1960, Comparison of Belgian Congo and synthetic "normal" samples: Nat. Bur. Standards (USA) Mtg. of the Advisory Committee for Standard Materials and Methods of Measurement (1960) 37 p. (unpublished).

Tetley, N. W., et al, 1976, A comparison of K-Ar and Rb-Sr ages in rapidly cooled igneous rocks (Preprint): 3 p.

Wetherill, G. W., 1966, Radioactive decay constants and energies, in S. Clark, ed., Handbook of physical constants: Geol. Soc. America Mem. 97, p. 513-519.

—— 1975, Radiometric chronology of the early solar system: Ann. Rev. Nuclear Science, v. 25, p. 283-328.

Pre-Cenozoic Phanerozoic Time Scale—Computer File of Critical Dates and Consequences of New and In-Progress Decay-Constant Revisions[1]

RICHARD LEE ARMSTRONG[2]

Abstract A computer file of K-Ar, Rb-Sr, and U-Pb dates that provide constraints on the pre-Cenozoic Phanerozoic time scale has been created. New data appear slowly and thus the file size grows at the rate of only a few percent per year. The time scale presented at the 1974 International Meeting for Geochronology, Cosmochronology, and Isotope Geology in Paris is not in conflict with the data added since then.

Precise chronologic subdivision of the Cretaceous is difficult because even the most optimistic uncertainties in the dates are greater than the duration of some stages. Nevertheless the stages of the Upper Cretaceous have been calibrated reasonably well.

Subdivision of the remainder of the Mesozoic and Paleozoic systems cannot be done precisely and objectively from geochronometric data. Important boundary dates must be derived from interpolation between points with experimental and geologic uncertainties of at least a few percent.

Efforts to obtain additional data for Lower Cretaceous to Upper Permian and Devonian and older rocks should be given special priority.

One potential source of confusion in time-scale calibration, and conversely in the assignment of geologic age to rocks that have been dated, is the use of different decay constants by different laboratories and by the same laboratory at different times. Recently adopted values for uranium decay constants have the effect of reducing previously published U-Pb dates by about 1%. Proposed new decay constants for potassium (K) would increase previously published Phanerozoic K-Ar dates by about 2% in the case of western literature, and would reduce dates published by the eastern European countries by about 2.5%. The suggested new decay constant for rhubidium (Rb) is a compromise between values previously used. Most western and all eastern European Rb-Sr dates would be reduced by about 2%. Dates from western labs that have used the 47×10^9-year half life would be increased about 3.5%.

Revised time scales will reflect these changes. Because most calibration points are K-Ar dates the net effect is a 1 to 2% increase in the ages assigned to time-scale boundaries for scales published by geologists from western countries, and an approximate 2% reduction for scales published in eastern Europe. A discrepancy exists between time scales used in the two groups of countries. The eastern European scale is younger than the western one because of the greater use of glauconite K-Ar dates by the eastern Europeans. In contrast western emphasis is on dated volcanic rocks coupled with skepticism toward glauconite and whole-rock K-Ar dates.

Introduction

For the past several years we have endeavored to maintain an up-to-date file of critical data that constrain the calibration for the pre-Cenozoic part of the Phanerozoic time scale. The Cenozoic was discussed at length by Berggren (1972) and Berggren and Van Couvering (1974). Further refinements are in progress (R. F. Zakrzewskii, written commun., 1974; G. S. Odin, personal commun., 1974; G. S. Odin, 1975). The uncertainties remaining are small, on the order of ±1 m.y. for most stage boundaries. The

[1] Manuscript received, December 14, 1976.

[2] University of British Columbia, Vancouver, B.C., Canada, V6T 1W5.

Article Identification Number: 0149-1377/78/SG06-0007/$03.00/0

Mesozoic and Paleozoic time-scale uncertainties are, in contrast, large (more than 10 m.y. in places), and the available information has been accumulating steadily. The situation up to approximately 1963 was reviewed extensively in the first Phanerozoic time-scale publication (Harland et al, 1964) and updated to about 1969 by Lambert (1971) in the second Phanerozoic time-scale publication (Harland and Francis, 1971).

Data Base

Much of our critical data comes from the Phanerozoic time-scale publications; the remainder has been gleaned from literature not covered by the previous reviews. The information for each critical date has been condensed on a single computer card (format contained in Table 1). If available information cannot be reduced to a few simple and absolute statements then the particular item is of little use in refined time-scale calibration. The computer-readable form of the file allows automatic recalculation and plotting of the data, or of selected subsets of data. Items up to number 366 come from the Phanerozoic time-scale publications and retain the original numbering system. Items omitted from this review are either Cenozoic, and thus of no concern to us, or pre-Cenozoic dates that provide no useful constraint. The latter items are listed in Table 2, together with the reason for their omission—excessive stratigraphic or analytical uncertainty, inadequate information, or obvious discordance indicating daughter product loss or sample contamination. The nearly 260 dates currently in the file are listed in Table 3. Nearly half are from the literature sources listed in Table 4. Literature sources for items numbered below 405 may be found in the two Phanerozoic time-scale (PTS) publications.

Output

The items in the file are plotted graphically with coordinates of time (x) and stratigraphic succession (y). The stratigraphic scale is an integer array that has a specified order (the geologic time scale as defined stratigraphically and paleontologically) but no quantitative significance. The stratigraphic-scale subdivisions used are listed in Table 5. There is no absolute time unit implied in the steps on this scale, but organic evolution provides a regulating mechanism that gives units of broadly similar length. Different symbols (shown in Fig. 1) are used to distinguish the types of material dated and the nature of the constraint provided.

FIG. 1—Legend for symbols on Figure 2 and ideal-world time curve to illustrate fit of data sought in drawing curve.

Decay Constants

To review the data, all results should be recomputed using a single set of decay constants. A variety of constants, none of which is universally accepted, has been used at various times. New values for uranium decay constants (Jaffe et al, 1971) generally have been accepted. These values reduce previously published U-Pb dates by about 1%. Proposed decay constants for potassium would increase previously published Phanerozoic dates by about 2% in the case of western literature (Beckinsale and Gale, 1969), and would reduce dates published by eastern European countries by about 2.5% (Afanas'yev and Zykov, 1975). Although the potassium constants proposed by western and eastern geologists appear somewhat different, the dates calculated from the same analytic data are virtually identical for the Phanerozoic. The suggested new decay constant for rubidium (Armstrong, 1974; Huster, 1974; Neumann and Huster, 1974;

Table 1. Computer file Input-Card format and code

Columns	Content	Code
1 to 4	Item Number (I 4, right adjust)	
5 to 44	Name and/or comment	
45	Location	1 = Europe 2 = U.S.S.R. 3 = S. America 4 = Africa 5 = Australia, New Zealand 6 = U.S.A. 7 = Canada 8 = Other
46	Material Analysed	U = Uraninite B = Biotite H = Hornblende S = Sanidine F = Feldspar W = Whole Rock G = Glauconite C = Clay Z = Zircon I = Whole Rock Isochron N = Unknown
47	Geologic Setting	A = Pegmatite B = Granitic Pluton C = Intermediate Pluton D = Mafic Pluton V = Tuff W = Rhyolitic (Felsic) Volcanic X = Basaltic (Mafic) Volcanic S = Sediment M = Metamorphic Rock T = Time Scale Point (published scales)
48	Quality Code	$\emptyset$ = Acceptable D = Dubious
49 to 50	Decay Constant - Method K-Ar Rb-Sr U-Pb	A1 = (4.72/.584 or .585/1.19) = (4.76/.589/1.18) A2 = U.S.S.R. values (4.72/.557/1.19) A3 = New (4.905/.575/1.18) = (4.963/.581/1.167) A4 = New U.S.S.R. (4.72/.5747/1.18) A5 = (4.72/.584/1.18) A6 = λ not reported S1 = 1.39 S2 = 1.47 S3 = 1.42 S4 = λ not reported U1 = (.154/.971 or .972/137.7) U2 = New (.155125/.98485/137.88) U3 = λ not reported
51 to 55	Date (F 5.1)	
56 to 60	Reported Error (σ) (F 5.1)	
61 to 64	(A) Upper Stratigraphic Limit (I4)	Code given in Table 5 [0240 used to indicate empty category – a point will not be plotted on the output.]
65 to 68	(B) Stratigraphic Age (I4)	
69 to 72	(C) Lower Stratigraphic Limit (I4)	

Usually B *or* A and C are given, not A, B, and C. In the case of volcanic ash, lava, glauconite, etc. dates it is usually possible to specify B. Intrusive rocks provide (C), the stratigraphic age of the youngest rocks cross cut, metamorphosed, or deformed by the emplacement event and (A), the age of overlying sediments or sediments containing erosion products of the solidified pluton. The information provided by a date on metamorphic rocks is of similar character–limits, never a precise stratigraphic age.

Afanas'yev and Zykov, 1975) lies between the values used previously. Most western and all eastern European Rb-Sr dates would be reduced by about 2%. Dates from western labs that have been using the 47×10^9-year half life would be increased by about 3.5%.

Revised time scales will reflect these changes. Because most dates used are by K-Ar, the net effect on the time-scale boundaries is essentially the same as the effect on K-Ar dates—a slight increase in western time-scale boundary ages and a reduction of boundary ages for previously published eastern European time scales.

Results

The time-scale data file, plotted on Figure 2 using recently proposed decay constants, is subject to revision if the present values undergo further change. Versions based on other sets of constants can be prepared readily. All versions appear similar. The only noticeable change is a small systematic shift of most points as a consequence of the different K-decay constants used.

Caution should be observed. The presentation of data in Table 6 is only a recomputation, using revised decay constants, of time-scale items as originally presented in the literature cited. We do not discuss the problems and confidence limits associated with each date; that is a matter for the reviews of the successive systems. Allocation of time to stages within systems is a discipline that involves detailed knowledge of biostratigraphy and relative thickness relations, examples of which can be found in Boucot (1975), Obradovich and Cobban (1975), and Van Hinte (1976a, b). This is the procedure for refinement with a precision that geochronometric data do not yet provide.

An ideal time scale would be a series of horizontal steps that agree with all items plotted (shown schematically on Figure 1). With insufficient data the time scale is not well constrained. The common discordance, due to daughter product loss, will displace dates toward lower values. This situation is chronic for glauconite dates and common to many other types, so that the scale should be constructed at the maximum values consistent with minimum violation of other available constraints.

Although volcanic rocks usually are correlated more precisely with the geologic time scale, many of the critical dates for time-scale calibration come from intrusive igneous rocks. The reasons for this are practical, relating to the stability of geochrono-

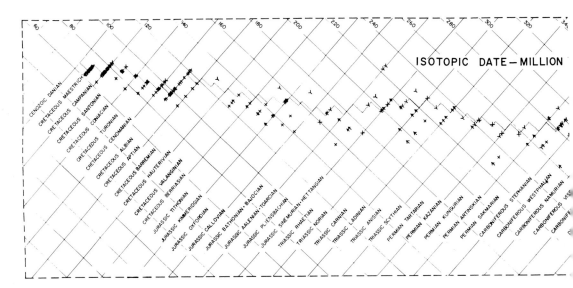

FIG. 2–Geologic time scale based on data file of Table 3. New decay constants for K, Rb, and U were used to recompute all dates plotted. Possible version of scale shown by dotted line. Various symbols explained in Figure 1 show constraints used to define position of time scale—control is

metric systems. Volcanic rocks are more unstable chemically and typically more porous and fractured; therefore probability of closed-system behavior is very much less than for massive, holocrystalline, plutonic rocks. Nevertheless, despite less satisfactory stratigraphic brackets, dates for intrusive rocks are important for any attempt at time-scale calibration.

The dotted line on Figure 2 is a plausible interpretation of the data. The graphic display of constraints shows the variable quality of the time scale. Where control is scarce the curve is carried through as smoothly as possible. The resulting discordances are few and nearly all are explainable as daughter product loss. The weakest parts of the calibration are Lower Cretaceous to Upper Permian, and Devonian and older. The Upper Cretaceous is unusually well dated (Obradovich and Cobban, 1975) and illustrates the fact that stratigraphic subdivisions are not of uniform duration. The remainder of the time scale could be equally variable, but the data simply do not exist to establish this. Even the designation of system boundaries may involve considerable interpolation, and discussion of stage lengths may be futile.

Comparisons (Table 6)

Where comparisons are made using the same decay constants, the proposed time scale is different from the 1964 Phanerozoic time scale in several ways. The bases of the Cenozoic and Upper Cretaceous have shifted to slightly younger values; the bases of the Cambrian, Ordovician, Silurian, and Permian are virtually unchanged; and the bases of the Cretaceous, Jurassic, Triassic, Carboniferous, and Devonian are shifted to older values. The older and newer Russian time scales (Afanas'yev and Zykov, 1975) are virtually the same when recalculated with the decay constants used in Figure 2. They are similar to the 1964 time scale only back to the base of the Upper Cretaceous and at the base of the Triassic. Elsewhere, the Russian scale consistently assigns younger dates to the system boundaries. This is largely due to the heavier emphasis placed on glauconite and whole-rock K-Ar dates by Russian geochronologists, as contrasted with western emphasis on dated volcanic and plutonic rocks (K-Ar and Rb-Sr on mineral separates and whole-rock Rb-Sr isochrons).

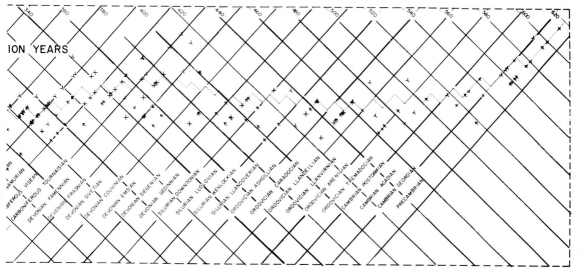

good in parts, and obviously poor in others. Decay constants used: $\lambda U^{238} = 0.155125 \times 10^{-9}\,yr^{-1}$; $K^{40}/K = 1.18 \times 10^{-4}$ atom ratio; $K^{40}\lambda_\beta = 4.905 \times 10^{-10}\,yr^{-1}$; $K^{40}\lambda_e = 0.575 \times 10^{-10}\,yr^{-1}$; $Rb^{87}\lambda = 1.42 \times 10^{-11}\,yr^{-1}$.

TABLE 2. Items in Phanerozoic Time Scale books that were excluded from the present datafile. Cenozoic items are not listed.*

Item	Grounds for Exclusion from Plot	Item	Grounds for Exclusion from Plot
004	not a tight limit, dtl (mm)	187	dubious significance (auth. bi)
007	not a tight limit (i)	188	dubious strat age limit (mm)
009	not a tight limit, dtl (i)	189	Pb- α (v)
010	obsolete, incorrect dates (i)	190	inaccurate strat bracket (i)
011	uncertain interpretation, dtl (i)	193	inaccurate strat bracket (v)
016	dtl (gl)	194	dubious date (i)
019	dtl (gl)	195	discordant data (gl)
032	dtl (il)	196	weak strat control (i)
033	dtl (i)	205	contaminated (gl)
034	dtl (sed U-Pb)	210	dtl (gl)
035	uncertain interp. dtl (i)	214	dtl (gl)
041	not a tight limit (i)	216	contaminated (gl)
043	not a tight limit (mm)	218	dtl (gl)
044	dtl (i)	221	dtl (gl)
048	doubtful significance (il)	222	dtl (gl)
052	dtl (gl)	223	dtl (gl)
067	not a tight limit (i)	224	contaminated (gl)
071	dtl (i)	225	strat age uncertain (gl)
072	dtl (gl)	231	dtl (gl)
078	dtl (gl)	232	dtl (gl)
091	not a tight limit (i)	234	dtl (gl)
092	dubious strat age (i)	235	dtl (gl)
099	Ar dtl (i)	236	dtl (gl)
100	dtl (il)	238	dtl (gl)
121	inaccurate date (i) ReOs	240	dtl (gl)
158	dtl (gl)	241	contaminated (gl)
159	dtl (gl)	319	strat age uncertain (i)
162	dubious strat age (i) Whin sill better dated	320	strat age uncertain (mm) dtl
164	dtl (gl)	321	insufficient information (i)
165	dtl (gl)	323	not restrictive (i)
166	dtl (gl)	324	not restrictive (i)
167	dtl (gl)	325	not restrictive (i)
168	dtl (gl)	326	not restrictive (i)
169	dtl (gl)	327	not restrictive (i)
170	dubious strat age (i)	330	strat age uncertain (i)
175	dtl (i)	331	Pb- α (i)
179	dtl (gl)	334	strat age uncertain (i)
180	dtl (gl)	337	strat age uncertain (i)
182	Ar dtl, Ca date inaccurate (syl)	342	dtl (v)
184	dubious strat age (U ore)	349	dtl (v)

*Legend: metamorphic rock = **mm**; intrusive rock = **i**; volcanic rock = **v**; glauconite = **gl**; illite = **il**; sylvite = **syl**; stratigraphic = **strat**; date too low = **dtl**.

Table 3. Computer datafile. Consecutive. First digit is continuation code, thus 1360 follows 0360 as a multiple listing.

```
LIST OF SAMPLES
MATERIAL ANALYSED                              METHOD, CONSTANTS
PUBLISHED DATE + ERROR (+000.0 IF NOT GIVEN)   LOWER STRAT LIMIT
UPPER STRAT LIMIT                              STRATIGRAPHIC AGE
```

```
0001  KAP FRANKLIN GRANITE                         CANADA
BIOTITE              INTRUSIVE GRANITE
393.0  012.0         ARGON 1.19/4.72/0.584
DEVONIAN GIVETIAN                    DEVONIAN GIVETIAN

0002  CHATTANOOGA SHALE                            USA
CLAY/SHALE          SEDIMENTARY
350.0  010.0         URANIUM 0.154 0.971 CLU
                                     DEVNIAN FAMENNIAN

0003  BENTONITES IN GRAND GREVE FORMATION          CANADA
                     VULCANIC TUFFACEOUS
385.0  015.0         ARGON 1.19/4.72/0.584
DEVNIAN GEDINNIAN                    DEVNIAN GEDINNIAN

0005  CALAIS GRANITE                               USA
BIOTITE              INTRUSIVE GRANITE
404.0  008.0         ARGON 1.19/4.72/0.584
DEVONIAN FAKENIAN                    DEVONIAN GEDINNIAN

0006  SHAP GRANITE                                 EUROPE
BIOTITE              INTRUSIVE GRANITE
391.0  007.0         ARGON 1.19/4.72/0.584
CARBONIFEROUS TOURNAISIAN            SILURIAN DOWNTONIAN

0006  DARTMOOR GRANITE                             EUROPE
BIOTITE              INTRUSIVE GRANITE
295.0  006.0         ARGON 1.19/4.72/0.584
PERMIAN SAKMARIAN                    CARBONIFEROUS STEPHANIAN

0012  MARSHALLTOWN FORMATION                       USA
GLAUCONITE          SEDIMENTARY
069.0  004.0         ARGON 1.19/4.72/0.584
                                     CRETACEOUS CAMPANIAN

0013  NAVESINK FORMATION                           USA
                     SEDIMENTARY
065.0  000.0         STRONTIUM LAMBDA=1.47
                                     CRETACEOUS MAESTRICHTIAN

0024  SHALE IN PENNSYLVANIAN LIMESTONE             USA
CLAY/SHALE          SEDIMENTARY
295.0  000.0         ARGON 1.19/4.72/0.584
                                     CARBONIFEROUS WESTPHALIAN

0030  UPPER KLITUNG SERIES              AUSTRALIA AND NEW ZEALAND
BIOTITE              VULCANIC DACITE RHYOLITE
278.0  000.0         ARGON 1.19/4.72/0.584
                                     CARBONIFEROUS STEPHANIAN

0031  HARZBURGER GABBRO                            EUROPE
BIOTITE              INTRUSIVE BASALT GABBRO
327.0  000.0         ARGON 1.19/4.72/0.584
CARBONIFEROUS STEPHANIAN             CARBONIFEROUS VISEAN

0042  VIRE CARCLLES GRANITE                        EUROPE
BIOTITE              INTRUSIVE GRANITE
553.0  000.0         ARGON 1.19/4.72/0.584
CAMBRIAN GEORGIAN                    PRECAMBRIAN

0045  DRAMMEN GRANITE                              EUROPE
BIOTITE              INTRUSIVE GRANITE
259.0  007.0         ARGON 1.19/4.72/0.584
                                     PERMIAN ARTINSKIAN

0046  OSLO NORDMARKITE                             EUROPE
ZIRKON              INTRUSIVE GRANITE
255.0  000.0         URANIUM 0.154 0.971 CLL
                                     PERMIAN ARTINSKIAN

0047  BUTTERLY DOLOMITE                            USA
CLAY/SHALE          SEDIMENTARY
437.0  000.0         ARGON 1.19/4.72/0.584
                                     ORDOVICIAN ARENIGIAN

0049  LOWER GREENSAND BARGATE BEDS                 EUROPE
GLAUCONITE          SEDIMENTARY
110.0  000.0         ARGON 1.19/4.72/0.584
                                     CRETACEOUS APTIAN

0050  LOWER GREENSAND HYTHE BEDS                   EUROPE
GLAUCONITE          SEDIMENTARY
115.0  000.0         ARGON 1.19/4.72/0.584
                                     CRETACEOUS APTIAN

0051  GAULT                                        EUROPE
GLAUCONITE          SEDIMENTARY
094.0  000.0         ARGON 1.19/4.72/0.584
                                     CRETACEOUS ALBIAN

0053  SYLVINITE                                    USSR
WHOLE ROCK          SEDIMENTARY
240.0  005.0         ARGON 1.19/4.72/0.557 USSR
                                     PERMIAN KUNGURIAN

0054  RIPLEY FORMATION                             USA
GLAUCONITE          SEDIMENTARY
069.0  000.0         ARGON 1.19/4.72/0.584
                                     CRETACEOUS MAESTRICHTIAN

0055  SHINGLOGWE URANINITE                         AFRICA
URANINITE           SEDIMENTARY
620.0  020.0         URANIUM 0.154 0.971 OLD
                                     PRECAMBRIAN

0056  UPPER GREENSAND                              EUROPE
GLAUCONITE          SEDIMENTARY
096.0  000.0         ARGON 1.19/4.72/0.584
                                     CRETACEOUS ALBIAN

0057  EMSCHER                                      EUROPE
GLAUCONITE          SEDIMENTARY
087.0  000.0         ARGON 1.19/4.72/0.584
                                     CRETACEOUS CONIACIAN

0058  BOCHUMER GREENSAND                           EUROPE
GLAUCONITE          SEDIMENTARY
084.0  000.0         ARGON 1.19/4.72/0.584
                                     CRETACEOUS TURONIAN

0059  SOESTER GREENSAND                            EUROPE
GLAUCONITE          SEDIMENTARY
079.0  000.0         ARGON 1.19/4.72/0.584
                                     CRETACEOUS TURONIAN

0060  GLAUCONITIC SANDSTONE                        EUROPE
GLAUCONITE          SEDIMENTARY
102.0  000.0         ARGON 1.19/4.72/0.584
                                     CRETACEOUS APTIAN

0061  ESSENER GREENSAND                            EUROPE
GLAUCONITE          SEDIMENTARY
086.0  000.0         ARGON 1.19/4.72/0.584
                                     CRETACEOUS TURONIAN

0062  GLAUCONITIC SANDSTONE                        EUROPE
GLAUCONITE          SEDIMENTARY
081.0  000.0         ARGON 1.19/4.72/0.584
                                     CRETACEOUS CAMPANIAN
```

0063 LES TUFS DE BRASSAC   EUROPE
     VOLCANIC TUFFACEOUS
     STRONTIUM LAMBDA=1.47
BIOTITE   008.0
288.0   CARBONIFEROUS STEPHANIAN

0064 VERAYA TIER   USSR
     SEDIMENTARY
GLAUCONITE 000.0   ARGON 1.19/4.72/0.557 USSR
308.0   CARBONIFEROUS WESTPHALIAN

0065 PATERSON TOSCANITE   AUSTRALIA AND NEW ZEALAND
     VOLCANIC DACITE RHYOLITE
BIOTITE   000.0   ARGON 1.19/4.72/0.584
298.0   CARBONIFEROUS STEPHANIAN

0066 LOWER KUTTUNG LAVAS   AUSTRALIA AND NEW ZEALAND
     VOLCANIC DACITE RHYOLITE
BIOTITE   000.0   ARGON 1.19/4.72/0.584
327.0   CARBONIFEROUS NAMURIAN

0068 BERKELY LATITE   AUSTRALIA AND NEW ZEALAND
     VOLCANIC DACITE RHYOLITE
FELDSPAR 000.0   ARGON 1.19/4.72/0.584
252.0   PERMIAN KUNGURIAN

0069 STANTHORPE GRANITE   AUSTRALIA AND NEW ZEALAND
     INTRUSIVE GRANITE
BIOTITE   000.0   ARGON 1.19/4.72/0.584
223.0   PERMIAN KUNGURIAN
TRIASSIC RHAETIAN

0070 BOISDALE HILLS GRANITE   CANADA
     INTRUSIVE GRANITE
BIOTITE   000.0   ARGON 1.19/4.72/0.584
518.0   CAMBRIAN ACADIAN

0073 GLAUCONITIC SANDSTONE HANOVER EUROPE
     SEDIMENTARY
GLAUCONITE 000.0   ARGON 1.19/4.72/0.584
139.0   JURASSIC TITHONIAN

0075 SHASTA BALLY BATHOLITH   USA
     INTRUSIVE INTERMEDIATE
BIOTITE   000.0   ARGON 1.19/4.72/0.557 USSR
134.0   JURASSIC KIMMERIDGIAN
CRETACEOUS BARREMIAN

0076 QTZ-DIORITE CR GABBRO INTRUSION, CALF USA
     INTRUSIVE GRANITE
BIOTITE   000.0   ARGON 1.19/4.72/0.557 USSR
143.0   JURASSIC KIMMERIDGIAN

0077 GLAUCONITIC SANDSTONE OBERPFALZ EUROPE
     SEDIMENTARY
GLAUCONITE 000.0   ARGON 1.19/4.72/0.584
136.0   JURASSIC OXFORDIAN

0089 DIORITE PLUTON TALKEETNA MOUNTAINS USA
     INTRUSIVE PEGMATITE
BIOTITE   000.0   ARGON 1.19/4.72/0.584
169.0   JURASSIC PLIENSBACHIAN
JURASSIC BATHONIAN BAJOCIAN

0090 INTRUSIONS WESTERN GEORGIA U.S.S.R USSR
     INTRUSIVE GRANITE
BIOTITE   000.0   ARGON 1.19/4.72/0.557 USSR
173.0   JURASSIC BATHONIAN BAJOCIAN

0093 CREETOWN GRANITE   EUROPE
     INTRUSIVE GRANITE
BIOTITE   000.0   ARGON 1.19/4.72/0.584
390.0   SILURIAN WENLOCKIAN

0094 CHATTANOOGA SHALE   USA
     VOLCANIC TUFFACEOUS
BIOTITE   006.0   ARGON 1.19/4.72/0.584
340.0   DEVONIAN FRASNIAN

0095 SNOBS CREEK RHYODACITE   AUSTRALIA AND NEW ZEALAND
     VOLCANIC DACITE RHYOLITE
BIOTITE   000.0   ARGON 1.19/4.72/0.584
350.0   DEVONIAN FAMENNIAN

0096 KINEO VOLCANIC SEQUENCE   USA
     VOLCANIC DACITE RHYOLITE
     STRONTIUM ISOCHRON
     STRONTIUM LAMBDA=1.39
375.0   000.0   DEVONIAN EMSIAN

0097 SNOWY RIVER GRANITE   AUSTRALIA AND NEW ZEALAND
     INTRUSIVE GRANITE
BIOTITE   000.0   ARGON 1.19/4.72/0.584
394.0   SILURIAN LUDLOVIAN

0098 BEAR RIVER GRANITE   CANADA
     INTRUSIVE GRANITE
BIOTITE   000.0   ARGON 1.19/4.72/0.584
370.0   DEVONIAN EMSIAN
CARBONIFEROUS VISEAN

0099 B NICTAUX GRANITES   CANADA
     INTRUSIVE GRANITE
     STRONTIUM LAMBDA=1.39
BIOTITE   000.0
386.0   DEVONIAN EMSIAN
CARBONIFEROUS VISEAN

0116 LOWER PART OF ASHINSK SERIES USSR
     SEDIMENTARY
GLAUCONITE 000.0   ARGON 1.19/4.72/0.557 USSR
573.0   PRECAMBRIAN

0117 LAMINARITES BEDS AND PROBLE EQUIVALENTS USSR
     SEDIMENTARY
GLAUCONITE 000.0   ARGON 1.19/4.72/0.557 USSR
595.0   PRECAMBRIAN

0118 UKSK BEDS OF KARATAU SERIES USSR
     SEDIMENTARY
GLAUCONITE 000.0   ARGON 1.19/4.72/0.557 USSR
615.0   PRECAMBRIAN

0119 GRANITE DE CHATEAU-CHINON   EUROPE
     INTRUSIVE GRANITE
     STRONTIUM LAMBDA=1.47
BIOTITE   010.0
300.0   CARBONIFEROUS STEPHANIAN

0120 SAKMARIAN LIMESTONE   USSR
     SEDIMENTARY
GLAUCONITE 000.0   ARGON 1.19/4.72/0.557 USSR
274.0   PERMIAN SAKMARIAN

0156 CASTRO DAIRE GRANITE   EUROPE
     INTRUSIVE GRANITE
     STRONTIUM LAMBDA=1.47
ZIRCON   007.0
282.0   PERMIAN ARTINSKIAN

0156 CARTERS LIMESTONE EGGLESTON LIMESTONE FM USA
     VOLCANIC TUFFACEOUS
ZIRCON   003.0   URANIUM 0.154 0.571 CLD
447.0   ORDOVICIAN CARADOCIAN

0157 BENTONITES IN CHASMOPS LIMESTONE EUROPE
     VOLCANIC TUFFACEOUS
BIOTITE   004.0   ARGON 1.19/4.72/0.584
444.0   ORDOVICIAN CARADOCIAN

0160 PITCHBLENDE,CHINLE FORMATION USA
     SEDIMENTARY
URANINITE 000.0   URANIUM 0.154 0.571 CLD
210.0   TRIASSIC CARNIAN

0163 MARDU DEPOSIT   USSR
     SEDIMENTARY
GLAUCONITE 000.0   ARGON 1.19/4.72/0.584
453.0   ORDOVICIAN ARENIGIAN

0171 VOSGES GRANITES   EUROPE
     INTRUSIVE GRANITE
     STRONTIUM LAMBDA=1.47
BIOTITE   000.0
320.0   CARBONIFEROUS STEPHANIAN

0172 UPPER VISEAN TUFFS FRANCE   EUROPE
     VOLCANIC TUFFACEOUS
     STRONTIUM LAMBDA=1.47
BIOTITE   000.0
328.0   CARBONIFEROUS VISEAN

0203 HARMON SHALE
BIOTITE   VOLCANIC TUFFACEOUS   CANADA
117.0   004.0   ARGON 1.19/4.72/0.584
                   CRETACEOUS ALBIAN

0204 MOWRY SHALE
BIOTITE   VOLCANIC TUFFACEOUS   USA
096.0   002.0   ARGON 1.19/4.72/0.584
                   CRETACEOUS ALBIAN

0207 GLAUCONITE IN SENONIAN SEDIMENTS SARATOV
GLAUCONITE   SEDIMENTARY   USSR
079.0   000.0   ARGON 1.19/4.72/0.557
                   CRETACEOUS SANTONIAN

0208 CONIACIAN SEDIMENTS
GLAUCONITE   SEDIMENTARY   USSR
078.0   000.0   ARGON 1.19/4.72/0.557
                   CRETACEOUS CONIACIAN

0209 CENOMANIAN SANDSTONE
GLAUCONITE   SEDIMENTARY   EUROPE
093.0   000.0   ARGON 1.19/4.72/0.557 USSR
                   CRETACEOUS CENOMANIAN

0211 BUKANSKOYE DEPOSIT
GLAUCONITE   SEDIMENTARY   USSR
103.0   000.0   ARGON 1.19/4.72/0.557 USSR
                   CRETACEOUS CENOMANIAN

0212 ALBIAN DEPOSITS
GLAUCONITE   SEDIMENTARY   USSR
103.0   000.0   ARGON 1.19/4.72/0.557 USSR
                   CRETACEOUS ALBIAN

0213 APTIAN SEDIMENTS
GLAUCONITE   SEDIMENTARY   USSR
107.0   000.0   ARGON 1.19/4.72/0.557 USSR
                   CRETACEOUS APTIAN

0215 NEOKOM
GLAUCONITE   SEDIMENTARY   USSR
136.0   000.0   ARGON 1.19/4.72/0.557 USSR
                   CRETACEOUS VALANGINIAN

0217 GRANODIORITE BAJA CALIFORNIA
ZIRCON   INTRUSIVE GRANITE   MISCELLANEOUS
103.0   000.0   URANIUM 0.154 0.971 CLD
                   CRETACEOUS MAESTRICHTIAN   CRETACEOUS ALBIAN

0219 MANNVILLE FORMATION
GLAUCONITE   SEDIMENTARY   CANADA
108.0   000.0   ARGON 1.19/4.72/0.584
                   CRETACEOUS ALBIAN

0220 LOWER CRETACEOUS SEDIMENTS
GLAUCONITE   SEDIMENTARY   USSR
103.0   000.0   ARGON 1.19/4.72/0.584 USSR
                   CRETACEOUS ALBIAN

0227 BENTONITE IN CRETACEOUS SHALE
BIOTITE   VOLCANIC TUFFACEOUS   USA
090.0   000.0   ARGON 1.19/4.72/0.584
                   CRETACEOUS CENOMANIAN

0227 UPPER GREENSAND
GLAUCONITE   SEDIMENTARY   EUROPE
091.0   000.0   ARGON 1.19/4.72/0.584
                   CRETACEOUS ALBIAN

0228 ALBIAN SEDIMENTS
GLAUCONITE   SEDIMENTARY   EUROPE
097.0   000.0   ARGON 1.19/4.72/0.584
                   CRETACEOUS ALBIAN

0173 GRANITE DE GLEN-SU-COAL
BIOTITE   VOLCANIC CALCITE PHYLLITE   EUROPE
334.0   007.0   STRONTIUM LAMBDA=1.47
                   CARBONIFEROUS VISEAN

0174 MERCYMIAN GRANITES ERZGEBIRGE MTS
BIOTITE   INTRUSIVE GRANITE   EUROPE
350.0   000.0   ARGON 1.19/4.72/0.557 USSR
PERMIAN SAKMARIAN   CARBONIFEROUS VISEAN

0176 WHIN SILL
WHOLE ROCK   INTRUSIVE BASALT GABBRO   EUROPE
295.0   019.0   ARGON 1.19/4.72/0.584
PERMIAN SAKMARIAN   CARBONIFEROUS NAMURIAN

0177 SANDRINGHAM SANDS
GLAUCONITE   SEDIMENTARY   EUROPE
131.0   000.0   ARGON 1.19/4.72/0.584
                   CRETACEOUS BERRIASIAN

0178 SANDRINGHAM SANDS
GLAUCONITE   SEDIMENTARY   EUROPE
132.0   004.0   ARGON 1.19/4.72/0.584
                   JURASSIC TITHONIAN

0181 SEDIMENTARY ROCKS
GLAUCONITE   SEDIMENTARY   USSR
097.0   000.0   ARGON 1.19/4.72/0.557 USSR
                   CRETACEOUS APTIAN

0183 MURRAY SHALE
GLAUCONITE   SEDIMENTARY   USA
284.0   050.0   STRONTIUM LAMBDA=1.39
                   CAMBRIAN GEORGIAN

0185 KESSYUSSE BEDS LF ALDAN SERIES
GLAUCONITE   SEDIMENTARY   USSR
550.0   000.0   ARGON 1.19/4.72/0.557 USSR
                   CAMBRIAN GEORGIAN

0186 GRANITE WICHITA MTS
ZIRCON   INTRUSIVE GRANITE   USA
523.0   000.0   URANIUM 0.154 0.971 CLD
CAMBRIAN POTSDAMIAN   PRECAMBRIAN

0191 MILLHOUSE BASALT SILL
WHOLE ROCK   INTRUSIVE BASALT GABBRO   EUROPE
322.0   012.0   ARGON 1.19/4.72/0.584
                   CARBONIFEROUS NAMURIAN

0192 ESSEXITE PORPHYRITE NORWAY
BIOTITE   VOLCANIC ANDESITIC BASALTIC   EUROPE
284.0   000.0   ARGON 1.19/4.72/0.584
                   PERMIAN SAKMARIAN

0198 JAMAICA GRANODIORITE
BIOTITE   INTRUSIVE INTERMEDIATE   MISCELLANEOUS
066.0   005.0   STRONTIUM LAMBDA=1.39
CENOZOIC DANIAN   CRETACEOUS MAESTRICHTIAN

0200 KNEEHILLS TUFF ZONE
BIOTITE   VOLCANIC TUFFACEOUS   CANADA
066.0   000.0   ARGON 1.19/4.72/0.584
                   CRETACEOUS MAESTRICHTIAN

0201 BEARPAW SHALE
BIOTITE   VOLCANIC TUFFACEOUS   CANADA
074.0   000.0   ARGON 1.19/4.72/0.584
                   CRETACEOUS CAMPANIAN

0202 CRO-SNAST VOLCANICS
SANIDINE   VOLCANIC CALCITE PHYLLITE   CANADA
102.0   003.0   ARGON 1.19/4.72/0.584
                   CRETACEOUS CENOMANIAN

0345 KOJTASH GRANITE   INTRUSIVE GRANITE   USSR
BIOTITE   000.0   ARGON 1.19/4.72/0.557 USSR
268.0
PERMIAN SAKMARIAN

0346 YATYRGVARTA INTRUSIONS   USSR
BIOTITE   003.0   ARGON 1.19/4.72/0.557 USSR
290.0
TRIASSIC SCYTHIAN

0347 FISSET BROOK FM LAKE AINSLIE VOLCANICS   CANADA
STRONTIUM ISOCHRON   VOLCANIC DACITE RHYOLITE
379.0   017.0   STRONTIUM LAMBDA=1.39
CARBONIFEROUS TOURNAISIAN   DEVONIAN FAMENNIAN

0348 MIEDZYGORZ GLAUCONITIC SANDSTONE   USSR
GLAUCONITE   SEDIMENTARY
492.0   000.0   ARGON 1.19/4.72/0.557 USSR
ORDOVICIAN TREMADOCIAN

0350 BALLANTRAE IGNEOUS COMPLEX   EUROPE
BIOTITE   INTRUSIVE BASALT GABBRO
475.0   000.0   ARGON 1.19/4.72/0.584
ORDOVICIAN CARADOCIAN   ORDOVICIAN ARENIGIAN

0351 BAIL HILL VOLCANICS   USSR
                 VOLCANIC ANDESITIC BASALTIC
BIOTITE   ARGON 1.19/4.72/0.584
445.0   000.0   ORDOVICIAN CARADOCIAN

0352 HOLYROOD GRANITE   INTRUSIVE GRANITE   CANADA
STRONTIUM ISOCHRON   STRONTIUM LAMBDA=1.47
574.0   011.0   PRECAMBRIAN
CAMBRIAN GEORGIAN

0353 MASSACHUSETTS GRANITES   USA
STRONTIUM ISOCHRON   INTRUSIVE GRANITE
569.0   004.0   STRONTIUM LAMBDA=1.39
CAMBRIAN GEORGIAN   PRECAMBRIAN

0354 CERBEREAN VOLCANICS   AUSTRALIA AND NEW ZEALAND
BIOTITE   VOLCANIC DACITE RHYOLITE
362.0   006.0   ARGON 1.19/4.72/0.584
DEVONIAN FAMENNIAN

0355 EASTPORT AND HEDGEHOG VOLCANICS   USA
STRONTIUM ISOCHRON   VOLCANIC DACITE RHYOLITE
411.0   005.0   STRONTIUM LAMBDA=1.39
DEVONIAN GEDINNIAN

0356 VOSGES GRANITES   INTRUSIVE GRANITE   EUROPE
STRONTIUM ISOCHRON   STRONTIUM LAMBDA=1.47
308.0   007.0   CARBONIFEROUS WESTPHALIAN
CARBONIFEROUS STEPHANIAN

0357 GYRANDA VOLCANICS   AUSTRALIA AND NEW ZEALAND
BIOTITE   VOLCANIC DACITE RHYOLITE
239.0   000.0   ARGON 1.19/4.72/0.584
TRIASSIC SCYTHIAN

0358 TRIASSIC GRANITES   AUSTRALIA AND NEW ZEALAND
STRONTIUM ISOCHRON   INTRUSIVE GRANITE
218.0   016.0   STRONTIUM LAMBDA=1.47
JURASSIC AALENIAN TOARCIAN   TRIASSIC LADINIAN

0359 WHIN SILL CF 176   EUROPE
WHOLE ROCK   INTRUSIVE BASALT GABBRO
295.0   006.0   ARGON 1.19/4.72/0.584
PERMIAN SAKMARIAN   CARBONIFEROUS NAMURIAN

0360 A BRITISH BASALT BARROW HILL   EUROPE
WHOLE ROCK   INTRUSIVE BASALT GABBRO
308.0   010.0   ARGON 1.19/4.72/0.584
CARBONIFEROUS WESTPHALIAN

0361 PREDAZZO GRANITE   INTRUSIVE GRANITE   EUROPE
STRONTIUM ISOCHRON   STRONTIUM LAMBDA=1.47
230.0   000.0   TRIASSIC LADINIAN

0229 SANTONIAN SEDIMENTS   EUROPE
GLAUCONITE   SEDIMENTARY
083.0   003.0   ARGON 1.19/4.72/0.584
CRETACEOUS SANTONIAN

0230 ALBIAN SEDIMENTS   EUROPE
GLAUCONITE   SEDIMENTARY
094.0   000.0   ARGON 1.19/4.72/0.584
CRETACEOUS ALBIAN

0233 CLEARWATER FORMATION   CANADA
GLAUCONITE   SEDIMENTARY
115.0?   003.0   ARGON 1.19/4.72/0.584
CRETACEOUS ALBIAN

0237 UPPER ALBIAN SEDIMENTS   USSR
GLAUCONITE   SEDIMENTARY
114.0   000.0   ARGON 1.19/4.72/0.584
CRETACEOUS ALBIAN

0239 SENONIAN SEDIMENTS   USSR
GLAUCONITE   SEDIMENTARY
079.0   000.0   ARGON 1.19/4.72/0.557 USSR
CRETACEOUS SANTONIAN

0252 LOWER GAULT   EUROPE
GLAUCONITE   SEDIMENTARY
098.0   003.0   ARGON 1.19/4.72/0.584
CRETACEOUS ALBIAN

0322 FOUREN DEPOSIT   USSR
GLAUCONITE   SEDIMENTARY
125.0   000.0   ARGON 1.19/4.72/0.557 USSR
CRETACEOUS VALANGINIAN

0428 MAGADAN BATHOLITH   INTRUSIVE GRANITE   USSR
BIOTITE   ARGON 1.19/4.72/0.557 USSR
134.0   000.0   CRETACEOUS ...

0335 GRANITIC INTRUSIONS   USSR
                 INTRUSIVE GRANITE
WHOLE ROCK   ARGON 1.19/4.72/0.557 USSR
094.0   000.0
CRETACEOUS TURONIAN

0336 VOLCANIC SEDIMENTARY SERIES MORTIN   USSR
                 VOLCANIC DACITE RHYOLITE
WHOLE ROCK   ARGON 1.19/4.72/0.557 USSR
105.0   000.0   CRETACEOUS ALBIAN

0338 SEMEITAU LAVAS   USSR
                 VOLCANIC DACITE RHYOLITE
WHOLE ROCK   ARGON 1.19/4.72/0.557 USSR
255.0   015.0   PERMIAN TARTARIAN

0339 KEREGETASS VOLCANICS   USSR
BIOTITE   VOLCANIC DACITE RHYOLITE
330.0   010.0   ARGON 1.19/4.72/0.557 USSR
CARBONIFEROUS NAMURIAN

0340 TASKULUK TRACHYDACITE   USSR
WHOLE ROCK   VOLCANIC DACITE RHYOLITE
310.0   015.0   ARGON 1.19/4.72/0.557 USSR
CARBONIFEROUS STEPHANIAN

0341 KYZYLKIY FIELDSITE   USSR
BIOTITE   INTRUSIVE GRANITE
287.0   010.0   ARGON 1.19/4.72/0.557 USSR
PERMIAN ARTINSKIAN

0342 KUTAN RHYOLITE   USSR
WHOLE ROCK   VOLCANIC TUFFACEOUS
245.0   000.0   ARGON 1.19/4.72/0.557 USSR
PERMIAN KAZANIAN

0344 PEREDOVOY CONGLOMERATE   USSR
                 INTRUSIVE GRANITE
BIOTITE   ARGON 1.19/4.72/0.557 USSR
288.0   000.0
PERMIAN SAKMARIAN

0363 KNEEHILLS BENTONITES CANADA
BIOTITE VOLCANIC TUFFACEOUS
065.5 001.0 ARGON 1.19/4.72/0.584
CRETACEOUS MAESTRICHTIAN

0364 TRICERATOPS ZONE BENTONITES CANADA
BIOTITE VOLCANIC TUFFACEOUS
064.0 001.0 ARGON 1.19/4.72/0.584
CRETACEOUS MAESTRICHTIAN

0365 BEARPAW BENTONITES CANADA
BIOTITE VOLCANIC TUFFACEOUS
071.0 001.0 ARGON 1.19/4.72/0.584
CRETACEOUS CAMPANIAN

0366 GUICHON CREEK BATHOLITH CANADA
FELDSPAR INTRUSIVE GRANITE
200.0 002.0 STRONTIUM LAMBDA=1.39
JURASSIC PLIENSBACIAN TRIASSIC CARNIAN

0406 HELLHOLE WILLIAMS GRYW USA
WHOLE ROCK METAMORPHIC
124.0 000.0 ARGON 1.19/4.72/0.584
CRETACEOUS VALANGINIAN

0407 KRIVOKLAT-ROKYCANY VOLC EUROPE
STRONTIUM ISOCHRCN VCLCANIC DACITE RHYOLITE
474.0 005.0 STRONTIUM LAMBDA=1.47
ORDOVICIAN TREMADCCIAN

0408 VILASUND GRANITE EUROPE
STRONTIUM ISOCHRCN INTRUSIVE GRANITE
447.0 006.0 STRONTIUM LAMBDA=1.39
SILURIAN LUDLCVIAN SILURIAN LLANDOVERIAN

0409 BROOKS RA SILLS USA
BIOTITE INTRUSIVE BASALT GABBRO
194.0 004.0 ARGON 1.19/4.72/0.584
JURASSIC SINEMUR-HETTANGIAN TRIASSIC NORIAN

0410 ELKHORN VOLC USA
HORNBLENDE VOLCANIC DACITE RHYOLITE
078.0 002.0 ARGON 1.19/4.72/0.584
CRETACEOUS CAMPANIAN

0411 CALEDONIAN PLUTONS EUROPE
STRONTIUM ISOCHRON INTRUSIVE INTERMEDIATE
474.0 000.0 STRONTIUM LAMBDA=1.39
ORDOVICIAN CARADOCIAN ORDOVICIAN LLANDELLIAN

0412 CASHEL-LOUGH WEELAUN PLUTON EUROPE
ZIRCON INTRUSIVE BASALT GABBRO
510.0 010.0 URANIUM 0.154 0.971 OLD
ORDOVICIAN LLANVIRNIAN

0413 MASSIF CENTRAL EUROPE
BIOTITE INTRUSIVE GRANITE
347.0 000.0 STRONTIUM LAMBDA=1.47
CARBONIFEROUS VISEAN

0414 STAPPOGIEDE SLATE EUROPE
STRONTIUM ISOCHRON SEDIMENTARY
515.0 007.0 STRONTIUM LAMBDA=1.39
ORDOVICIAN TREMADOCIAN

0415 ESQUIBEL ISLAND USA
HORNBLENDE VOLCANIC TUFFACEOUS
432.9 003.3 ARGON 1.18/4.905/0.575 NEW
SILURIAN LLANDOVERIAN

0416 BAY OF ISLANDS OBDUCTION CANADA
HORNBLENDE INTRUSIVE BASALT GABBRO
460.0 005.0 ARGON 1.19/4.72/0.584
ORDOVICIAN LLANVIRNIAN ORDOVICIAN ARENIGIAN

0417 AZERBAIJAN KAZAKH CITY USSR
BIOTITE INTRUSIVE GRANITE
086.0 005.0 ARGON 1.19/4.72/0.557 USSR
CRETACEOUS CAMPANIAN CRETACEOUS SANTONIAN

0418 BELORUSSIA USSR
GLAUCONITE SEDIMENTARY
090.0 004.0 ARGCN 1.19/4.72/0.557 USSR
CRETACEOUS CENOMANIAN

0419 ABKHAZIA MT BZYB USSR
WHOLE ROCK VOLCANIC ANDESITIC BASALTIC
164.0 006.0 ARGON 1.19/4.72/0.557 USSR
JURASSIC BATHONIAN BAJOCIAN

0420 KAZAKHSTAN AKCHATAU USSR
BIOTITE INTRUSIVE GRANITE
293.0 015.0 ARGON 1.19/4.72/0.557 USSR
PERMIAN SAKMARIAN CARBONIFEROUS STEPHANIAN

0421 KAZAKHSTAN AKCHATAU USSR
ZIRCON INTRUSIVE GRANITE
287.0 010.0 URANIUM 0.154 0.971 OLD
PERMIAN SAKMARIAN CARBONIFEROUS STEPHANIAN

0422 KAZAKHSTAN AKCHATAU USSR
BIOTITE INTRUSIVE GRANITE
291.0 020.0 STRONTIUM LAMBDA=1.39
PERMIAN SAKMARIAN CARBONIFEROUS STEPHANIAN

0423 DERGUNOVKA VEREYAN HORIZ USSR
GLAUCONITE SEDIMENTARY
308.0 010.0 ARGON 1.19/4.72/0.557 USSR
CARBONIFEROUS NAMURIAN

0424 KAZAKHSTAN CENTRAL USSR
BIOTITE INTRUSIVE GRANITE
355.0 010.0 ARGON 1.19/4.72/0.557 USSR
CARBONIFEROUS TOURNAISIAN DEVONIAN FRASNIAN

0425 KURSKIY DISTRICT USSR
GLAUCONITE SEDIMENTARY
366.0 010.0 ARGCN 1.19/4.72/0.557 USSR
DEVONIAN FRASNIAN

0426 PAKISTAN SALT RANGE USSR
GLAUCONITE SEDIMENTARY
530.0 020.0 ARGON 1.19/4.72/0.557 USSR
CAMBRIAN ACADIAN

0427 UKRAINE DNIEPER RIVEF USSR
GLAUCONITE SEDIMENTARY
590.0 015.0 ARGON 1.19/4.72/0.557 USSR
PRECAMBRIAN

0428 BULGARIA SWINETCH MTN USSR
GLAUCONITE SEDIMENTARY
106.0 004.0 ARGON 1.19/4.72/0.557 USSR
CRETACEOUS ALBIAN

0429 N CACASUS USSR
GLAUCONITE SEDIMENTARY
110.0 005.0 ARGON 1.19/4.72/0.557 USSR
CRETACEOUS ALBIAN

0430 N CACASUS USSR
GLAUCONITE SEDIMENTARY
128.0 005.0 ARGON 1.19/4.72/0.557 USSR
CRETACEOUS VALANGINIAN

0431 BULGARIA WRATZA MTN USSR
GLAUCONITE SEDIMENTARY
145.0 005.0 ARGON 1.19/4.72/0.557 USSR
JURASSIC CALLCVIAN

```
0432  BULGARIA SOFIA                              SEDIMENTARY                           USSR
GLAUCONITE  140.0   006.0     ARGON 1.19/4.72/0.557  USSR
JURASSIC PLIENSBACIAN

0433  N CACASUS SILLS          VOLCANIC ANDESITIC BASALTIC                 USSR
WHOLE ROCK  175.0   000.0     ARGON 1.19/4.72/0.557  USSR
JURASSIC PLIENSBACIAN

0434  AKSAHUT RIVER            VOLCANIC ANDESITIC BASALTIC                 USSR
WHOLE ROCK  255.0   000.0     ARGON 1.19/4.72/0.557  USSP
PERMIAN KAZANIAN

0435  N CACASUS BI GRANITES    INTRUSIVE INTERMEDIATE                      USSR
HORNBLENDE  270.0   000.0     ARGON 1.19/4.72/0.557  USSR
PERMIAN KUNGURIAN

0436  N CACASUS DAMJT RIVER    VOLCANIC ANDESITIC BASALTIC                 USSR
WHOLE ROCK  292.0   000.0     ARGON 1.19/4.72/0.557  USSR
CARBONIFEROUS WESTPHALIAN

0437  LIBA RIVER LIPARITE      VOLCANIC DACITE RHYOLITE                    USSR
WHOLE ROCK  245.0   000.0     ARGON 1.19/4.72/0.557  USSR
TRIASSIC SCYTHIAN

0438  KALDOMIN COMPLEX         INTRUSIVE GRANITE                           USSR
BIOTITE     299.0   000.0     ARGON 1.19/4.72/0.557  USSR
CARBONIFEROUS WESTPHALIAN

0439  KALDOMIN COMPLEX         INTRUSIVE GRANITE                           USSR
BIOTITE     291.0   000.0     STRONTIUM LAMBDA=1.39
CARBONIFEROUS WESTPHALIAN

0440  N CACASUS AKSAHUT RIVER / VORONEZ BAS   VOLCANIC ANDESITIC BASALTIC  USSR
WHOLE ROCK  345.0   010.0     ARGON 1.19/4.72/0.557  USSR
DEVONIAN FAMENNIAN

0441  N CACASUS DAMJT RIVER    INTRUSIVE GRANITE                           USSR
HORNBLENDE  365.0   020.0     ARGON 1.19/4.72/0.557  USSR
DEVONIAN FRASNIAN

0442  CENT KAZAKHSTAN SARY-ADYP   INTRUSIVE GRANITE                        USSR
BIOTITE     355.0   010.0     ARGON 1.19/4.72/0.557  USSR
DEVONIAN FRASNIAN

0443  N CACASUS UGUM RANGE     INTRUSIVE GRANITE                           USSR
BIOTITE     355.0             ARGON 1.19/4.72/0.557  USSR
DEVONIAN FRASNIAN

0444  MORRISON FORMATION VOLCANISM   VOLCANIC TUFFACEOUS                   USA
BIOTITE     150.0   005.0     ARGON 1.19/4.72/0.584
JURASSIC KIMMERIDGIAN

0445  BASE OF GREAT VALLEY SEQUENCE CALIF   INTRUSIVE BASALT GABBRO        USA
HORNBLENDE  155.0   00000     ARGON 1.19/4.72/0.584
JURASSIC TITHONIAN

0446  INDONESIAN TIN BELT      INTRUSIVE GRANITE                           MISCELLANEOUS
STRONTIUM ISOCHRON  217.0   005.0   STRONTIUM LAMBDA=1.39
TRIASSIC RHAETIAN

0447  INDONESIAN TIN BELT      INTRUSIVE GRANITE                           MISCELLANEOUS
HORNBLENDE  214.0   004.0     ARGON 1.18/4.72/0.584
TRIASSIC RHAETIAN


0448  KATAHDIN BATHOLITH       INTRUSIVE GRANITE                           USA
STRONTIUM ISOCHRON  395.0   016.0   STRONTIUM LAMBDA=1.39
DEVONIAN SIEGENIAN

0449  NAVARRO FM NCA           SEDIMENTARY                                 USA
GLAUCONITE  68.5    00000     STRONTIUM LAMBDA=1.39
CRETACEOUS MAESTRICHTIAN

0450  N AM K BENTONITE 2       VOLCANIC TUFFACEOUS                         USA
BIOTITE     067.4   000.7     ARGON 1.19/4.72/0.584
CRETACEOUS MAESTRICHTIAN

0451  N AM K BENTONITE 3       VOLCANIC TUFFACEOUS                         USA
BIOTITE     068.5   000.7     ARGON 1.19/4.72/0.584
CRETACEOUS MAESTRICHTIAN

0452  N AM K BENTONITE 4       VOLCANIC TUFFACEOUS                         USA
BIOTITE     071.5   000.7     ARGON 1.19/4.72/0.584
CRETACEOUS CAMPANIAN

0453  N AM K BENTONITE 5       VOLCANIC TUFFACEOUS                         USA
BIOTITE     072.6   000.7     ARGON 1.19/4.72/0.584
CRETACEOUS CAMPANIAN

0454  N AM K BENTONITE 5       VOLCANIC TUFFACEOUS                         USA
FELDSPAR    071.3   000.7     ARGON 1.19/4.72/0.584
CRETACEOUS CAMPANIAN

0455  N AM K BENTONITE 6       VOLCANIC TUFFACEOUS                         USA
BIOTITE     072.2   000.7     ARGON 1.19/4.72/0.584
CRETACEOUS CAMPANIAN

0456  N AM K BENTONITE 7       VOLCANIC TUFFACEOUS                         USA
BIOTITE     072.8   000.7     ARGON 1.19/4.72/0.584
CRETACEOUS CAMPANIAN

0457  N AM K BENTONITE 8       VOLCANIC TUFFACEOUS                         USA
BIOTITE     077.9   000.8     ARGON 1.19/4.72/0.584
CRETACEOUS CAMPANIAN

0458  N AM K BENTONITE 9       VOLCANIC TUFFACEOUS                         USA
BIOTITE     077.5   000.8     ARGON 1.19/4.72/0.584
CRETACEOUS CAMPANIAN

0459  N AM K BENTONITE 10      VOLCANIC TUFFACEOUS                         USA
BIOTITE     078.2   000.8     ARGON 1.19/4.72/0.584
CRETACEOUS CAMPANIAN

0460  N AM K BENTONITE 11      VOLCANIC TUFFACEOUS                         USA
BIOTITE     082.5   000.8     ARGON 1.19/4.72/0.584
CRETACEOUS SANTONIAN

0461  N AM K BENTONITE 13      VOLCANIC TUFFACEOUS                         USA
BIOTITE     086.8   000.9     ARGON 1.19/4.72/0.584
CRETACEOUS CONIACIAN

0462  N AM K BENTONITE 14      VOLCANIC TUFFACEOUS                         USA
BIOTITE     088.9   000.9     ARGON 1.19/4.72/0.584
CRETACEOUS TURONIAN

0463  N AM K BENTONITE 15      VOLCANIC TUFFACEOUS                         USA
BIOTITE     091.3   000.9     ARGON 1.19/4.72/0.584
CRETACEOUS CENOMANIAN
```

0464 N AM K BENTONITE 16   VOLCANIC TUFFACEOUS   USA
BIOTITE   ARGON 1.19/4.72/0.584
092.1   CRETACEOUS CENOMANIAN

0465 N AM K BENTONITE 17   VOLCANIC TUFFACEOUS   USA
BIOTITE   ARGON 1.19/4.72/0.584
095.3   CRETACEOUS ALBIAN

0466 N AM K BENTONITE 17   VOLCANIC TUFFACEOUS   USA
FELDSPAR   ARGON 1.19/4.72/0.584
095.3   CRETACEOUS ALBIAN

0467 HORNERSTOWN FM NJ-MD   SEDIMENTARY   USA
GLAUCONITE   ARGON 1.19/4.72/0.584
061.0   CENOZOIC DANIAN

0468 RED BANK - NAVESINK FMS NJ-MD SEDIMENTARY   USA
GLAUCONITE   ARGON 1.19/4.72/0.584
063.1   CRETACEOUS MAESTRICHTIAN

0469 MOUNT LAUREL FM NJ-MD   SEDIMENTARY   USA
GLAUCONITE   ARGON 1.19/4.72/0.584
083.3   CRETACEOUS CAMPANIAN

0470 MERCHANTVILLE FM NJ-MD   SEDIMENTARY   USA
GLAUCONITE   ARGON 1.19/4.72/0.584
081.0   CRETACEOUS CAMPANIAN

0471 ST GEORGE PLUTON   INTRUSIVE GRANITE   CANADA
STRONTIUM ISOCHRON   STRONTIUM LAMBDA=1.39
394.0   DEVONIAN GEDINNIAN

0472 ST GEORGE - ST STEVEN PLUTONS INTRUSIVE GRANITE   CANADA
HORNBLENDE   ARGON 1.19/4.72/0.584
402.0   DEVONIAN GEDINNIAN

0473 FLATHEAD SANDSTONE   SEDIMENTARY   USA
GLAUCONITE   STRONTIUM LAMBDA=1.39
542.0   CAMBRIAN ACADIAN

0474 FLATHEAD SANDSTONE   SEDIMENTARY   USA
STRONTIUM ISOCHRON   STRONTIUM LAMBDA=1.39
555.0   CAMBRIAN ACADIAN

0475 GUICHON CREEK BATHOLITH   INTRUSIVE GRANITE   CANADA
BIOTITE   ARGON 1.19/4.72/0.584
200.0   JURASSIC PLIENSBACHIAN   TRIASSIC CARNIAN

0476 MT MONZONI GRANITE   INTRUSIVE GRANITE   EUROPE
BIOTITE   STRONTIUM LAMBDA=1.47
230.0   TRIASSIC LADINIAN

0477 TRIASSIC-JURASSIC LAVAS   VOLCANIC ANDESITIC BASALTIC   USA
WHOLE ROCK   ARGON 1.19/4.72/0.584
202.0   JURASSIC SINEMUR-HETTANGIAN

0478 CUDDY MTN GRANITE   INTRUSIVE INTERMEDIATE   USA
HORNBLENDE   ARGON 1.19/4.72/0.584
206.0   JURASSIC SINEMUR-HETTANGIAN

0479 TALKEETNA MTNS GRANITES   INTRUSIVE INTERMEDIATE   USA
BIOTITE   ARGON 1.19/4.72/0.584
166.0   JURASSIC OXFORDIAN   JURASSIC AALENIAN TOARCIAN

0480 FANGS GRANITE   INTRUSIVE INTERMEDIATE   EUROPE
BIOTITE   ARGON 1.19/4.72/0.584
150.0   JURASSIC KIMMERIDGIAN

0481 FANGS GRANITE   INTRUSIVE INTERMEDIATE   EUROPE
BIOTITE   STRONTIUM LAMBDA=1.47
150.0   JURASSIC KIMMERIDGIAN

0482 ARISAIG VOLCANICS   VOLCANIC DACITE RHYOLITE   USA
STRONTIUM ISOCHRON   STRONTIUM LAMBDA=1.39
430.0   SILURIAN LLANDOVERIAN   ORDOVICIAN THEMAIUCIAN

0483 RODEZ STRAIT   VOLCANIC DACITE RHYOLITE   EUROPE
BIOTITE   ARGON 1.19/4.72/0.584
285.0   PERMIAN SAKMARIAN   CARBONIFEROUS STEPHANIAN

0484 SWISS JURASSIC   SEDIMENTARY   EUROPE
GLAUCONITE   GC3.C
145.0   JURASSIC OXFORDIAN

0485 SWISS JURASSIC   SEDIMENTARY   EUROPE
GLAUCONITE   ARGON 1.19/4.72/0.584
135.5   JURASSIC KIMMERIDGIAN

0486 VINE CARCLLES   INTRUSIVE GRANITE   EUROPE
NOT SPECIFIED   URANIUM 0.154 0.971 OLD
570.0   PRECAMBRIAN
CAMBRIAN GEORGIAN

0487 PENOBSCOT FORMATION   VOLCANIC DACITE RHYOLITE   USA
STRONTIUM ISOCHRON   STRONTIUM LAMBDA=1.39
460.0   ORDOVICIAN CARADOCIAN

0488 RED BEACH GRANITES   INTRUSIVE GRANITE   USA
STRONTIUM ISOCHRON   STRONTIUM LAMBDA=1.35
400.0   DEVONIAN FRASNIAN

0489 MM TORBROOK FM   METAMORPHIC   CANADA
WHOLE ROCK   ARGON 1.19/4.72/0.584
390.0   DEVONIAN SIEGENIAN

0490 BEEMERVILLE CPLX   INTRUSIVE GRANITE   USA
BIOTITE   ARGON 1.19/4.72/0.584
437.0   SILURIAN LLANDOVERIAN   ORDOVICIAN CARADOCIAN

0491 BEEMERVILLE CPLX   INTRUSIVE GRANITE   USA
BIOTITE   STRONTIUM LAMBDA=1.47
430.0   SILURIAN LLANDOVERIAN   ORDOVICIAN LANACELIAN

0492 MID PORTLANDIAN   SEDIMENTARY   EUROPE
GLAUCONITE   ARGON 1.19/4.72/0.584
129.0   JURASSIC TITHONIAN

0493 LOW ALBIAN   SEDIMENTARY   EUROPE
GLAUCONITE   ARGON 1.19/4.72/0.584
100.0   CRETACEOUS ALBIAN

```
0494 LUW ALBIAN             G 429I                          EUROPE
GLAUCONITE                  SEDIMENTARY
106.3                       ARGON 1.19/4.72/0.584
                              CRETACEOUS ALBIAN

0495 BASAL CENOMANIAN GL-0                                  EUROPE
GLAUCONITE      001.8       SEDIMENTARY
091.5                       ARGON 1.19/4.72/0.584
                              CRETACEOUS CENOMANIAN

0496 UPPER ALBIAN           G 288A                          EUROPE
GLAUCONITE                  SEDIMENTARY
097.1           003.3       ARGON 1.19/4.72/0.584
                              CRETACEOUS ALBIAN

0497 UPPER ALBIAN           G 286A                          EUROPE
GLAUCONITE      004.0       SEDIMENTARY
096.0                       ARGON 1.19/4.72/0.584
                              CRETACEOUS ALBIAN

0498 BASAL CENOMANIAN       G 277A                          EUROPE
GLAUCONITE      002.3       SEDIMENTARY
093.0                       ARGON 1.19/4.72/0.584
                              CRETACEOUS CENOMANIAN

0499 UPPER ALBIAN           G 273A                          EUROPE
GLAUCONITE      005.3       SEDIMENTARY
095.7                       ARGON 1.19/4.72/0.584
                              CRETACEOUS ALBIAN

0500 MID TO UPPER ALBIAN    G 272A                          EUROPE
GLAUCONITE                  SEDIMENTARY
094.9           003.7       ARGON 1.19/4.72/0.584
                              CRETACEOUS ALBIAN

0501 ALBIAN                 G 466A                          EUROPE
GLAUCONITE      006.1       SEDIMENTARY
097.2                       ARGON 1.19/4.72/0.584
                              CRETACEOUS ALBIAN

0502 BASAL CENOMANIAN       G 412A                          EUROPE
GLAUCONITE                  SEDIMENTARY
088.9           002.8       ARGON 1.19/4.72/0.584
                              CRETACEOUS CENOMANIAN

0503 OSLO REGION PLUTONS                                    EUROPE
STRONTIUM ISOCHRON  007.C   INTRUSIVE INTERMEDIATE
274.0                       STRONTIUM LAMBDA=1.39
PERMIAN ARTINSKIAN            PERMIAN SAKMARIAN

0504 TALKMA PASS C DIORITE                                  CANADA
HORNBLENDE      006.0       INTRUSIVE GRANITE
197.0                       ARGON 1.19/4.72/0.584
                              JURASSIC SINEMUR-HETTANGIAN

0505 TUFF NEAR MCASON CREEK                                 CANADA
HORNBLENDE                  VOLCANIC TUFFACEOUS
173.0                       ARGON 1.19/4.72/0.584
                              JURASSIC SINEMUR-HETTANGIAN

0506 Q MONZONITE S OF MAHTZEHTZEL MTN                       CANADA
HORNBLENDE      008.0       INTRUSIVE GRANITE
195.0                       ARGON 1.19/4.72/0.584
                              JURASSIC SINEMUR-HETTANGIAN

0507 Q MONZONITE N OF TOPLEY LANDING                        CANADA
HORNBLENDE      009.0       INTRUSIVE GRANITE
205.0                       ARGON 1.19/4.72/0.584
                              JURASSIC SINEMUR-HETTANGIAN

0508 SASK CRETACEOUS                                        CANADA
SANIDINE        003.0       VOLCANIC TUFFACEOUS
080.0                       ARGON 1.19/4.72/0.584
                              CRETACEOUS CAMPANIAN

0509 SASK CRETACEOUS                                        CANADA
SANIDINE        003.0       VOLCANIC TUFFACEOUS
087.0                       ARGON 1.19/4.72/0.584
                              CRETACEOUS SANTONIAN

0510 SASK CRETACEOUS                                        CANADA
SANIDINE        003.0       VOLCANIC TUFFACEOUS
090.0                       ARGON 1.19/4.72/0.584
                              CRETACEOUS TURONIAN

0511 GULPEN CHALK NETHERLANDS                               EUROPE
GLAUCONITE      003.0       SEDIMENTARY
072.4                       ARGON 1.18/4.72/0.584
                              CRETACEOUS CAMPANIAN

0512 GULPEN CHALK NETHERLANDS                               EUROPE
GLAUCONITE      003.0       SEDIMENTARY
066.7                       ARGON 1.18/4.72/0.584
                              CRETACEOUS MAESTRICHTIAN

0513 GULPEN CHALK NETHERLANDS                               EUROPE
GLAUCONITE      003.0       SEDIMENTARY
069.9                       ARGON 1.18/4.72/0.584
                              CRETACEOUS MAESTRICHTIAN

0514 GULPEN CHALK NETHERLANDS                               EUROPE
GLAUCONITE      003.0       SEDIMENTARY
070.8                       ARGON 1.18/4.72/0.584
                              CRETACEOUS MAESTRICHTIAN

0515 CRAIE BLANCH BELGIUM                                   EUROPE
GLAUCONITE      003.0       SEDIMENTARY
073.8                       ARGON 1.19/4.72/0.584
                              CRETACEOUS CAMPANIAN

0516 CRAIE BLANCH BELGIUM                                   EUROPE
GLAUCONITE      003.0       SEDIMENTARY
070.5                       ARGON 1.19/4.72/0.584
                              CRETACEOUS CAMPANIAN

0517 ORPHAN KNOLL                                           CANADA
GLAUCONITE      003.C       SEDIMENTARY
094.0                       ARGON 1.19/4.72/0.584
                              CRETACEOUS CENOMANIAN

0518 ORPHAN KNOLL                                           CANADA
GLAUCONITE      003.C       SEDIMENTARY
096.0                       ARGON 1.19/4.72/0.584
                              CRETACEOUS ALBIAN

0519 ISHIGUALASTO-ISCHIGUALA BASIN                          SOUTH AMERICA
WHOLE ROCK      005.L       INTRUSIVE BASALT GABBRC
227.0                       ARGON 1.19/4.72/0.584
TRIASSIC CARNIAN              TRIASSIC LATINIAN

0520 PUESTC VIEJC FORMATION                                 SOUTH AMERICA
WHOLE ROCK      004.0       INTRUSIVE BASALT GABBRC
242.0                       ARGON 1.19/4.72/0.584
TRIASSIC ANISIAN              TRIASSIC SCYTHIAN

0521 STATE CIRCLE SHALE                                     AUSTRALIA AND NEW ZEALAND
STRONTIUM ISOCHRON          SEDIMENTARY
455.0           015.0       STRONTIUM LAMBDA=1.39
                              SILURIAN LLANDOVERIAN

0522 SILURIAN DACITE                                        AUSTRALIA AND NEW ZEALAND
STRONTIUM ISOCHRON  004.0   VOLCANIC DACITE RHYOLITE
438.0                       STRONTIUM LAMBDA=1.39
                              SILURIAN DOWNTONIAN

1006 B          SNAP GRANITE                                EUROPE
BIOTITE         011.0       INTRUSIVE GRANITE
397.0                       STRONTIUM LAMBDA=1.47
CARBONIFEROUS TOURNAISIAN     SILURIAN OLENTONIAN
```

1300 B BRITISH BASALT NEW BRIDGE   EUROPE
WHOLE ROCK   016.0   INTRUSIVE BASALT GABBRO
313.0   ARGON 1.19/4.72/0.584
                 CARBONIFEROUS WESTPHALIAN

1361 B PRECAZIC GRANITE   EUROPE
BIOTITE   000.0   INTRUSIVE GRANITE
231.0   ARGON 1.19/4.72/0.584
                 TRIASSIC LADINIAN

2136 C CARTERS LIMESTONE EGGLESTON LS FM   USA
BIOTITE   007.0   VOLCANIC TUFFACEOUS
471.0   STRONTIUM LAMBDA=1.39
                 ORDOVICIAN CARADOCIAN

2198 C JAMAICA GRANODIORITE   MISCELLANEOUS
NOT SPECIFIED   INTRUSIVE INTERMEDIATE
004.0   005.0   URANIUM 0.154 0.971 CLU
CENOZOIC DANIAN   CRETACEOUS MAESTRICHTIAN

2354 C CERBERAN VOLCANICS   AUSTRALIA AND NEW ZEALAND
STRONTIUM ISOCHRON   VOLCANIC DACITE RHYOLITE
367.0   022.0   STRONTIUM LAMBDA=1.47
                     DEVONIAN FAMENNIAN

2360 C BRITISH BASALT LITTLE WENLOCK   EUROPE
WHOLE ROCK   017.0   VOLCANIC ANDESITIC BASALTIC
334.0   ARGON 1.19/4.72/0.584
                 CARBONIFEROUS WESTPHALIAN

3300 D BRITISH BASALT BURNTISLAND   EUROPE
WHOLE ROCK   004.0   VOLCANIC ANDESITIC BASALTIC
338.0   ARGON 1.19/4.72/0.584
                 CARBONIFEROUS WESTPHALIAN

4300 E BRITISH BASALT ARTHURS SEAT   EUROPE
WHOLE ROCK   007.0   VOLCANIC ANDESITIC BASALTIC
347.0   ARGON 1.19/4.72/0.584
                 CARBONIFEROUS VISEAN

5300 F BRITISH BASALT LAMPSIE FELS   EUROPE
WHOLE ROCK   006.0   VOLCANIC ANDESITIC BASALTIC
359.0   ARGON 1.19/4.72/0.584
                 CARBONIFEROUS TOURNAISIAN

1038 B DARTMOOR GRANITE   EUROPE
BIOTITE   000.0   INTRUSIVE GRANITE
27..0   STRONTIUM LAMBDA=1.47
PERMIAN SAKMARIAN   CARBONIFEROUS STEPHANIAN

1093 B CREETOWN GRANITE   EUROPE
BIOTITE   016.0   INTRUSIVE GRANITE
399.0   STRONTIUM LAMBDA=1.47
                 SILURIAN WENLOCKIAN

1156 B CARTERS LIMESTONE EGGLESTON LS FM   USA
BIOTITE   005.0   VOLCANIC TUFFACEOUS
420.0   ARGON 1.19/4.72/0.584
                 ORDOVICIAN CARADOCIAN

1198 B JAMAICA GRANODIORITE   MISCELLANEOUS
BIOTITE   005.0   INTRUSIVE INTERMEDIATE
067.0   ARGON 1.19/4.72/0.584
CENOZOIC DANIAN   CRETACEOUS MAESTRICHTIAN

1322 B EGGREY DEPOSIT   USSR
GLAUCONITE   000.0   SEDIMENTARY
136.0   ARGON 1.19/4.72/0.557
                 JURASSIC TITHONIAN

1354 B CERBERAN VOLCANICS   AUSTRALIA AND NEW ZEALAND
STRONTIUM ISOCHRON   VOLCANIC DACITE RHYOLITE
355.0   015.0   STRONTIUM LAMBDA=1.47
                     DEVONIAN FAMENNIAN

1356 B TRIASSIC GRANITES   AUSTRALIA AND NEW ZEALAND
BIOTITE   002.0   INTRUSIVE GRANITE
218.0   ARGON 1.19/4.72/0.584
JURASSIC AALENIAN TOARCIAN   TRIASSIC LADINIAN

Table 4. Phanerozoic Time Scale supplemental reference list

Item	Citation	Item	Citation
406	Suppe, 1973	476	Borsi, et al, 1969
407	Vidal, et al, 1973	477, 478	Armstrong and Besancon, 1970;
408	Gee and Wilson, 1974		Cornet, et al, 1973; Bruce, 1970;
409	Tailleur, 1974	479	Grantz, et al, 1963
410	Tilling, et al, 1968;	480, 481	Borsi, et al, 1966
	Robinson, et al, 1968	482	Fullagar and Bottino, 1968
411	Dewey and Pankhurst, 1970	483	Fuchs, et al, 1970
412	Pidgeon, 1969	484, 485	Gygi and McDowell, 1970
413	Vialette, 1966	486	Pasteels, 1970
414	Sturt, et al, 1975	487	Boucot, et al, 1972
415	Lanphere, et al, 1976	488	Spooner and Fairbairn, 1970
416	Dallmeyer and Williams, 1975	489	Reynolds, et al, 1973
417-443	Afanas'yev, 1970	490, 491	Proko and Hargraves, 1973;
444	Armstrong and Suppe, 1973		Zartman, et al, 1967
445	Lanphere, 1971	492-494	Odin, 1975
446, 447	Priem, et al, 1975a, b	495-502	Jurignet, et al, 1975
448	Naylor, et al, 1974	503	Heier and Compston, 1969
449	Harris and Bottino, 1974	504-507	Wanless, et al, 1974
450-466	Obradovich and Cobban, 1975	508-510	Williams and Baadsgaard, 1974
467-470	Owens and Sohl, 1973	511-516	Priem, et al, 1975a, b
471, 472	Pajari, et al, 1974	517, 518	Van Hinte, et al, 1975
473, 474	Chaudhuri and Brookins, 1969	519, 520	Valencio, et al, 1975
475	White, et al, 1967	521, 522	Bofinger, et al, 1970

References Cited

Afanas'yev, G. D., 1970, Certain key data for the Phanerozoic time-scale: Eclogae Geol. Helvetiae, v. 63, p. 1-7.
—— and S. I. Zykov, 1975, Phanerozoic geochronological time scale in the light of significantly new decay constants: Moscow, Nayka, 100 p.
Armstrong, R. L., 1974, Proposal for simultaneous adoption of new U, Th, Rb, and K decay constants for calculation of radiometric dates (abs.): Internat. Mtg. Geochronology, Cosmochronology, and Isotope Geology (Paris), abs. vol.
—— and J. Besancon, 1970, A Triassic time scale dilemma—K-Ar dating of Upper Triassic mafic igneous rocks, eastern U.S.A. and Canada, and post-Upper Triassic plutons, western Idaho, U.S.A.: Eclogae Geol. Helvetiae, v. 63, p. 15-28.
—— and W. G. McDowall, 1974, Proposed refinement of the Phanerozoic time scale (abs.): Internat. Mtg. Geochronology, Cosmochronology, and Isotope Geology (Paris), abs. vol.
—— and J. Suppe, 1973, Potassium-argon geochronometry of Mesozoic igneous rocks in Nevada, Utah, and southern California: Geol. Soc. America Bull., v. 84, p. 1375-1392.
Beckinsale, R. D., and N. H. Gale, 1969, A reappraisal of the decay constants and branching ratio of ^{40}K: Earth and Planetary Sci. Letters, v. 6, p. 289-294.
Berggren, W. A., 1972, A Cenozoic time-scale—some implications for regional geology and paleobiogeography: Lethaia, v. 5, p. 195-215.
—— and J. A. Van Couvering, 1974, The late Neogene: biostratigraphy, geochronology, and paleoclimatology of the last 15 million years in marine and continental sequences: Palaeogeography, Palaeoclimatology, and Palaeoecology, v. 16, p. 1-216.
Bofinger, V. M., W. Compston, and B. L. Gulson, 1970, A Rb-Sr study of the Lower Silurian State Circle Shale, Canberra, Australia: Geochim. et Cosmochim. Acta, v. 34, p. 433-445.
Borsi, S., et al, 1966, Age stratigraphique et radiometrique Jurassique superieur d'un granite des zones internes des Hellenides: Rev. Geographie Phys. Geologie Dyn., Ser. 2, v. 8, pt. 4, p. 279-287.
—— et al, 1969, Isotopic age measurements of the M. Monzoni intrusive complex: Mineralog. et Petrog. Acta, v. 14, p. 171-183.
Boucot, A. J., 1975, Evolution and extinction rate controls: Amsterdam, Elsevier, 427 p.
—— et al, 1972, Staurolite zone Caradoc (Middle-Late Ordovician) age, Old World province brachiopods from Penobscot Bay, Maine: Geol. Soc. America Bull., v. 83, p. 1953-1960.

TABLE 5. Breakdown of geologic time scale used for datafile and for plotting on Figure 2. First column is numerical sequence of zones, second is coding used on data-input cards.

#	Code	Period	Stage	#	Code	Period	Stage
1	000	PRECAMBRIAN		30	074	PERMIAN	KAZANIAN
2	011	CAMBRIAN	GEORGIAN	31	075	PERMIAN	TARTARIAN
3	012	CAMBRIAN	ACADIAN	32	101	TRIASSIC	SCYTHIAN
4	013	CAMBRIAN	POTSDAMIAN	33	102	TRIASSIC	ANISIAN
5	021	ORDOVICIAN	TREMADOCIAN	34	103	TRIASSIC	LADINIAN
6	022	ORDOVICIAN	ARENIGIAN	35	104	TRIASSIC	CARNIAN
7	023	ORDOVICIAN	LLANVIRNIAN	36	105	TRIASSIC	NORIAN
8	024	ORDOVICIAN	LLANDEILIAN	37	106	TRIASSIC	RHAETIAN
9	025	ORDOVICIAN	CARADOCIAN	38	111	JURASSIC	SINEMURIAN HETTANGIAN
10	026	ORDOVICIAN	ASHGILLIAN	39	112	JURASSIC	PLIENSBACHIAN
11	031	SILURIAN	LLANDOVERIAN	40	113	JURASSIC	AALENIAN TOARCIAN
12	032	SILURIAN	WENLOCKIAN	41	114	JURASSIC	BATHONIAN BAJOCIAN
13	033	SILURIAN	LUDLOVIAN	42	115	JURASSIC	CALLOVIAN
14	034	SILURIAN	DOWNTONIAN	43	116	JURASSIC	OXFORDIAN
15	041	DEVONIAN	GEDINNIAN	44	117	JURASSIC	KIMMERIDGIAN
16	042	DEVONIAN	SIEGENIAN	45	118	JURASSIC	TITHONIAN
17	043	DEVONIAN	EMSIAN	46	121	CRETACEOUS	BERRIASIAN
18	044	DEVONIAN	COUVINIAN	47	122	CRETACEOUS	VALANGINIAN
19	045	DEVONIAN	GIVETIAN	48	123	CRETACEOUS	HAUTERIVIAN
20	046	DEVONIAN	FRASNIAN	49	124	CRETACEOUS	BARREMIAN
21	047	DEVONIAN	FAMENNIAN	50	125	CRETACEOUS	APTIAN
22	051	CARBONIFEROUS	TOURNAISIAN	51	126	CRETACEOUS	ALBIAN
23	052	CARBONIFEROUS	VISEAN	52	131	CRETACEOUS	CENOMANIAN
24	061	CARBONIFEROUS	NAMURIAN	53	132	CRETACEOUS	TURONIAN
25	062	CARBONIFEROUS	WESTPHALIAN	54	133	CRETACEOUS	CONIACIAN
26	063	CARBONIFEROUS	STEPHANIAN	55	134	CRETACEOUS	SANTONIAN
27	071	PERMIAN	SAKMARIAN	56	135	CRETACEOUS	CAMPANIAN
28	072	PERMIAN	ARTINSKIAN	57	136	CRETACEOUS	MAESTRICHTIAN
29	073	PERMIAN	KUNGURIAN	58	221	CENOZOIC	DANIAN

Table 6. Time Scales (age to base)

	Kulp, 1961	Harland et al 1964 (PTS-1)	Lambert, 1971 (PTS-2)	Armstrong & McDowall 1974 (PTS decay constants)	1974 (revised decay constants)	Afanas'yev & Zykov, 1975 (new constants)
Cenozoic	63	70	65	64	65	66
Upper Cretaceous	110	100	95	≥ 94	≥ 96	
Lower Cretaceous	135	136	~135	140	143	132
Jurassic	181	~193	~200	208	212	185
Triassic	~230	225	~240	≥242	≥247	235
Permian	280	280	280	284	289	280
Upper Carboniferous	310			335	341	
Lower Carboniferous	345	345	370	360	367	345
Devonian	405	395	~415	409	416	400
Silurian	~425	~435	~445	≥436	≥446	435
Ordovician	500	~500	~515	≥500	≥509	490
Cambrian	≈600	~570	~590	~564	~575	570

Ed. Note: PTS is abbreviation for Phanerozoic Time Scale.

Bruce, W. R., 1970, Geology, mineral deposits, and alteration of parts of the Cuddy Mountain District, western Idaho: PhD thesis, Oregon State Univ., 165 p.

Chaudhuri, S., and D. G. Brookins, 1969, The isotopic age of the Flathead Sandstone (Middle Cambrian), Montana: Jour. Sed. Petrology, v. 39, p. 364-366.

Cornet, B., A. Traverse, and N. G. McDonald, 1973, Fossil spores, pollen and fishes from Connecticut indicate early Jurassic age for part of the Newark Group: Science, v. 182, p. 1243-1247.

Dallmeyer, R. D., and H. Williams, 1975, 40Ar/39Ar ages from the Bay of Islands metamorphic aureole: Their bearing on the timing of Ordovician ophiolite obduction: Canadian Jour. Earth Sci., v. 12, p. 1685-1690.

Dewey, J. F., and R. J. Pankhurst, 1970, The evolution of the Scottish Caledonides in relation to their isotopic age pattern: Royal Soc. Edinburgh Trans., v. 68, p. 361-389.

Fuchs, Y., F. Leutwein, and J.-L. Zimmermann, 1970, Etude geochronologique et geochimique des roches volcaniques du Stephanien et Autunien du Detroit de Rodes: Acad. Sci. (Paris) Comptes Rendus, Ser. D, v. 270, p. 2415-2417.

Fullagar, P. D., and M. L. Bottino, 1968, Geochronology of Silurian and Devonian age volcanic rocks from northeastern North America: 23d Int. Geol. Cong. (Prague), Proc., v. 6, p. 17-32.

Gee, D. G., and M. R. Wilson, 1974, The age of orogenic deformation in the Swedish Caledonides: Am. Jour. Sci., v. 274, p. 1-9.

Grantz, A., et al, 1963, Potassium-argon and lead-alpha ages for stratigraphically bracketed plutonic rocks in the Talkeetna Mountains, Alaska: U.S. Geol. Survey Prof. Paper 475 B, p. 56-59.

Gygi, R. A., and F. W. McDowell, 1970, Potassium-argon ages of glauconites from a biochronologically dated Upper Jurassic sequence of northern Switzerland: Eclogae Geol. Helvetiae, v. 63, p. 111-118.

Harland, W. B., and E. H. Francis, ed., 1971, The Phanerozoic time scale: a supplement: Geol. Soc. London Spec. Pub. 5, 356 p.

—— A. G. Smith, and B. Wilcock, eds., 1964, The Phanerozoic time-scale: Geol. Soc. London Quart. Jour., v. 120 supp., 458 p.

Harris, W. B., and M. L. Bottino, 1974, Rb-Sr study of Cretaceous lobate glauconite pellets, North Carolina: Geol. Soc. America Bull., v. 85, p. 1475-1478.

Heier, K. S., and W. Compston, 1969, Rb-Sr isotopic studies of the plutonic rocks of the Oslo region: Lithos, v. 2, p. 133-145.

Huster, E., 1974, The half-life of natural rubidium 87 measured as difference between the isotopes Rb 87 and Rb 85 (abs.): Int. Mtg. Geochronology, Cosmochronology, and Isotope Geology (Paris), abs. vol.

Jaffey, A. H., et al, 1971, Decay constants for U: Phys. Rev. Ser. C, v. 4, p. 1889.

Juignet, P., J. C. Hunziker, and G. S. Odin, 1975, Datation numerique du passage Albien-Cenomanien en Normandie: Acad. Sci. (Paris) Comptes Rendus, Ser. D, v. 280, p. 379-382.

Kulp, J. L., 1961, Geologic time scale: Science, v. 133, p. 1105-1114.

Lambert, R. St. J., 1971, The pre-Pleistocene Phanerozoic time scale—a review, in The Phanerozoic time-scale—a supplement: Geol. Soc. London Spec. Pub. 5, p. 9-31.

Lanphere, M. A., 1971, Age of the Mesozoic oceanic crust in the California Coast Ranges: Geol. Soc. America Bull., v. 82, p. 3209-3212.

—— M. Churkin, Jr., and G. D. Eberlein, 1976, Radiometric age of the *Monograptus cyphus* graptolite zone in southeastern Alaska—an estimate of the Ordovician-Silurian boundary: Geol. Mag., v. 114, p. 15-24.

Naylor, R. S., R. Hon, and P. D. Fullagar, 1974, Age of the Katahdin batholith, Maine (abs.): Geol. Soc. America Abs. with Programs, v. 5, p. 59.

Neumann, W., and E. Huster, 1974, The half-life of ^{87}Rb measured as difference between the isotopes of ^{87}Rb and ^{85}Rb: Zeitschrift. Physik, v. 270, p. 121-127.

Obradovich, J. D., and W. A. Cobban, 1975, A time-scale for the late Cretaceous of the western interior of North America: Geol. Assoc. Canada Spec. Paper 13, p. 31-54.

Odin, G. S., 1975, Les glauconies; constitution formation, ages: Thèse Doctorat d'Etat, Paris, 278 p.

Owens, J. P., and N. F. Sohl, 1973, Glauconites from New Jersey - Maryland coastal plain: their K-Ar ages and application in stratigraphic studies: Geol. Soc. America Bull., v. 84, p. 2811-2838.

Pajari, G. E., Jr., et al, 1974, The age of the Acadian deformation in southwestern New Brunswick: Canadian Jour. Earth Sci., v. 11, p. 1309-1313.

Pasteels, Paul, 1970, Uranium-lead radioactive ages of monazite and zircon from the Vire-Carolles Granite (Normandy)—a case of zircon-monazite discrepancy: Eclogae Geol. Helvetiae, v. 63, p. 231-237.

Pidgeon, R. T., 1969, Zircon U-Pb ages from the Galway granite and the Dalradian Connemara, Ireland: Scottish Jour. Geology, v. 5, p. 375-392.

Priem, H. N. A., et al, 1975a, Isotope geochronology in the Indonesian tin belt: Geologie en Mijnbouw, v. 54, p. 61-70.

—— et al, 1975b, Isotopic dating of glauconites from the Upper Cretaceous in Netherlands and Belgian Limburg: Geologie en Mijnbouw, v. 54, p. 205-207.

Proko, M. S., and R. B. Hargraves, 1973, Paleomagnetism of the Beemerville (New Jersey) alkaline complex: Geology, v. 1, p. 185-186.

Reynolds, P. H., E. E. Kublick, and G. K. Muecke, 1973, Potassium-argon dating of slates from the Meguma Group, Nova Scotia: Canadian Jour. Earth Sci., v. 10, p. 1059-1067.

Robinson, G. D., M. R. Klepper, J. D. Obradovich, 1968, Overlapping plutonism, volcanism, and tectonism in the Boulder batholith region, western Montana: Geol. Soc. America Mem. 116, p. 557-576.

Spooner, C. M., and H. W. Fairbairn, 1970, Relation of radiometric age of granitic rocks near Calais, Maine, to the time of Acadian orogeny: Geol. Soc. America Bull., v. 81, p. 3663-3670.

Sturt, B. A., I. R. Pringle, and D. Roberts, 1975, Caledonian nappe sequence of Finnmark, northern Norway, and timing of orogenic deformation and metamorphism: Geol. Soc. America Bull., v. 86, p. 710-718.

Suppe, John, 1973, Geology of the Leech Lake Mountain - Ball Mountain region, California: Univ. Calif. Pubs. Geol. Sci., v. 107, 82 p.

Tailleur, I. L., 1974, Triassic - Jurassic time boundary—new data from northern Alaska: in prep.

Tilling, R. I., M. R. Klepper, and J. D. Obradovich, 1968, K-Ar ages and time span of emplacement of the Boulder batholith, Montana: Am. Jour. Sci., v. 266, p. 671-689.

Valencio, D. A., J. E. Mendia, and J. F. Vilas, 1975, Paleomagnetism and K-Ar ages of Triassic igneous rocks from the Ischigualasto-Ischichuca basin and Puesto Viejo Formation, Argentina: Earth and Planetary Sci. Letters, v. 26, p. 319-330.

Van Hinte, J. E., 1976a, A Cretaceous time scale: AAPG Bull., v. 60, p. 498-516.

—— 1976b, A Jurassic time scale: AAPG Bull., v. 60, p. 489-497.

—— J. A. S. Adams, and D. Perry, 1975, K-Ar age of Lower-Upper Cretaceous boundary at Orphan Knoll (Labrador Sea): Canadian Jour. Earth Sci., v. 12, p. 1484-1491.

Vialette, Y., 1966, Granitisation hercynienne dans le massif central francais; *in* Interpretation geologique des mesures effectuees au spectrometre de masse dans le domaine de la geochronologie absolue; Colloques Int. (Nancy, France): Centre Natl. Recherche Sci,Paris, p. 163-176.

Vidal, P., et al, 1973, Contribution to the age of the Cambrian-Ordovician boundary; radiometric age of volcanics of the Krivoklat-Rokycany zone (Bohemian massif), (abs.): E.C.O.G. III, Oxford, 4-8 Sept. 1973.

Wanless, R. K., et al, 1974, Age determinations and geological studies; K-Ar isotopic ages, Report 12: Canada Geol. Survey Paper 74-2, 72 p.

White, W. H., et al, 1967, Isotopic dating of the Guichon batholith, B.C.: Canadian Jour. Earth Sci., v. 4, p. 677-690.

Williams, G. D., and H. Baadsgaard, 1974, Potassium-argon dates and upper Cretaceous biostratigraphy in eastern Saskatchewan: Geol. Assoc. Canada Spec. Paper 13, p. 417-426.

Zartman, R. E., et al, 1967, K-Ar and Rb-Sr ages of some alkalic intrusive rocks from central and eastern United States: Am. Jour. Sci., v. 265, p. 848-870.

Applicability of the Rubidium-Strontium Method to Shales and Related Rocks[1]

UMBERTO G. CORDANI[2], KOJI KAWASHITA[2] and ANTONIO THOMAZ FILHO[3]

Abstract From the critical interpretation of many examples of rubidium-strontium (Rb-Sr) dating of shales and related rocks, two main conclusions can be drawn: (1) Isochron diagrams of whole rock samples of pelitic sediments, originated from the same depositional environment, in many cases allow the determination of the age of sedimentation. In such cases, a uniform dispersion of clastic material within the basin must be inferred. (2) Isochron diagrams for subsystems of the fine fraction (insoluble and soluble materials) can indicate almost always the time of the diagenesis, anchi- or epi-metamorphism suffered by each sample. In such cases, Sr isotopic homogenization among clay minerals and soluble material occur.

Several examples for the above referred concepts were found in the literature, and others were produced in the São Paulo Geochronology Laboratory on more than 300 samples from the following sedimentary units: Trombetas Formation (Silurian, marine); Botucatu Formation (Jurassic, continental); Rio do Rasto Formation (Permian, continental); Irati Formation (Permian, marine); Rio Bonito Formation (Permian, marine); Sepotuba Formation (Cambrian, marine); Estancia Formation (Cambro-Ordovician, marine); and Bambui Group (late Precambrian, marine).

Introduction

The general opinion of geochronologists, in the last 15 years, regarding the applicability of the Rb-Sr method to shales has changed several times from optimistic to pessimistic. Apparently, the tendency is currently pessimistic because radiometric ages for sedimentary material are considered to be reliable only when authigenic minerals such as glauconite are used.

The main difficulty concerning the interpretation of Rb-Sr results for sediments and sedimentary rocks is the detrital component, which introduces great uncertainty with respect to the Sr^{87}/Sr^{86} ratio. For instance, Dasch (1969), and Biscaye and Dasch (1971) demonstrated clearly that clay minerals in marine environments do not equilibrate isotopically with seawater strontium.

Summarizing previous investigations, Faure and Powell (1972) concluded that it is highly improbable that shales could satisfy the basic assumptions for application of the

[1] Manuscript received, January 28, 1977.

[2] Instituto de Geociencias, Univ. of São Paulo CX. P. 20899-São Paulo.

[3] Cenpes, Petrobras-Ilha do Fundão-Quadra 7 Rio de Janeiro.

We wish to acknowledge the Brazilian Oil Company, Petrobrás, and its chief of exploration, C. W. M. Campos, for granting us permission to make use of their samples and geologic data.

We are indebted to M. Berenholc and M. S. M. Mantovani, and to the technical personnel of the Geochronology Research Center at USP, for help during the analytical work.

Geologists, M. A. M. de Oliveira and L. P. Quadros, of Petrobrás, and Dr. Bonhomme of the Strasbourg group, contributed helpful discussions. The latter was also responsible for most of the clay mineralogy determinations at Strasbourg. T. R. Fairchild kindly corrected the original English version.

The Brazilian National Council for Scientific and Technological Development (CNPq) partially supported this research, through several scientific grants awarded to the São Paulo Laboratory.

Article Identification Number: 0149-1377/78/SG06-0008/$03.00/0

Rb-Sr method. Nevertheless, they pointed out that in many cases, Rb-Sr whole rock isochrons on shales did indicate concordant results, within experimental errors, with the known ages of deposition (Compston and Pidgeon, 1962; Whitney and Hurley, 1964; Allsopp and Kolbe, 1965; Faure and Kovach, 1969; Chaudhuri and Brookins, 1969; Bofinger et al, 1970).

The geochronologists of the University of Strasbourg have been working with clay minerals for several years with significant results. They have demonstrated that, in certain environments in which aggradation or neoformation processes occur, there is a complete strontium homogenization due to exchanges and equilibration with the strontium in the water (Bonhomme and Clauer, 1972; Clauer and Bonhomme, 1974). However, such environments are fairly uncommon in nature, and most shales would not satisfy the criteria required by the Strasbourg group for yielding significant results in terms of their stratigraphic ages.

A few years ago at the Geochronology Research Center of the University of São Paulo (USP), we decided to look further into the problem to establish the real potentiality of the method, and to attempt to enlarge its field of application. The Brasilian Oil Company, Petrobrás, made available to us its collection of core samples, and enabled us to use their stratigraphic data. Because of this, we could select many very good samples with well-established stratigraphic position and ages varying from Paleozoic to early Mesozoic. The results of more than 300 Rb-Sr determinations (also included in the papers of Kawashita (1972) and Thomaz (1976) are considered in this paper, as well as some other determinations carried out on Brazilian shales or extracted from the pertinent international literature.

In this work, for the sake of clarity, our main ideas and conclusions regarding the Sr isotopic behavior in pelitic sediments will be expressed first, followed by our proposed models for the interpretation of Rb-Sr diagrams in terms of ages of sedimentation and of later geologic events affecting the sediments. The available examples will be dealt with in a subsequent section.

Considerations on The Evolution of The Sr Isotopic Composition of Sedimentary Material

Compston and Pidgeon (1962) were pioneers in dating shales by the Rb-Sr isochron method. Dealing with marine shales, they commented upon the possibility of the mixture of detrital and authigenic material being uniform within the entire sedimentary unit, producing an initial Sr^{87}/Sr^{86} ratio slightly higher than the seawater value. This idea was later followed by other investigators, but could not be conclusively demonstrated or properly explained.

Chemical changes within the depositional environment acting upon the strontium isotopic composition of the clay fraction have been considered in several important works:

1. Bonhomme et al (1964) made interpretations on the behavior of clay minerals during degradation and neoformation. They focussed on transformational processes and discussed the isotopic homogenization of clays during diagenesis and anchimetamorphism.

2. Faure and Chaudhuri (1967) verified that the Nonesuch Shale did not equilibrate isotopically with the water during sedimentation and suggested that strontium isotopic homogenization occurred with connate water during diagenesis.

3. Working on modern sediments of the Pacific Ocean, Clauer et al (1975) clearly verified ionic exchanges between the connate waters and the clay fraction, which represent the first steps in an isotopic homogenization process.

Acid leaching, in order to separate components with different Rb-Sr ratios, was used by several authors, such as Faure and Chaudhuri (1967), Bofinger et al (1968). This technique was followed by Kawashita (1972) and Thomaz (1976), because it was observed that significant interpretations could be obtained from the resultant isochron diagrams.

Two main concepts can be derived from interpretation of the whole of the available data regarding Rb-Sr dating of shales and related rocks. The first concept concerns the possibility of the detection of sedimentation age, and the second is related to subsequent phenomena affecting the sediment, such as diagenesis and/or metamorphism.

The Problem of Age of Sedimentation and the Process of Mechanical Dispersion of Detrital Material

Many cases are quoted in the specialized literature in which whole-rock isochron diagrams of shales may indicate the approximate age of sedimentation. We are convinced that if some criteria are followed regarding the selection of samples (based on their mineral composition and rubidium and strontium contents), and if it can be ascertained that all samples are derived from the same depositional environment, the sedimentation age can be obtained in practice from any kind of unmetamorphosed pelitic rock.

To us, this is due to a mechanical dispersion of the detrital material in the depositional basin, leading to a uniform strontium isotopic composition for whole-rock samples a few centimeters in size. However, the material will not be isotopically homogeneous. For instance, there may be small mica or feldspar crystals with high Sr^{87} content mixed together with montmorillonitic material with low Sr^{87} content. However, the pelitic sediment, as a whole, will present a fairly uniform bulk isotopic composition; moreover, the initial uniform strontium isotopic composition of the sediment is not enough to allow the application of the Rb-Sr isochron method. It is also necessary that a convenient span in the Rb/Sr ratio occur within the depositional environment at the same time and independently of the mechanism of dispersion of the detrital material. Some evidence that these premises actually hold, due to the geochemically different behavior of rubidium and strontium, are presented later in this paper.

Figures 1, 2, and 3 are examples taken from the literature which illustrate the mechanism of uniform dispersion of clastic material in depositional environment.

The first example is taken from Dasch (1969) and includes the measurements of carbonate-free fractions of recent sediment samples from the bottom of the North Atlantic, in the Sargasso Sea, and in the region passed over by the Gulf Stream current (Fig. 1). The Sr^{87}/Sr^{86} ratios are fairly uniform, with values between 0.72 and 0.73, whereas the Rb/Sr ratios vary by a factor of 4, ranging between 0.4 and 1.5.

The second example (Fig. 2) exhibits the data for the western South Atlantic, in the region of the Argentine basin and adjacent areas, taken from Biscaye and Dasch (1971). This also deals with carbonate-free fractions of recent sediment samples. The Sr^{87}/Sr^{86} values are consistently between 0.705 and 0.710, but the Rb/Sr ratios vary by a factor of 10, ranging between 0.1 and 1.0.

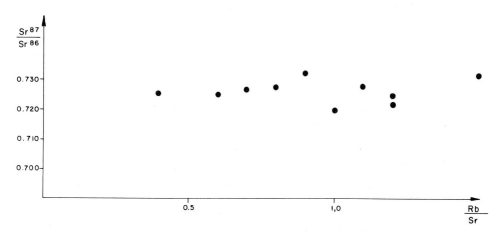

FIG. 1—Measurements of carbonate-free fractions of recent sediment samples from the bottom of the North Atlantic, in the Sargasso Sea, as in the vicinity of the Gulf Stream Current (from Dasch, 1969).

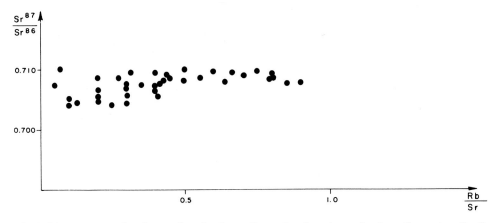

FIG. 2—Measurements of carbonate-free fractions of recent sediment samples from the western South Atlantic in the area of the Argentine basin (from Biscaye and Dasch, 1971).

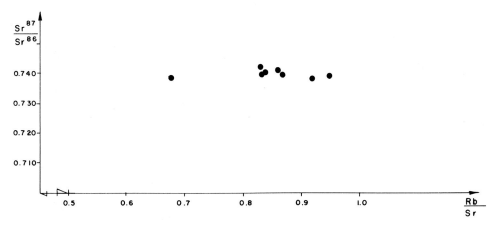

FIG. 3—Measurements of whole-rock samples of recent sediments from Lake Superior (from Hart and Tilton, 1966).

Figure 3 includes the data obtained by Hart and Tilton (1966) for a few whole-rock samples of recent sediments from Lake Superior, a large continental depositional basin. The Sr^{87}/Sr^{86} ratios are quite similar with values slightly below 0.74, and the Rb/Sr ratios are between 0.67 and 1.0, a variation which is not large, but adequate for obtaining significant interpretation in isochron diagrams.

We are aware that to demonstrate clearly the validity of this proposition, many more measurements on recent clays are needed from all kinds of depositional sites. The three quoted examples refer to very large depositional environments but the indicated values for their Sr^{87}/Sr^{86} parameters are distinctive and are clearly related to the provenance of the material. The high value for the Lake Superior clays can be related to the rivers which bring material from the old provinces of the Canadian Shield. The intermediate value for North Atlantic clays is probably related to the denudation of North and Central America; and the lower value for the South Atlantic in the region south of the Rio de La Plata can be associated with material brought from the Argentinian rivers, whose ultimate source is the young and mainly volcanogenic province of the Andes.

We are convinced that any geologic process leading to the formation of a shale, independent of its depositional site (open sea, epicontinental sea, lake, alluvial plain) in principle can produce the kind of uniform dispersion referred to above by mechanical mixing of clastic material. Because of the extremely large number of particles below 2μ, which is involved in a bulk sample a few centimeters in size, we think that a final uniformity within the depositional unit is more likely to occur than not. Moreover, the independence of Rb and Sr contents, regarding the overall Sr^{87}/Sr^{86} ratio, can be achieved in any part of the general processes of rock alteration, transportation by solution and/or suspension, or finally by chemical exchanges within the depositional basin just after sedimentation.

Age of Diagenetic and Metamorphic Processes and the Significance of Sample Isochrons

In several papers, published mainly by the Strasbourg group, it was demonstrated that clay minerals are open in various degrees to chemical exchanges. Isotopic homogenizations of strontium between the connate fluids and many of the argillaceous phases is common, but depends upon the intensity of the diagenetic and metamorphic processes. In this paper, the concept of diagenesis is applied in its broader significance, including all the mechanisms leading to the lithification of a sediment just after deposition, as well as any physical or chemical transformations occurring thereafter, in weak environmental conditions outside the field of metamorphic reactions.

The occurrence of such isotopic equilibration, after which each argillaceous phase seems to act as a closed system, makes the fine fraction of any sediment suitable to be studied in a way analogous to a policyclic metamorphic rock for which mineral isochrons can be obtained. These mineral isochrons would indicate the age of the last thermal episode and the new initial Sr^{87}/Sr^{86} ratio produced in the rock during that event.

The problem is different for a shale, where pure clay phases cannot be adequately separated from each other. However, at least two very convenient fractions can be separated by acid leaching: a soluble one (LCH) with a low Rb/Sr ratio, and a residue (RES) with a high Rb/Sr ratio. Acid leaching is commonly performed only in the fine fraction (particles below 2μ or 4μ), previously separated from the bulk sample by normal pipetting techniques.

In an isochron diagram, the analytical points for the soluble material and the residue will define a mixing line, here defined as *Sample isochron*, where the point for the fine fraction shall be located. Whereas samples from the same depositional units will exhibit, in most cases, a series of parallel mixing lines, we conclude that they are geologically meaningful and that their slope is dependent on the time at which the clay minerals closed (in relation to rubidium and strontium exchanges). This can logically be associated with the end of the last diagenetic and(or) metamorphic process undergone by the sedimentary unit.

Proposed Models for Interpretation on Rb-Sr Diagrams in Pelitic Rocks

It may be inferred from the previous section that shales are potentially adequate to be studied by the Rb-Sr method. In whole-rock unmetamorphosed samples, they can be used to obtain a stratigraphic age; and after separation of the fine-fraction system into soluble and insoluble subsystems, shales can be used for age determination of subsequent diagenetic and anchi- or epi-metamorphic processes. Considering the results obtained in the São Paulo Laboratory, with those already published in the specialized literature, we are proposing a series of models for interpretation of Rb-Sr ages of pelitic rocks.

These models are ideal examples of isochron diagrams in which all the cogenetic whole-rock samples were subjected to ideal conditions for the mechanism of uniform dispersion of clastic material, and therefore exhibited the same initial Sr^{87}/Sr^{86} ratio. The resultant whole-rock isochrons indicate the sedimentation age, at least for the unmetamorphosed units. In addition, the models suggest a subsequent event of strontium homogenization, acting upon the fine fraction of each sample. The lines which join the subsystems LCH (soluble) and RES (insoluble) in the proposed models are the sample isochrons, as defined previously, and are used to determine the age of the homogenization phenomena.

We will discuss each of the conceived models, in terms of the mineralogy of the pelitic rock concerned, and of the intensity of the environmental conditions leading to the isotopic homogenization of strontium. Figure 4 shows all the proposed models, and Figure 5 is the legend for all the isochron diagrams shown in this paper. It is convenient to emphasize that the proposed models are applicable only to shales, which are defined as predominantly argillaceous sedimentary rocks bearing only small-sized detrital fragments that are not larger than silt or very fine sand particles.

Isochronic Model I
(Isotopic Homogenization in the Fine Fraction System at Hand Specimen Level)

Model I characterizes a pelitic unit in which low-temperature diagenetic processes produced strontium isotopic homogenization in the fine fraction system within volumes of a few cubic centimeters. The chemical exchange between phases at sizes below 2μ is limited to small parts of the sediment isolated from each other, which behave like closed systems and gain a particular strontium isotopic composition. In the isochron diagram, the sample isochrons are parallel lines separated from each other according to differences in the initial Sr^{87}/Sr^{86} ratios; their slope indicates the age of the diagenetic event.

The analytical points of the whole-rock samples are located well outside and above each respective sample isochron, since the mineralogical composition of the fine fraction is considerably different from that of the whole rock sample. The latter includes many detrital phases which are not open to chemical exchanges as are the clay minerals, within conditions of the low energy diagenetic field. The whole rock isochron line indicates the sedimentation age.

Isochronic Model II
(Isotopic Homogenization in the Whole Rock Sample at the Hand Specimen Level)

This model refers to the pelitic unit in which low temperature diagenetic processes produced strontium isotopic homogenization in the whole-rock sample within volumes of a few cubic centimeters. Again (as in Model I), the sample isochrons are parallel, but this time they include the analytical points of the whole-rock samples.

This model occurs when detrital material (grains larger than 2μ) is not quantitatively important or is composed essentially of sterile minerals with respect to strontium content, such as quartz. Once again, the whole-rock isochron indicates the sedimentation age.

Isochronic Model III
(Isotopic Homogenization in the Fine Fraction System, at the Level of the Lithostratigraphic Unit)

In this model, the postdepositional diagenetic or metamorphic processes have been intense enough to induce a generalized strontium isotopic homogenization within the unit, with widespread chemical exchanges within the clay minerals. All the analytical points for the fine fraction systems are aligned along the same isochron line, together with all the LCH and RES subsystems. Even the whole-rock samples are somewhat affected by chemical exchanges, and their analytical points do not define an isochron line; in the diagram, some of them are dislocated towards the younger isochron line.

It is important that a large quantity of clay minerals open to chemical exchanges exists in this case. The shifting of the whole-rock samples towards the isochron line depends on the relative amount of argillaceous material in the system. Model III occurs where a fairly strong diagenetic or metamorphic event takes place long after sedimentation.

Isochronic Model IV
(Isotopic Homogenization in the Whole-Rock Sample at the Level of the Lithostratigraphic Unit)

Model IV characterizes a pelitic unit affected by a diagenetic or metamorphic event strong enough to produce a generalized strontium isotopic homogenization within the unit. In this case, the detrital components coarser than 2μ also participate in the chem-

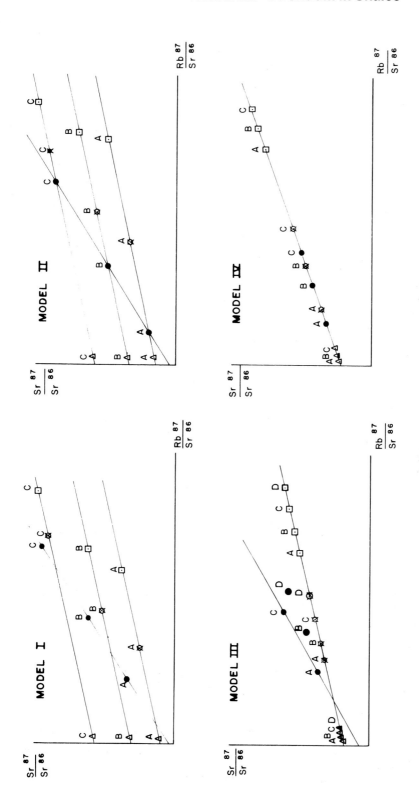

FIG. 4—Four models for isochron diagrams on shales, proposed in this paper.

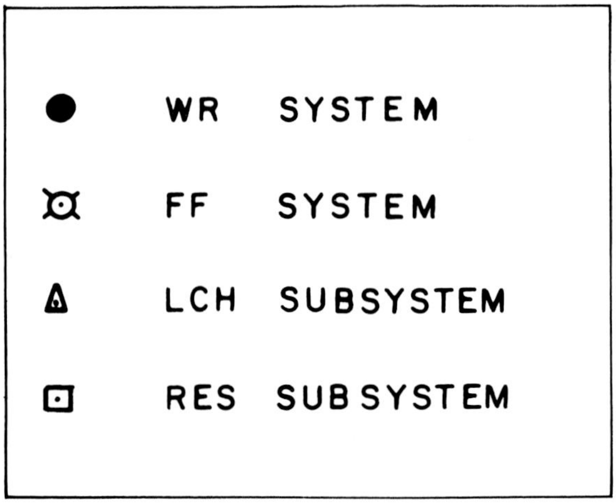

●	WR	SYSTEM
✪	FF	SYSTEM
◮	LCH	SUBSYSTEM
⊡	RES	SUBSYSTEM

FIG. 5—Legend for all isochron diagrams proposed in this
paper (Fig. 4, and 6 to 21).

ical exchanges, and only one isochron line is defined. Along this line, all the analytical points of whole-rock and fine-fraction, as well as LCH and RES subsystems, will be displayed.

Model IV occurs generally where the sediment has been placed in environmental conditions already within the field of anchi- or epi-metamorphic reactions.

Relations Between Models

In principle, Models I, II, and III occur commonly in shales affected by weak thermodynamic processes within the field of diagenesis. Model IV, although correlatable to stronger events already within metamorphism, can occur occasionally in shales that are only diagenetic. For this to happen it is sufficient that the rock consist essentially of clay minerals open to ionic exchanges, and that interstitial fluids within the system induce wide circulation of ions. In this case, we would reach an extreme case of Model II, in which all the samples would bear the same initial Sr^{87}/Sr^{86} ratio.

In general, shales contain strontium-bearing detrital minerals. In weak thermodynamic conditions, isochronic Model I will be developed. Increasing pressure and temperature conditions will lead to Model II if the quantitative amount of detrital material larger than 2μ is not very significant. A tendency towards Model III with increasing pressure and temperature conditions will occur in rocks with a large amount of clay minerals open to ionic exchanges, and(or) with a significant amount of large detrital components.

Models II and III grade towards Model IV with increasing conditions of pressure and temperature. In both cases, the tendency is towards strontium isotopic homogenization even at the level of the phases with larger dimensions. Within the field of epimetamorphism, or in more extreme pressure and temperature conditions, Model IV will be the only possibility for shale-like metamorphic rocks.

Case Examples—A Discussion

It must be pointed out, at this point, that the analytical work carried out at the São Paulo Laboratory was performed largely before we established our interpretative ideas, and consequently before we determined definitive technical procedures, field sampling criteria, and criteria for selection of samples based on mineralogy and chemistry. For this reason, the set of data is not uniform. In some cases, only whole rocks or fine fractions were analyzed. At other times, only two or three samples were investigated. In addition, some cases explained here were taken directly from already published work.

Following up the ideas expressed in the first sections, the interpretations will be succint. Only information directly bearing on the interpretations will be given. For further information, the readers are referred to the bibliography. In the data presented, the Rb^{87} constant of $1.47 \times 10^{-11} y^{-1}$ was employed. Moreover, all the indicated Kubler indices of crystallinity of illites were obtained at the Strasbourg Laboratory, in which the lower boundaries for diagenesis and anchimetamorphism are fixed at 5.75 and 3.50, respectively.

Trombetas Formation

The Trombetas Formation (Kawashita, 1972; Thomaz, 1976) is the lowest unit of the Amazon basin sequence and includes four members, one of which, the Pitinga member, is made up of laminated black shales from which the analyzed samples were collected. The stratigraphic age of this formation is Early Silurian (Daemon and Contreiras, 1971).

The study samples came from three different wells, located a few hundreds of kilometers apart. From one of them (UI-2-AM, see Thomaz, 1976) were collected samples AT-125, AT-127, AT-130, AT-133, AT-135, and AT-136, for which sample isochrons were obtained. Their mineralogy included quartz, kaolinite, illite-muscovite, interstratified illite-montmorillonite, and montmorillonite. Only one of the samples (AT-133) presented a significant amount of feldspar. The Kubler index presented values between 6.2 and 7.9, well within the diagenetic field, and the Esquevin index (between 0.33 and 0.49) indicated aluminous illite, of detrital character. The detrital origin of the clays was already revealed by the presence of kaolinite.

Three of the sample isochrons traced in Figure 6 are subparallel (AT-125, AT-135, and AT-136), with very high initial Sr^{87}/Sr^{86} ratios and an age around 160 m.y. (Jurassic). One of the sample isochrons is younger, and the other is older. The Mesozoic sample isochrons, in this lower Paleozoic unit, indicate a strontium isotopic homogenization much younger than sedimentation, probably associated with the Mesozoic tectonic and volcanic activity that affected the Amazon basin when the Atlantic Ocean formed.

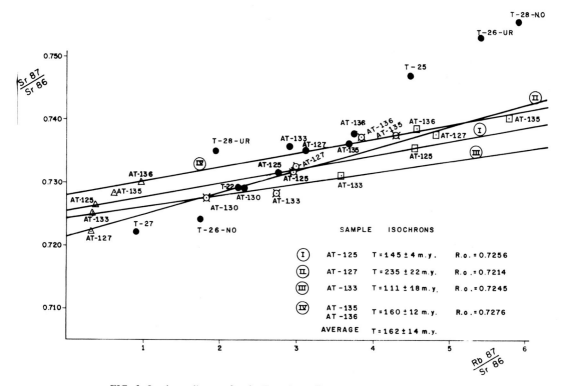

FIG. 6–Isochron diagram for the Trombetas Formation, Amazon basin.

If point AT-133 were excluded, the errorchron traced through the other AT analytical points would indicate an age of 420 ± 34 m.y. (which is concordant with the paleontologic record) and an initial Sr^{87}/Sr^{86} ratio equal to 0.7148. Most whole-rock analytical points are located slightly above their respective sample isochron, defining Model I from the weak diagenetic event undergone by the samples.

A virtually parallel Silurian errorchron, with a much higher initial ratio, can be traced through the T-25, T-26-UR, and T-28-NO analytical points, perhaps indicating the existence of a simultaneous but different depositional environment.

Rio do Rasto Formation

The Rio do Rasto Formation (Thomaz, 1976; Thomaz et al, 1976) of the Paraná basin consists of interbedded sandstones, siltstones, and shales, deposited under continental flood plain environmental conditions. Its stratigraphic age is Late Permian, with good control, as quoted by Barberena and Daemon (1974).

The analyzed samples came from a single well and exhibited a fairly uniform mineral composition: some detrital grains including quartz, feldspars and micas, clay minerals, mainly montmorillonite and illite, register a Kubler index between 6.2 and 7.8, well within the diagenetic field. The Esquevin index between 0.15 and 0.30 indicated magnesian illites, suitable to ionic exchanges.

The mineralogic similarity of the samples produced an isochron diagram (Fig. 7) in which all the whole-rock, as well as the fine fraction analytical points are very close. The sample isochrons are virtually superposed, and only one was traced, as a reference. Its age was determined as 211 ± 12 m.y., with Sr^{87}/Sr^{86} equal to 0.7106. It is interpreted as the time of the diagenetic event. The whole-rock points are located near the sample isochrons, or slightly above, indicating Model I, or Model II.

An errorchron traced through the whole-rock analytical points (Thomaz, 1976) indicated a value of 228 ± 9 m.y., which is indeed consistent with the inferred stratigraphic age.

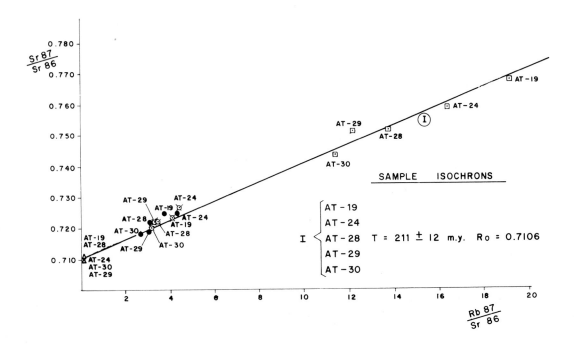

FIG. 7–Isochron diagram for the Rio do Rasto Formation, Paraná basin.

Botucatu Formation

The Botucatu Formation (Thomaz, 1976; Thomaz et al, 1976) of the Paraná basin is made up of eolian and fluvial sandstones, in which some thin argillaceous siltstones and some argillites are interbedded. The stratigraphic age of this formation is poorly controlled; its lower limit is indicated by the overlying basaltic lava flows of Early Cretaceous age (Cordani and Vandoros, 1967).

A few very thin argillaceous levels containing quartz, feldspars, micas, montmorillonite, and illite were selected for analytical work. It must be emphasized that these are the least desirable conditions for an attempt at age dating: continental sediments, very thin argillaceous layers, presence of detrital feldspars and micas.

Nevertheless, in the diagram of Figure 8, some meaningful interpretations can be made:

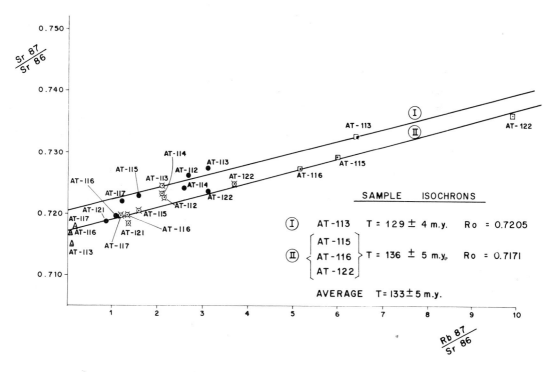

FIG. 8—Isochron diagram for the Botucatu Formation, Paraná basin.

1. The sample isochrons for three of the analyzed specimens (AT-115, AT-116, and AT-122) are virtually identical, with an age of 136 ± 5 m.y. and Sr^{87}/Sr^{86} equal to 0.7171. The sample isochron for AT-113 indicates a similar age, but a higher value for the initial ratio. The sample isochrons indicated that the age of the strontium isotopic homogenization phenomena within the Botucatu Formation simultaneously occurred with the outpouring of the overlying Serra Geral basaltic rocks in the Early Cretaceous.

2. Two of the whole-rock analytical points (AT-116, AT-122) lie on the sample isochrons, and two others are slightly above. This indicates the coexistence of Models I and II and a weak diagenetic event. Because of the low illite content, it was not possible to determine the Kubler index of crystallinity for the analyzed samples.

3. The whole-rock samples are much too dispersed in the diagram, preventing the tracing of an isochron or errorchron line. We believe this is because of changes in the depositional conditions of the interbedded argillites, which are separated by thick sandstone layers. Thomaz et al (1976) interpreted the marked tendency to alignment

of points AT-117, AT-115, AT-112, and AT-113 as being significant, defining an isochron with 197 ± 3 m.y. age and Sr^{87}/Sr^{86} equal to 0.7184. This was taken as a good estimate of the age of sedimentation, supposed to be Late Triassic.

Irati Formation

The Irati Formation (Thomaz, 1976; Thomaz et al, 1976), generally less than 100 m thick, extends over the entire Paraná basin, and is composed predominantly of black shales assumed to be euxinic and associated with limestones and chert in the northern part of the basin. Daemon and Quadros (1970) indicated a Late Permian stratigraphic age for the unit.

Six samples from the same well (LI-1-SP, Thomaz, 1976) were selected for analytical work. Two of them (AT-50 and AT-51) are composed essentially of quartz, montmorillonite, interstratified illite-montmorillonite, illite, chlorite, and some feldspar and gypsum. One of them is made up almost entirely of montmorillonite (AT-46), another is an argillaceous dolomite (AT-45), and two others (AT-48 and AT-48a) are evaporites composed essentially of anhydrite. The two Kubler indices for the shales were different (7.5 and 5.0) and only one of them was within the diagenetic field. The Esquevin indexes of 0.31 and 0.38 indicated magnesian illites open to chemical exchanges.

Figure 9 shows the isochron diagram for the Irati Formation. The two sample isochrons are parallel, indicating an age of about 180 m.y. (Jurassic). The whole-rock samples in both cases were virtually aligned with the sample isochrons, indicating the occurrence of Model II.

Samples AT-45, AT-48, and AT-48a, very low in rubidium content, produced Sr^{87}/Sr^{86} results very close to the seawater strontium value, 0.709. The whole-rock isochron age, traced through points AT-48, AT-48a, AT-50, and AT-51, which are very close stratigraphically, resulted in 256 ± 19 m.y., concordant with the paleontologic control.

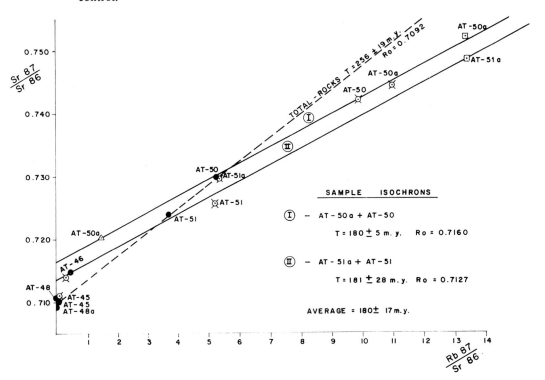

FIG. 9–Isochron diagram for the Irati Formation, Paraná basin.

Estrada Nova Formation

The Estrada Nova Formation (Thomaz, 1976; Thomaz et al, 1976) in the Paraná basin, lies conformably over the previously mentioned Irati Formation. Its palynological record defines a Kazanian (Late Permian) age (Daemon and Quadros, 1970). Its main lithology consists of siltstones, interbedded with silty shales, with lenses of limestones, and chert. From analysis of the sedimentary structures it is considered to have formed within large tidal plains.

The analyzed samples came from the well MC-1-SC, and included two dolomites (AT-66 and AT-67) and six silty shales, in which montmorillonite and chlorite predominate. Quartz, feldspars, and micas were identified as detrital fragments, and in two of the samples (AT-64 and AT-83) aragonite was present. Illite in some of the samples exhibited a Kubler index close to 6.0, within the diagenetic field, and an Esquevin index between 0.23 and 0.30, indicating magnesian illites open to chemical exchanges.

Acid leaching of those samples was not performed. However, in the isochron diagram of Figure 10, the lines connecting the whole-rock samples with the respective fine fractions are virtually subparallel, and suggest that the samples were affected by strontium isotopic homogenization at the hand specimen level (Model II). It is interesting to note that the resulting age for this inferred homogenization event (about 180 m.y.) is practically identical to that displayed by the Irati Formation, a fact which strengthens the validity of the age number in terms of geologic significance.

Although the analytical points for the total rock samples are somewhat dispersed in the diagram, an errorchron can be traced, with an age of 243 ± 14 m.y. and initial Sr^{87}/Sr^{86} ratio of 0.7102. Once again, this calculated age is compatible with the paleontologic record of the formation.

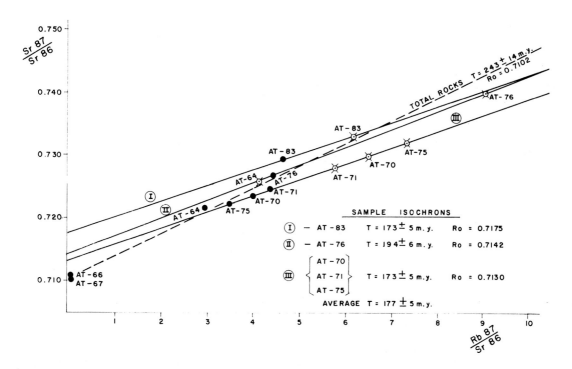

FIG. 10—Isochron diagram for the Estrada Nova Formation, Paraná basin.

Ponta Grossa Formation

The Ponta Grossa Formation (Kawashita, 1972; Thomaz et al, 1976) belongs to the lower sequence of the Paraná basin, and its paleontologic record (Lange, 1967) indicates a Middle Devonian age for the unit. Essentially it comprises interbedded shales and siltstones formed in a neritic to infraneritic marine environment. The samples analyzed by Kawashita (1972) were collected in two wells (MO-1-PR and MO-2-PR) located a few kilometers apart. They are composed of illite, kaolinite, and chlorite, and by quartz, micas, and some feldspar as detrital components. The Kubler index determined for one sample was 4.2, already within the field of anchimetamorphism.

In the isochron diagram of Figure 11, the six whole-rock points exhibit considerable dispersion, but the points related to the fine fractions are aligned, suggesting Model III. The LCH (soluble) analytical points lie on the fine fraction alignment, but at least one of the RES points (T-19) is located significantly above, indicating that it escaped the strontium isotopic homogenization phenomenon. The age of this event, 173 ± 16 m.y., much younger than the sedimentation age, is once again similar to that previously mentioned as having affected the Irati and Estrada Nova Formations. It also coincides with the onset, in the Jurassic, of the widespread post-Paleozoic tectonism related to the opening of the South Atlantic, which in the Paraná basin caused, among other effects, the reactivation of the Ponta Grossa arch.

A reference isochron traced as an upper envelope for the whole-rock samples, plus the analytical point for the residue T-19, would indicate a Devonian age.

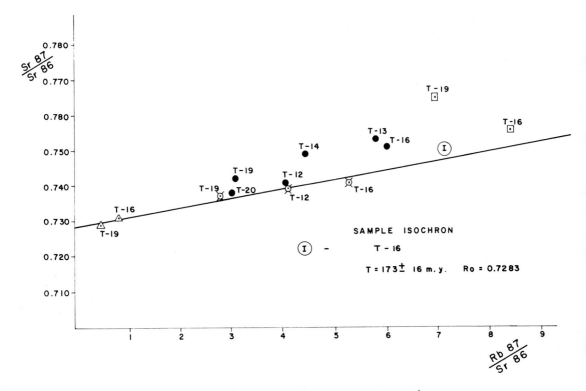

FIG. 11—Isochron diagram for the Ponta Grossa Formation, Paraná basin.

Tubarão Group, Rio Bonito Formation

The Rio Bonito Formation (Kawashita, 1972; Thomaz et al, 1976) of the Tubarão Group, in the Paraná basin, is made up mainly of fine sandstones deposited in a neritic marine environment. Silty shales occur interbedded in the central part of the sequence.

The stratigraphic age based on palynology was determined by Daemon and Quadros (1970) as Late Permian.

Three samples were analyzed by Kawashita (1972) from the TV-4-SC well, and the whole-rock analytical points as well as one fine fraction are plotted in Figure 12. These samples are silty shales, containing mainly illite and kaolinite, and some detritic fragments, essentially quartz and micas. In one of the samples, the Kubler index was determined, and resulted around 4.0, within the field of anchimetamorphism. Esquevin indices were between 0.24 and 0.32, indicating magnesian illites, open to chemical exchanges.

In the isochron diagram, a reference isochron can be traced through the three whole-rock points, with an age of 273 ± 34 m.y. and an initial Sr^{87}/Sr^{86} ratio of 0.7115. Because of the large quoted experimental error, the value is still within the limits of the stratigraphic age.

No acid leaching experiments were performed on these samples. However, the only available point for the fine fractions, dislocated to the whole-rock isochron line, already excludes the possibility of the attribution of Model IV to the Rio Bonito Formation.

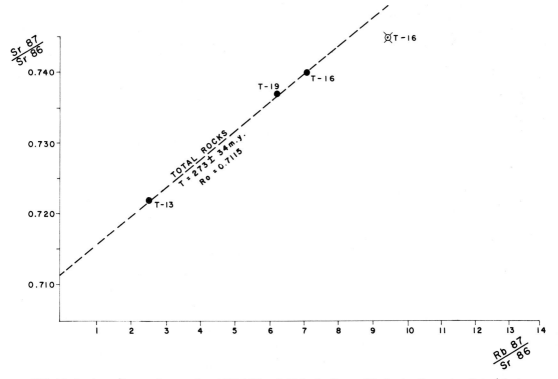

FIG. 12–Isochron diagram for samples of TV-4-SC well, Tubarão Group, Rio Bonito Formation, Paraná basin.

Sepotuba Formation

The unfossiliferous shales of the Sepotuba Formation (Cordani et al, in preparation) belong to the Alto Paraguay Group, which is considered to be a postorogenic unit of the late Precambrian Brazilian Cycle. They overlie discordantly the sediments of the Jangada and Araras Groups, as well as the metasediments of the Paraguay-Araguayia geosyncline.

The analyzed samples are shales collected within the same general outcrop. They are composed essentially of anchimetamorphic illite, with a Kubler index of 5.3, and an Esquevin index of 0.40.

Figure 13 is the isochron diagram for the Sepotuba Formation. The whole-rock analytical points are aligned, and define an isochron line with an age of 547 ± 5 m.y. (Cambrian) and an initial Sr^{87}/Sr^{86} ratio of 0.711. This value is taken as representing the age of the sedimentation.

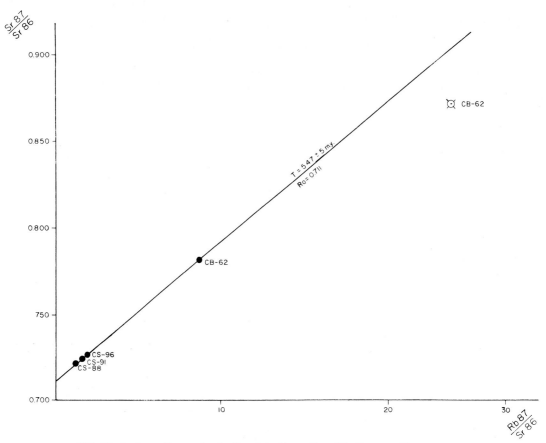

FIG. 13–Isochron diagram for the Sepotuba Formation, Alto Paraguay Group.

Because no acid leaching experiments were performed, it is impossible to classify the Sepotuba Formation according to the proposed models. However the position of the fine fraction CB-62 on the diagram already precludes the classification as Model IV.

Estancia Formation

The unmetamorphosed and unfossiliferous sediments of the Estancia Formation (Brito Neves et al, in preparation), in northeast Brazil, are also considered to be post-orogenic deposits of the late Precambrian Brazilian Cycle. They overlie discordantly the metasediments of the Vaza Barris Group, and the basement complex of the Salvador Craton.

The analyzed samples are shales and siltstones, and even fine-grained sandstones, collected in three different outcrops in the vicinity of Lagarto (State of Sergipe) within a distance of 30 km from one another. Illite predominates among the clay minerals, and quartz and feldspars were detected as the main detrital components. The Kubler indices of the illites (from 3.0 to 4.3) were near the boundary between anchi- and epi-metamorphism, and the Esquevin indices (around 0.30) indicated magnesian illites.

Figure 14 brings the isochron diagram for the Estancia Formation, in which all the points (whole-rock and fine fraction samples) are aligned along an isochron with 459 ± 15 m.y. and a Sr^{87}/Sr^{86} ratio equal to 0.7161. This pattern can be related to Model IV, characteristic of units subjected to an anchimetamorphic episode. The obtained age, Ordovician, is assumed to represent this event, younger than the sedimentation age.

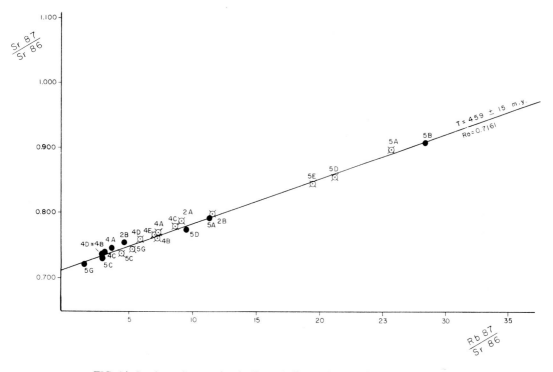

FIG. 14—Isochron diagram for the Estancia Formation, northeastern Brazil.

Bambui Group

The Bambui Group (Bonhomme, 1976; Amaral and Kawashita, 1967) covers a very large area, in eastern Brazil, overlying most of the São Francisco Craton and a great part of the Salvador Craton. The unit is essentially unmetamorphosed and only slightly folded, except at the borders of the tectonic units where it has been strongly folded and metamorphosed.

Limestones and siltstones predominate within the Bambui Group. The interbedded shales exhibit illite as the main clay mineral, which results in generally high Rb/Sr ratios, very favorable for Rb-Sr age measurements. Stromatolites were found in the southern part of the unit which suggests a late Precambrian age of at least 900 m.y. (Cloud and Dardenne, 1973). Due to the marked lithologic similarities throughout the outcrop areas, we are inclined to accept the idea that the Bambui Group was deposited in a reasonably short time interval, at any rate not longer than a few hundred million years. Consequently, the depositional ages should not differ greatly from place to place within the unit.

Samples for Rb-Sr studies were collected in several places. We refer here to the examples of Lages do Batata, Januaria, and João Pinheiro (Bonhomme, 1976), as well as Vazante (Amaral and Kawashita, 1967).

Figure 15 illustrates the analytical points for Lages do Batata, located within a stable part of the Salvador Craton, in central Bahia State. The shales are made up

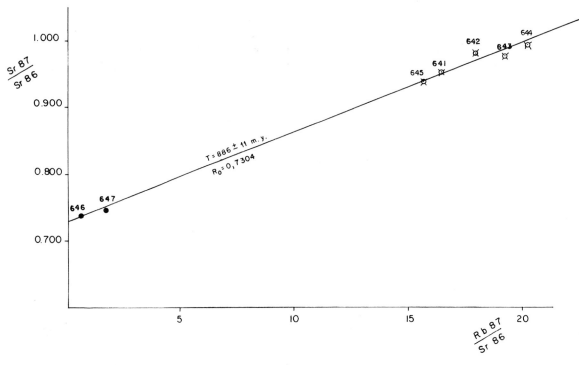

FIG. 15–Isochron diagram for the Bambui Group, Lages do Batata.

essentially of illite, with Kubler indices around 9.0, well within the diagenetic field, and Esquevin indices between 0.35 and 0.42, indicating aluminous illite. Only the fine fractions and two associated carbonate rocks (points 646 and 647) were analyzed.

The obtained isochron, with an age of 886 ± 11 m.y., must be taken as related to a diagenetic age, and consequently represents a minimum value for the deposition of the Bambui Group. The very high initial Sr^{87}/Sr^{86} ratio of 0.7304 indicates that isotopic homogenization affected even the carbonates, at the level of the lithologic unit, defining Model IV of strontium evolution.

The samples of Januaria and João Pinheiro are located within a stable part of the São Francisco Craton in northern Minas Gerais State. They were collected from very thin, millimetric shale lenses interbedded with limestones. Essentially pure illite makes up the Januaria samples, and illite and vermiculite predominate in the João Pinheiro samples (633 and 634). Sample 640 represents the main carbonate lithology of Januaria. The Kubler indices of the illites are around 6.7 for the Januaria samples, and around 5.4 for João Pinheiro. This latter value is already within the field of anchimetamorphism. In all cases, the Esquevin index was higher than 0.30, indicating aluminous illites. Once again, only the fine fractions were analyzed.

In the isochron diagram of Figure 16, an age much younger than in the previous example was determined. The fine fractions are all aligned, defining an isochron with 619 ± 17 m.y., and an initial Sr^{87}/Sr^{86} ratio of 0.7084. This case seems to define the existence of a strong episode that led to a complete strontium isotopic homogenization at least for the fine fractions, which occurred at Januaria during diagenetic environmental conditions, and at João Pinheiro already during anchimetamorphism. The general age around 600 m.y. is well known in the Brazilian territory, as it can be associated with strong tectonism in the orogenic units of the Late Precambrian Brazilian Cycle.

It is apparent that the effects of this tectonism were felt well inside the supposedly stable regions, at distances of several hundreds of kilometers from the orogenic belts.

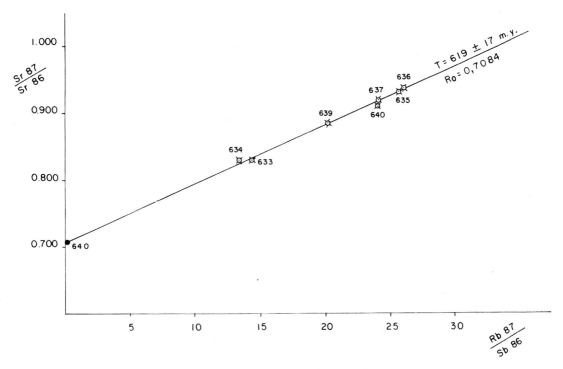

FIG. 16–Isochron diagram for the Bambui Group, at Januária and João Pinheiro, northern Minas Gerais.

Although whole-rock analyses were not performed, the Bambui sediments at Januaria and João Pinheiro seem to have been affected by Model III or Model IV types of strontium evolution.

At Vazante, Minas Gerais State, the Bambui Group was perceivably affected by the Late Precambrian tectonism, along the western border of the São Francisco Craton. There, the shales contain essentially illite, with a Kubler index of around 4.0 (well inside the field of anchimetamorphism), and an Esquevin index between 0.15 and 0.30 (indicating magnesian illites, open to chemical exchanges).

Figure 17, taken from Amaral and Kawashita (1967), shows the isochron diagram for whole-rock samples of the Bambui Group at Vazante. The points lie very close to the reference isochron of 600 m.y., with an extremely high Sr^{87}/Sr^{86} ratio of 0.83. It is evident that a large scale strontium isotopic homogenization occurred at the level of the formation, thus characterizing Model IV.

These three isochron diagrams related to rocks of the Bambui Group are good examples of the complex evolution suffered by this unit in its different outcrop regions. Its minimum age is around 900 m.y., the time at which a well-defined diagenetic episode occurred at Lages do Batata. In addition, a late diagenetic and(or) anchimetamorphic episode associated with the strong regional tectonism related to the Brazilian Cycle (around 600 m.y.) apparently affected the entire region covered by the Bambui Group in Minas Gerais and is well exposed in the isochron diagrams.

Other Examples

Several examples of published work regarding Rb-Sr dating of shales can be found in the international literature. Many conclude that the obtained isochron lines are geologically meaningful and approach the depositional ages, such as Compston and Pidgeon (1962) Allsopp and Kolbe (1965), Bonhomme et al (1965), Bofinger and Compston (1967), Faure and Chaudhuri (1967) and others. However, these studies were carried out by several different laboratories with different techniques. Thus, their

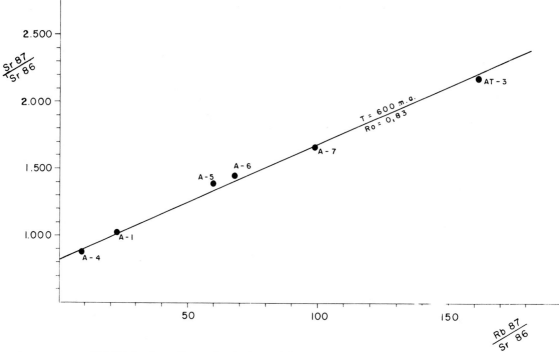

FIG. 17–Isochron diagram for the Bambui Group, at Vazante, Minas Gerais.

results can not be interpreted directly in terms of our models for strontium evolution within sediments because of the lack of significant data such as analyses on fine fraction systems or subsystems resulting from acid leaching.

We reproduce here, in Figures 18 to 21, four examples in which an interpretation in terms of our proposed models can be made:

1. Figure 18 is taken from Bonhomme et al (1966) and refers to shales of the Erquy and Kerity Formations (France). The fine fraction points are aligned, defining an isochron of 373 ± 12 m.y. The whole-rock points are located above the line, and characterize our Model III. The age obtained for the reference isochron connecting the whole-rock analytical points (Cambrian) apparently can be taken as the sedimentation age, considering the available paleontologic data.

2. Figure 19 concerns the Anse du Veryac'h Series (France) (Bonhomme et al, 1968), which is controlled by paleontologic data indicating a Siluro-Ordovician age. In the diagram, the fine fractions are aligned along an isochron with 308 ± 19 m.y. (Upper Carboniferous). In our opinion, once again Model III can be characterized, and the younger isochron is indicative of diagenetic transformations affecting the sediment during a strong regional tectonic activity. The older reference isochron traced in Figure 19 (passing close to four whole-rock analytical points) is concordant with the stratigraphic Siluro-Ordovician age assigned to the unit.

3. Figure 20 illustrates the data obtained by Bofinger et al (1968) for the Ordovician shales of the Great Artesian basin of Australia. The Bofinger group performed acid leaching directly on the whole-rock samples without extracting the fine fractions, but their results can be interpreted in terms of our Model IV with minor adaptations. The indicated isochron age, 448 m.y., refers to a diagenetic event which must have occurred shortly after the deposition of the unit in Ordovician time.

4. Figure 21 is taken from Clauer and Bonhomme (1970), and is related to shales of the Steige and Villé Series (France). Whole-rock and fine-fraction samples are aligned along an isochron with an age of 358 ± 4 m.y. The samples are slightly metamorphosed and characterize our Model IV.

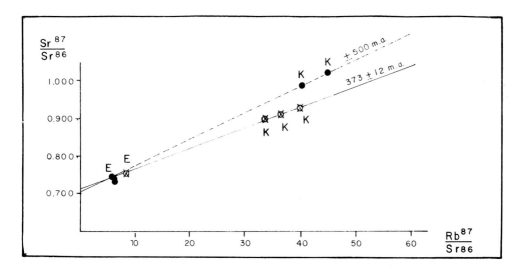

FIG. 18–Isochron diagram for the Erquy and Kerity Formations France (taken from Bonhomme et al, 1966).

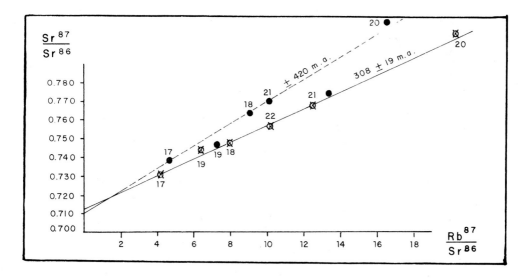

FIG. 19–Isochron diagram for the Anse du Veryac'h Series, France (taken from Bonhomme et al, 1968).

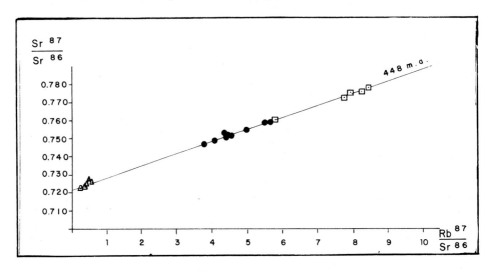

FIG. 20—Isochron diagram for the Great Artesian basin, Australia (taken from Bofinger et al, 1968).

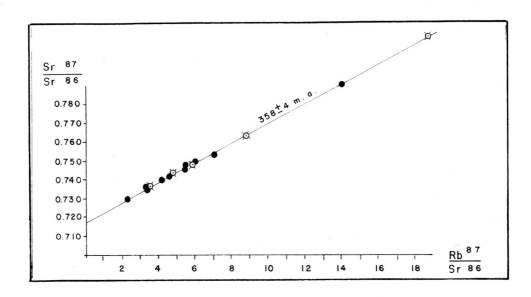

FIG. 21—Isochron diagram for the Steige and Villé Series, France (taken from Clauer and Bonhomme, 1970).

Final Considerations

We consider the Rb-Sr method as potentially capable of offering interesting and geologically meaningful age data when applied to pelitic rocks. In our opinion, the study is still in its beginning, and many improvements and(or) insights can be expected for the future regarding interpretations of the geologic history of a sedimentary rock in terms of the clay mineralogy, role of detrital material, and role of the chemical components.

We do not claim that our proposed models for strontium evolution within shales are applicable in all cases. With additional information available, we think that they could probably be improved and that other alternative models could be revealed.

More work is necessary to demonstrate the effectiveness of the proposed mechanism of uniform dispersion of detrital material in the depositional basins and to see if this geologic process is as widespread in nature as we think it may be. For instance, it was observed, in the given examples, that the whole-rock isochron of shales, even when indicating a reasonable value in terms of sedimentation age, always exhibits a considerable scattering of points. This effect could be exclusively attributed to the mechanism of uniform dispersion, since variations in the initial Sr^{87}/Sr^{86} ratios within the depositional site can be expected. These variations may be about 0.005, judging from Figures 1 through 3.

In order to check these possibilities, Rb-Sr analytical work will be performed on recent sediments of variable mineralogy and from different environments. Such studies, some of which are presently in progress at the São Paulo Laboratory, will produce insights regarding the limitation of the proposed models and the criteria to be established for field collecting and for sample selection in the laboratory, based on mineralogy, rubidium and strontium chemistry, and determination of other parameters such as the Kubler and Esquevin indices of illite.

We anticipate that the Rb-Sr isochron method will be of great use in establishing the depositional age of shales and related rocks. Perhaps not for the Phanerozoic, for which fossils remain the more powerful and precise method, but for the Precambrian in whose sedimentary rocks fossils are not frequently encountered, or when encountered do not have adequate stratigraphic value. For instance, Rb-Sr seems to be the most appropriate tool for global correlation of Precambrian flat-lying sedimentary sequences which exist on all continents.

Finally, we consider that the examples supplied here sufficiently demonstrate the geologic meaning of the sample isochrons performed on the fine fractions and their respective subsystems produced by acid leaching. This matter seems to be extremely important when dealing with the history of a sediment in times subsequent to its deposition.

In the study carried out by Thomaz (1976), data regarding the evolutionary stage of the organic matter contained within the sediment were included. A correlation exists between the index of thermal alteration, hydrocarbon varieties, diagenetic facies of organic matter, and maximum paleotemperature attained by the sediment (Quadros, 1975). The determination of sample isochrons is potentially capable of indicating the epoch of maturation of the organic matter, a potentiality which has great implications for the processes of oil formation and oil migration in sedimentary sequences. We do not have to emphasize the importance of additional research along these lines to future oil prospection.

References Cited

Allsopp, H. L., and P. Kolbe, 1965, Isotopic age determinations on the Cape Granite and intruded Malmesbury sediments, Cape Peninsula, South Africa: Geochim. Cosmochim. Acta, v. 29, p. 1115-1130.

Amaral, G., and K. Kawashita, 1967, Determinacao da idade do Grupo Bambui pelo metodo Rb-Sr: 21st Cong. Bras. Geol. (Curitiba) Anais, p. 214-217.

Barberena, M. C., and R. F. Daemon, 1974, A primeira ocorrencia anphibia Labyrinthondontia na Formacao Rio do Rasto. Implicacoes geocronologicas e estratigraficas: 28th Cong. Bras. Geol., (Porto Alegre) Anais, RS.

Biscaye, P. E., and E. J. Dasch, 1971, The rubidium, strontium, strontium-isotope system in deep-sea sediments; Argentine Basin: Jour. Geophys. Research, v. 76, no. 21, p. 5087-5096.

Bofinger, V. M., and W. Compston, 1967, A reassessment of the age of the Hamilton Group, New York and Pennsylvania, and the role of inherited radiogenic Sr^{87}: Geochim. Cosmochim. Acta, v. 31, p. 2353-2359.

—— —— and B. L. Gulson, 1970, A Rb-Sr study of the Lower Silurian Stage Circle Shale, Canberra, Australia: Geochim. Cosmochim. Acta, v. 34, p. 433-445.

—— —— and M. J. Vernon, 1968, The application of acid leaching to the Rb-Sr dating of a Middle Ordovician shale: Geochim. Cosmochim. Acta, v. 32, p. 823-833.

Bonhomme, M., 1976, Mineralogie des fractiones fines et datations Rubidium-Strontium dans le Groupe Bambui (MG, Bresil): Rev. Bras. Geociencias, v. 6, no. 4.

—— and N. Clauer, 1972, Possibilites d'utilisation stratigraphique des datations directes rubidium-strontium sur les mineraix et les roches sedimentaires: Colloque sur les Methodes et Tendances de la Stratigraphie: Bur. Recherches Geol. Minieres Mem., v. 77, p. 943-950.

—— J. Lucas, and G. Millot, 1964, Signification des determinations isotopiques dans la geochronologie des sediments: Sciences de la Terre, t X, v. 3-4, p. 539-565.

—— P. Vidal, and J. Cogne, 1968, Determination de l'age tectonique de la serie ordovicienne et silurienne de l'Anse du Veryac'h (Presq'ile de Crozon, Finistere): Serv. Carte Geol. Als. Lorr. Bull., v. 21, no. 4, p. 249-252.

—— F. Weber, and R. Favre-Mercuret, 1965, Age par la methode rubidium-strontium des sediments du Bassin de Franoeville: Serv. Carte Geol. Als. Lorr. Bull., v. 18, no. 4, p. 243-252.

—— J. Cogne, F. Leuiwein, and J. Sonet, 1966, Donnees nouvelles sur l'age des series rouges du golfe normanno-breton: Acad. Sci. (Paris) Comptes Rendus Ser. T, v. 262, p. 606-609.

Brito Neves, B. B., and K. Kawashita, in preparation, Estudo Geocronologico do Grupo Estancia pelo Metodo Rb-Sr.

Chaudhuri, S., and D. G. Brookins, 1969, The Rb-Sr whole-rock age of the Stearns Shale (Lower Permian), Eastern Kansas, before and after acid leaching experiments: Geol. Soc. America Bull. v. 80, p. 2605-2610.

Clauer, N., and M. Bonhomme, 1970, Homogeneisation isotopique du strontium entre les schistes de Steige et la serie de Ville (Vosges) pendant la phase bretonne de l' orogenese hercynienne: Acad. Sci. (Paris) Comptes Rendus, Ser. T, v. 271, p. 1844-1847.

—— —— 1974, Isotopic homogeneisation of strontium in clays: Int. Mtg. Geochronology, Cosmochronology and Isotope Geology (Paris), p. 26-31.

—— M. Hoffert, D. Grimaud, and G. Millot, 1975, Composition isotopique du strontium d'eaux interstitielles extraites de sediments recents: un argument en faveur de l'homogeneisation isotopique des mineraux argileux: Geochim. Cosmochim. Acta, v. 39, p. 1579-1582.

Cloud, P., and M. Dardenne, 1973, Proterozoic age of the Bambui Group in Brazil: Geol. Soc. America Bull., v. 84, no. 5, p. 1673-1676.

Compston W., and R. T. Pidgeon, 1962, Rubidium-Strontium dating of shales by the total-rock method: Jour. Geophys. Research, v. 67, no. 9, p. 3493-3502.

Cordani, U. G., and P. Vandoros, 1967, Basaltic rocks of the Parana Basin in J. J. Bigarella, G. D. Becker and I. D. Pinto, eds., Problems in Brazilian Gondwana geology: p. 207-231.

—— K. Kawashita, and S. M. M. Mantovani, in preparation, Idade Rb-Sr de folhelhos do Grupo Alto Paraguai.

Daemon, R. F., and C. J. A. Contreiras, 1971, Zoneamento palinologico da Bacia do Amazonas: 25th Cong. Bras. Geol. (Sao Paulo) Anais, v. 3, p. 79-92.

—— and L. O. Quadros, 1970, Bioestratigrafia do Neopaleozoico da Bacia do Parana: 24th Cong. Bras. Geol., (Brasilia) Anais, p. 359-412.

Dasch, E. J., 1969, Strontium isotopes in weathering profiles, deep-sea sediments, and sedimentary rocks: Geochim. Cosmochim. Acta, v. 33, p. 1521-1552.

Esquevin, J., 1969, Influence de la composition chimique des illites sur leur cristallinite: Centre Rech. Pau-SNPA Bull., v. 3, no. 1, p. 147-153.

Faure, G., and S. Chaudhuri, 1967, The geochronology of the Keweenawan rocks of Michigan and origin of the copper deposits: Final Report, Grants GP-3090, GA-470, The National Science Foundation, Washington, 41 p.

—— and J. Kovach, 1969, The age of the Gunflint Iron Formation of the Animikie Series in Ontario, Canada: Geol. Soc. America Bull., v. 80, p. 1725-1736.

—— and J. C. Powell, 1972, Strontium isotope geology: Minerals, Rocks and Inorganic Materials, Mon. Ser. Theoretical and Experimental Studies, v. 5, 188 p.

Hart, S. R., and G. R. Tilton, 1966, The isotope geochemistry of strontium and lead in Lake Superior sediments and water: Geoph. Mon., 10, The Earth Beneath the Continents, p. 127-137.

Kawashita, K., 1972, O metodo Rb-Sr em rochas sedimentares-aplicacao para as Bacias do Parana e Amazonas: PhD Thesis, Instituto de Geociencias da Universidade de Sao Paulo, 111 p.

Kubler, B., 1966, La cristallinite de l'illite et les zones tout a fait superieures du metamorphisme:

Etages tectoniques, Colloque de Neuchatel, p. 105-122.

Lange, F. W., 1967, Subdivisao bioestratigrafica e revisao da coluna siluro - devoniana da Bacia do Baixo Amazonas: Atas do Simposio sobre a Biota Amazonica, v. 1 (geociencias), p. 215-326.

Quadros, L. P., 1975, Organopalinologia na prospeccao de petroleo: Bol. Tec. Petrobras, v. 18, no. 1, p. 3-11.

Thomas Filho, A., 1976, Potencialidade do Metodo Rb-Sr para Datacao de Rochas Sedimentares Argilcsas: PhD Thesis, Instituto de Geociencias da Universidade de Sao Paulo, 128 p.

—— U. G. Cordani, and K. Kawashita, 1976, Aplicacao do Metodo Rb-Sr na Datacao de Rochas Sedimentares Argilosas da Bacia do Parana: 29th Cong. Bras. Geol., (Belo Horizonte) Anais (in press).

Whitney, P. R., and P. M. Hurley, 1964, The problem of inherited radiogenic strontium in sedimentary age determinations: Geochim. Cosmochim. Acta, v. 28, p. 425-436.

Potassium-Argon Isotopic Dating Method and Its Application to Physical Time-Scale Studies[1]

IAN McDOUGALL[2]

Abstract Advantages and disadvantages of the potassium-argon (K-Ar) isotopic dating method are discussed in relation to its application to the accurate calibration of the physical time scale for the Phanerozoic. The method is best applied to the dating of igneous rocks but authigenic glauconite in sedimentary rocks also is amenable to dating. In physical time-scale studies, the ideal situation is that of volcanic rocks datable from sequences for which there is good stratigraphic and biostratigraphic control. Ages determined on intrusive rocks in a biostratigraphically well-controlled sequence are useful in providing a younger limit to the age of the sediments. As with all dating methods, it is of utmost importance to measure ages on several samples in known stratigraphic relation in order to evaluate and test the assumptions underlying the K-Ar method, in particular those involving closed-system behavior and the possible effect of later events.

Introduction

In chronology two events are either simultaneous or one precedes the other. Using this principle, we can order events uniquely, without reference to the numerical aspects of time. For example, successive layers of sedimentary rock are a chronologic series. Using this approach, with the evolutionary change of species traced through fossils in rocks, geologists of the 19th century were able to organize a relative time scale for the Phanerozoic.

This relative geologic time scale has been progressively developed and refined. Correlation is effected throughout the world mainly by biostratigraphic means. Calibrating this relative scale in time units (years) has been attempted from almost the earliest days of modern geology, perhaps because a physical time scale provides the foundation for estimates of rates of geologic processes. In addition such a time scale permits estimates of elapsed time.

At the turn of the century there was much controversy about the age of the earth—Kelvin's preferred estimate of about 20 m.y. was hotly disputed by many geologists who argued that much more time was involved. Kelvin's estimates, based on the assumption of an initially molten earth that has been cooling ever since, were nullified when Becquerel discovered radioactivity which later was shown to be an additional source of heat. And of course the discovery of radioactivity made possible the development of isotopic dating methods. Indeed, in the first decade of this century it was shown that some rocks at the earth's surface were many hundreds of millions of years

[1] Manuscript received, October 4, 1976.

[2] Research School of Earth Sciences, Australian National University, Canberra, A.C.T. 2600, Australia.

The incremental-heating experiment on hornblende was carried out by E. Farrar in this laboratory, and I thank him for permission to use the results for illustrative purposes. The Australian Institute of Nuclear Science and Engineering facilitated the irradiation.

Article Identification Number: 0149-1377/78/SG06-0009/$03.00/0

old. It was not until the 1950s however, that development of the K-Ar and Rb-Sr methods, as well as further development of the U- and Th-to-Pb methods, were sufficiently advanced that dating of rocks especially chosen because their position in the relative time scale was well known, became practical on a reasonably large scale. Between about 1955 and 1965, a lot of effort was directed toward organizing a physical time scale; summary papers appeared (Holmes, 1947, 1959; Kulp, 1961; Harland et al, 1964) which formed an evolutionary sequence of progressively better correlations and age assignments as the biostratigraphic and physical age data increased.

The physical time scale as presently known was derived by applying all the dating techniques mentioned, but for various reasons results by the K-Ar method predominate. Because there is considerable interest in constructing even more precise and accurate time scales, it seems appropriate to review the K-Ar method and its usefulness for time-scale purposes.

Sampling for Time-Scale Studies

For time-scale studies it is desirable to measure isotopic ages on samples that are controlled as precisely as possible in the relative geologic time scale. The ideal situation perhaps is that of an undeformed, unmetamorphosed sequence consisting of volcanic rocks, amenable to isotopic dating, interbedded with sediments which contain a diagnostic fossil assemblage that enables the position on the relative time scale to be determined with precision. Deformed and/or metamorphosed sequences are much less favorable because of possible disturbance of the isotopic dating systems during such events. Tuffaceous deposits interbedded with fossil-bearing sediments also can be very useful for time-scale studies, provided that the tuffs contain separable minerals suitable for isotopic dating and provided that contamination of the tuff by older material is absent or can be eliminated.

Glauconite has been used extensively for time-scale studies. It is virtually the only mineral suitable for direct isotopic dating of sediments. This authigenic, micalike mineral, however, loses radiogenic argon at relatively low temperatures ($<150°C$) so that measured ages commonly are demonstrably too young. Glauconite can form by degradation of detrital mica so there is always the risk that some inherited daughter product may be present, causing the measured ages to be too old. Thus, results from glauconite dating must be carefully authenticated.

Age measurements on intrusive rocks can be of some value in the construction of the physical time scale because the ages provide a younger limit to the age of the rocks intruded. Where an intrusive body has been eroded and sediments deposited, an age on the intrusive rocks gives a maximum physical age for the overlying sediments. If the time gaps between sediment deposition, emplacement of an intrusive body, subsequent erosion, and deposition of more sediment are sufficiently short, then isotopic ages on such intrusive bodies can be of considerable value for time-scale purposes.

It should be noted that several different stratigraphic or zonal schemes comprise the relative time scale. The importance of obtaining sufficient physical age data to calibrate each scheme independently already has been demonstrated by the discovery, through isotopic dating, that correlations between the vertebrate-bearing continental succession in North America and marine sequences were considerably in error (Berggren, 1969; Page and McDougall, 1970).

Potassium-Argon Dating Method

The potassium-argon (K-Ar) dating method is now a rather well known and widely applied technique of isotopic age determination. Useful general references include Dalrymple and Lanphere (1969) and York and Farquhar (1972). The method is based on the naturally occurring ^{40}K isotope, which has an atomic abundance of 1.167 x 10^{-4} relative to total K (Garner et al, 1975), and has a half life of about 1,250 m.y. The isotope ^{40}K has a dual decay; about 89.5% of all decays produce ^{40}Ca by emission of an electron, and the remainder of the decays (about 10.5%) produce ^{40}Ar by orbital electron capture and emission of a gamma-ray. The K-Ar method depends upon the fact that argon, a noble gas, normally is lost from magma during eruption, crystallization, and cooling and that, subsequently, radiogenic argon produced from the decay of ^{40}K will be trapped within the rock. Thus we are dealing with an accumulative type of isotopic clock. The basic equation is as follows:

$$t = \frac{1}{\lambda} \ln \left(1 + \frac{\lambda}{\lambda_e} \frac{^{40}Ar^*}{^{40}K}\right),$$

where t = time since crystallization, λ = the ^{40}K decay constant, λ_e = partial decay constant for the ^{40}K to ^{40}Ar branch, ^{40}Ar* = radiogenic argon content, and ^{40}K = amount of this isotope present in the sample.

A rock containing 1% total K generates about 4×10^{-8} cc NTP ^{40}Ar*/g in 1 m.y. Such small amounts of argon can be measured precisely using mass spectrometric techniques. Potassium is measured by flame photometry, atomic absorption, or neutron-activation techniques. In principle it is possible to measure K-Ar ages on suitable samples older than about 1 m.y. with a precision of about 1%. With the development of digital readout systems, data acquisition from a mass spectrometer directly on-line with a computer is becoming common, resulting in increased precision of the measurement of isotopic ratios. Nevertheless, uncertainties in the calibrations of the mass spectrometer, the tracers (spikes), and standards for K and Ar probably are the limiting factors in precision and accuracy of K-Ar measurements at present, and these uncertainties will vary from laboratory to laboratory. Lanphere and Dalrymple (1967) showed that K-Ar measurements on an 81-m.y.-old muscovite (P-207) measured in 24 different laboratories gave a standard deviation of 2.6%. Thus, it is of some importance to realize that interlaboratory differences exceeding 3% may not be uncommon, even using material that technically is not difficult to measure.

As with all dating methods certain assumptions must be met if a K-Ar age measurement is to give a valid estimate of age.

1. At the time of crystallization of the rock all preexisting radiogenic argon must be lost. Clearly if some radiogenic argon is incorporated at the time of crystallization, the measured age will be greater than the true age. This assumption appears to be valid for subaerially erupted volcanic rocks and rocks intrusive at relatively shallow depths, provided that no older, xenolithic material is present within the rock. Rocks that have crystallized under high pressure, deep within the crust or mantle or in the deep oceans, may have incorporated radiogenic argon at the time of crystallization.

2. To yield a valid estimate of age the rock must have remained a closed system since crystallization. In other words, there should be no loss or gain of K or radiogenic Ar since crystallization and cooling except for those changes produced by the decay of ^{40}K to ^{40}Ar. Failure of this assumption is perhaps the most common cause of incorrect estimation of K-Ar ages, and generally leads to a calculated age that is too young, because loss of radiogenic argon by diffusion is the most common form of open-system behavior.

3. Any nonradiogenic argon present in the sample has the isotopic composition of atmospheric argon, or the appropriate composition can be recognized and used in the calculations.

4. Because ^{40}K comprises only about one atom in every 8,600 atoms of K, it is usual to measure the total K and assume that the ^{40}K/K ratio is constant in nature. For terrestrial materials, this assumption appears to be met (Garner et al, 1975) although Verbeek and Schreiner (1967) reported a variation of up to 3.5% in the ^{39}K/^{41}K ratio across a sharp intrusive contact between a granitic body and an amphibolite.

Apart from the requirement that K and Ar be measured precisely and accurately, the three most important aspects in K-Ar dating are: (1) the choice of appropriate material for dating; (2) the assessment as to whether basic assumptions have been fulfilled; and (3) a good understanding of the geology of the area. Careful petrographic examination of all samples prior to dating is desirable because it will assist in making a judgement as to whether an age measurement on a sample is likely to relate to the time of original crystallization or a subsequent event, such as metamorphism. Of equal importance is that microscopic study of a thin section enables the observer to decide which minerals can be separated for age measurement and whether the sample can be used for a whole-rock determination. There is a great deal of empirical evidence available about which minerals are those most useful for K-Ar dating. The list includes hornblende, biotite, muscovite, monoclinic potassium feldspar, and plagioclase. Whole-rock samples, particularly of fine-grained volcanic rocks, are commonly used for K-Ar dating; an age measurement on a whole rock naturally reflects the argon-retention characteristics of the component minerals. The most reliable ages are obtained on fresh, unaltered whole rocks that are well crystallized. Altered whole-rock samples and mineral separates commonly yield aberrant young ages because of loss of radiogenic argon by diffusion. Volcanic glass, or whole rocks containing volcanic glass, must be used with caution for K-Ar dating because glass readily loses radiogenic argon.

For a proper assessment of the K-Ar method, it is essential to measure a number of samples in known stratigraphic relation to each other. In the ideal situation, measurements will be made on several minerals and/or rock types, so that the relative consistency of the results will enable a judgement as to whether the assumptions have been met. For example, if K-Ar ages on biotite and sanidine phenocrysts in a volcanic rock agree with each other and with ages of whole-rock samples from the same volcanic sequence, we would be very confident that the ages measured have geologic significance. In an undisturbed sequence this would relate to the crystallization and cooling of the volcanic rocks. If K-Ar ages on biotite and hornblende from an intrusive rock disagree, and the biotite gives a significantly younger age than the hornblende, slow cooling may be suspected, or else the intrusion may have been metamorphosed mildly so that some radiogenic argon was lost from the biotite, but none from the more retentive hornblende. The interpretation of both concordant and discordant data must rely heavily on the known geology—this point cannot be overemphasized.

Another useful approach in assessing data is to plot results from samples regarded as cogenetic on isochron-type diagrams such as the K versus radiogenic Ar diagram (Funkhouser et al, 1966) or the $^{40}Ar/^{36}Ar$ versus $^{40}K/^{36}Ar$ diagram (McDougall et al, 1969). Ideally, the points will lie on a single line, the slope of which is proportional to age. Samples thought to be cogenetic, but showing discordant K-Ar ages when calculated using the usual assumptions, sometimes may form a linear array on an isochron diagram. A good example of this behavior was given by Hayatsu and Carmichael (1970) who demonstrated that a sequence of rocks in Newfoundland deposited in the Cambrian was reset in the Devonian, and that the argon then incorporated in the samples differed considerably from that of atmospheric argon.

$^{40}Ar/^{39}Ar$ Method

The $^{40}Ar/^{39}Ar$ method is a variant of the K-Ar method first suggested by Sigurgeirsson (1962) and independently by Merrihue (1965). The first results were given by Merrihue and Turner (1966) on meteorite samples, and the technique soon was extended to terrestrial rock samples (Mitchell, 1968; York and Berger, 1970; Dalrymple and Lanphere, 1971; McDougall and Roksandic, 1974).

The basis of the $^{40}Ar/^{39}Ar$ technique is the conversion of a proportion of the ^{39}K in a sample to ^{39}Ar by an n,p reaction involving fast neutrons. The sample is irradiated in a nuclear reactor together with a standard sample of known K-Ar age; the standard sample acts as a monitor of the fast neutron dose received by the unknown. Subsequent to the irradiation the argon is extracted from the samples and analysed isotopically, from which the ratio of the radiogenic ^{40}Ar, derived from the decay of ^{40}K through geologic time, can be determined relative to the ^{39}Ar generated from the ^{39}K during irradiation by fast neutrons. By comparison of the ratio determined for the standard and the unknown, an age is calculated for the unknown.

Total fusion $^{40}Ar/^{39}Ar$ ages are those determined by measuring the argon extracted by directly fusing the sample after the irradiation; such ages should be indistinguishable from conventional K-Ar ages. Advantages of this technique are that sample inhomogeneity is of no consequence because the age is determined on a single aliquot of sample, less material is required, and the determination of absolute abundances of K and Ar is not necessary.

The greatest advantage of the $^{40}Ar/^{39}Ar$ technique, however, is that, instead of melting the sample directly, the Ar can be released in stages by stepwise heating, and the gas evolved analyzed to provide a series of apparent ages on the one sample. This approach, which is coming into wider use, is known as the stepwise or incremental-heating method. The technique relies upon the diffusion characteristics of the radiogenic ^{40}Ar and the ^{39}Ar in the sample. In the ideal case, where the sample has remained a closed system since crystallization, the $^{40}Ar/^{39}Ar$ age calculated for each gas fraction released during stepwise heating will be identical (Fig. 1). If the sample has been disturbed subsequent to its crystallization, a more complex age spectrum is likely and in some circumstances can provide greater insight to the geologic history of the sample.

The concordant ages for the hornblende give an ideal age spectrum (Fig. 1), and Figure 2 shows these same data plotted on an isochron or correlation diagram, the properties of which have been discussed by Brereton (1972). The plateau age, obtained by averaging the calculated age from each step, and the age derived from the slope of the line in Figure 2 are virtually indistinguishable from one another and the conventional K-Ar age of 199.6 ± 2.1 m.y. (Roddick and Farrar, 1971). This hornblende is unlikely to have lost any of its radiogenic Ar subsequent to crystallization—a simple model of crystallization with closed system behavior is indicated. Such an interpretation could not have been made on the basis of an individual K-Ar or total fusion $^{40}Ar/^{39}Ar$ age. The intercept on the y axis in Figure 2 is 294.1 ± 3.1, indistinguishable from the atmospheric $^{40}Ar/^{36}Ar$ ratio, indicating that the contaminating nonradiogenic Ar, indeed, has atmospheric Ar composition.

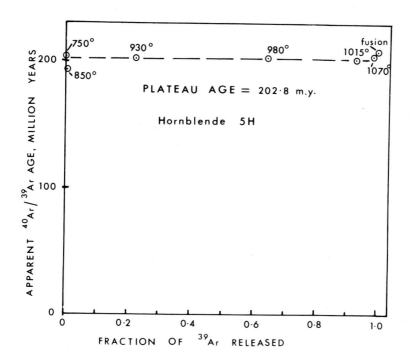

FIG. 1—Apparent $^{40}Ar/^{39}Ar$ age measured on gas fractions released at temperatures indicated are plotted against fraction of ^{39}Ar released. Data are for hornblende 5H.

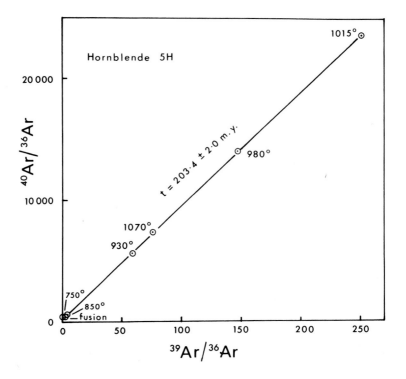

FIG. 2—Data derived from isotopic analysis of argon extracted at temperatures indicated for hornblende 5H plotted on isochron diagram. Slope of line, determined by a least-squares procedure, is proportional to age, and intercept on y axis is 294.1 ± 3.1, indistinguishable from atmospheric $^{40}Ar/^{36}Ar$ ratio.

In the case of a sample from which some loss of radiogenic ^{40}Ar has occurred, a lower apparent age is likely for the gas released at low temperatures, with the higher temperature fractions yielding a consistent age, which may correspond to the time of crystallization (Fig. 3). Such a pattern is expected because the radiogenic Ar that has diffused from the mineral will have originated from the least retentive sites, whereas distribution of ^{39}Ar, generated from ^{39}K, is expected to be uniform throughout the mineral. More complex age spectra are indicative of a more complex geologic history, but caution must be used in making detailed interpretations of the spectra.

The $^{40}Ar/^{39}Ar$ incremental-heating method has proved to be of considerable value in identifying undisturbed or minimally disturbed samples. The method has not been used extensively in physical time-scale studies, but is likely to come into wider use because of the additional information that it yields.

Decay Constants

In setting up a physical time scale, it is desirable to use the most precise and accurate decay-constant values available. The constants generally used in calculating K-Ar ages are those recommended by Aldrich and Wetherill (1958), although different decay constants have been used in eastern European countries. Beckinsale and Gale (1969) summarized the physical measurements of the specific beta and gamma activities of natural K from which the decay constants are derived. Recently Garner et al (1975) redetermined the percent atomic abundance of ^{40}K as 0.01167 ± 0.0004 and this figure, combined with specific activities recommended by Beckinsale and Gale (1969), yields the following values: $\lambda_\beta = 4.962 (\pm0.009) \times 10^{-10}y^{-1}$; $\lambda_e + \lambda_e' = 0.581 (\pm0.004) \times 10^{-10}y^{-1}$; and $\lambda = 5.543 (\pm0.010) \times 10^{-10}y^{-1}$. These may be regarded as

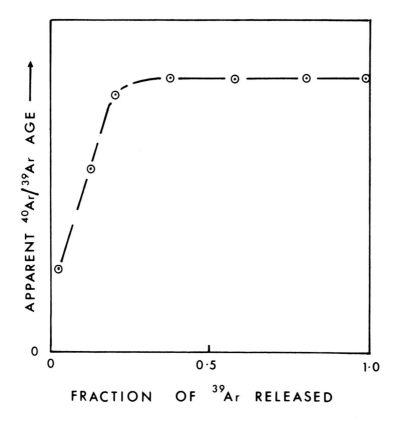

FIG. 3—Diagrammatic age spectrum illustrating loss of radiogenic Ar from least retentive sites in sample. Plateau age is age of crystallization of sample.

the most precise estimates for the decay constants of ^{40}K that are available. The calculation of Phanerozoic K-Ar ages using these revised constants gives values that are 1.6% greater at 600 m.y., and 2.6% greater at 1 m.y. when compared with those calculated using the Aldrich and Wetherill (1958) constants.

Conclusions

The K-Ar isotopic dating method is particularly useful for measuring the age of igneous rocks that have not been geologically disturbed since crystallization and cooling. The method can be applied to rocks ranging in age from Precambrian through Pleistocene. Because of its applicability to a wide range of time and mineral and rock types the K-Ar method has been used extensively in physical time-scale studies. Such studies are best made on suitable volcanic rocks interbedded with sediments containing a definitive fossil assemblage, so that the position in the relative geologic time scale is controlled.

Only the freshest mineral and whole-rock samples should be used for K-Ar dating; altered and poorly crystallized materials commonly yield relatively low ages because of loss of radiogenic argon. For each study, sufficient measurements should be made, preferably on a variety of materials, so that the assumptions underlying the method can be properly assessed. A K-Ar age measurement on an isolated sample may be more misleading than informative because it is not possible to test the basic assumptions on the basis of a single measurement. Where practicable it is valuable, of course, to make age measurements using more than one decay scheme. The overall assessment and interpretation of the geochronologic data, whether obtained for time-scale or other purposes, should be made with the fullest possible knowledge of the geology of the region.

References Cited

Aldrich, L. T., and G. W. Wetherill, 1958, Geochronology by radioactive decay: Ann. Rev. Nuclear Sci., v. 8, p. 257-298.

Beckinsale, R. D., and N. H. Gale, 1969, A reappraisal of the decay constants and branching ratio of ^{40}K: Earth and Planetary Sci. Letters, v. 6, p. 289-294.

Berggren, W. A., 1969, Cenozoic chronostratigraphy, planktonic foraminiferal zonation and the radiometric time scale: Nature, v. 224, p. 1072-1075.

Brereton, N. R., 1972, A reappraisal of the $^{40}Ar/^{39}Ar$ stepwise degassing technique: Royal Astron. Soc. Geophys. Jour., v. 27, p. 449-478.

Dalrymple, G. B., and M. A. Lanphere, 1969, Potassium-argon dating: San Francisco, W. H. Freeman and Co., 258 p.

—— —— 1971, $^{40}Ar/^{39}Ar$ technique of K-Ar dating; a comparison with the conventional technique: Earth and Planetary Sci. Letters, v. 12, p. 300-308.

Funkhouser, J. G., I. L. Barnes, and J. J. Naughton, 1966, Problems in the dating of volcanic rocks by the potassium-argon method: Volcanol. Bull., v. 29, p. 709-718.

Garner, E. L., et al, 1975, Absolute isotopic abundance ratios and the atomic weight of a reference sample of potassium: Nat. Bur. Standards (USA) Research Jour., v. 79A, p. 713-725.

Harland, W. B., A. Smith, and B. Wilcock, eds., 1964, The Phanerozoic time scale: Geol. Soc. London Quart. Jour., v. 120 supp., 458 p.

Hayatsu, A., and C. M. Carmichael, 1970, K-Ar isochron method and initial argon ratios: Earth and Planetary Sci. Letters, v. 8, p. 71-76.

Holmes, A., 1947, The construction of a geological time-scale: Glasgow Geol. Soc. Trans., v. 21, p. 117-152.

—— 1959, A revised geological time scale: Edinburgh Geol. Soc., Trans., v. 17, p. 183-216.

Kulp, J. L., 1961, Geologic time scale: Science, v. 133, p. 1105-1114.

Lanphere, M. A., and G. B. Dalrymple, 1967, K-Ar and Rb-Sr measurements on P-207, the USGS interlaboratory standard muscovite: Geochim. et Cosmochim. Acta, v. 31, p. 1091-1094.

McDougall, I., and Z. Roksandic, 1974, Total fusion $^{40}Ar/^{39}Ar$ ages using Hifar reactor: Geol. Soc. Australia Jour., v. 21, p. 81-89.

—— H. A. Polach, and J. J. Stipp, 1969, Excess radiogenic argon in young subaerial basalts from the Auckland volcanic field, New Zealand: Geochim. et Cosmochim. Acta, v. 33, p. 1485-1520.

Merrihue, C. M., 1965, Trace element determinations and potassium-argon dating by mass spectroscopy of neutron-irradiated samples (abs.): Am. Geophys. Trans. (EOS), v. 46, p. 125.

—— and G. Turner, 1966, Potassium-argon dating by activation with fast neutrons: Jour. Geophys. Research, v. 71, p. 2852-2857.

Mitchell, J. G., 1968, The argon-40/argon-39 method for potassium-argon age determination: Geochim. et Cosmochim. Acta, v. 32, p. 781-790.

Page, R. W., and I. McDougall, 1970, Potassium-argon dating of the Tertiary f_{1-2} stage in New Guinea and its bearing on the geological time-scale: Am. Jour. Sci., v. 269, p. 321-342.

Roddick, J. C., and E. Farrar, 1971, High initial argon ratios in hornblendes: Earth and Planetary Sci. Letters, v. 12, p. 208-214.

Sigurgeirsson, T., 1962, Age dating of young basalts with the potassium-argon method: Univ. Iceland, Physics Lab. Rept.

Verbeek, A. A., and G. D. L. Schreiner, 1967, Variations in ^{39}K:^{41}K ratio and movement of potassium in a granite-amphibolite contact region: Geochim. et Cosmochim. Acta, v. 31, p. 2125-2133.

York, D., and G. W. Berger, 1970, $^{40}Ar/^{39}Ar$ age determinations on nepheline and basic whole rocks: Earth and Planetary Sci. Letters, v. 1, p. 333-336.

—— and R. M. Farquhar, 1972, The earth's age and geochronology: Oxford, Pergamon Press, 178 p.

Results of Dating Cretaceous, Paleogene Sediments, Europe[1,3]

G. S. ODIN[2]

Abstract Since 1970, a new set of experiments and research on the dating of glauconites has been underway in different laboratories of Europe. The quality of the results seems to be due to a rigorous preselection of the outcrops, levels, and pellets before dating. This work, preliminary to an isotopic analysis, permits an easier interpretation of the apparent ages obtained.

From the glauconites chosen in Europe it is possible to test the local horizontal and vertical reproducibility and accuracy of the chronometer. After interlaboratory analysis and internal verification (Rb-Sr and K-Ar isochrons with 10 points or more) it appears that a carefully selected sediment sample gives an apparent age equivalent to other samples of the same stratigraphic level. The accuracy is lower with the Rb-Sr method.

Comparison of results obtained by both radiometric methods or with different chronometers (high temperature - low temperature) does not show an important systematic difference.

As far as it can be concluded with the present dates, it is clear that most boundaries proposed since 1964 must be changed for Cretaceous and Tertiary stages. There are three reasons: (1) we know that some of the oldest results and reasoning taken into account were not adequate; (2) we have more results on better known samples; and (3) the decay constants are nearer their probable value today.

Apparent ages proposed are essentially obtained from glauconites and must be completed and compared with more apparent age data on known high-temperature minerals.

For more than a century, the stratigraphic position of European basin sediments has been studied. These investigations have led to the construction of the stratigraphic column. A maximum number of isotopic dates on these stratigraphically well-determined horizons is needed to correlate the stratigraphic column with a numeric time scale.

Glauconite as a geochronometer is available in many sediments. Specific precautions have been determined and are observed. For a few years, different laboratories have undertaken preliminary isotopic studies for a better understanding of glauconite.

Introduction

This paper presents the numerous results obtained during 5 years in different laboratories. The principal part of the isotopic measurements presented was done at

[1] Manuscript received, January 6, 1977.

[2] Lab de Géochimie, Université P. & M. Curie, Paris, France.

[3] This work was possible due to the generous help of the team of Geochronology of Bern, especially J. C. Hunziker who lent me his apparatus for one or two months each year and E. Jager, the leader who encouraged, helped, and facilitated the work.

Numerous field workers helped me with the samples: D. Curry, M. Gulinck and P. Juignet primarily but also, J. Alvinerie, G. Bignot, A. Blondeau, P. Cotillon, ᴴ. Gamble, C. Lorenz and B. Pomerol.

Some data have been obtained at Orsay (with H. Bellon) and Strasbourg (with M. Bonhomme and N. Clauer); at Brussels, many data were obtained by the Rb-Sr method with E. Keppens and P. Pasteels. Discussions with scientists at Hannover (H. Kreuzer) and Amsterdam were highly appreciated. Since 1974, the International Geological Correlations Program Project 133 permits yearly exchanges of ideas, data, and samples, and a coordination of the research highly useful for all. Since July 1976, the author has been the leader of this project.

Article Identification Number: 0149-1377/78/SG06-0010/$03.00/0

the University of Bern. But a few other laboratories in Europe are working on the same subject and the comparison and exchange of all results and some samples are important factors for success of this common research. One cannot report in detail all the preliminary experimentations and stratigraphic or sedimentologic studies.

One may find here the essential conclusions and a corresponding bibliography on three kinds of work: (1) the principles of choice of the best glauconites are given; (2) the reproducibility, "exactness," and accuracy are discussed, utilizing recent works; and (3) a study of the boundaries of some stages is done as a contribution to a more reliable numeric time scale.

Preliminary Choice of the Geochronometer

Polevaya et al (1961) insisted on the necessary purification of green grains before using for dating purposes, while Gygi and McDowell (1970) gave a method of purification, and Obradovich (1964) gave isotopic results on the ultrasonic agitation in a water medium. More recently, Thompson and Hower (1973), Gramann et al (1975), and Pasteels et al (1976) proposed new data on the preliminary treatments, independent of the writer's works summarized in Odin (1975). From the preceding works and more general considerations on sediment dating (Hurley et al, 1963; Bonhomme et al, 1965; Hurley, 1966; Odin, 1973a), it is possible to propose a specific scheme of choice for the use of the best glauconites.

Isotopic measurements must be done at the level, sample, and fraction of green grains richest in potassium. In most cases a glauconite with less than 5.5 to 6% of the element potassium is not suitable. This is our *a priori* criteria to distinguish bad glauconites from those that may provide reliable age data. It is evident at first that great care must be used in the field. The method of glauconite fractionation is less known (Fig. 1). This *fractionation* uses essentially the grain size (Aubry and Odin,

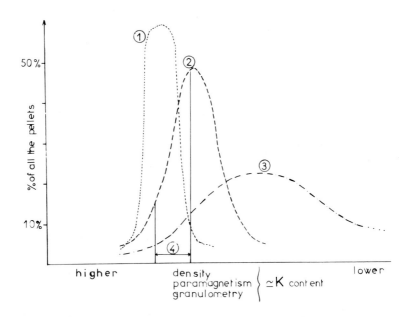

FIG. 1—Method of choice of the "best green grains" The aim of fractionation is to select the potassium-richest grains (more evolved). The bigger, heavier, more attractable grains consist of a better micaceous mineral. But grains that are too large, too heavy or too attractable are not pure. More evolved glauconite (1) is more homogeneous and the curve is to the left side. A less evolved glauconite (3) is not useful for dating purpose. In a middle evolved glauconite (2) I do not choose the middle part of the curve (more abundant), but only the left (4).

1973; Odin et al, 1975), paramagnetism, and density (Shutov et al, 1970; Priem et al, 1975). A rapid method of verification of the suitability of glauconites was given in Odin and Hunziker (1974) and Odin (1975, p. 49) using the fact that powder X-ray diffractograms give a good appreciation of the mineralogic composition of the grains. Grains must be monomineral (their componants free from oxides, hydroxides, or phyllites which are not of the 10 Å family) and the phyllites must have less than 10 to 15% of expandable layers.

If cleaning is necessary, ultrasonic treatment is generally used with different chemical cleaners—acidic (Pasteels et al, 1976) or basic (Kreuzer et al, 1973). The result is a potassium enrichment and a loss of common strontium, both things being useful, but one must find the boundary between the cleaning and the weathering of the samples.

The necessity of abundant potassium in grains is well understood. We know that this element is the key which closed the chronometer, sticking the layers of the micaceous mineral to each other. Obradovich and Cobban (1975) gave a good example of the influence of the potassium content on the apparent ages of biotites; the question is the same here. Another problem with low-potassium glauconites is that the pellets may keep inherited argon. This was shown clearly in the recent glauconites which lie in the Guinean Gulf (Giresse and Odin, 1973; Fig. 2). Apparent ages may be higher but they may also be lower because ionic exchange between the grains and the environment is easier. This is shown by the cationic exchange capacity which is low when the potassium is abundant and high when the potassium content is low (Manghnani and Hower, 1964; McRae and Lambert, 1968; Cimbalnikova, 1971). However the so-called systematic slow enrichment in potassium during time as proposed by Hurley et al (1960), Hower (1961), and Hurley (1966) is not proved. Obradovich (1964) showed on the contrary that this hypothesis is not good. Evernden et al (1961) and many

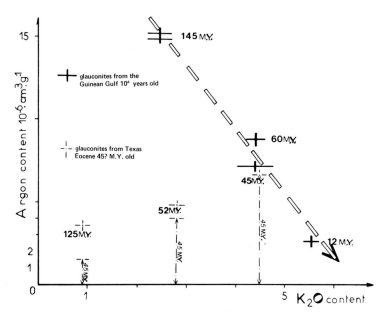

FIG. 2—Apparent ages of low-potassium glauconites. Few Guinean Gulf grains give very high ages (original data after Giresse, unpublished). With potassium enrichment, initial time "zero" is lowered. These glauconites are still "living." They are opened but there is not yet equilibrium with atmospheric argon. We find other cases in the literature. Ghosh (1972) gave a good example of Texas Eocene glauconites. Owens and Sohl (1973) also showed ages too high for low-potassium glauconites and it is possible that the Cenomanian glauconite of Van Hinte et al (1975) is another example of a K-Ar age too high for the same reason.

workers after them proposed a systematic loss of argon during time. It has not been proved that this phenomenon exists for potassium-rich glauconites and that the effects are important in our sedimentation basins where sediments are not deeply buried. Argon is well preserved in evolved glauconites.

Experiences on the reference material, glauconite GL-0, have shown that even at $180°$ C under high vacuum, one cannot see the loss of radiogenic argon. In selected cases of quiet basins older than Cambrian, such as the Russian and Australian shields, Soviet authors and Webb et al (1963) have probably found correct ages on glauconites. The case is different for deeply buried glauconites with low-potassium content (Evernden et al, 1961).

Glauconite Dating, Reliability

Presentation

Localization of samples is given in Figure 3. Most of the samples are from fresh outcrops because it was difficult to have a large quantity of sample in a core. In many cases, biostratigraphic knowledge is from the stratigrapher who collected the sample. The apparent ages commonly may be compared with older data compiled in the Phanerozoic time scale (Harland et al, 1964) or more recent data.

We must take into account the last data given by (Baadsgaard et al, 1964). There are examples of systematic erroneous results in the literature. One may observe that the first work of Evernden et al (1961) assigned older ages than Obradovich (1964) on the same samples. Personal communication with Curtis indicated that there were problems with potassium measurements in the first work. Other works on European glauconites also gave systematic ages that are too old (Odin, 1973b). The problem was in the argon measurements and all these ages must be abandoned. After this bad experience I did the argon measurements with systematic controls, by reference materials, or by working in different laboratories.

As the more recent numeric time scales for the Paleogene are based on these results (Berggren, 1972), one can say that modifications are still necessary.

We must compare data obtained by K-Ar and Rb-Sr methods. During the International Meeting in Paris, 1974, new self-consistent decay constants were proposed

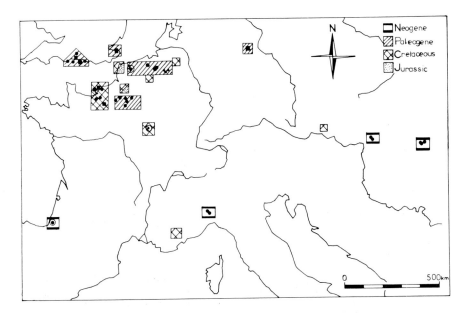

FIG. 3–Location of European samples. Details are published for the Neogene in Odin et al (1975) and for the Paleogene in Odin et al (in press).

(Armstrong, 1974; McDougall, 1974). All results are calculated, or recalculated, with the following physical constants:

$$\lambda^{40}K \text{ total} = 0.575 \cdot 10^{-10} \cdot y^{-1};$$
$$^{40}K/K \text{ total} = 1.18 \cdot 10^{-4} \text{ (atomic)}^4;$$
$$\lambda^{87}Rb = 1.42 \cdot 10^{-11} \cdot y^{-1}.$$

Reproducibility of Glauconites

The best ways to test how glauconites are satisfying reproducibility during time-scale studies are either to date a single glauconite horizon at several localities, or to date numerous glauconitic beds in a well-defined stratigraphic sequence (modified after Obradovich and Cobban, 1975). I present here the two approaches that I have followed.

Two consecutive levels are very rich in green grains in the Paris basin at the top of Albian and at the base of Cenomanian (Juignet et al, 1975). An isochrone can be drawn with the recent Rb-Sr results of Hunziker (unpublished) on seven Cenomanian glauconites, two argillaceous fractions (smectites), and one carbonate (Fig. 4). A representation of all K-Ar data is also given graphically—one line for the base of the Cenomanian, another for the upper Albian (Fig. 5). This is a good test of reproducibility. Many other examples can be found comparing the apparent ages obtained by the same or different authors on different potassium-rich glauconites of the same level (Odin, 1975)—Lutetian, upper Portlandian, etc.

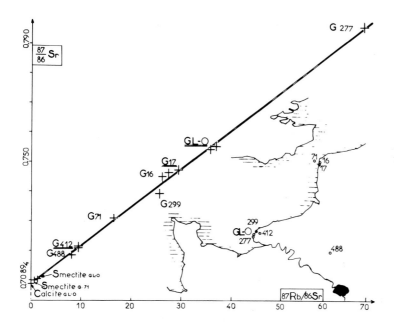

FIG. 4—Rb-Sr isochrone line for eight glauconites from base of Cenomanian (after work by Hunziker); location of the samples is given. Most are taken on cliffs and almost all points are on a line. Initial isotopic ratio calculated is the same as that in the calcite from GL-O. Three samples have been measured twice (underlined). Data obtained on the argillaceous fractions are on the isochrone line. This is not always the case for inherited clay minerals (Bonhomme et al, 1970).

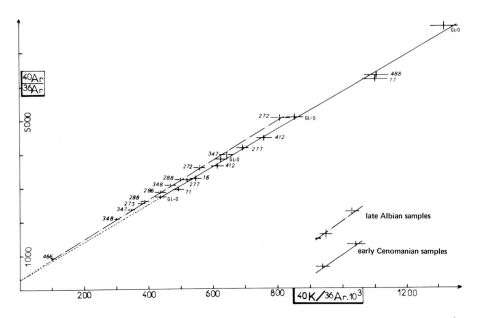

FIG. 5—K-Ar results on late Albian and early Cenomanian samples. Uncertainties are given for each sample (a few have been measured twice for argon). Reproducibility of the chronometer is well illustrated by both glauconitic levels. "Initial argon ratio" is normal (the same value as the atmospheric one). All samples come from the Paris basin (Fig. 4). A measurement on GL-O smectite gave an "age" of 145 ± 9 m.y. which is not on the line.

The second approach is well illustrated by the results obtained on Eocene sediments of Great Britain, Belgium, and France (Odin et al, in press). Figure 6 shows an abstract of these results.

The numerous apparent ages obtained on the boundaries of the Lutetian stage (a rare case) permit us to define their age with accuracy (Fig. 7). One cannot say that the data are completely accurate because the genesis of the chronometer is complex and requires a long time (Odin, 1975). Owens and Sohl (1973) gave us another example of regional *reproducibility* on Maryland glauconites.

Progress in dating glauconites also can be seen by comparing the diagram of data obtained in 1964 along the admitted scale (Obradovich, 1964; Hurley, 1966) and the present results for the Paleogene (Fig. 8). The difference is due to the preliminary choice of the samples and green grains.

Exactness and Reliability

There are two ways to prove these qualities: (1) to compare apparent ages obtained by two different methods of isotopic dating or (2) with two different chronometers of equivalent horizons.

Geochemical behaviors of potassium, rubidium, and strontium ions and argon atoms are very different in micaceous minerals due to their respective properties (Hower, 1961; Kulp and Engels, 1963). A possible alteration will have different consequences on the apparent ages (Odin et al, 1974a). Comparable ages by both methods will prove an absence of exchange between the chronometer and its environment and the probable exactness of the apparent ages. The first comparisons have been made by Hurley et al (1960). With the decay constants proposed by Armstrong (1974)[4], data from basal Cenomanian give us useful examples.

---

[4] The new dates concerning the K-Ar method proposed at Sydney during the meeting: $\lambda^{40}K = 0.5811 \cdot 10^{-10} \cdot y^{-1}$ and $^{40}K/K = 1.167 \cdot 10^{-4}$ (atomic)—give the same results in our field of work.

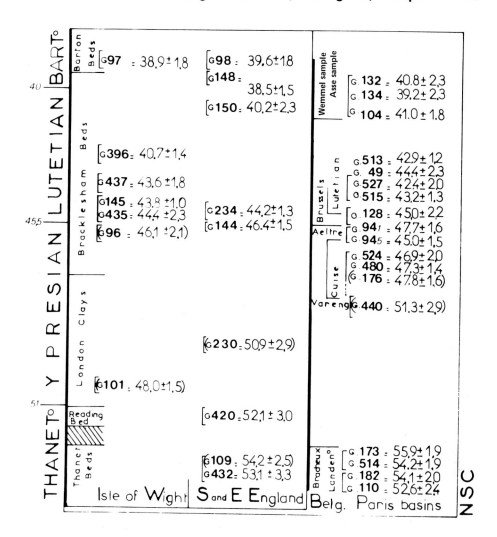

FIG. 6—Vertical reproducibility of glauconites. Apparent ages given in megayears with analytical uncertainties. Succession of samples is same in the field and in the figure. Parallelism between basins is well known however equivalence between different Paleocene facies is questionable. Dates in parentheses concern glauconites with less than 5.7% K -6.9% K_2)- (but more than 5.25).

We find 94.0 ± 1.6 m.y. for the K-Ar isochrone age and 90.3 ± 1.2 for the Rb-Sr isochrone age ($\lambda^{87}Rb = 1.42 \cdot 10^{-11} \cdot y^{-1}$). The Rb-Sr age is younger (approximately 4%). Other examples were published by Odin et al (1974b) for Oligocene-Miocene glauconites. A few glauconites of the same level show equivalent apparent ages by both methods. In another formation, Rb-Sr apparent ages are clearly older by 10 to 20%, but the geochemical history of these glauconitic sands is not yet known.

A third example was reported by Elewaut et al (1976) and a study is underway at Brussels with Keppens. Seven samples from the Bartonian and Lutetian Stages of England gave equivalent apparent ages of ± 3 to 4% by both methods, whereas seven samples from the Bande Noire of Belgium gave equivalent ages after the necessary selection.

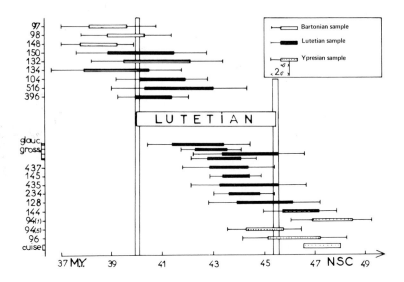

FIG. 7–Boundaries of Lutetian (example of accuracy)–45.5 ± 0.5; 40.0 ± 0.5 are after glauconites of European sedimentary basins. The σ and 2σ uncertainties are given for each sample. Some samples are more or less at boundary of two stages and are shown in black and white. Both boundaries can be estimated at plus or minus half a megayear. This possibility is not frequent with glauconites.

Taking into account the uncertainties of the measurements and the quality of the glauconites the apparent ages are commonly (but not always) equivalent. The differences are sometimes systematic in a horizon. They are moderate in the glauconites of Europe. The older ages are not always from Rb-Sr data; this proves that a systematic loss of argon by diffusion is not the rule. The question is different in other basins. In the very old formations of Australia (Precambrian), McDougall et al (1965) showed us an example where temperatures higher than in our basins are frequent. However with the new standard constants for recalculations, a third of the glauconites show the same apparent age with both methods. In the other cases the K-Ar apparent age is lower (5 to 11%). There is only one glauconite with more than 5.8% potassium and this glauconite gives the same age by both methods.

The second way to test the reliability is the comparison between different geochronometers. For the lower Maastrichtian and Cenomanian, there is agreement between data from Obradovich and Cobban (1975) on bentonite minerals of North America and those from Priem et al (1975) and ours on glauconites. For the Oligocene-Miocene boundary the apparent ages from the work of Kreuzer et al (1973) on German glauconites agreed well with those of Turner (1970) on volcanic rocks, with data obtained in the Bergell Massif and with the compilation by Afanas'yev (1970), and Afanas'yev and Zykov (1975). There commonly are problems of correlation in these comparisons.

As a general rule one can say that at this time the so-called too low ages of glauconites are specific cases. What is proved is that glauconites can be used to give a reproducible age in the sediments.

It is known today that the apparent age of a potassium-rich glauconite is given with an uncertainty of at least 10^6 years due to the complex and long genesis of the chronometer. The initial time of accumulation of radiogenic argon is probably late, just before the deposit of the overlying sediments, in most cases. It is true that questionable ages are obtained in particular studies. This is not a reason to take into account *a priori* only the older apparent ages of the glauconites from the literature. The result is that the proposed time scale is "too old" for numerous boundaries determined since 1964.

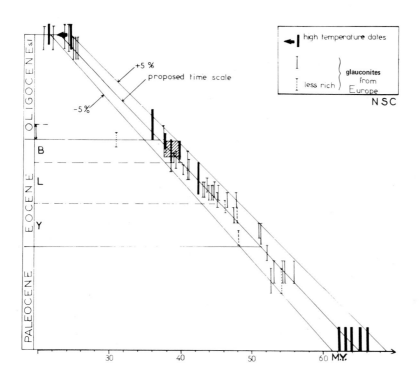

FIG. 8—Dates concerning the Paleogene. Most of data are in the ± 5% interval. This is a better result compared with 1964 Phanerozoic time scale on the same chronometer. Available ages on well-correlated high temperature minerals show good agreement. Stratigraphic situation of each sample is shown by a bar. Eocene-Oligocene boundary is not the same for all authors. Both possible positions are given. Apparent ages on glauconites are those given by Kreuzer et al (1973), Gramann et al (1975), Odin et al (1975), and Odin et al (in press). High temperature dates from Evernden et al (1964), Folinsbee et al (1966), Page and McDougall (1970), Afanas'yev (1970), Ghosh (1972), and Obradovich and Cobban (1975).

The Numerical Time Scale

Comparing our data on Cretaceous strata of Europe with those of the literature (Fig. 9), we propose a numerical time scale in which all considerations on the mean duration of stages or biozones or sedimentation rates are excluded.

Figure 10, which gives some extrapolations, shows that Turonian, Coniacian, and Santonian are three very short stages. We must add that Obradovich (personal commun., 1974) thinks that the base of the Albian is at least at -112 m.y. The Early Cretaceous is less known but the presented age data (Fig. 9B) permit useful conclusions. The stages are also of varied duration. It seems best not to give data for boundaries without new measurements.

So, we can see from Figure 10 that there is still much work to do to complete and verify the proposed ages.

For comparison, other scales are given in Figure 10. The scales of Gill and Cobban (1966) and Kauffman (1970) gave the first modern view of the scale of the Late Cretaceous stages because they abandoned the idea of equal stages.

A scale for the Paleogene (after apparent ages given essentially by glauconites) is illustrated in Figure 11.

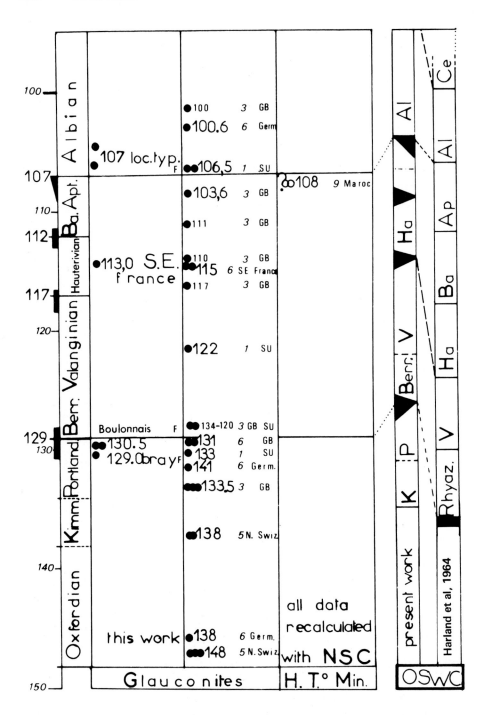

FIG. 9–Data available for Cretaceous (on glauconites of Europe compared with some recent apparent ages of high temperature minerals on right side) in megayears. Samples from Austria (Aus), Belgium (Bel), France (F), Great Britain (G.B.), Germany (Germ), Netherlands Limburg (Lim), Soviet Union (S.U.), Northern Switzerland (N. Swiz) are included. Published data are taken from works of Afanas'yev (1970) (1), Dodson et al (1964) (3), Folinsbee et al (1963; 1966) (4), Gygi and McDowell (1970) (5), Obradovich (1964) (6), Obradovich and Cobban (1975) (7), Priem

et al (1975) (8) or unpublished Bonhomme (2), Thuizat (9) and this work partly after Odin (1975). Uncertainties are generally near ± 4% and diminish when measurements are done on two or more samples (one circle by sample). When glauconite is potassium poor, apparent age is given in small type. Apparent ages are given with new standardized constants (NSC), and with old standardized western constants (OSWC) used in Phanerozoic time scale 1964 for an actual comparison in the columns.

    A. In the Lower Cretaceous a systematic difference is observed for all boundaries. This is due to another evaluation of Lower Cretaceous boundaries. In the Phanerozoic time scale (P.T.S.) they are both clearly too old. Presented data show that Portlandian-Perriasian boundary is near 130 m.y. not more after glauconites which seem as reliable as others.

    B. These Upper Cretaceous data show an age of 95 ± 1 m.y. for the base of Cenomanian and 73-72 for base of Maastrichtian. If base of Campanian is at 84 m.y. in Europe as in North America, Turonian, Coniacian, and Santonian are very short stages. That conclusion cannot be obtained considering the number of biozones. There are very young ages obtained from potassium-rich glauconites from Germany. This is probably due to samples from bore holes for various reasons. In this figure as in the preceding one we use only the data given by the richest glauconites.

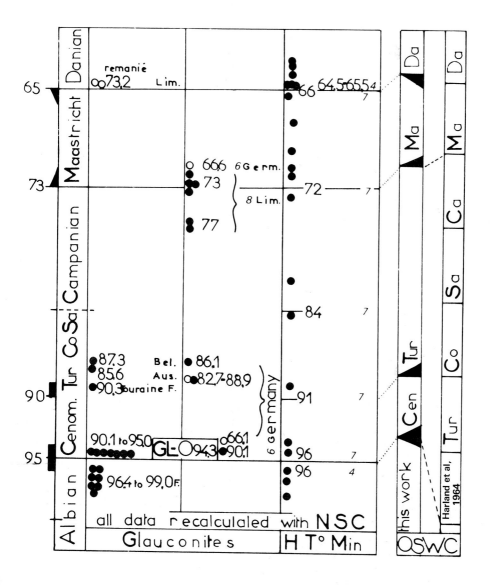

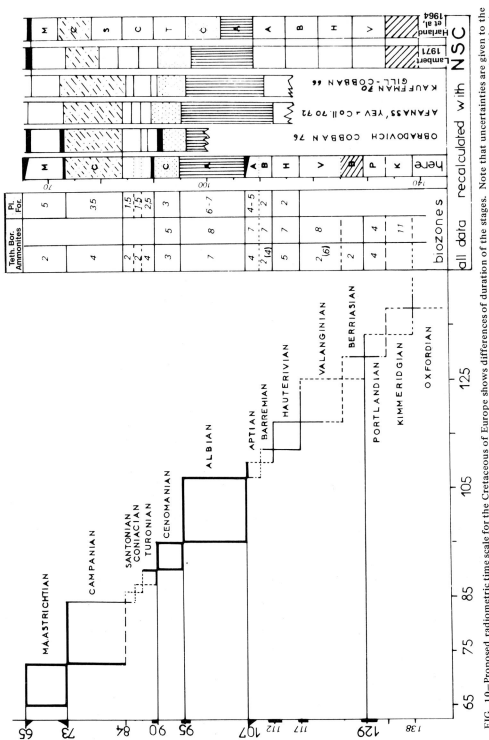

FIG. 10—Proposed radiometric time scale for the Cretaceous of Europe shows differences of duration of the stages. Note that uncertainties are given to the left side. This modern view of the time scale is compared with more classical views after Harland et al (1964) and Lambert (1971) drawn on the right side. An evolution is clear also for the Albo-Cenomanian boundary—younger than in 1964. I think we must accept the same evaluation in knowledge of the Jurassic-Cretaceous boundary. Note also that there is no constant ratio between duration and number of biozones for each stage (Tethyan and Boreal Ammonite zones and Planktonic Foraminifera estimated after a compilation by Van Hinte (1976). Biozone duration seems a bad criteria in this case to predict accurately age of a boundary. Concerning Cretaceous-Tertiary boundary, data by Folinsbee and colleagues gave ages between 63 and 64 m.y. (OSWC), the mean of which gave 65 m.y. (NSC). Some authors prefer on older age. Then we can propose an age of 65-66 m.y. (NSC).

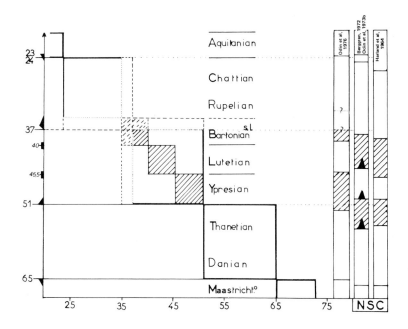

FIG. 11—Proposed radiometric time scale for the Paleogene in Europe. Proposed ages given in megayears and uncertainties drawn on the left side. We have no data concerning the Rupelian-Chattian and Danian-Thanetian boundaries in Europe. After data on Silberberg and underlying beds (Gramann et al, 1975), base of the Rupelian stage is younger than 37 m.y. probably between 34.5 and 36.5 m.y. Taking into account duration of biozones which seem much more regular in Paleogene than in Cretaceous formations, we obtain a maximum age of 33 m.y. instead of 37 (after an idea of Curry in Odin et al, in press). More data are hoped and I still prefer radiometric age data today. Depending on the chosen data, other boundaries are at 23 or 24 m.y., 51 or less for a short lower Eocene, 52 or more if we prefer to include Woolwich beds in the Ypresian, and 65 or 66 m.y. as it is discussed in Figure 9B. Boundaries of the Lutetian are given according to Figure 7 assuming that this stage includes the biozones N.P. 14 and N.P. 15 only. Base of the Barton beds is younger. Other numeric scales are given on the right side after Harland et al (1964), Berggren (1972), and Odin (1973b) which gave maximum ages only and which must be abandoned now for analytic purposes considering age data of Odin et al (in press).

Conclusion

The recently obtained data on sedimentology and isotopic geochemistry of glauconites permit a better use of this chronometer. If some problems can be observed in the interpretation of the results, for all the chronometer, it remains possible to date with reliability a sediment after a specific study of the geochemical environment. The numerous new dates summarized here lead to a better knowledge of the time scale which must be improved considering that some of the dates from the literature are overemphasized.

References Cited

Afanas'yev, G. D., 1970, Certain key data for the Phanerozoic time-scale: Eclogae Geol. Helvetiae, v. 63, no. 1, p. 1-7.
—— and J. I. Zykov, 1975, Echelle geochronologique du Phanerozoique: Moscow, Nauka, p. 1-100.
Armstrong, R. L., 1974, Proposal for simultaneous adoption of new U, Th, Rb, and K decay constants for calculation of radiometric dates: Int. Mtg. Geochronology Cosmochronology and Isotope Geology (Paris), Abst.

Aubry, M. P., and G. S. Odin, 1973, Sur la nature mineralogique du verdissement des craies: formation d'une phyllite apparentee aux glauconies en milieu semi-confine poreaux: Soc. Geol. Normandie Bull., v. 61, p. 11-22.

Baadsgaard, H., et al, 1964, Limitations of radiometric dating, *in* Geochronology in Canada: Royal Soc. Canada, Spec. Pub. 8, p. 20-38.

Berggren, W. A., 1969, Cenozoic chronostratigraphy, planktonic foraminiferal zonation and the radiometric time scale: Nature, v. 224, no. 5224, p. 1072-1075.

—— 1972, A Cenozoic time-scale; Some implications for regional geology and paleobiogeography: Lethaia, v. 5, p. 195-215.

Bonhomme, M., N. Clauer, and G. S. Odin, 1970, Resultats preliminaires de datations rubidium-strontium sur des sediments glauconieux dans le Paleogene d'Angleterre: Alsace-Lorraine Serv. Carte Geol. Bull., v. 23, no. 3-4, p. 209-213.

—— J. Lucas, and G. Millot, 1965, Signification des determinations isotopiques dans la geochronologie des sediments: Sci. Terre, v. 10 no. 2-4, p. 539-565.

Cimbalnikova, A., 1971, Cation exchange capacity of glauconites: Cas. Mineral. Geol., v. 16, no. 1, p. 15-21.

Dodson, M. H., et al, 1964, Glauconite dates from the Upper Jurassic and Lower Cretaceous: Geol. Soc. London Quart. Jour., v. 120, p. 145-158.

Elewaut, E., et al, 1976, K-Ar and Rb-Sr radiometric ages on glauconites from calcareous nanno-plancton zones NP 14 to NP 21 in North Sea Basins: 4th E.C.O.G. (Amsterdam) Abst., p. 31.

Evernden, J. F., et al, 1961, On the evaluation of glauconite and illite for dating sedimentary rocks by the K-Ar method: Geochim. Cosmochim. Acta, v. 23, p. 78-99.

—— et al, 1964, Potassium-argon dates and the Cenozoic mammalian chronology of North America: Am. Jour. Sci., v. 262, p. 145-198.

Folinsbee, R. E., H. Baadsgaard, and G. L. Cumming, 1963, Dating of volcanic ash beds (bentonites) by the K-Ar method: Nuclear Geophysics, Nuclear Sci. Ser. Rept., v. 38, p. 70-82.

—— —— —— 1970, Geochronology of the Cretaceous-Tertiary boundary of the Western plains of North America: Eclogae Geol. Helvetiae, v. 63, no. 1, p. 91.

—— et al, 1965, Late Cretaceous radiometric dates for the Cypress hills of western Canada, *in* Cyprus Hills plateau, Alberta and Saskatchewan: Alberta Soc. Petrol. Geol., 15th Ann. Field Conf. Guidebook, p. 162-174.

Ghosh, P. K., 1972, Use of bentonites and glauconites in potassium 40/argon 40 dating in Gulf Coast stratigraphy: PhD Thesis, Rice Univ., 136 p.

Gill, J. R., and W. A. Cobban, 1966, The red bird section of the Upper Cretaceous Pierre Shale: U.S. Geol. Survey Prof. Paper, 398A, 73 p.

Giresse, P., and G. S. Odin, 1973, Nature mineralogique et origine des glauconies du plateau continental du Gabon et du Congo: Sedimentology, v. 20, no. 4, p. 457-488.

Gramann, F., et al, 1975, K-Ar ages of Eocene to Oligocene glauconitic sands from Helmstedt and Lehrte (N. W. Germany): Newsletter Stratigraphy, v. 4, p. 71-86.

Gygi, R. A., and F. W. McDowell, 1970, Potassium-argon ages of glauconites from a biochronologically dated Upper Jurassic sequence of Northern Switzerland: Ecologae Geol. Helvetiae, v. 63, no. 1, p. 111-118.

Harland, W. B., A. G. Smith, and B. Wilcock, 1964, The Phanerozoic time scale—A symposium: Geol. Soc. London, Quart. Jour., v. 120, supp., 458 p.

Hower, J., 1961, Some factors concerning the nature and origin of glauconite: Am. Mineralogist, v. 46, p. 313-334.

Hurley, P. M., 1966, K-Ar dating of sediments, *in* O. A. Schaeffer and J. Zahringer, compilers, Potassium argon dating: New York, Springer-Verlag, p. 134-151.

—— et al, 1960, Reliability of glauconite for age measurement by K-Ar and Rb-Sr methods: AAPG Bull., v. 44, no. 11, p. 1793-1808.

—— et al, 1963, K-Ar age values in pelagic sediments of the North Atlantic: Geochim. Cosmochim. Acta, v. 27, no. 4, p. 393-399.

Juignet, P., J. C. Hunziker, and G. S. Odin, 1975, Datation numerique du passage Albien-Cenomanien en Normandie; etude preliminaire par la methode a l'argon: Acad. Sci. (Paris) Comptes Rendus, v. 280, p. 379-382.

Kauffman, E. G., 1970, Population systematics, radiometrics and zonation. A new biostratigraphy *in* Correlation by fossils: North Am. Paleont. Conf. Proc., p. 612-666.

Kreuzer, H., et al, 1973, K-Ar dates of some glauconites of the Northwest German Tertiary basin: Fortschr. Mineral., v. 50, no. 3, p. 94-95.

Kulp, J. L., and S. Engels, 1963, Discordances in K-Ar and Rb-Sr isotopic ages in radio-active dating: Int. Atom. Energy Agency (Athens), Symposium, p. 919.

Lambert, R. St. J., 1971, The Pre-Pleistocene Phanerozoic time-scale *in* the Phanerozoic time-scale, a supplement, Geol. Soc. London, Spec. Pub. 5, p. 33-34.

McDougall, I., 1974, The present status of the decay constants, *in* Report of activities of the

I.U.G.S. subcommission of Geochronology: Int. Mtg. Geochronology, Cosmochronology, Isotope Geology (Paris) Proc., p. 1-5.

——— et al, 1965, Isotopic age determinations on Precambrian rocks of the Carpentoria region (Australia): Geol. Soc. Australia Jour., v. 12, p. 67-90.

McRae, S. G., and J. L. M. Lambert, 1968, A study of some glauconites from Cretaceous and Tertiary formations in southeast England: Clay Miner., v. 7, no. 4, p. 431-440.

Manghnani, M. H., and J. Hower, 1964, Glauconite cation exchange capacities and infrared spectra: Am. Mineralogist, v. 49, p. 586-598.

Obradovich, J., 1964, Problems in the use of glauconite and related minerals for radioactivity dating: PhD Thesis, Univ. California, Berkeley, 160 p.

——— and W. A. Cobban, 1975, A time scale for the late Cretaceous of the western interior of North America: Geol. Assoc. Canada, Spec. Paper, 13, p. 31-54.

Odin, G. S., 1973a, Specificite des datations radiometriques dans les bassins sedimentaires europeens au Mesozoique et au Cenozoique: Rev. Geographie Phys. Geologie Dynam., v. 15, no. 5, p. 523-532.

——— 1973b, Resultats de datations radiometriques dans les series sedimentaires du Tertiaire de l'Europe occidentale: Rev. Geographie Phys. Geologie Dynam., v. 15, no. 3, p. 317-330.

——— 1975, De glauconiarum, constitutione, origine, aetateque: These Doct. Etat, Paris, 280 p.

——— and J. C. Hunziker, 1974, Etude isotopique de l'alteration naturelle d'une formation a glauconie (methode a l'argon). Contrib. Mineral. Petrol., v. 48, p. 9-22.

——— D. Curry, and J. C. Hunziker, (in press), Radiometric dating by glauconites from N. W. Europe and the time-scale of the Palaeogene: Geol. Soc. London Quart. Jour.

——— J. C. Hunziker, and C. R. Lorenz, 1975, Age radiometrique du Miocene inferieur en Europe occidenale et centrale: Geol. Rundsch., v. 64, no. 2, p. 570-592.

——— et al, 1974a, Influence of natural and artificial weathering on glauconite K-Ar and Rb-Sr apparent ages: Int. Mtg. Geochronology, Cosmochronology, Isotope Geology (Paris) Proc.

——— et al, 1974b, Analyse radiometrique de glauconies par les methodes au strontium et a l'argon; l'Oligo-Miocene de Belgique: Soc. Belge Geol. Bull., v. 83, p. 35-48.

Owens, J. P., and N. F. Sohl, 1973, Glauconites from New Jersey-Maryland Coastal plain, their K-Ar ages and application in stratigraphic studies: Geol. Soc. America Bull., v. 84, p. 2811-2838.

Page, R. W., and I. McDougall, 1970, Potassium-argon dating of the Tertiary f. 1-2 stage in New Guinea and its bearing on the geological time-scale:Am. Jour. Sci., v. 269, p. 321-342.

Pasteels, P., P. Laga, and E. Keppens, (1976),   Essai d'application de la methode radiometrique au strontium aux glauconies du Neogene; le probleme du traitement de l'echantillon avant l'analyse: Acad. Sci. (Paris) Comptes Rendus. , v. 282, p. 2029-2032.

Polevaya, N. I., G. A. Murina, and G. A. Kazakov, 1961, Glauconite in absolute dating: New York Acad. Sci., Annals, v. 91, no. 2, p. 298-310.

Priem, H. N. A., et al, 1975, Isotopic dating of glauconites from the Upper Cretaceous in Netherlands and Belgian Limburg: Geologie Mijnbouw, v. 54, no. 3-4, p. 205-207.

Shutov, V. D., et al, 1970, Crystallochemical heterogeneity of glauconite as depending on the conditions of its formation and post-sedimentary change: Int. Clay Conf. (Madrid) Proc. I, p. 327-339.

Thompson, G. R., and J. Hower, 1973, An explanation for low radiometric ages from glauconite: Geochim. Cosmochim. Acta, v. 37, p. 1473-1491.

Turner, D. L., 1970, Potassium-argon dating of Pacific Coast Miocene Foraminiferal stages: Geol. Soc. America Bull., Spec. Paper, 124, p. 91-129.

Van Hinte, J. E., 1976, A cretaceous time scale: AAPG Bull., v. 60, p. 498-516.

——— J. A. S. Adams, and D. Perry, 1975, K-Ar of Lower-Upper Cretaceous boundary at Orphan Knoll (Labrador Ocean): Canadian Jour. Earth Sci., v. 12, p. 1484-1491.

Webb, A. W., I. McDougall, and J. A. Cooper, 1963, Retention of radiogenic argon in glauconites from Proterozoic sediments, Northern Territory Australia: Nature, v. 199, p. 270-271.

Isotopic Ages and Stratigraphic Control of Mesozoic Igneous Rocks in Japan[1]

KEN SHIBATA[2], TATSURO MATSUMOTO[3], TAKERU YANAGI[3], and REIKO HAMAMOTO[3]

Abstract Mesozoic igneous rocks from Southwest Japan are critically reviewed in terms of isotopic and stratigraphic ages, and new K-Ar and Rb-Sr age results on the Cretaceous granitic and volcanic rocks from the Kitakami Mountains and some other areas are presented and discussed. Cretaceous granites from the western half of the Inner Zone of Southwest Japan are bounded between the Albian Shimonoseki Subgroup and the Campanian and Maestrichtian Izumi Group. The K-Ar ages of the granites generally agree with the stratigraphic evidence, and indicate that the granites were uplifted and eroded soon after the intrusion. Granites and related Cretaceous formations become younger eastwards, suggesting that a site of plutonism, associated with a site of volcanism and sedimentation, migrated eastwards at a rate of 2.6 cm/year. The isotopic ages of granites from Amami-oshima and those of the Funatsu granite also indicate a rapid uplift and erosion subsequent to the granite intrusion.

The Miyako and Taro granites from the Kitakami Mountains, which are stratigraphically bracketed between the Neocomian and the upper Aptian, give Rb-Sr whole-rock isochron ages of 121 ± 6 m.y. and 128 ± 12 m.y. respectively, and agree with the stratigraphic evidence. Seven mineral samples of the Miyako granite give an average K-Ar age of 113 ± 3 m.y., suggesting that the Aptian-Albian boundary may be slightly younger than 113 m.y. Cretaceous volcanic rocks from the Kitakami Mountains give K-Ar whole rock ages of 93-119 m.y., all slightly younger than stratigraphically estimated. Ages of the Harachiyama and Kanaigaura Formations are equal to those of the Miyako granite, and indicate either the contact effect of the granite or contemporaneity with the granite. Lower ages of the Niitsuki and Yamadori Formations may be attributed to the alteration. Volcanic rocks of the Sennan acid pyroclastic rocks in Kinki and those of the Mifune Group in Kyushu yield much younger K-Ar whole rock ages than estimated. A biotite from tuff in the Middle Yezo Group in northwest Hokkaido is dated at 91.4 ± 2.4 m.y., which agrees with the stratigraphically assigned Turonian age.

Introduction

Granitic rocks compose approximately 12% of the Japanese Islands, and are more widely exposed in the Inner Zone of Southwest Japan (Fig. 1). Recently, potassium-argon (K-Ar) age determinations on Japanese granites have been done extensively, with the result that most of the granites are Late Cretaceous to early Tertiary in age, and that K-Ar ages of the granites indicate a characteristic zonal arrangement corre-

[1] Manuscript received, September 27, 1976.
[2] Geological Survey of Japan, Kawasaki 213, Japan.
[3] Department of Geology, Kyushu University, Fukuoka 812, Japan.

We thank T. Yoshida, F. Takizawa and Y. Teraoka of the Geological Survey of Japan for helpful discussions and critical review of the manuscript; N. Isshiki, K. Ono and K. Okumura of the Survey for assistance in petrographic description of volcanic rocks; K. Tanaka and T. Nozawa of the Survey for valuable information.

We are also indebted to S. Kanisawa of Tohoku University, H. Kanaya and Y. Horikawa of the Survey, Y. Miyata of Kyushu University, and members of the Research Group for the Sennan Group, for providing samples and useful information, and to S. Uchiumi and T. Nakagawa of the Survey for technical assistance in K-Ar dating.

Article Identification Number: 0149-1377/78/SG06-0011/$03.00/0

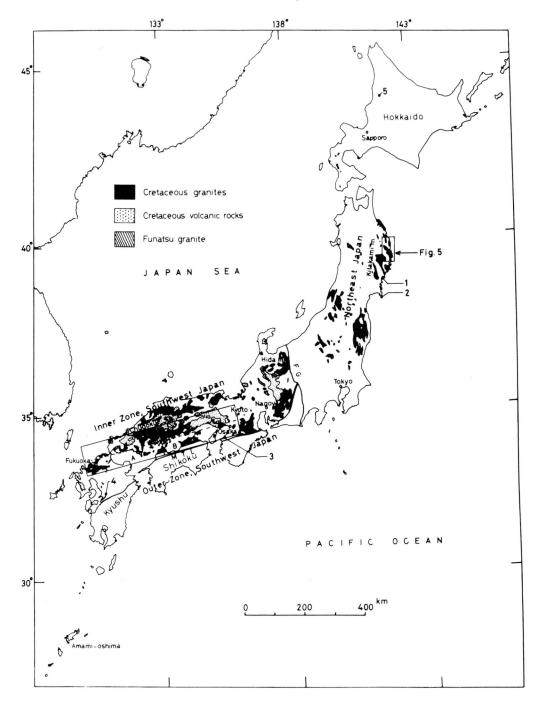

FIG. 1—Map showing distribution of Mesozoic igneous rocks in the Japanese Islands and sample localities of volcanic rocks. 1—Kanaigaura and Niitsuki Formations, 2—Yamadori Formation, 3—Sennan acid pyroclastic rocks, 4—Mifune Group, 5—Middle Yezo Group.

sponding to the petrographic and tectonic provinces (Kawano and Ueda, 1967; Shibata, 1968). Generally, it has been considered that granites in the Japanese Islands are not well controlled stratigraphically except for a few cases, but close evaluation of the relation between the ages of granites and their stratigraphic control has not been made so far.

Associated with granites, volcanic rocks also are exposed widely in the Inner Zone of Southwest Japan. They are mostly pyroclastic rocks with interbedded lavas and clastic rocks, and some of them are well defined stratigraphically. Few age determinations on these rocks have been made, because the rocks are more or less altered or weathered and are thought to be unsuitable for dating.

In an attempt to evaluate any geochronological data in Japan that are useful for the calibration of the Phanerozoic time scale, we will review, in the first part of this paper, the relationship between the isotopic ages of the Mesozoic granites in Southwest Japan and the biostratigraphic evidence of the related Mesozoic strata. In the latter half, we will present and discuss new K-Ar and rubidium-strontium (Rb-Sr) age determinations on the Cretaceous granitic and volcanic rocks from the Kitakami Mountains and some other areas.

Throughout this study isotopic ages are discussed by reference to a time scale prepared by Armstrong and McDowall (1974). The ages of stage boundaries in their time scale are recalculated by the decay constants of this paper: $^{40}K\lambda\beta = 4.72 \times 10^{-10}$ /y; $\lambda_e = 0.584 \times 10^{-10}$ /y; $^{40}K/K = 0.0119$ atom %; $^{87}Rb\lambda = 1.47 \times 10^{-11}$/y, and consequently they are lower than those of the original values, by a factor of 1.023.

Isotopic Ages and Related Stratigraphic Evidence

Cretaceous System Biostratigraphy

Late Mesozoic granites are widely exposed in the Inner Zone of Southwest Japan for an area measuring about 800 km long by a width of about 120 km (Fig. 1). Reliable fossil evidence for the estimation of the period of the granite intrusion is available in the western half of this area (A and B, Fig. 1). Therefore the comparison between the biostratigraphic evidence and the K-Ar ages of the granites is made in the western half.

The Kwanmon Group exposed in North Kyushu and West Chugoku is an accumulation of clastic sediments and andesitic volcanic materials in a continental basin. The group is divided into two subgroups with a disconformity. The lower half is the Wakino Subgroup and the upper half the Shimonoseki Subgroup. In the Wakino Subgroups, lacustrine black shale predominates with some sandstone intercalation and small bodies of limestone. A few beds of tuff also occur. The thickness is about 1,200 m in North Kyushu. However, in West Chugoku, it lacks its lower half and hence its thickness diminishes to about 600 m (Hase, 1960). Zones of *Brotiopsis wakinoensis* and *B. kobayashii-Viviparus onogonensis* are well traced in the subgroup and correlated provisionally to the upper Neocomian (Ota, 1960).

The Shimonoseki Subgroup rests disconformably on the Wakino Subgroup, but sometimes overlaps onto the older basement. In North Kyushu, the Shimonoseki Subgroup outcrops on the northwestern side of the Wakino Subgroup. However, in West Chugoku, the Shimonoseki Subgroup overlaps towards the east onto the older basement (Hase, 1960). This subgroup is made up of conglomerate, sandstone, shale, tuff, tuff breccia, andesite, dacite, and rhyolite. Among them, volcanic materials of andesitic composition predominate. The thickness is about 2,300 m. The age is estimated to be the Aptian to Albian (Matsumoto, 1963; Murakami, 1974). Fossils are few: *Nippononia* (?) *obsoleta* (Hase, 1960) and *Estherites kyongsanensis poucillneata* (Kusumi, 1961). Therefore this age estimation leaves some doubt.

The Shunan Group lies disconformably or unconformably on the Kwanmon Group and is distributed sporadically in West Chugoku (Murakami and Matsusato, 1970). This group consists of rhyolitic to dacitic pyroclastic rocks, mainly welded, with a subordinate amount of andesitic rocks. The thickness varies from place to place and is estimated to be 500 to 2,000 m. No fossil is reported. Ichikawa et al (1968) and Murakami (1974) claimed that the type of eruption changed from a central one in the Shimonoseki Subgroup to a fissure eruption in the Shunan and Abu Groups.

The Abu Group is exposed extensively in West Chugoku and overlies almost flatly the Kwanmon and Shunan Groups and much older basement rocks. It consists mainly of rhyolitic to rhyodacitic pyroclastic materials, sometimes with intercalation of andesitic to dacitic pyroclastics, rhyolitic lavas and lacustrine sediments (Murakami, 1974). Much of the pyroclastics is welded. The thickness amounts to 2,000-3,000 m. The Abu Group has not been found in North Kyushu. Plant fossils that suggest the Upper Cretaceous were reported from the Abu Group (Takahashi, 1959).

Small-scale and local intrusions of granitic rocks into the Shimonoseki Subgroup occurred before the accumulation of the Shunan Group. One of them is the Deba acid igneous complex composed of granodiorite and plagiophyre with aplitic dykes (Murakami, 1960). The outcrop has an extent of 4 by 2 km. Also small-scale intrusions of plutonic rocks of intermediate composition occurred in the Shunan Group before the accumulation of the Abu Group. They all have petrographic characters indicating the solidification at shallow depth in the crust (Murakami, 1974). The large-scale batholithic intrusion of granitic rocks occurred after the accumulation of the Abu Group. The Kwanmon, Shunan, and Abu Groups around the granites were affected by thermal metamorphism.

In East Chugoku and West Kinki, intrusion of the batholithic granites occurred after the accumulation of the Aioi Group (Kishida and Wadatsumi, 1967) which is very similar in various petrographic aspects to the Abu Group and affected by the thermal metamorphism around the granites. Thickness of the Aioi Group is estimated to be 1,350 to 2,850 m. Plant fossils of the Asuwa flora were reported by Matsuo (1964), who once claimed that the Asuwa flora is much akin to the middle Cenomanian-Turonian flora in Sakhalin, but later he considered that the age of the Asuwa flora may possibly be much younger (Matsuo, 1975).

The Izumi Group in East Shikoku lies directly on the southern part of the batholithic granites. It overlaps eastwards onto the granites. The outcrop extends along the Median Tectonic Line for 200 km with a width of about 10 km. It is made up of alternate conglomerates, sandstones, and shales of various thicknesses. The sediments have the characters of turbidite facies. The strata dip moderately to the south or southeast with a synclinal structure, the axis of which is nearly parallel to the Median Tectonic Line and plunges east-northeastwards. Acidic volcanic rocks, porphyries, and granites predominate among the constituent clastic materials and hence indicate that these rocks extensively occupied the source area. A result of paleocurrent direction analyses gave the longitudinal orientation to the west (Tanaka, 1965; Nishimura, 1976).

Since ammonites and inocerami are not uncommon in shaly members (and even sandstones and conglomerates sometimes contain marine fossils), the zones of *Inoceramus orientalis, I. balticus, I. shikotanensis,* and *I. (?) awajiensis* are recognized successively and hence correlated with the Campanian and the Maestrichtian (Matsumoto, 1963; Suyari, 1973).

Paleomagnetic Evidence

Fossil evidence for the Aioi and Abu Groups is so scarce and inconclusive that the age estimation has to be made by indirect evidence. All of the 34 samples from 10 localities of the welded tuff at the upper part of the Aioi Group have a reversal of the natural remanent magnetization (Sasajima and Shimada, 1966). Kono et al (1974) took samples from the whole sequence of the Aioi Group and showed that all of the reliable data have the reversal, although half of the samples were unstable to demagnetization. The recent success of Larson and Pitman (1972) in worldwide correlation of the Mesozoic marine magnetic anomalies enables us to estimate the age of the Aioi Group. The reversal may possibly be correlated to that of 82 to 85 m.y. without serious conflict. This correlation is in harmony with the plant fossil evidence previously stated. It is worth noting that, if this correlation is correct, the accumulation of the Aioi Group occurred during a very short period of the reversal—a few m.y.

All samples of the Abu Group, except those from one locality in the lower part, have normal polarity (Sasajima and Shimada, 1966). The exceptional reversal may await further study, because the direction of the natural remanent magnetization differs largely from those of other Cretaceous samples.

Samples from the lower half of the Shimonoseki Subgroup also have the normal polarity (Sasajima and Shimada, 1966). As stated earlier, this group is estimated stratigraphically to be the Aptian to Albian, the span of which is defined as 120 to 95 m.y. according to the time scale of Armstrong and McDowall (1974). However, in the geomagnetic reversal time scale of Larson and Pitman (1972) frequent reversals occurred during a period from 120 to 111 m.y. Therefore we conclude that the Shimonoseki Subgroup is the Albian. The Abu and Shunan Groups are similar in petrographic aspects and also in types of eruption to the Aioi Group, and hence they accumulated possibly at a rate as rapidly as the Aioi Group. In addition, there is no distinctive structural difference between the Kwanmon and the Abu Groups (Murakami, 1974). Volcanic rocks change in composition successively from andesitic ones in the Shimonoseki Subgroup to rhyolitic ones in the Abu Group. The facts suggest that these three groups are successive without a significantly large time interval between, although the Shunan Group seems to be local. We therefore conclude that the Abu and Shunan Groups are older than the Turonian and possibly the upper Albian to Cenomanian.

Granite K-Ar Ages

Potassium-argon ages of the granites from the Inner Zone of Southwest Japan are plotted in Figure 2. Only the biotite K-Ar ages from the areas denoted A and B in Figure 1 are used. The data sources are Aldrich et al (1962), Shibata and Karakida (1965), Kawano and Ueda (1966), and Shibata (1968).

The abscissa of Figure 2 is a distance measured along the dotted line in Figure 1 from the west end to the east end. Figure 2 indicates that the K-Ar age band with a width of about 15 m.y. varies systematically with the distance. The age band becomes younger at a rate of 3.8×10^4 years/km. The histograms of the K-Ar ages are shown in Figure 3 for the areas denoted A and B in Figure 1. Each of these two areas has an east-west length of 280 km. The K-Ar ages of two areas have a bimodal pattern with two peaks similar to each other. An age difference between the two peaks is the same in both areas, and is 8 m.y.

K-Ar Ages and Stratigraphic Evidence Comparison

The time range of the Cretaceous granitic rocks and stratigraphic units in the Inner Zone of Southwest Japan, based on the biostratigraphic and paleomagnetic evidence as discussed above is shown in Figure 2 by reference to the time scale of Armstrong and McDowall (1974). Stratigraphically the lower limit of the granites in the Inner Zone of Southwest Japan is bounded by the Shimonoseki Subgroup and the Abu Group in the western half (area A), and by the Aioi Group in the eastern half (area B), whereas the upper limit is bounded by the Izumi Group. As seen in Figure 2, the relationship between the K-Ar ages of the granites and the time range of the related stratigraphic units is generally in agreement with a few exceptional points scattered outside the upper and lower boundaries. The agreement may support the time scale of Armstrong and McDowall (1974) in a general way. It may further suggest that the K-Ar ages of the granites in the Inner Zone of Southwest Japan represent the time of intrusion and that areas were rapidly uplifted and eroded after the intrusion of the granites. The latter idea is in agreement with the thick and rapid sedimentation of the Izumi Group.

Figures 2 and 3 indicate that the K-Ar ages of the granites become younger eastwards from North Kyushu. This fact was briefly mentioned by Kawano and Ueda (1967) and Nozawa (1970), and was discussed in more detail by Teraoka (1977). In addition, Figure 2 clearly indicates that the Cretaceous groups associated with the granites become younger eastwards, too. These facts indicate that the granitic intrusion and subsequent uplift of land occurred first in North Kyushu and then the site of uplift migrated eastwards preceded by violent volcanism and followed by rapid marine sedimentation. The sites of contemporaneous sedimentation, intrusion and uplift, and volcanism cover a distance of about 300 km from west to east. The migration rate calculated from Figure 2 is 2.6 cm/year.

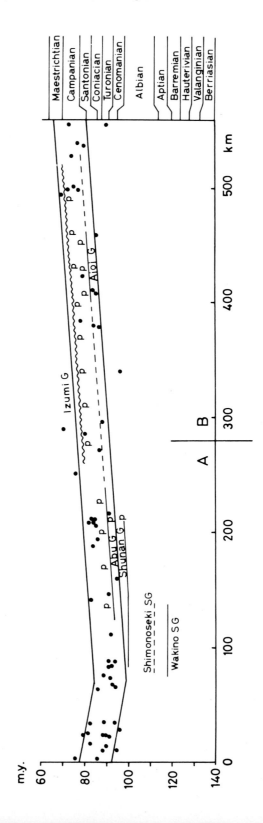

FIG. 2–K-Ar ages of granites in the Inner Zone of Southwest Japan (for areas A and B in Figure 1, closed circles), and stratigraphic positions of related Cretaceous groups. "P" indicates a time-spatial position of plutonism stratigraphically defined.

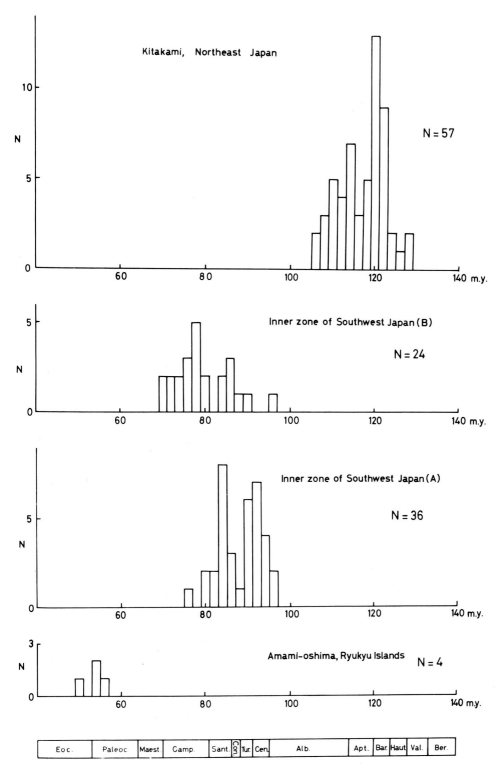

FIG. 3—Histograms of K-Ar ages for granites in the Kitakami Mountains and Southwest Japan. N is the number of dates.

Amami-oshima Granites

Small stocks of granites are intruded into folded sedimentary rocks in Amami-oshima of the Ryukyu Islands (Fig. 1). An ammonite fossil indicating the Upper Cretaceous (probably Turonian) was found from the eastern part of this island (Matsumoto et al, 1966), where the Ohgachi Formation is exposed. Clasts of granitic rocks are found in conglomerate of the Wano Formation, which outcrops to the east of the Ohgachi Formation with a fault contact (Ishida, 1969). From the Wano Formation, a few specimens of Eocene nummulites were discovered by Ishida (1969).

Potassium-argon ages on biotite of the granite stocks reported by Shibata and Nozawa (1966) range from 49 to 56 m.y., and are plotted in Figure 3. As the Eocene is defined to be from 53.5 to 37.5 m.y. (Berggren, 1972), these K-Ar ages indicate that the uplift and erosion took place soon after the granitic intrusion.

Funatsu Granite

The Funatsu granite is exposed in the Hida Mountains in Central Japan (Fig. 1), closely associated with the Hida metamorphic rocks. It is intruded into Carboniferous-age strata, and is overlain by the Tetori Group of the Middle Jurassic to Lower Cretaceous. The Tetori Group is exposed in several isolated areas in the Hida Mountains, and is composed of conglomerate, sandstone, and shale, with the sandstone almost always arkosic. It is divided into three subgroups. The Kaizara Shale, a member within the lower subgroup, contains ammonites of the lower to upper Callovian; *Oppelia* aff. *O. subradiata*, *Grossouvria* cf. *G. subtilis*, and *Neuqueniceras yokoyamai* (Sato, 1962). Approximately 1,000 m of strata exist between the Kaizara Shale and the basement, therefore the age of the base of the Tetori Group has been tentatively assigned to the Bajocian. The basal conglomerate of the Tetori Group contains abundant clasts of gneiss and granite which are petrographically identified with the Hida gneiss and the Funatsu granite respectively, suggesting that the materials were supplied mainly from the area of gneiss and granite.

Isotopic dating of the Funatsu granite has been done as a part of the geochronologic study of the Hida metamorphic terrain (Kuno et al, 1960; Kawano and Ueda, 1966; Ishizaka and Yamaguchi, 1969; Shibata et al, 1970). Figure 4 shows the histogram of isotopic ages of the Funatsu granite. There is a sharp peak at 170 to 180 m.y. by U-Pb, Rb-Sr, and K-Ar ages. The concordant isotopic ages clearly indicate that the age of 170 to 180 m.y. is the time of intrusion for the Funatsu granite. A few lower biotite ages by K-Ar method are considered as the result of weathering or alteration.

According to the time scale of Armstrong and McDowall (1974), the Callovian-Bathonian boundary is 162 m.y., and that of the Bajocian-Aalenian is 173 m.y. Since the base of the Tetori Group is supposed to be the Bajocian, the isotopic ages of the Funatsu granite suggest a rapid uplift and erosion of the granite after intrusion. This suggestion is in agreement with the sedimentary facies of the Tetori Group.

There exists another Jurassic group, called the Kuruma Group, on the northeastern side of the Tetori Group. The Kuruma Group overlies the Paleozoic strata and metamorphic rocks of the Hida marginal belt, of which the age is determined to be 300 to 350 m.y. (Shibata and Nozawa, 1968; Shibata et al, 1970). The middle part of the Kuruma Group is correlated to the upper Pliensbachian to upper Toarcian by ammonites (Sato, 1955, 1962). In marked contrast to the Tetori Group, the Kuruma Group contains no clasts of gneiss (Kobayashi et al, 1957), although those of granitic rocks are abundant in the uppermost formation of the group. The evidence suggests that the Funatsu granite and the Hida gneiss were not exposed during most of the period when the Kuruma Group was deposited, and that the uplift and erosion of the Funatsu granite occurred towards the end of the Kuruma period, presumably in the Aalenian age (~175 m.y.). The sharp peak of ages at 170 to 180 m.y. for the Funatsu granite supports this idea.

Geochronology of Cretaceous Igneous Rocks

There are wide exposures of Cretaceous granitic rocks in the Kitakami Mountains (Fig. 1). The K-Ar age determinations have been extensively done on these rocks (Shibata and Miller, 1962; Kawano and Ueda, 1965), and their ages are mostly in a

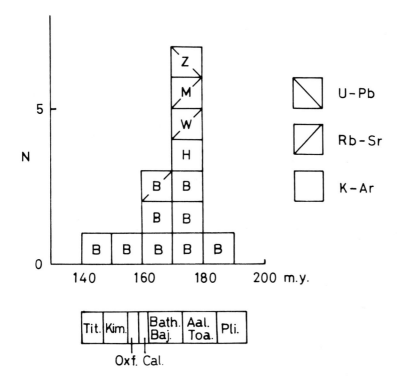

FIG. 4—Histogram of isotopic ages for the Funatsu granite, Hida Mountains.
B—biotite; M—muscovite; H—hornblende; Z—zircon; W—whole rock.

range of 100 to 125 m.y. with peaks at 114 and 120 m.y. (Fig. 3). Among many granitic masses in the Kitakami Mountains, the Miyako and Taro granites are stratigraphically well defined. We therefore made further geochronologic study on these rocks by K-Ar and Rb-Sr methods. In addition, we carried out K-Ar age determinations on Cretaceous volcanic rocks from the Kitakami Mountains, Southwest Japan and Hokkaido, of which the geologic ages are relatively well known.

Analytical Procedures

Rubidium-strontium whole rock analyses were done at Kyushu University, whereas Rb-Sr analyses on biotite and K-Ar analyses were done at the Geological Survey of Japan. The Rb-Sr analytical procedures are similar to those described in Yanagi (1975). Briefly, rubidium and strontium concentrations for whole rock samples were measured by isotope dilution on the Hitachi RMU-5G mass spectrometer, and $^{87}Sr/^{86}Sr$ ratios were determined on unspiked samples using the JEOL-05RB mass spectrometer. All $^{87}Sr/^{86}Sr$ ratios were normalized to $^{86}Sr/^{88}Sr$ ratio equal to 0.1194. Replicate analyses of the A and E standard at Kyushu University yielded an average ratio of 0.70802 ± 0.00014 (2σ). Errors in $^{87}Sr/^{86}Sr$ ratios for whole rock samples are indicated at 2σ level and are all within ± 0.03%, whereas those for biotite are estimated to be ± 0.1%. Errors in rubidium and strontium concentrations are estimated to be within ± 2%. The whole rock isochrons were calculated by the least-square method of York (1966), taking account of an error of ± 3% for $^{87}Rb/^{86}Sr$ ratio and a variable error (0.006 to 0.025%) associated with each $^{87}Sr/^{86}Sr$ ratio. The errors in ages are given at 2σ level. The error in Rb-Sr biotite ages is estimated to be ± 5%.

The analytical method for K-Ar dating is essentially the same as that described in Shibata (1968) except that potassium contents were determined by the atomic absorption analysis. Errors in ages are indicated at 1σ level. The decay constants used in this paper are: ^{87}Rb $\lambda = 1.47 \times 10^{-11}$/y; ^{40}K $\lambda\beta = 4.72 \times 10^{-10}$/y; $\lambda_e = 0.584 \times 10^{-10}$/y; ^{40}K/K = 0.0119 atom %. The Rb-Sr and K-Ar analytical results are given in Tables 1 and 2 respectively.

Miyako and Taro Granites

The Miyako and Taro granites are distributed in the coastal district of the northern Kitakami Mountains (Fig. 5). The Miyako granite is exposed in an area about 50 km in length, and composed mainly of granodiorite and tonalite (Ishihara and Suzuki, 1974; Yoshii and Katada, 1974). The Taro granite is situated to the east of the northern part of the Miyako granite, and is a rather small mass composed mainly of granodiorite. Both granites are intruded into the Harachiyama Formation and overlain by the Miyako Group, and considered to be stratigraphically the most closely limited granites in Japan as explained in the following.

The Miyako Group is exposed at several small and isolated areas along the Rikuchu coast of the northern Kitakami Mountains (Fig. 5). The group consists mainly of calcareous, shelly, and conglomeratic sandstones, and its basal conglomerate contains boulders of granites and volcanic rocks derived from the Harachiyama Formation (Hanai et al, 1968; Shimazu et al, 1970). The total thickness is estimated to be about 200 m. The Hiraiga Formation, which represents the middle part of the Miyako Group, is characterized by the occurrence of guide ammonites such as *Diadochoceras nodosocostatiforme*, *Eodouvilleiceras matsumotoi* and *Valdedorsella akuschaensis*, and well correlated to the upper part of the upper Aptian. The Aketo Formation, the uppermost unit of the Miyako Group, is the lower Albian by the occurrence of *Pseudoleymeriella hataii* and then *Douvilleiceras mammilatum* (Obata, 1977).

Table 1.   Rb-Sr analytical data for the Miyako and Taro granites

Sample No.	Rock type	Rb (ppm)	Sr (ppm)	^{87}Rb/^{86}Sr	^{87}Sr/^{86}Sr
Miyako granite					
107	tonalite	31.97	996.2	0.0929	0.70409 ± 5
113	granodiorite	91.54	532.3	0.4977	0.70475 ± 5
114	granodiorite	106.3	447.4	0.6877	0.70504 ± 8
115	granite	149.9	254.3	1.706	0.70693 ± 4
140	tonalite	65.93	476.9	0.4001	0.70459 ± 6
141	granodiorite	91.97	397.5	0.6697	0.70514 ± 4
142	tonalite	41.73	637.2	0.1895	0.70420 ± 17
550	granodiorite	60.82	679.9	0.2589	0.70431 ± 5
830	tonalite	61.77	467.5	0.3824	0.70459 ± 6
2708	biotite (ton.)	347.0	11.68	86.03	0.8589
141	biotite (granod.)	503.9	12.91	113.0	0.8875
Taro granite					
802	granodiorite	77.67	298.0	0.7544	0.70546 ± 16
809	granite	55.22	169.0	0.9457	0.70585 ± 8
816	granite	73.88	191.5	1.117	0.70601 ± 6
817	quartz diorite	32.30	362.6	0.2578	0.70428 ± 8
821	gabbro	15.86	408.4	0.1124	0.70396 ± 6
826	granodiorite	91.29	124.3	2.126	0.70763 ± 5

Table 2. K-Ar Ages of the Miyako Granite and Cretaceous Volcanic Rocks

Sample No.	Rock	Mineral	K_2O (%)	^{40}Ar rad $(10^{-6} ccSTP/g)$	Atmospheric ^{40}Ar (%)	Age (m.y.)
Miyako granite						
2601	tonalite	biotite	8.85, 8.83	32.7	16.2	109 ± 4
		hornblende	0.818, 0.826	3.24	70.8	116 ± 8
		K-feldspar	14.37	53.1	31.4	109 ± 4
2708	tonalite	biotite	8.66, 8.75	33.9	19.9	114 ± 4
		hornblende	0.419, 0.415	1.59	81.5	112 ± 12
0406	granodiorite	biotite	8.10, 8.17	32.3	27.8	116 ± 4
		hornblende	0.437	2.01	88.1	134 ± 25
			0.435, 0.429	1.89	81.4	128 ± 14
				2.00	80.4	135 ± 14
					av.	132 ± 11
144	granodiorite	hornblende	1.04	4.16	76.9	117 ± 10
Cretaceous volcanic rocks						
1027	tuff (Harachiyama F.)	whole rock	1.64, 1.68	6.73	43.6	119 ± 5
1034	andesite (Harachiyama F.)	whole rock	1.42	5.69	53.8	118 ± 5
1083	dacite (Harachiyama F.)	whole rock	2.35	9.14	35.1	114 ± 4
70032	basalt (Kanaigaura F.)	whole rock	0.610, 0.611	2.49	73.7	119 ± 9
1402	andesite (Niitsuki F.)	whole rock	0.526	1.66	76.5	93.0 ± 7.7
1903p	amphibolite (Yamadori F.)	hornblende	0.847, 0.825 0.854	2.99	74.0	104 ± 8
19A-8	rhyolite (Sennan G.)	whole rock	2.87, 2.89	5.73	51.3	59.4 ± 2.4
M-2	tuff (Mifune G.)	whole rock	0.916, 0.927	1.39	72.7	44.9 ± 3.2
M-4	tuff (Mifune G.)	whole rock	0.445, 0.459	0.836	87.7	55.1 ± 9.2
680	tuff (Middle Yezo G.)	biotite	7.30, 7.27	22.4	16.8	91.0 ± 2.8
			7.26	22.6	51.6	91.8 ± 3.8
					av.	91.4 ± 2.4

The Harachiyana Formation is the uppermost formation of the Rikuchu Group (Sugimoto, 1969), and is exposed widely, together with other formations in the group, along the Rikuchu coast extending over 100 km (Fig. 5). The formation is composed of acid and intermediate volcanic rocks and clastic rocks, and the total thickness is estimated to be over 2,000 m (Shimazu et al, 1970; Sugimoto, 1974). The Rikuchu Group is known to have been folded and faulted before the intrusion of granites and is overlain by the Miyako Group (Shimazu et al, 1970). No fossils have been found to date from the Harachiyama Formation, but the conformably underlying Omoto Formation contains plant fossils of the Lower Cretaceous "Ryoseki flora." The formation also contains molluscan fossils including *Pterotrigonia hokkaidoana,* which indicates the Lower Cretaceous and younger (Sugimoto, 1974). Accordingly, the Omoto Formation is assigned to the Lower Cretaceous, probably lower Neocomian. Therefore, the Harachiyama Formation may possibly be assigned to the upper Neocomian, and the geologic age of the Miyako and Taro granites are bounded between the upper Aptian and the Neocomian. The stratigraphic position of the Miyako and Taro granites in relation to the Cretaceous formations is shown in Table 3.

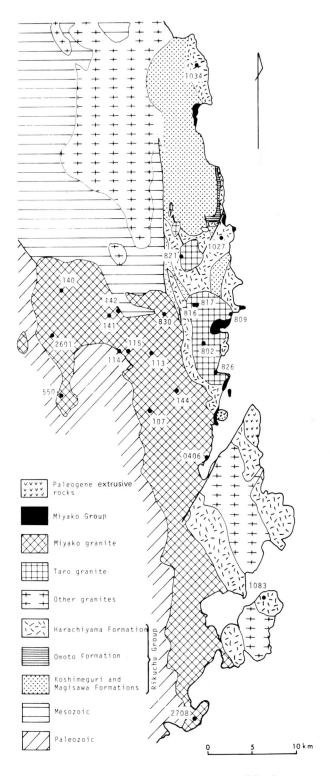

FIG. 5–Generalized geological map of the Rikuchu coast, northern Kitakami Mountains, showing sample localities (based on the map by Shimazu et al, 1970).

Table 3. Correlation of the Lower Cretaceous formations in the Kitakami mountains

Age (m.y.)	Stage		Rikuchu coast	Kesennuma district	Oshika Peninsula
114	Albian	U / M / L	Miyako G.		
120	Aptian	U / L	Miyako and Taro granites (121,128,113)	Qz monzonite	Qz diorite
124	Barremian	Neocomian	Harachiyama F. (114–119)	Oshima F.	Yamadori F. (104)
128	Hauterivian		Omoto F.	Kanaigaura F. (119)	Niitsuki F. (93)
133	Valanginian			Isokusa F.	Ayukawa F.
133	Berriasian		Koshimeguri F.	Kogoshio F.	
140	Tithonian		Magisawa F.		Oginohama F.

Rikuchu G.

Numbers in parentheses are isotopic ages (m.y.) obtained in this study.

Prior to this work, four K-Ar dates ranging from 112 to 123 m.y. were reported for the Miyako and Taro granites (Shibata and Miller, 1962; Kawano and Ueda, 1965). In order to define more precisely the isotopic ages of the Miyako and Taro granites, we analyzed nine and six whole rock samples of the Miyako and Taro granites respectively by the Rb-Sr method, and analyzed eight minerals separated from five samples of the Miyako granite by the K-Ar method. In addition, Rb-Sr analyses were made on two biotites from the Miyako granite. The Rb-Sr analytical results are shown in Table 1.

The whole rock samples of the Miyako granite have rather low rubidium contents and high strontium contents; the highest strontium value amounting to 1,000 ppm, and so $^{87}Rb/^{86}Sr$ ratios are relatively low. Nevertheless, nine samples define an isochron of 121 ± 6 m.y. with an initial $^{87}Sr/^{86}Sr$ ratio of 0.70389 ± 0.00005 (Fig. 6). It is remarkable that the Miyako granite, exposed in an elongated area (measuring about 50 km from north to south) is so well homogenized isotopically. Considering the geologic evidence and a low initial ratio, the age of 121 ± 6 m.y. is safely considered to represent the time of intrusion for the Miyako granite. The age is within the Barremian according to the time scale prepared by Armstrong and McDowall (1974),

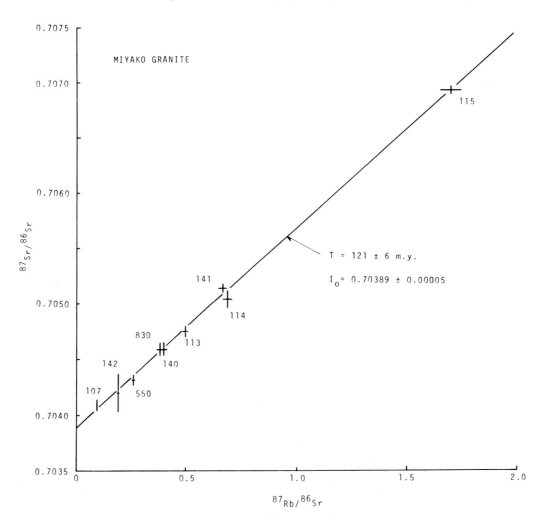

FIG. 6–Rb-Sr whole rock isochron plot for the Miyako granite.

and is in agreement with the geologic evidence that the Miyako granite is intruded into the Harachiyama Formation of the upper Neocomian. The Rb-Sr analytical data for the two biotites from the Miyako granite are also given in Table 1, and the ages are calculated to be 122 ± 6 m.y. (2708) and 110 ± 6 m.y. (141); both are slightly discordant to each other, but the older age is identical to the whole rock age.

The Rb-Sr data for the whole rock samples of the Taro granite are plotted in Figure 7. The data points do not define a good isochron and two points are significantly off the best-fit line. Excluding these points (802 and 809), the line yields an age of 128 ± 12 m.y. with an initial $^{87}Sr/^{86}Sr$ ratio of 0.70378 ± 0.00015. The age is apparently older than that of the Miyako granite, but a larger error makes it impossible to affirm that the age difference is real. If all the samples of the Miyako granite and those of the Taro granite excluding the above-mentioned two are combined, an isochron age of 124 ± 6 m.y. is obtained. This age is nearly equal to that of the Miyako granite, and is more realistic for the age of the Taro granite. The Taro granite contains abundant basic inclusions of various sizes, and the lack of the perfect isochron plot may reflect the inhomogeneity of the whole rock system to some extent.

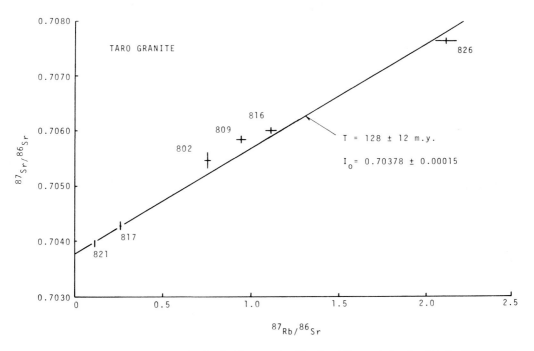

FIG. 7–Rb-Sr whole rock isochron plot for the Taro granite. The fitted line excludes the data for 802 and 809.

The K-Ar mineral ages for the Miyako granite are shown in Table 2 together with other ages for Cretaceous volcanic rocks. Except for one hornblende, all minerals separated from the Miyako granite yield ages of 109 to 117 m.y. with an average of 113 ± 3 m.y. Coexisting minerals in each rock are concordant with each other within experimental errors, and even an age of K-feldspar is not significantly younger than that of coexisting biotite or hornblende. All K-Ar mineral ages for the Miyako granite are younger than the whole rock age of 121 m.y., and the average age of 113 m.y. is also slightly younger than that. If new decay constants of ^{40}k: $\lambda_\beta = 4.962 \times 10^{-10}$/y, $\lambda_e = 0.581 \times 10^{-10}$/y, $^{40}K/K = 0.01167$ atom %, are used for the K-Ar age and $\lambda = 1.42 \times 10^{-11}$/y is used for the Rb-Sr age, the average mineral and the whole rock ages are calculated to be 116 and 125 m.y. respectively. This time difference may be interpreted to indicate the cooling period of the Miyako granite after intrusion. A hornblende gives a K-Ar age of 132 m.y., which is older than the whole rock age. The host rock is not different from others geologically or petrographically, and the co-existing biotite gives an age of 116 m.y. At present we only assume that the age might be apparently too old due to incorporation of excess ^{40}Ar in hornblende.

As stated before, the Miyako and Taro granites are unconformably covered by the Miyako Group of which the middle part is well correlated to the upper part of the upper Aptian. This evidence closely defines the upper limit of the granites. The basal conglomerate of the Miyako Group, the Raga Formation, contains abundant boulders of granites which are petrographically identified with the Miyako and Taro granites. It is natural to consider that the sedimentation of the Miyako Group started after the Miyako and Taro granites were cooled below a temperature at which the radiogenic argon was retained. This leads to a conclusion that the Miyako Group must be younger than 113 ± 3 m.y.; the mean mineral age of the Miyako granite. There still exist approximately 100 m of strata between the Miyako granite and the bed that has been assigned to the upper part of the upper Aptian. Thus the age of the upper part of the upper Aptian must be 113 ± 3 m.y. at the maximum, and probably slightly younger than this age.

According to the time scale of Armstrong and McDowall (1974), the Aptian-Albian boundary is 114 m.y. if the constants of this paper are used. Our age results on the Miyako granite indicate that this boundary must be younger than 113 ± 3 m.y., and this may be in agreement with the above time scale making allowance for the uncertainty in age determination. However, we tentatively estimate that the Aptian-Albian boundary may be slightly younger than 113 m.y.

Harachiyama Formation

There remains some ambiguity in using the age of plutonic rocks for the time scale work even if the rocks are closely defined stratigraphically. The best samples for this purpose are volcanic rocks interbedded in well-defined strata. In the Kitakami Mountains, characteristic volcanic rocks are intercalated in the Lower Cretaceous sedimentary series. They are called the Harachiyama Formation in the Rikuchu coast (of which the details are already described), the Kanaigaura and Niitsuki Formations in the Kesennuma district, and the Yamadori Formation in the Oshika Peninsula. Fortunately, these volcanic rocks are fairly well defined stratigraphically, and are thought to be generally contemporaneous to each other (Onuki, 1969). We therefore determined K-Ar ages on some of these rocks. The stratigraphic sequence and correlation of the Lower Cretaceous formations in the Kitakami Mountains are given in Table 3.

The Harachiyama Formation, although nonfossiliferous and composed mostly of volcanic rocks, is bounded by the upper Barremian-lower Aptian Miyako granite for the upper limit and by the lower Neocomian Omoto Formation for the lower limit. Therefore the age of the Harachiyama Formation is assigned to the upper Neocomian. In some places the formation is intruded by the Miyako and Taro granites and thermally metamorphosed, and hence we carefully selected the samples having no effect of contact under the microscope. Three whole rock samples were dated and the results are shown in Table 2. Dacite tuff (1027) and andesite lava (1034) give ages of 119 and 118 m.y. respectively, which are similar to the whole rock Rb-Sr age of the Miyako granite, whereas dacite lava (1083) gives an age of 114 m.y., which is slightly younger than the previous two and equal to the mineral ages of the Miyako granite.

The last sample is rather close to the granite outcrop (1 km), and the age might be reduced by the contact effect although no such effect is observed petrographically. As stated earlier, the Harachiyama Formation is assigned to the upper Neocomian of which the time span is defined to be 128 to 120 m.y. according to Armstrong and McDowall (1974), and the older ages of the Harachiyama Formation 118 to 119 m.y., might be barely equal to the geologic age. As the Harachiyama Formation has been subjected to a low-grade metamorphism of prehnite-pumpellyite to greenschist facies (Moriya, 1969), it is not possible to obtain samples without slight alteration in this district. In fact epidote and chlorite are commonly observed in the dated samples. It is therefore difficult to tell whether the whole rock ages of the Harachiyama Formation represent the true age of extrusion. Geologically, a significant time gap is evident involving tectonic movement and plutonism between the Miyako Group and the Rikuchu Group, the event of which has been called the Oshima orogeny (Kobayashi, 1941). Then, the age of approximately 120 m.y. for the Harachiyama Formation may be slightly younger than geologically estimated. This may have resulted from the contact effect of the granite although the possibility that the Harachiyama Formation could be nearly contemporaneous with the Miyako granite remains.

Kanaigaura and Niitsuki Formations

Both formations are distributed in the Kesennuma district of the southern Kitakami Mountains, and composed of andesitic lavas and pyroclastic rocks (Kambe and Shimazu, 1961; Fig. 1). The Kanaigaura Formation overlies the Isokusa Formation with an unconformity and grades upwards into the Oshima Formation in its uppermost part. The Isokusa Formation contains characteristic ammonites of the Berriasian to Valanginian; *Berriasella* sp., *Thurmanniceras isokusense, Kilianella* sp., and *Olcostephanus* sp. The lowermost part of the Isokusa Formation contains *Substeueroceras* sp., which is correlated to the upper Tithonian. Therefore, the Isokusa Formation is assigned to the upper Tithonian to middle-upper Valanginian (Sato, 1959; Takahashi, 1973). The Oshima Formation contains *Crioceratites ishiharai*, which can be assigned to the Hauterivian (Obata and Matsumoto, 1977). Therefore the geologic age of the Kanaigaura Formation is estimated to be Hauterivian, defined to be 128 to 124 m.y. (Armstrong and McDowall, 1974). The Niitsuki Formation, although without known direct contact with the Cretaceous formation, is considered to be contemporaneous with the Kanaigaura Formation. The Niitsuki Formation is intruded by a quartz monzonite having probably the same age as the Miyako granite.

A whole rock sample of basalt (70032) from the Kanaigaura Formation gives a K-Ar age of 119 ± 9 m.y., which is equal to those of the Harachiyama Formation (Table 2). However, a large error makes it difficult to evaluate precisely the agreement with the geologic evidence. In addition, the sample has probably been involved in a metamorphism of prehnite-pumpellyite facies because epidote, chlorite, and prehnite are observed, and it is again difficult to decide whether the obtained age represents the time of extrusion. In spite of this fact, it is certain that the argon loss due to the metamorphism, if any, occurred during a short period soon after the extrusion, since the age of the basalt is as old as the Miyako granite.

A whole rock sample of andesite (1402) from the Niitsuki Formation gives a K-Ar age of 93.0 ± 7.7 m.y., significantly younger than that of the Kanaigaura Formation. The sample has been subjected to a rather severe alteration, probably of hydrothermal character, and the younger age is attributed to this alteration.

Yamadori Formation

The Yamadori Formation is exposed in the Oshika Peninsula, southernmost of the Kitakami Mountains (Fig. 1). It is composed of pyroclastic rocks and lavas having andesitic or basaltic composition (Takizawa et al, 1974). It covers the Ayukawa Formation with an unconformity. The middle part of the Ayukawa Formation contains ammonites such as *Thurmanniceras* sp., *Kilianella* sp. and *Berriasella* sp. (Takizawa, 1970), and is assigned to the Berriasian to lower Valanginian. About 1,000 m of strata are between these ammonite-bearing horizons and the base of the Yamadori Formation. Considering the moderate rate of sedimentation and the time gap of the unconformity, at least 3 m.y. may have elapsed before the extrusion of the Yamadori Formation, and its base may probably be the Hauterivian or younger. Although the Yamadori Formation is not directly contacted by the Cretaceous granite, in some places it is thermally metamorphosed probably by a quartz diorite, which is exposed to the northwest of this formation and is dated to be 120 m.y. (Kawano and Ueda, 1965).

A K-Ar age of 104 ± 8 m.y. is obtained on hornblende separated from a lithic fragment in lapilli tuff of the Yamadori Formation (Table 2). The age is younger than that of the above-mentioned quartz diorite or the Miyako granite, and probably does not indicate a real age of extrusion. The hornblende sample is separated from a hornblendite (lithic fragments in lapilli tuff) and is fairly fresh associated only with small amounts of apatite and iron ore. It is difficult, at present, to explain an apparently low age of the Yamadori Formation. It might be related to the low grade metamorphism of prehnite-pumpellyite facies observed in the Yamadori Formation.

Sennan Acid Pyroclastic Rocks

The Sennan acid pyroclastic rocks are exposed in an elongated area to the south of Osaka (Fig. 1), extending approximately 20 km in the east-west direction and composed predominantly of welded tuff of rhyolite to rhyodacite composition, intercalated with small amounts of clastic rocks and lavas according to the research group for the Sennan Group (in preparation). The total thickness attains 3,000 m. The rocks are unconformably overlain by the Izumi Group of the Campanian and the Maestrichtian, therefore the age must be older than about 75 m.y.

A whole rock sample of rhyolite lava (19A-8) from the lowermost part of the Sennan rocks gives a K-Ar age of 59.4 ± 2.4 m.y. (Table 2), which is considerably younger than the geologically estimated age. The sample appears slightly altered because sericite is noticed under the microscope. The alteration might be the cause of the low age reading. Another possibility is that the argon loss occurred during a tectonic movement of a later period, which is related to the Median Tectonic Line, because the rocks are situated close to the line.

Mifune Group

The Mifune Group is exposed on the western foot of the Aso volcano in Central Kyushu (Fig. 1), and is composed of three units; basal conglomerate which overlies crystalline schists and the Upper Permian strata, the lower formation consisting mainly of sandstone and shale and containing bivalves and gastropods, and the upper formation consisting predominantly of red beds with intercalated acid tuff (Matsumoto, 1939). The group was formerly correlated roughly with the upper Albian to Turonian (Matsumoto, 1954), but recently the lower formation was defined more closely to the middle to upper Cenomanian based on the occurrence of *Inoceramus concentricus costatus* and *Eucalycoceras* sp. cf. *E. spathi* (Tamura et al, 1974). The upper limit of the Mifune Group is defined by the age of the Gankaizan Formation, which overlies disconformably the Mifune Group at its westernmost part, and which is assigned to the lower Santonian because of *Inoceramus amakusensis* (Tamura and Tashiro, 1966). Thus the upper formation of the Mifune Group is older than the Santonian, and is supposed to be the Turonian.

The K-Ar age determinations were carried out on two whole rock samples of tuff from the upper formation of the Mifune Group. Sample M-2, biotite rhyolite pumice tuff, gives an age of 44.9 m.y., whereas M-4, vitric biotite rhyolite tuff, gives an age of 55.1 m.y. (Table 2). These two ages are considerably younger than stratigraphically estimated, because the Santonian is defined to be 86 to 81 m.y. (Armstrong and McDowall, 1974). The whole rock samples are fairly fresh and no remarkable alteration is observed under the microscope. Therefore it is quite difficult to interpret the lower ages, yet possible causes will be considered. The Mifune Group is developed close to the probable western extension of the Median Tectonic Line, and the lower ages might be related to a tectonic movement after the sedimentation, though its geologic evidence is not yet known. An alternative interpretation is that the tuffs have been thermally affected by the adjacent younger volcanic rocks represented by the Funano-yama andesite. However, the most probable interpretation is that the whole rock samples of tuff from the Mifune Group are not retentive of argon and thus not useful for K-Ar dating.

Cretaceous Tuff from the Middle Yezo Group

The Cretaceous Yezo Group is extensively developed in the central zone of Hokkaido. Although layers of tuff are intercalated in various horizons of a thick sequence of sandstone and shale, samples suitable for dating have not yet been obtained. During a more detailed study of the Cretaceous System of the Obirashibe area (Tanabe et al, 1977), a tuff containing fresh biotite flakes was found from the Middle Yezo Group. Calcareous nodules containing *Inoceramus teshioensis* and ammonites were obtained near the outcrop of the tuff. They indicate the upper Turonian. Although the nodules were taken from the debris and not from the outcrop of the tuff-bearing formation, they are probably derived from nearby beds judging from the topography and rock facies. Unless the tuff-bearing bed is separated from the fossiliferous bed by faulting, this tuff can also be assigned to the upper Turonian.

Biotite separated from the tuff gives a K-Ar age of 91.4 ± 2.4 m.y. According to the time scale of Armstrong and McDowall (1974), the Turonian is 91 to 88 m.y. The age of 91.4 m.y. is close to the Cenomanian-Turonian boundary, thus it is slightly older than the geologically estimated age. However, the precise correlation is difficult because of the ambiguity in stratigraphic relation between the tuff and fossiliferous beds, and also because of the error associated with the age. It may be said, at least, that this biotite age is generally in agreement with the geologic age. As another alternative, the tuff might be correlated to the predominant tuffite of the lower Turonian sequence, which contains *Inoceramus labiatus* and *Fagesia* sp. and is exposed more widely in the northeastern part of the Obirashibe area.

Summary

1. The relationship between the isotopic ages of the Mesozoic granites in Southwest Japan and the related biostratigraphic evidence is critically reviewed. Paleomagnetic data are used to evaluate the stratigraphic ages of Cretaceous pyroclastic rocks. The Cretaceous granites from the western half of the Inner Zone of Southwest Japan are bounded by the Albian Shimonoseki Subgroup and probably upper Albian to Cenomanian Abu Group for the lower limit, and by the Campanian and Maestrichtian Izumi Group for the upper limit. The volcanic activity represented by the Abu-Aioi Group is in the lower part of this time range. The K-Ar ages of the granites are mostly within this range when comparing to the time scale of Armstrong and McDowall (1974). The general agreement between the K-Ar and stratigraphic ages roughly supports their time scale, and suggests that the K-Ar ages represent the time of intrusion for the granites, followed by rapid uplift and erosion. Furthermore, it is made clear that both the granites and the Cretaceous formations that bound the granites become younger eastwards. This fact indicates that the sites of contemporaneous sedimentation, plutonism, and volcanism, which are arranged from west to east for about 300 km, migrated eastwards at a rate of 2.6 cm/year.

The isotopic ages of granites in Amami-oshima of the Ryukyu Islands and those of the Funatsu granite in the Hida Mountains are generally in harmony with the stratigraphic evidence of overlying strata, and suggest again a rapid uplift and erosion of granites subsequent to the intrusion.

2. The Rb-Sr and K-Ar age determinations have been done on the Cretaceous granites and volcanic rocks from the Kitakami Mountains, where their stratigraphic relation to the Cretaceous System is relatively well known. The Miyako and Taro granites, which are thought to be the most closely controlled granites in Japan and bounded between the Neocomian and the upper Aptian, give Rb-Sr whole rock isochron ages of 121 ± 6 m.y. and 128 ± 12 m.y. respectively, which is in agreement with the stratigraphic evidence that the granites are intruded into the Harachiyama Formation of possibly the upper Neocomian. Seven K-Ar mineral ages of the Miyako granite range from 109 to 117 m.y. with an average age of 113 ± 3 m.y., and suggest that the Aptian-Albian boundary in the middle of the biostratigraphically well controlled Miyako Group may be slightly younger than 113 m.y. Cretaceous volcanic rocks in the Kitakami Mountains yield the K-Ar whole rock ages of 93 to 119 m.y. The ages of the Harachiyama Formation and the Hauterivian Kanaigaura Formation are 114 to 119 m.y.; equal to the ages of the Miyako granite, and may reflect the thermal effect of granites, although the possibility of contemporaneity of the volcanic rocks with the granites cannot be ruled out. The lower ages of the Niitsuki and Yamadori Formations may be related to the alteration.

Volcanic rocks from the Sennan acid pyroclastic rocks in the Kinki district and from the Mifune Group in Kyushu yield much younger K-Ar whole rock ages than geologically estimated. The alteration or weathering, a tectonic movement in a later period, and the non-retentivity of the samples for argon, are considered as the possible causes of younger ages. A biotite separated from tuff of the Middle Yezo Group in northwestern Hokkaido gives an age of 91.4 ± 2.4 m.y., which is slightly older than but still agrees with the assigned stratigraphic age—Turonian.

3. In conclusion, a careful examination of the isotopic age results and related stratigraphic evidence on the Mesozoic igneous rocks in Japan reveals that some of the age data can be used for the calibration of the Phanerozoic time scale, and that they are generally in harmony with the time scale of Armstrong and McDowall (1974).

APPENDIX: Rock Type, Locality of Samples

Miyako Granite

K-107*	Biotite-hornblende tonalite	39-38.9, 141-53.4**
K-113	Biotite-hornblende granodiorite	39-42.3, 141-53.4
K-114	Biotite-hornblende granodiorite	39-42.5, 141-50.6
K-115	Biotite granite	39-42.6, 141-51.2
K-140	Hornblende-biotite tonalite	39-45.8, 141-46.7
K-141	Biotite-hornblende granodiorite	39-44.3, 141-50.2
K-142	Biotite-hornblende tonalite	39-44.7, 141-50.8
K-550	Hornblende-biotite granodiorite	39-39.7, 141-46.3
70830	Hornblende-biotite tonalite	39-44.5, 141-54.5
72080406	Hornblende-biotite granodiorite	39-36.4, 141-57.7
72102601	Hornblende-biotite tonalite	39-42.8, 141-45.2
72102708	Hornblende-biotite tonalite	39-21.5, 141-57.1
K-144	Biotite-hornblende granodiorite	39-40.2, 141-55.4

Taro Granite

70802	Hornblende-biotite granodiorite	39-43.1, 141-57.5
70809	Biotite granite	39-44.7, 141-59.6
70816	Hornblende-biotite granite	39-45.3, 141-56.8
70817	Hornblende quartz diorite	39-45.3, 141-56.9
70821	Biotite-bearing hornblende gabbro, Kurumihata mass	39-47.7, 141-55.5
70826	Biotite granodiorite	39-41.3, 141-58.3

Cretaceous Volcanic Rocks

741027	Dacite sandy tuff, Harachiyama Formation	39-48.4, 141-57.8
741034	Basic andesite, Harachiyama Formation	39-59.2, 141-56.7
741083	Quartz dacite, Harachiyama Formation	39-28.5, 142-02.3
70032	Basalt, Kanaigaura Formation	38-52.2, 141-37.4
73-1402	Altered andesite, Niitsuki Formation	38-52.6, 141-32.6
NI70031903p	Amphibolite fragment in lapilli tuff, Yamadori Formation	38-16.8, 141-30.8
741019A-8	Rhyolite, Sennan acid pyroclastic rocks	34-22.8, 135-27.5
M-2	Biotite rhyolite pumice tuff, Mifune Group	32-44.0, 130-49.2
M-4	Vitric biotite rhyolite tuff, Mifune Group	32-45.2, 130-49.4
KU-680	Tuff, Saku Formation, Upper Yezo Group	44-01.8, 141-54.4

*Full sample number, only the last three or four digits are used in the text.
**Latitude (N) and longitude (E).

References Cited

Aldrich, L. T., et al, 1962, Radiometric ages of rocks: Carnegie Inst. Wash. Year Book, v. 61, p. 234-239.

Armstrong, R. E., and W. G. McDowall, 1974, Proposed refinement of the Phanerozoic time scale (abs.): Int. Mtg. Geochronology, Cosmochronology and Isotope Geology (Paris) Abs. Vol.

Berggren, W. A., 1972, A Cenozoic time-scale—some implications for regional geology and paleobiogeography: Lethaia, v. 5, p. 195-215.

Hanai, T., I. Obata, and I. Hayami, 1968, Notes on the Cretaceous Miyako Group: Natl. Sci. Mus. Mem. (Tokyo), no. 1, p. 20-28.

Hase, A., 1960, The Late Mesozoic formations and their molluscan fossils in west Chugoku and north Kyushu, Japan: Hiroshima Univ. Jour. Sci., ser. C, v. 3, p. 281-342.

Ichikawa, K., et al, 1968, Late Mesozoic igneous activity in the inner side of southwest Japan, in Pacific Geology: Tokyo, Tsukiji Shokan Publ. Co., v. 1, p. 97-118.

Ishida, S., 1969, Wano formation (Eocene) in Amami-oshima, Ryukyu Islands, Japan: Geol. Soc. Japan Jour., v. 75, p. 141-156.

Ishihara, S., and Y. Suzuki, 1974, Modal compositions, in Cretaceous granitic rocks in the Kitakami mountains, petrography and zonal arrangement: Japan Geol. Surv. Rept., no. 251, p. 23-42.

Ishizaka, K., and M. Yamaguchi, 1969, U-Th-Pb ages of sphene and zircon from the Hida metamorphic terrain, Japan: Earth and Planetary Sci. Letters, v. 6, p. 179-185.

Kambe, N., and M. Shimazu, 1961, The geological sheet map "Kesennuma," scale 1:50,000, and its explanatory text: Japan, Geol. Surv., 73 p.

Kawano, Y., and Y. Ueda, 1965, K-A dating on the igneous rocks in Japan (II), granitic rocks in Kitakami massif: Japan. Assoc. Mineral. Petrol. Econ. Geol. Jour., v. 53, p. 143-154.

—— —— 1966, K-A dating on the igneous rocks in Japan (V), granitic rocks in Southwest Japan: Japan. Assoc. Mineral. Petrol. Econ. Geol. Jour., v. 56, p. 191-211.

—— —— 1967, K-Ar dating on the igneous rocks in Japan (VI), granitic rocks, summary: Japan. Assoc. Mineral. Petrol. Econ. Geol. Jour., v. 57, p. 177-187.

Kishida, K., and K. Wadatsumi, 1967, Volcanostratigraphy of the Himeji acid volcano-plutonic complex, studies of the Late Mesozoic igneous rocks in Kinki, Southwest Japan, Part I, in T. Sudo, ed., Prof. H. Shibata Mem. Vol., p. 241-255.

Kobayashi, T., 1941, The Sakawa orogenic cycle and its bearing on the origin of the Japanese Islands: Tokyo Imp. Univ. Fac. Sci. Jour., ser. 2, v. 5, p. 219-578.

—— et al, 1957, On the Lower Jurassic Kuruma group: Geol. Soc. Japan Jour., v. 63, p. 182-194.

Kono, M., M. Ozima, and K. Wadatsumi, 1974, Paleomagnetism and K-Ar ages of Himeji volcanics, in M. Kono, ed., Rock magnetism and paleogeophysics: Japan Rock Magn. Paleogeophysics Res. Group, v. 2, p. 45-49.

Kono, H., et al, 1960 Potassium-argon dating of the Hida metamorphic complex, Japan: Japan. Jour. Geol. Geogr., v. 31, p. 273-278.

Kusumi, H., 1961, Studies on the fossil estherids: Hiroshima Univ. Geol. Rept., no. 7, p. 1-88.

Larson, R. L., and W. C. Pitman, III, 1972, Worldwide correlation of Mesozoic magnetic anomalies, and its implications: Geol. Soc. America Bull., v. 83, p. 3645-3662.

Matsumoto, T., 1939, Geology of Mifune district, Kumamoto Prefecture, Kyushu, with special reference to the Cretaceous system: Geol. Soc. Japan Jour., v. 46, p. 1-12.

—— (ed.) 1954, The Cretaceous system in the Japanese Islands: Japan Soc. Promotion Sci. Res., 324 p.

—— 1963, The Cretaceous, in F. Takai, T. Matsumoto, and R. Toriyama, eds., Geology of Japan: Univ. Tokyo Press, p. 99-128.

—— H. Ishikawa, and S. Yamaguchi, 1966, A Mesozoic ammonite from Amami-oshima: Paleontol. Soc. Japan Trans. Proc., n.s., no. 62, p. 234-241.

Matsuo, H., 1964, On the late Cretaceous flora in Japan: Kanazawa Univ. Ann. Sci., v. 1, p. 39-65.

—— 1975, A few evidences of the climatic conditions of the Neophyta in the Innerside of Honshu, Japan: Kanazawa Univ. Ann. Sci., v. 12, p. 73-90.

Moriya, S., 1969, Low-grade metamorphic rocks of the Sanriku coast, northeastern Japan: Japan Assoc. Mineral. Petrol. Econ. Geol. Jour., v. 62, p. 55-65.

Murakami, N., 1960, Cretaceous and Tertiary igneous activity in western Chugoku: Yamaguchi Univ. Sci. Rept., v. 11, p. 21-126.

—— 1974, Some problems concerning Late Mesozoic to early Tertiary igneous activity on the inner side of Southwest Japan, in Pacific Geology: Tokyo, Tsukiji Shokan Publ. Co., v. 8, p. 139-151.

—— and H. Matsusato, 1970, Intrusive volcanic breccias in the Late Mesozoic Zenjoji-yama formation in West Chugoku and their possible relevance to the formation of cauldron structure: Japan Assoc. Mineral. Petrol. Econ. Geol. Jour., v. 64, p. 73-94.

Nishimura, Y., 1976, Petrography of the Izumi sandstones in the east of the Sanuki mountain range, Shikoku, Japan: Geol. Soc. Japan Jour., v. 82, p. 231-240.

Nozawa, T., 1970, Isotopic ages of late Cretaceous acid rocks in Japanese Islands, summary and notes in 1970: Geol. Soc. Japan Jour., v. 76, p. 493-518.

Obata, I., 1977, The Miyakoan (Aptian and Albian) on the Pacific coast of Northeast Japan, *in* R. A. Reyment, ed., IGCP Project Mid-Cretaceous events, Mus. d'Hist. Nat. Nice Ann. Rept.

—— and T. Matsumoto, 1977, Correlation of the Lower Cretaceous formations in Japan: Kyushu Univ. Dept. Geol. Sci. Rept., v. 12.

Onuki, Y., 1969, Geology of the Kitakami massif, Northeast Japan: Tohoku Univ. Inst. Geol. Paleont. Contr., no. 69. 239 p.

Ota, Y., 1960, The zonal distribution of the non-marine fauna in the Upper Mesozoic Wakino Subgroup: Kyushu Univ. Fac. Sci. Mem., ser. D, v. 9, p. 187-209.

Sasajima, S., and M. Shimada, 1966, Paleomagnetic studies of the Cretaceous volcanic rocks in Southwest Japan, an assumed drift of the Honshu island: Geol. Soc. Japan Jour., v. 72, p. 503-514.

Sato, T., 1955, Les ammonites recueillies dans le groupe de Kuruma, nord du Japon central: Paleontol. Soc. Japan Trans. Proc., n.s., no. 20, p. 111-118.

—— 1959, Presence du Berriasien dans la stratigraphie du plateau de Kitakami (Japon septentrional): Soc. Geol. Fr. Bull., 6e ser., v. 8, p. 585-599.

—— 1962, Etudes biostratigraphiques des ammonites du Jurassique du Japon: Soc. Geol. Fr. Mem., n.s., v. 41, no. 94, p. 1-122.

Shibata, K., 1968, K-Ar age determinations on granitic and metamorphic rocks in Japan: Japan Geol. Surv. Rept., no. 227, 71 p.

—— and Y. Karakida, 1965, Potassium-argon ages of the granitic rocks from northern Kyushu: Geol. Surv. Bull, v. 16, p. 443-445.

—— and J. A. Miller, 1963, Potassium-argon ages of granitic rocks from the Kitakami highlands: Japan Geol. Surv. Bull., v. 16, p. 443-445.

—— and J. A. Miller, 1962, Potassium-argon ages of granitic rocks from the Kitakami highlands: Japan Geol. Surv. Bull., v. 13, p. 709-711.

—— —— 1968, K-Ar age of Omi schist, Hida mountains, Japan: Japan Geol. Surv. Bull., v. 19, p. 243-246.

—— —— and R. K. Wanless, 1970, Rb-Sr geochronology of the Hida metamorphic belt, Japan: Canadian Jour. Earth Sci., v. 7, p. 1383-1401.

Shimazu, M., K. Tanaka, and T. Yoshida, 1970, Geology of the Taro district: Japan Geol. Surv. Quadr. Ser., 54 p.

Sugimoto, M., 1969, Geology of the Omoto-Tanohata district, Outer Kitakami belt, Northeast Japan: Tohoku Univ. Inst. Geol. Paleont. Contr., no. 70, 22 p.

—— 1974, Stratigraphical study in the Outer belt of the Kitakami massif, Northeast Japan: Tohoku Univ. Inst. Geol. Paleont. Contr., no. 74, 48 p.

Suyari, K., 1973, On the lithofacies and the correlation of the Izumi group of the Asan mountain range, Shikoku: Tohoku Univ. Sci. Rept. 2nd ser., sp. vol., no. 6, p. 489-495.

Takahashi, E., 1959, Floral change since the Mesozoic age in western Honshu, Japan: Yamaguchi Univ. Sci. Rept., v. 10, p. 181-237.

Takahashi, H., 1973, The Isokusa Formation and its Late Jurassic and Early Cretaceous ammonite fauna: Tohoku Univ. Sci. Rept. 2nd ser. sp. vol., no. 6, p. 319-336.

Takizawa, F., 1970, Ayukawa Formation of the Ojika Peninsula, Miyagi Prefecture, Northeast Japan: Japan Geol. Surv. Bull., v. 21, p. 567-578.

—— N. Isshiki, and M. Katada, 1974, Geology of the Kinkasan district: Japan Geol. Surv. Quadr. Ser., 62 p.

Tamura, M., and M. Tashiro, 1966, Upper Cretaceous system south of Kumamoto: Kumamoto Univ. Fac. Educ. Mem., no. 14, sec. 1, p. 24-35.

—— M. Matsumura, and T. Matsumoto, 1974, On the age of the Mifune group, central Kyushu, Japan, with a description of ammonite from the group Kumamoto Univ. Fac. Educ. Mem., no. 23, sec. 1, p. 47-56.

Tanabe, K., et al, 1977, Stratigraphy of the Upper Cretaceous deposits in the Obira area, northwestern Hokkaido: Kyushu Univ., Dept. Geol. Sci. Rept., v. 12, p. 181-202.

Tanaka, K., 1965, Izumi group in the central part of the Izumi mountain range, Southwest Japan, with special reference to its sedimentary facies and cyclic sedimentation: Japan Geol. Surv. Rept., no. 212, 33 p.

Teraoka, Y., 1977, Cretaceous sedimentary basins in the Ryoke and Sambagawa belts, *in* K. Hide, ed., The Sambagawa belt: Hiroshima Univ. Press, p. 419-431.

Yanagi, T., 1975, Rubidium-strontium model of formation of the continental crust and the granite at the island arc: Kyushu Univ. Fac. Sci. Mem. ser. D, v. 22, p. 37-98.

York, D., 1966, Least-squares fitting of a straight line: Canadian Jour. Phys., v. 44, p. 1079-1086.

Yoshii, M., and M. Katada, 1974, Granitic rocks in the Northern Kitakami mountains, *in* Cretaceous granitic rocks in the Kitakami mountains, petrography and zonal arrangement: Japan Geol. Surv. Rept., no. 251, p. 8-22.

Isotopic Methods in Quaternary Geology[1]

VLADIMIR ŠIBRAVA[2]

Abstract Results achieved in different Quaternary sediments by different methods are discussed, especially with regard to the worldwide correlation of Quaternary sediments. The necessity of a mutual checking of isotopic dating results by other geologic, paleontologic, and geophysical methods is stressed. There is danger of overestimating some methods of absolute dating (in Quaternary geology) which has led in many cases to false stratigraphic conclusions. There is danger especially in conditions of continental sedimentation, where the Quaternary is characterized by the alternation of different genetic and lithologic types of deposits. Isotopic methods must be regarded only as one of the existing methods leading to the establishment of a geochronologic scale, considering the many existing breaks of sedimentations.

Introduction

The Quaternary, the youngest geologic period, covers about 1.8 m.y. According to the recommendations of the INQUA, the IUGS Stratigraphic Commission, and the decision of the 24th International Geological Congress at Montreal, 1972, the Plio-Pleistocene boundary is defined now at the base of the Mediterranean Calabrian by the first appearance of the psychrophilic species *Arctica islandica* and *Hyalinea baltica*. For determination of the base of the Quaternary and for its time-stratigraphic assignment, isotopic geochronometry (used either in conjunction with other geophysical methods or with classic stratigraphic methods) is being applied with greater frequency.

Survey of Isotopic Methods

The most widely used dating method in Quaternary geology is by carbon-14. Thousands of dates are available now. The present technical standard of radiocarbon dating makes possible carbon 14 dates to 50,000 years B.P., and exceptionally to 70,000 years which corresponds to a period equaling 10 half lives of C^{14}. Thus, the method covers the Holocene and a part of the upper Pleistocene. The disadvantage of its relatively short time range became obvious when attempts were made to determine the duration and the boundary of the last interglacial epoch, which is beyond the range of this method of dating.

A second geochronometric method applicable to the Quaternary is the potassium-argon (K-Ar) method that recently has been used in dating rock ages near the Pleistocene-Pliocene boundary and as a correlation method for rocks in paleomagnetic studies. Dating of more recent Pleistocene rocks continues to be a problem and is still restricted to several highly sophisticated laboratories. Even K-Ar dates around 10,000 years (Dalrymple, 1968) are known from literature, so that theoretically radiometric dates may supplement C^{14} dates. However, the two methods do not overlap in prac-

[1] Manuscript received, October 7, 1976.
[2] Geological Survey of Czechoslovakia, C.S.S.R.

Article Identification Number: 0149-1377/78/SG06-0012/$03.00/0

tice and are applicable only to the end (C^{14}) and beginning (K-Ar) of the Quaternary period. The major part of Quaternary time is virtually undated by these methods.

For this reason attention has been focused on methods using the disequilibrium of the uranium and thorium radioactive series; namely at the U^{234}/U^{238} and uranium, $Th/Io^{230}/U^{238}$ methods. The half lives of the U^{234} and Th^{230} isotopes (284,000 and 75,000 years) allow determinations to cover practically the entire Quaternary Period, but these methods are mainly convenient for the dating of marine and fresh-water sediments, respectively, and exceptionally useful for carbonate travertines and sediments of karst caves (U/Io). However, these methods are applicable to a limited extent only in genetically and lithologically varied Quaternary sediments.

Application of the age study by stable nonradiogenic isotopes (i.e. D [Deuterium], C^{13}, N^{15}, O^{18} and S^{34}) for Quaternary material is essentially the same as for material from earlier geologic periods. Of special importance now in Quaternary geology is the oxygen isotopic paleotemperature analysis (Urey, 1947), because Pleistocene time is affected by significant changes of climate. This method is gaining favor because of the increasing interest in the stratigraphy of Quaternary marine sediments, which are far better traceable in uninterrupted sequences than continental deposits. Rhythmic temperature variations in sea basins are reflected in the isotopic composition of oxygen in fossil forams of *Globigerina* oozes (that is in samples from regions not influenced by continental water containing the light oxygen isotope). For instance the Emiliani (1955) paleotemperature curve, based on foraminifers recovered from the first meters of deep sea cores has shown that during individual glacial intervals the seawater temperature dropped by 5 to 6°C. Later, it became obvious that Emiliani had undervalued the changes in the isotopic composition of seawater caused by the formation of huge ice sheets containing light oxygen isotopes. Consequently, water enriched in O^{18} isotope remained in the sea when the water level was significantly lowered. According to Shackleton's (1967) calculations, Emiliani's curve is a consequence of both factors, that is not only of the lowering of the water level in the sea but also of its enrichment by the O^{18} isotope. Actual temperature variations of seawater did not exceed 3°C.

In continental deposits, (stalagmites in karst caves), a similar or even a more pronounced variation of the O^{18} value is indicated. Also we have to take into account both the primary temperature factor and the changes in isotopic composition of vadose water from which the stalagmites were formed. Whereas seawater during the glacials was richer in O^{18} than standard mean ocean water (SMOW), the vadose water on the continents was depleted in O^{18}. After a similar correction for the isotopic composition of water, results indicate (when studying stalagmites) a variation of isotopic paleotemperatures which is greater than without correction (the opposite of Emiliani's marine samples).

The fact that vadose water became depleted in O^{18} in the Pleistocene is also furnished by the study of O^{18} and deuterium in hydrogeology. For instance, in studying artesian water of the Nubian sandstones in the eastern Sahara from O^{18} values (about -10 per mil of SMOW) it is determined that this water cannot be due to present-day rainfall in the Mediterranean region, because it is richer in O^{18} isotope (about -3 per mil). Degens, in Knetsch et al (1962), claimed that this is due to infiltrated water of the pluvial episode of the Pleistocene confirmed by a radiocarbon age of the water (30,000 years). Therefore, these methods (O^{18} and deuterium determinations) might be used also for the correlation of pluvial and interpluvial episodes with glacial and interglacial intervals.

Significance of Isotopic Methods

Radiometric dating is one of the important techniques for correlation of Quaternary sediments on an intercontinental and global scale where other methods are difficult to use. However, it should always be used in conjunction with other available methods such as paleontologic and geophysical techniques. The K-Ar method has brought good results in the correlation of the Plio-Pleistocene boundary. Savelli (1975) gave (in his survey of absolute ages from the time span of the Plio-Pleistocene boundary) 21 dates, 14 measured by the K-Ar method and 2 by the fission-track method.

The K-Ar method also was used in determining the 2.3 ± 0.2 m.y. age of the volcanic rocks of the Roccastrada (Grosseto, Italy), the 1.6 m.y. age of the basalts in Valros (southern France) lying between the Astian and Villafranchian (Evernden et al, 1964), and the age of anorthoclase from the well-known Olduvai Gorge in Tanzania.

The age of Olduvai Gorge determined by the fission-track method is 2.03 ± 0.23 m.y. Because of the remains of fossil Man at this locality, these results are some of the most important radiometric dates of the Quaternary. At Mt. Coupet in France (Savage and Curtis, 1970), the age of the basalt is 1.9 m.y., the age of the so-called "Cover-Basalt" in northern Israel (Siedner and Horowitz, 1974) is 1.7 to 2.0 ± 0.1 m.y., and the age of basalts from the western coast of New Zealand, Auckland Province, (Stipp et al, 1967) is 2.5 ± 0.1 m.y. The latter material is correlated with a widespread ice sheet (the earliest glaciation on New Zealand). A similar age is given by Savage and Curtis (1970) for pumice fragments in tuff containing the Roca Neyra fauna in southern France. A K-Ar date of 3.0 ± 0.1 m.y. for the Plio-Pleistocene boundary from Palos Verdes Hills in California is reported by Obradovich (1965). Dates older than 1.8 m.y. predate the presently accepted boundary. This boundary at 3 m.y. has been based on the study of fossil vertebrates especially by vertebrate paleontologists; K-Ar and fission track dates prior to 3.5 to 4.2 m.y. are from the late Pliocene.

Dates of marine sediments from deep-sea cores when supplemented by paleomagnetic methods (Shackleton, 1969; Shackleton and Opdyke, 1973) allow limited correlation of marine and terrestrial sediments. Oxygen isotope analyses and paleomagnetic measurements have provided supplemental results. In evaluating borehole Verna 28-238 in the Pacific, Shackleton and Opdyke (1973) correlated an uninterrupted sequence of sediments representing the past 870,000 years with the sequence of strata reported by Emiliani (1955) from the Caribbean Sea and Atlantic Ocean. Oxygen-carbon isotope analysis has yielded new data on climatic changes during the Cenozoic; Shackleton and Kenneth (1975a, b) proposed a rapid accumulation of the Antarctic continental ice sheet at the onset of the late Miocene and found that even the subsequent temperature increase during the late Miocene had not influenced the extent of the ice.

Even though oxygen and carbon isotope analysis of deep-sea cores has supplied many data on temperature changes on whose variations the stratigraphic subdivision of the Quaternary is based, the problem of the correlation of Quaternary deep-sea and continental sediments remains. Problems in correlation are due, not only to the geologic problems associated with the stratigraphy of continental and marine sediments and the methods used in their investigation (greater numbers of stratigraphic hiatuses in the continental sediments and different concepts of the classification of glacial and interglacial intervals), but also to the differences in the dating techniques and to the lack of a single dating method that would apply to sediments deposited under continental conditions during the entire Pleistocene and Holocene time span. In contrast, material for radiocarbon dating is lacking in suitable marine sediments. Hence, even though variations of climate in both marine and continental sediments have been proved by several methods, problems in the correlation of continental and marine sediments still remain, especially in middle and early Pleistocene time because of the absence of accurately defined boundaries between individual glacial and interglacial intervals. A unified stratigraphic time scale and the refinement of the cyclicity of paleoclimate in the Quaternary requires additional study.

Problems of Radiometric Dating

Introduction of new, exact radiometric dating techniques into Quaternary stratigraphy has meant a significant step forward in the study of this geologic period. On the other hand, as is often the case with new methods, there has been a tendency to overrate radiometric dating at the expense of other geologic or paleontologic methods and lithologic studies that are indispensable in every type of stratigraphic investigation.

Carbon-14 dating, applicable to the latest Pleistocene and Holocene time span, is the most frequently used radiometric dating technique in Quaternary geology. However, after it was introduced, some controversial dates appeared; they were due to one or more of the following—imperfect processing of the sample, inadequate sampling, distortion of a secondary character, and influences from the atmosphere such as enrichment with radiocarbon-14 from nuclear weapon tests. Hence, we find in the literature different ages are reported on the same beds from the same localities by different laboratories.

Data that were interpreted without having been sufficiently checked against data obtained by other methods have frequently led to erroneous stratigraphic conclusions which have persisted in the geologic literature. Apart from the inaccuracies of the carbon-14 analyses, a major inadequacy in interpretation has been the often-neglected fact that stratigraphic hiatuses exist in Quaternary continental sediments. As the interdisciplinary study of a number of key sections in Europe has shown, stratigraphic hiatuses are so frequent that we always have to bear in mind that isotopic dates apply to one definite horizon only and that the age of the underlying and overlying beds, respectively, and the time relation of these beds to the defined horizon have to be determined by using sedimentologic, paleontologic, or other types of analysis. Instances where relatively recent sediments are separated by a disconformity from appreciably earlier deposits have been sufficiently documented; but most erroneous conclusions in carbon-14 dating are due to overrating of its results at the expense of other regional methods and classical field observations.

Inconsistencies often appear even where dating earlier time intervals. Many paleomagnetic dates used in correlation studies and for the absolute age determination of the Quaternary, notably for areas where deep-sea cores have been recovered, have not been sufficiently checked against dates obtained using isotopic methods. This has led to different interpretations in the stratigraphic classification of the Quaternary. Olausson (1976) for instance, gives two Quaternary classifications, the so-called "long Quaternary chronology" and a "short Quaternary chronology," which are based on the extrapolated C^{14} scale and the correlation of Pleistocene stratigraphy and the chronology of the geomagnetic field reversals (Cox, 1969). According to the "short Quaternary chronology," the Brunhes-Matuyama (Paleomagnetic reversal) boundary is around 390,000 years and the late Pliocene - early Pleistocene boundary is at about 625,000 years; the writer favors the "short Quaternary chronology" and believes that the paleomagnetic methods and other data used in establishing the "long Quaternary chronology" are being erroneously interpreted.

Conclusion

Results of isotopic dating of the Quaternary thus far have significantly advanced the study of this geologic period. However, we must be careful not to rely too strongly on certain isotopic and other physical, geophysical, and physico-chemical methods. As yet, no single method has been developed that will solve intricate stratigraphic and chronologic relations by itself.

Classification of the Quaternary will continue to rely on interdisciplinary studies using all available methods, including regional geologic, paleontologic, petrographic, lithologic, and geochronologic studies. Disregard of such an approach may lead to erroneous results, such as we have witnessed in the past.

References Cited

Cox, A., 1969, Geomagnetic reversals; Science, v. 163, p. 237-245.

Dalrymple, G. B., 1968, Potassium-argon ages of Recent rhyolites of the Mono and Inyo Craters, California: Earth Planetary Sci. Letters, v. 3, p. 289-298.

Emiliani, C., 1955, Pleistocene temperatures: Jour. Geology, v. 63, p. 149-158.

Evernden, J., et al, 1964, Potassium-argon dates and the Cenozoic mammalian chronology of North America: Am. Jour. Sci., v. 262, p. 145-198.

Knetsch, G., et al, 1962, Untersuchungen an Grundwassern der Ost-Sahara: Geol. Rundsch., v. 52, p. 587-610.

Obradovich, J. D., 1965, Age of the marine Pleistocene of California: AAPG Bull., v. 49, p. 1087.

Olausson, E., 1976, Problems in Pleistocene correlations and datings: Report no. 3, IGCP Proj. 73/I/24 (Bellingham).

Savage, D. E., and G. H. Curtis, 1970, The Villafranchian Stage-age and its radiometric dating, in O. L. Bandy, ed., Radiometric dating and paleontologic zonation: Geol. Soc. America Spec. Pub. 124, p. 207-231.

Savelli, G., 1975, Radiometric ages relevant for the Plio-Pleistocene boundary and for the Villa-franchian Stage: IGCP, Proj. NO 73-I-41–Neogene - Quaternary Boundary Symp. (Bologna).

Shackleton, N. J., 1967, Oxygen isotope analyses and Pleistocene temperatures re-assessed: Nature, v. 215, p. 15-17.

—— 1969, The last interglacial in the marine and terrestrial records: Royal Soc. London Proc. Bull. 174, p. 135-154.

—— and J. P. Kennett, 1975a, Paleotemperature history of the Cenozoic and the initiation of Antarctic glaciation–Oxygen and carbon isotope analyses in DSDP Sites 277, 279 and 281, *in* J. P. Kennett et al, eds., Initial reports of the Deep Sea Drilling Project, v. 29.

—— —— 1975b, Late Cenozoic oxygen and carbon isotopic changes at DSDP Site 284–Implications for glacial history of the Northern Hemisphere and Antarctica, *in* J. P. Kennett et al, eds., Initial reports of the Deep Sea Drilling Project, v. 29.

—— and N. D. Opdyke, 1973, Oxygen isotope and paleomagnetic stratigraphy of equatorial Pacific core V28-238–Oxygen isotope temperatures and ice volumes on a 10^5 year and a 10^6 year scale: Quaternary Research (Wash. Univ. Quat. Res. Cent.), v. 3, no. 1, p. 39-55.

Siedner, G., and A. Horowitz, 1974, Radiometric ages of late Cenozoic basalts from northern Israel–chronostratigraphic implications: Nature, v. 250, no. 5461, p. 23-26.

Stipp, J. J., et al, 1967, K-Ar estimate of the Pliocene-Pleistocene boundary in New Zealand: Am. Jour. Sci., v. 265, p. 462-474.

Suggate, R. P., 1974, When did the last interglacial end?: Quaternary Research (Wash. Univ. Quat. Res. Cent.), v. 4, no. 3, p. 246-252.

Urey, H. C., 1947, The thermodynamic properties of isotopic substances: Chem. Soc. London Jour., Part 1, p. 562-581.

Status of the Boundary between Pliocene and Pleistocene[1]

K. V. NIKIFOROVA[2]

Abstract The boundary between the Neogene and Quaternary (Anthropogene) Systems has been the subject of discussion for many years. Its position has been discussed at almost all INQUA congresses since 1932 (2d International Conference, Association for Study of Quaternary Period in Europe, Leningrad), as well as at international geological congresses. At the 28th Session of IGC held at London (1948) it was proposed to draw the boundary between the Pliocene and Pleistocene at the first signs of climatic cooling in the marine section of the Italian Neogene. Calabrian marine deposits include the first appearance of northern immigrants in their faunas and their stratigraphical continental analogue, the Villafranchian, was adopted as the basal member of the Pleistocene. The 29th Session of IGC held in Algeria (1952) approved the resolutions of the 28th IGC.

At the 5th INQUA Congress at Madrid and Barcelona (1957) the Subcommission on the Pliocene-Pleistocene boundary was organized under the Commission on Nomenclature and Stratigraphy of the Quaternary. The Subcommission has submitted recommendations to every subsequent INQUA Congress but these were not approved by the Sessions of the International Geological Congress.

In 1972, the International Colloquium on the lower boundary of the Quaternary organized by the INQUA Subcommission was held in the Soviet Union. Decisions made at this Colloquium were approved at the 24th International Geological Congress held in Canada in 1972. These recommendations confirmed a proposal submitted to the 28th International Congress that Italy be the stratotype area for the boundary between the Neogene and Quaternary Systems, and the original definition that the base of the Pleistocene should be drawn in marine deposits at the lowermost level in the section at La Castella, Catanzaro, where *Hyalinea baltica* is recognized. In addition, the 1972 International Colloquium changed the correlation of the Calabrian with the subdivisions or biozones of Villafranchian continental deposits because, as proved by subsequent studies, the lower Villafranchian corresponds to the Astian, not the Calabrian. Where Calabrian analogues cannot be established easily, it was decided to use local subdivisions based on established stratotypes.

Introduction

Since the recommendations of the International Colloquium and 24th IGC in 1972, most specialists have agreed on the position of the boundary. Further studies dealing with an exact position of the boundary between the Neogene and Quaternary, and the correlation of upper Pliocene and Lower Quaternary deposits on a global scale have continued using paleontology, paleoclimatology, absolute age determinations, and paleomagnetism.

In 1974, the "Neogene-Quaternary boundary" was approved as a key project in the IGCP Program. The working group held its first meeting at Barcelona in September 1974, in conjunction with the INQUA Subcommission on the Plio-Pleistocene boundary. Some important Neogene and Quaternary sections were examined during excursions in Spain and France. The 2d Symposium of the IGCP Working Group and the

[1] Manuscript received, October 5, 1976.
[2] Geological Institute of the USSR Academy of Sciences, Moscow.

Article Identification Number: 0149-1377/78/SG06-0013/$03.00/0

INQUA Subcommission on Plio-Pleistocene boundary was held at Bologna in October, 1975. Excursions were held to the stratotype sections of Calabrian, Villafranchian, and the boundary between the Neogene and Quaternary. It was of great importance, as in accordance with the decisions of the INQUA International Congresses and International Geological Congresses (including its last Session held in 1972 at Montreal), Italy was adopted as a stratotype area for establishing the boundary between the Neogene and Quaternary.

At present the Pliocene–Pleistocene deposits of the ocean basins and continents of both the northern and the southern hemispheres are correlated with the deposits of these sections. This fact makes it necessary to discuss, first of all, the position of the Neogene-Quaternary boundary in Italy. The reports delivered by Selli (1975) and Azzaroli (1975) at the previous Symposium in Italy, and the excellent guide book of excursions, contributed greatly to an understanding of these relations.

Neogene-Quaternary Boundary Data

The history of the Calabrian stage and the position of the Neogene-Quaternary boundary in the Italian section can be briefly summarized as follows—the Calabrian stage was defined by Gignoux (1910, 1913) as the upper part of the Pliocene that overlay the Astian, and was overlain by the Sicilian (the beginning of the Quaternary). Later, Gignoux agreed to draw the lower boundary of the Quaternary system at the base of the Calabrian stage. Calabrian deposits were characterized by cold-water North Atlantic immigrant faunas. Gignoux suggested the Santa Maria di Catanzaro section (Calabria) as a stratotype. Emiliani, Mayeda, and Selli (1961) studied the section at La Castella (Crotone, Calabria) and carried out a paleotemperature analysis of deposits, using oxygen isotopes. The lower boundary of the Pleistocene is marked by the appearance of *Hyalinea baltica* and the section is boreal clay deposited under open-sea conditions (depth about 500 m) and rich in planktonic organisms. The study shows a decrease of surface paleotemperatures from the upper Pliocene to the Calabrian and the Calabrian shows abundant temperature oscillations.

Ruggieri (1961) showed that in some Calabrian sections *Arctica islandica* appears earlier than *Hyalinea baltica*, and suggested that the Neogene-Quaternary boundary be drawn on the first appearance of *H. baltica*. In 1965 he changed his opinion to conform with Gignoux (1913) and made the Calabrian the oldest part of the marine Pleistocene characterized by the appearance of *Arctica islandica* in the Mediterranean. *Hyalinea baltica* appears later in littoral and epineritic sections, and thus, cannot always be used to determine the base of the Calabrian.

Ruggieri et al (1975) suggested that the Calabrian stage be a period of time characterized by abundant *A. islandica* without *H. baltica*, that the Emilian stage be determined by the association of *A. islandica* and *H. baltica*, and finally, that the Sicilian stage be characterized by these two species and *Globorotalia truncatulinoides*.

Selli reported at the 8th Congress of INQUA (Denver, 1965) that all North Atlantic immigrants used for establishing the Pliocene-Pleistocene boundary in Italy are benthonic forms, and their first appearance was controlled by the environment and the ecology of each species. He suggested that the Neogene-Quaternary boundary be drawn at the first appearance of any coldwater species, whether it be *A. islandica, H. baltica,* or any other species, and that the marine Pleistocene be subdivided into two parts—preglacial and glacial. The beginning of the Pleistocene was estimated at 1.82 m.y. and the beginning of its glacial part at 0.82 m.y. These dates generally have been accepted.

During the last decade many geologists have worked in Italy, studying the sections at Santa Maria di Catanzaro, La Castella, Vrica, Santerno, and others.

The Santa Maria di Catanzaro section is not the best one for determination of the position of the Neogene-Quaternary boundary. It contains many missing intervals as a result of underwater erosion. *Arctica islandica* and *Hyalinea baltica* are present throughout the section and the fauna abounds in redeposited species. *Globorotalia truncatulinoides* was recognized 48 m below the base of the horizon "G" by Gignoux (1913) who considered this horizon the base of the Calabrian. Selli (1962, 1967) believed that horizon "G" is a turbidite (graded, parallel lamination), thus confirming its deep-sea origin. However, Sprovieri et al (1973) considered it a transgressive cal-carenite of shallow-water origin. An interpretation of the paleomagnetic inversion in the Santa Maria di Catanzaro section has not been made.

The La Castella section at Calabria is more suitable for determination of the Neo-gene-Quaternary boundary. The boundary is drawn at the base of the turbidite layer, where *Hyalinea baltica* and cold-water species of ostracods appear for the first time. Just below the boundary *Globigerinoides obliquus extremus* disappears and, 2 m lower, *Globorotalia truncatulinoides* appears. According to Nakagawa et al (1975), the boundary corresponds to the Olduvai (=Gilsa) event.

Analysis of benthonic forams shows that the basin was 500 m or more (up to 1,000 m) in depth. On the basis of other species of forams, ostracods, and nannoplankton, this section can be correlated worldwide.

All the stratigraphers of Italy agree to draw the Neogene-Quaternary boundary at the base of the Calabrian. This idea is supported by most geologists from other countries. But, bearing in mind the resolution of the IGC (London, 1948), this bound-ary should proceed from distinct climatic deterioration fixed in the sections of Italy. In this relation it is necessary to select such a section where the boundary would be marked by appearance of North Atlantic immigrants. An ideal stratotype section for the Neogene-Quaternary boundary should be continuous, homogenous, formed in the bathyal or deep-sea environment, rich in fossils and planktonic organisms, and correla-table with sections in other parts of the world. Section Vrica, south of Crotone, satisfies all these requirements and as a result of a decision of the 2d Symposium of the IGCP Working Group held in Italy in 1975, it is intended to study this section thoroughly and submit it as stratotypical at the 10th Congress of INQUA in England in 1977.

Calabrian-Villafranchian Correlation

At the 2d Symposium and during the excursions, problems of correlation of the marine Calabrian and the continental Villafranchian deposits of Italy were discussed. Studies by Azzaroli (1970, 1975) distinguished six faunal zones in the Villafranchian. These zones are (from the base upwards)—the Triversa and Montopoli (lower Villa-franchian); the Saint Vallier (middle Villafranchian); and the Olivola, Tasso, and Farneta (upper Villafranchian). All but the Saint Vallier zone are recognized in Italy. Fauna of the Saint Vallier zone that corresponds to an intense erosive phase has not yet been established in Italy.

On the basis of radiologic data, age limits of these zones range from 3.8 m.y. (Triversa zone) to 1 m.y. (Farneta zone). The Plio-Pleistocene boundary, according to Azzaroli, is drawn between the Montopoli and Saint Vallier zones, that is beneath the middle Villafranchian. However, this conclusion does not correspond to the radiologic age of the base of the Pleistocene (1.8 m.y. - Olduvai event), whereas the base of the middle Villafranchian is dated not younger than 2.5 m.y. The lower Pleistocene appears to be correlated more reliably with the upper Villafranchian—Olivola or even Tasso zone, where the Olivola fauna were correlated by Azzaroli with the fauna of Tiglian, which was correlated by Zagwijn (1974, 1975) with the uppermost parts of the middle Villafranchian.

General Remarks

Some available data on the Neogene-Quaternary boundary are available worldwide, and are summarized in Figure 1. Some of these data were published in Spain and others are included in papers of the 2d Symposium on the IGCP Project "Neogene-Quaternary boundary" that were published in Italy. Still other data have been pub-lished in papers of the 9th INQUA Congress held in New Zealand in 1973.

Some general theses and principles that must serve as the basis for establishing the Neogene-Quaternary boundary are as follows:

1. To confirm principal decisions of the 18th IGC held in London in 1948—(a) the Pliocene-Pleistocene (Neogene-Quaternary) boundary should be established in accor-dance with general stratigraphic principles and be based on changes in marine faunas; (b) a horizon chosen for the Pliocene-Pleistocene boundary should record the cooling

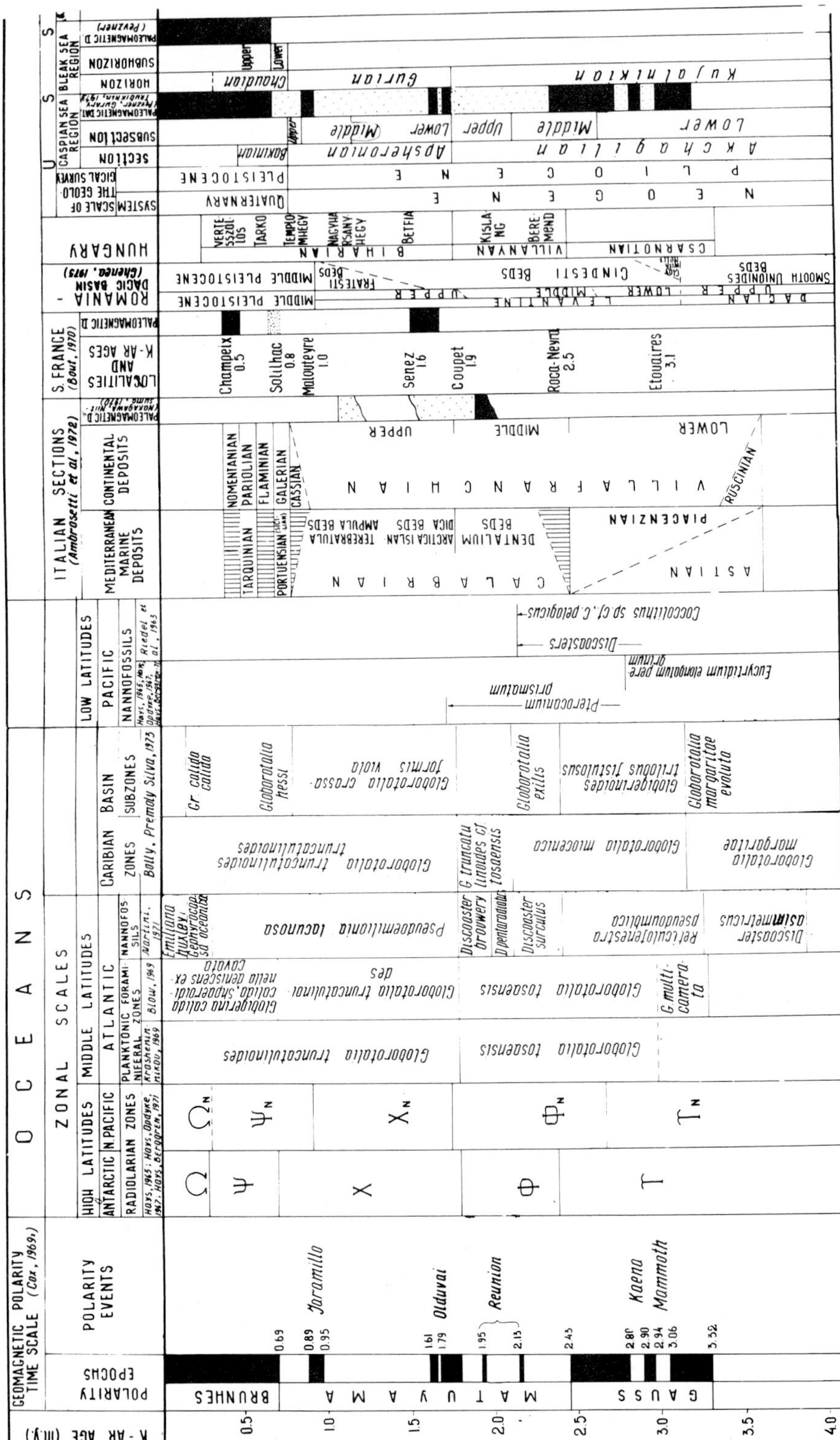

FIG. 1—Correlation diagram for the Neogene-Quaternary boundary.

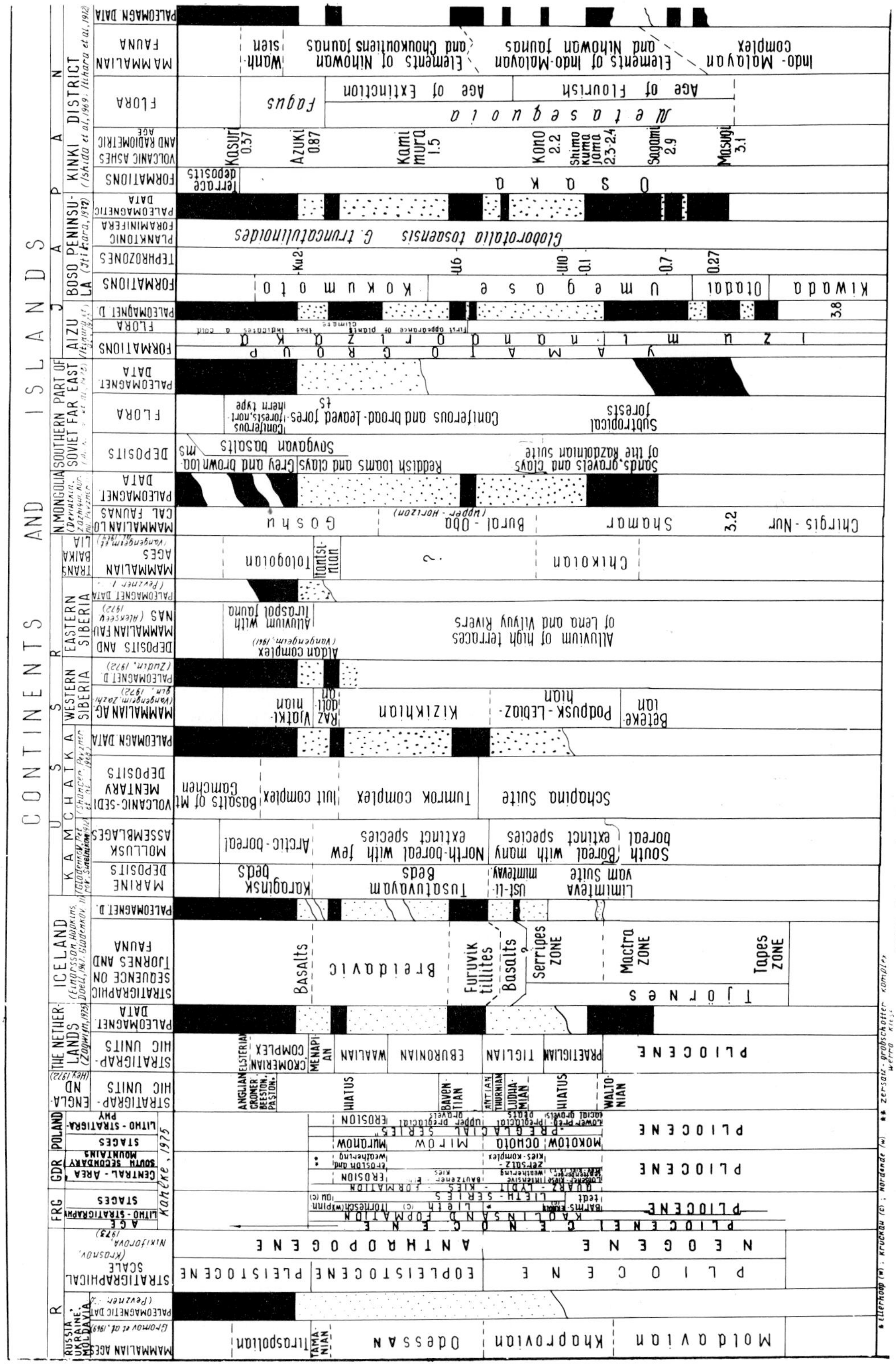

FIG. 1 (cont.)–Correlation diagram for the Neogene-Quaternary boundary.

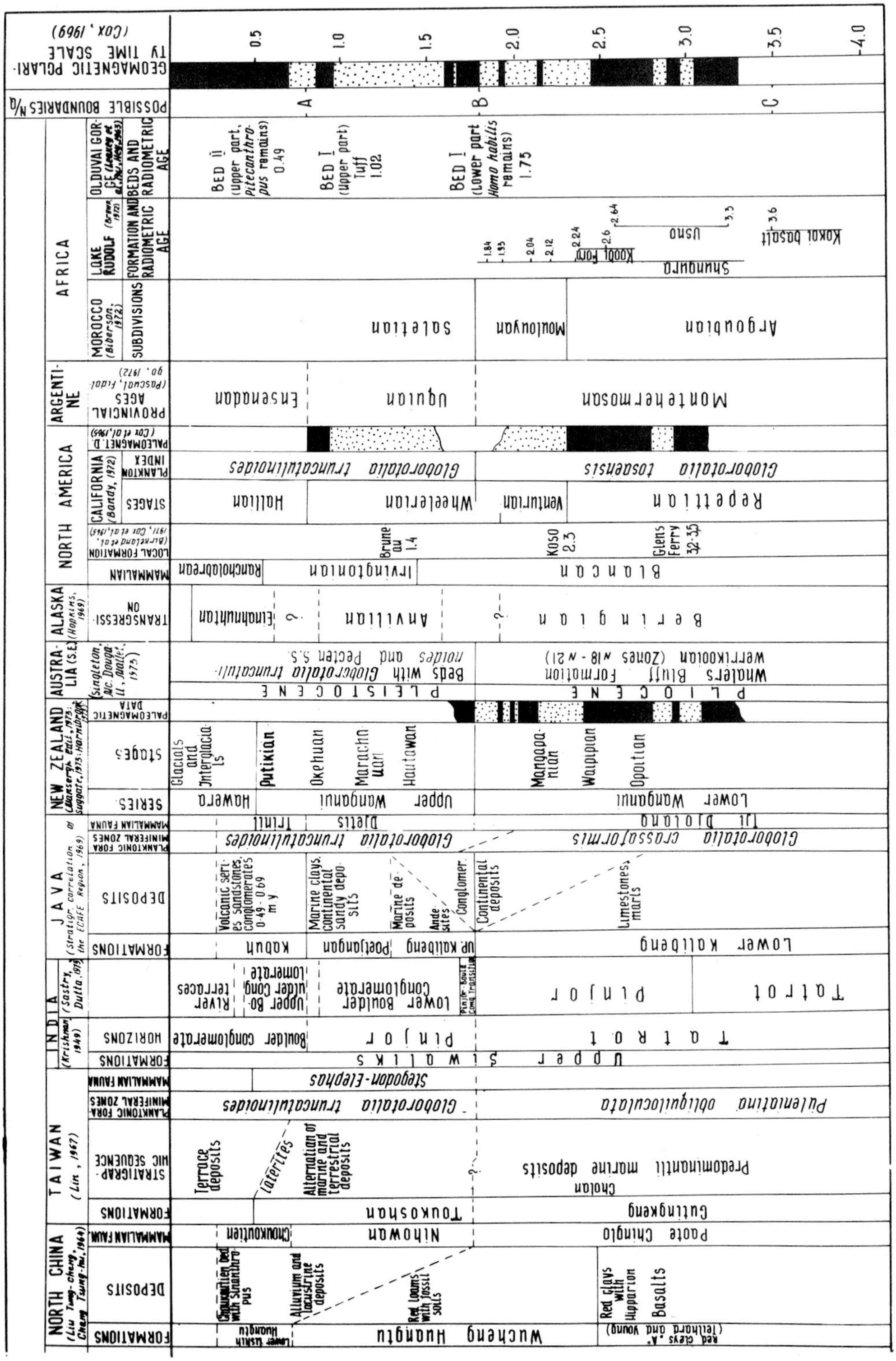

FIG. 1 (cont.)—Correlation diagram for the Neogene-Quaternary boundary.

of climate for the first time. The words "for the first time" should be omitted; (c) Italy is to serve as the stratotype area for establishing the boundary; and (d) the marine Calabrian Formation of Italy is the lowest unit of the Pleistocene.

2. If these theses are accepted, then these conclusions follow—(a) the position of the Neogene-Quaternary boundary should be determined in a continuous section of oceanic bottom sediments by all available methods—bio- and climato-stratigraphic, paleomagnetic, radiological, and paleotemperature. Most researchers believe that this is the lower boundary of the *Globorotalia truncatulinoides* zone coinciding with the Olduvai event, its radiological age being 1.85 m.y.; and (b) it is critical to study in detail the lower boundary of the Calabrian and its relation to the Neogene-Quaternary boundary. The suggestion of Selli (1975) that the Neogene-Quaternary boundary in the sections of Italy should be established by the first appearance of any cold-water immigrants (not only *Arctica islandica* or *Hyalinea baltica*) whould be adopted. The same boundary should also determine the base of the Calabrian. This will provide a stratotype common to both the Calabrian and the boundary. (I oppose dropping the determination of the stratotype area and the type section for establishing the Neogene-Quaternary boundary, as some researchers suggest).

3. I agree with Selli that we cannot draw the boundaries of any stratigraphic subdivision according to the first appearance of *Globorotalia truncatulinoides*, or the extinction of *Discoaster* in any local sections (including Italy) because the common time-stratigraphic subdivisions cannot be established through the local distribution of fossils.

4. As the Neogene-Quaternary boundary is isochronous, paleomagnetic and radiologic data should be taken into consideration for determining its position. Nevertheless, it would be wrong to rely on these data only without considering bio- and climato-stratigraphic criteria.

5. A detailed correlation with marine deposits is necessary for establishing the Neogene-Quaternary boundary in continental deposits. The boundary established in marine deposits should be taken as the standard. Data on paleomagnetic stratigraphy and radiologic age should be taken into account.

6. Determination of radiologic age and its place in the paleomagnetic scale will provide a reliable basis for global correlation of the exact horizon of the Neogene-Quaternary boundary in marine and continental deposits of Italy.

7. The position of the Neogene-Quaternary boundary in other countries and continents is determined, first of all, through working out the local detailed stratigraphic scales of the Pliocene and the Quaternary, correlating deposits occurring in various paleogeographical provinces with one another, and correlating marine with continental deposits.

Conclusion

The 3rd meeting of the IGCP Working Group on the Neogene-Quaternary boundary met jointly with the INQUA Subcommission on the Plio-Pleistocene boundary in Tokyo on May 20, 1976, during the First International Congress on Pacific Neogene Stratigraphy. Sections of Pliocene and Lower Quaternary deposits in the Boso Peninsula and in the Kinki and Kakegawa regions were examined and it was concluded: (1) that the Pliocene-Pleistocene sequences of Japan will be considered as key sections of the Neogene-Quaternary boundary for East Asia and the North Pacific (they can be correlated with the stratigraphic sections of the same age in Italy.); (2) that Italy is to remain the type area for the Plio-Pleistocene boundary, and the marine Calabrian formation of Italy is the key lower sequence of the Quaternary; (3) that the Neogene-Quaternary boundary as a time stratigraphic boundary has to be defined in a continuous section of oceanic bottom sediments by all available methods; and (4) that further micropaleontologic and paleomagnetic studies are necessary, particularly, in connection with the position of the *Globorotalia truncatulinoides* datum plane.

References Cited

Azzaroli, A., 1970, Villafranchian correlations based on large mammals: Giorn. Geol. (Bologna Mus. Geol.), v. 35, no. 1, p. 111-131.

——— 1975a, Communications, recommendations, conclusions–Group de Travail Limite Pliocene-Quaternaire (Villafranchien) 5th Congres du Neogene Meditteraneen, Lyon: (France) Bur. Recherches Geol. Minieres Mem., v. 78, no. 1.

——— 1975b, The Vilafranchian stage in Italy and the Plio-Pleistocene boundary, 2nd Symp. IGCP Work Group Neogene-Quaternary Boundary: Scientific Papers, (Bologna).

Emiliani, C., T. Mayeda, and R. Selli, 1961, Paleotemperature analysis of the Plio-Pleistocene section at La Castella, Calabria, southern Italy: Geol. Soc. America Bull., v. 72, p. 679-688.

Gignoux, M., 1910, Sur la classification du Pliocene et du Quaternaire dans l'Italie du sud: Acad. Sci. Comptes Rendus, v. 150, no. 1, p. 841-844.

——— 1913, Les formations marine pliocenes et quaternaires de l'Italie du Sud et de la Sicile: Lyon Univ. Ann., n.s. n. 1, 693 p.

Nakagawa, H., et al, 1975, Magnetic stratigraphy of Late Cenozoic stage in Italy and their correlatives in Japan, in T. Saito, L. Burckle, eds., Late Neogene epoch boundaries: Micropaleontology.

Ruggieri, G., 1961, Alcune zone biostratigrafiche del Pliocene e del Pleistocene italiano: Riv. Italiana Paleontologia e Stratigrafia, v. 62, no. 4, p. 405-417.

——— 1965, A contribution to the stratigraphy of the marine lower Quaternary sequence in Italy: Geol. Soc. America Spec. Paper 84, p. 141-151.

——— et al, in press, Un affioramento di Siciliano nel quadro della revisione del la stratigrafica del Pleistocene inferiore.

Selli, R., 1962, Le Quaternaire marin du versant Adriatique–Jonien de la peninsule Italienne: Quaternaria, v. 6, p. 391-413.

——— 1967, The Pliocene-Pleistocene boundary in Italian marine sections and its relationship to continental stratigraphies, in Progress in Oceanography: v. 4, p. 67-86.

——— 1975, The Neogene-Quaternary boundary in the Italian marine formations: 2nd Symp. IGCP Work Group Neogene-Quaternary Boundary, Scientific Papers (Bologna).

Sprovieri, R., S. D'Agostino, and E. Di Stefano, 1973, Giacitura del Calabriano nei dintorni di Catanzaro: Riv. Italiana Paleontologia e Stratigrafia, v. 79, n. 1, p. 127-140.

Zagwijn, W. H., 1974, The Plio-Pleistocene boundary in western and southern Europe: Boreas v. 3, no. 3, p. 75-97.

——— 1975, Variations in climate as shown by pollen analysis, especially in the lower Pleistocene of Europe: Geol. Jour. (Liverpool) Spec. Issue, no. 6, p. 137-152.

A Radiometric Time Scale for the Neogene of the Paratethys Region[1]

DIONYZ VASS[2] and GEVORG P. BAGDASARJAN[3]

Abstract The radiometric time scale of Paratethys Neogene is presented. Fifty-four radiometric ages of the volcanic rocks are taken into consideration. The majority of them are controlled mostly by marine biostratigraphy. The Egerian (Oligo-Miocene) and Eggenburgian (lower Miocene) are dated only by glauconite ages contradictory among themselves and the volcanic rock ages. As a result, glauconite ages are omitted in the time scale. The only existing volcanic rock age (21.9 m.y.) and three ages of the lower Karpathian (between 19.4 and 20.7 m.y.) are considered as the maximum age of the Ottnangian and the Karpathian. The base of the lower Badenian (Moravian) is estimated to 16.5 ± 0.5 m.y., the base of the upper Badenian (Kosovian) 15 ± 0.5 m.y., the base of the Sarmatian to 13.3 ± 0.5 m.y., and the base of the Pannonian to 10.5 to 11.0 ± 0.5 m.y.

Radiometric calibration of the Pontian, Dacian, and Rumanian has not been estimated because reliable radiometric dates with biostratigraphic control are not available. Two ages from the top of the Eastern Paratethys Pliocene are of 0.95 and 2.26 m.y. Probably the older one is closer to the actual age of the Pliocene-Pleistocene boundary.

Introduction

Paratethys is the name given to the Oligocene and mainly Neogene sedimentation area extending from the Western Alps foredeep to the Aral Sea (Fig. 1). The concept was primarily introduced by Laskarev (1924). The Paratethys area could be divided into Western Paratethys, Central Paratethys, and Eastern Paratethys.

The Central Paratethys extends from Bavaria, Germany, throughout the foredeep of the Alps and Carpathians to Moldavia, Rumania, including the basins on the inner side of the Carpathian orogenic mountains. It is rich in volcanic rocks forming the mountain ranges which are in close relationship to marine and continental sediments. For this reason the effort of the last 10 years to establish the radiometric time scale of Paratethys Neogene was focused on Central Paratethys.

The Central Paratethys Neogene is biostratigraphically divided into regional stages. The modern biostratigraphic division of the Central Paratethys Neogene is as follows (youngest to oldest): Rumanian, Dacian, Pontian, Pannonian, Sarmatian, Badenian, Karpathian, Ottnangian, Eggenburgian, and Egerian. A modern definition of stage may be found in Senes et al (1975).

Radiometric ages used as fundamental dates for the Paratethys Neogene time scale were obtained by K-Ar analysis of the volcanic rocks (lava flow and tuffs). Mostly whole rock samples were dated. One radiometric age was obtained by fission track analysis.

[1] Manuscript received, January 28, 1977.
[2] Geological Institute of D. Stur, Mlynska dolina 1, 809 40 Bratislava, CSSR.
[3] Geological Institute of Armenian Academy of Science, 375019 Barekamutian No. 24, Erivan, USSR.

Article Identification Number: 0149-1377/78/SG06-0014/$03.00/0

179

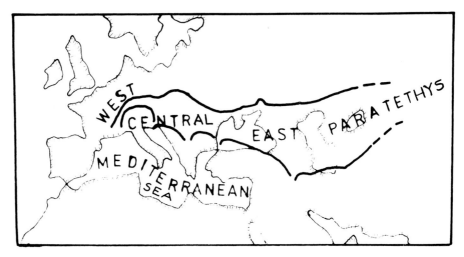

FIG. 1–Paratethys Neogene sedimentation area.

Fifty-four radiometric ages were used in the time scale construction. All are plotted in Figure 2 and listed in Table 1. More details concerning each age, geographic location, stratigraphic position, petrologic designation of the sample dated, method and decay constant originally used for analysis, evaluation of reliability, name of analyst, institution, and references are to be found in the appendix.

To facilitate and simplify a comparison of the time scale presented with the existing radiometric scales for the World Neogene, all K-Ar ages were calculated by means of the constants $\lambda_k = 0.584 \times 10^{-10} r^{-1}$; $\lambda_\beta = 4.72 \times 10^{-10} r^{-1}$. In most original papers the values of K-Ar ages were higher because the ages were calculated with a different constant, $\lambda_k = 0.557 \times 10^{-10} r^{-1}$.

Results

The Egerian, which is really a provincial stage of the Oligocene-Miocene boundary as well as the Eggenburgian has been dated by the radiometric ages of glauconite. Also some glauconite ages from the lowermost Ottnangian or uppermost Eggenburgian are reported.

Egerian

The Egerian provincial stage was originally dated by one age of 30.5 m.y., based on analysis of a glauconite from Eger (Odin et al, 1970, p. 221). Odin and Hunziker (1975, p. 144, 147) replaced the former datum by new radiometric ages obtained from the same glauconite–20.8 ± 1.3 m.y. and 19.6 ± 1.5 m.y. These are younger than the Oligocene-Miocene boundary in the modern time scale 23.5 m.y. (Selli, 1970), 22.5 m.y. (Turner, 1970; Berggren, 1969a, b, c, d, 1971a, b, c, 1972a, b; Berggren and Van Couvering, 1974), 23-24 m.y. (Thayer and Hammond, 1974; Ryan et al, 1974; Vass 1975). Neither the former nor the new datings are close to the actual age of the Egerian which is situated somewhere between those two extreme values.

Eggenburgian

Some authors recommend changing the former dating of glauconite from Bad Hall, Austria–25 m.y. (Evernden et al, 1961, p. 81, 99) to be replaced by another dating of (most likely the same glauconite horizon) 18.8 m.y. (Obradovich, 1964).

Ottnangian

Odin and Hunziker (1975, p. 142, 144) quoted an older dating of glauconite from Briglez, Romania–22.6 ± 1.2 m.y. but they abandoned it because it did not agree with

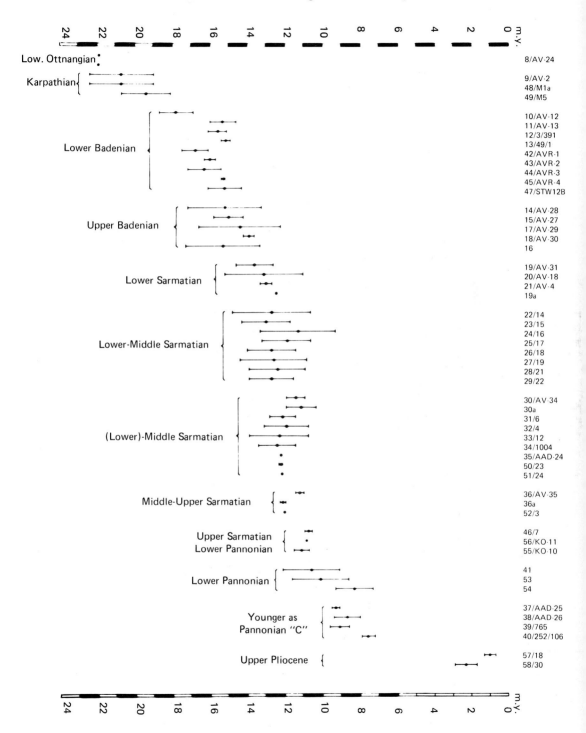

FIG. 2–Fundamental radiometric ages of the Parathys Neogene time scale (December 31, 1975).

Table 1. Review of Fundamental Radiometric Ages of Paratethys Neogene Time Scale.
(compiled by D. Vass, 1976)

Stratigraphic unit	Number	Sample	Site	Rock	Method	Radiometric age in $\lambda_K = 0.584 \times 10^{-1} y^{-1} / m.y$	Analyst
Akchagylian /Upp.Pliocene/	2	57/18	Dashbash[5]	andesito-	K-Ar/W.R./	0.95 ± 0.3	Rubinstein
		58/3	Trialeti[5]	-basalts	K-Ar/W.R./	2.26 ± 0.6	Rubinstein
Younger than Pannonian "c"	4	40/252/106	Bolshaja Beganj[3]	andesite	K-Ar/W.R./	7.6 ± 0.3	Bagdasarjan
		39/765	Sinjak Hill[3]	andesite	K-Ar/W.R./	9.1 ± 0.5	Bagdasarjan
		38/AAD-26	Vihorlat[1]	andesite	K-Ar/W.R./	8.7 ± 0.7	Bagdasarjan
		37/AAD-25	Vihorlat[1]	andesite	K-Ar/W.R./	9.3 ± 0.2	Bagdasarjan
Meotian / Lower Pannonian	3	54	Toloshi[5]	rhyodacite	K-Ar/W.R./	8.3 ± 1	Rubinstein
		53	Shabanebeli[5]	dacite	K-Ar/W.R./	10.1 ± 1.5	Rubinstein
		41	Saro[5]	rhyodacite	K-Ar/W.R./	10.6 ± 1.5	Rubinstein
Upp. Sarmatian / Lower Pannonian	3	56/KD-11	Stará Huta[1]	andesite	K-Ar/W.R./	10.8 ± 0	Bagdasarjan
		55/KO-10	Pstruša[1]	andesite	K-Ar/W.R./	11.1 ± 0.4	Bagdasarjan
		46/7	Kremnička[1]	rhyolite	K-Ar/W.R./	10.7 ± 0.2	Bagdasarjan
Upper-Middle Sarmatian	3	52/3	Vinné[1]	andesite	K-Ar/W.R./	12.05 ± 0.05	Kreuzer
		36a	Vinné[1]	andesite	K-Ar/W.R./	12.1 ± 0.15	Kreuzer
		36/AV-35'	Vinné[1]	andesite	K-Ar/W.R./	11.2 ± 0.25	Bagdasarjan
(Lower)/Middle Sarmatian	9	51/24	Zehna-Tuhriná[1]	andesite	K-Ar/W.R./	12.15 ± 0.05	Kreuzer
		50/23	Lesíček[1]	andesite	K-Ar/W.R./	12.2 ± 0.1	Kreuzer
		35/AAD-24	Tuhriná[1]	andesite	K-Ar/W.R./	12.2 ± 0	Kreuzer
		34/1004	Pelikan Hill[3]	rhyolite	K-Ar/W.R./	12.4 ± 1	Bagdasarjan
		33/12	Sharok Hill[3]	rhyolite tuff	K-Ar/W.R./	12.3 ± 1.6	Bagdasarjan
		32/4	Dlinnaja Hill[3]	obsidian	K-Ar	11.9 ± 1.2	Bagdasarjan
		31/6	Komárovce[1]	rhyolite	K-Ar/W.R./	12.1 ± 0.75	Bagdasarjan
		30a	Viničky[1]	rhyolite	F.T. /obsid./	11.1 ± 0.8	Repčok
		30/AV-34	Viničky[1]	rhyolite	K-Ar/W.R./	11.4 ± 0.5	Bagdasarjan
		29/22	Beregovo[3]	adularised		12.7 ± 1.2	Bojko et al.
		28/21				12.4 ± 1.4	-"-
		27/19				12.6 ± 1.8	-"-

Stage	Count	Sample No.	Locality	Rock type	Method	Age (Ma)	Author
Lower-Middle Sarmatian	8	26/18	Kosino[3]	and silici-fied tuffs		12,7 ± 1,3	-"-
		25/17	Kvasovo[3]			11,9 ± 1,3	-"-
		24/16				11,3 ± 2	-"-
		23/15				13,0 ± 1,3	-"-
		22/14				12,7 ± 2,1	-"-
Lower Sarmatian	4	19a	Ruskov[1]	andesite	K-Ar/W.R./	12,45 ± 0,05	Kreutzer
		19/AV-31	Ruskov[1]	andesite	K-Ar/W.R./	13,6 ± 1,0	Bagdasarjan
		21/AV-4	Sazdice[1]	rhyolite tuff	K-Ar/W.R./	13,0 ± 0,3	Bagdasarjan
		20/AV-18	Nižná Myšla[1]	andesite tuff	K-Ar/W.R./	13,1 ± 2,1	Bagdasarjan
Upper Badenian /Kosovian/	5	16	Malá Bara[1]	rhyolite	K-Ar/W.R./	15,3 ± 2	Tsonj
		18/AV-30	Královce[1]	rhyolite tuff	K-Ar/W.R./	13,9 ± 0,3	Bagdasarjan
		17/AV-29	Zamutov[1]	rhyolite	K-Ar/W.R./	14,4 ± 2	Bagdasarjan
		15/AV-27	Zatín[1]	andesite	K-Ar/W.R./	15,0 ± 0,8	Bagdasarjan
		14/AV-28	Žipov[1]	rhyolite tuff	K-Ar/W.R./	15,0 ± 2	Bagdasarjan
Lower Badenian /Moravian/	9	47/STW 12B	Weitendorf[4]	basalt	K-Ar/W.R./	15,2 ± 0,9	Lippolt
		45/AVR-4	Klause[4]	andesite	K-Ar/W.R./	15,3 ± 0,1	Bagdasarjan
		44/AVR-3	Klause[4]	andesite	K-Ar/W.R./	16,3 ± 0,9	Bagdasarjan
		43/AVR-2	Weitendorf[4]	basalt	K-Ar/W.R./	16,0 ± 0,3	Bagdasarjan
		42/AVR-1	Weitendorf[4]	basalt	K-Ar/W.R./	16,8 ± 0,75	Bagdasarjan
		13/49/1	Neresnica[3]	rhyodac.tuff	K-Ar/W.R./	15,2 ± 0,5	Bagdasarjan
		12/3/891	Novoselica[3]	rhyolite tuff	K-Ar/W.R./	15,6 ± 0,5	Bagdasarjan
		11/AV-13	Hrušov[1]	andesite	K-Ar/W.R./	15,3 ± 0,7	Bagdasarjan
		10/AV-12	Hrušov[1]	andesite	K-Ar/W.R./	16.45 ± 1.5	Bagdasarjan
Lower Karpathian	3	49/M5	Kisinocz-polytarnocz[2]	dacote	K-Ar/B./	19,4 ± 1,3	Balogh
		48/M1a	Magyarkút[1]	tuff	K-Ar /B./	20,7 ± 3,2	Balogh
		9/AV-2	Klenany[1]	glassy ash	K-Ar/W.R./	20,7 ± 1,7	Bagdasarjan
Lower Ottnangian	1	8/AV-24	Kalonda[1]	rhyolite tuff	K-Ar/W.R./	21,9 ± 0	Bagdasarjan
Total	54						

1 Czechoslovakia
2 Hungary
3 Transcarpathia, USSR
4 Austria
5 Georgia, USSR
F.T. = Fission Track
W.R. = Whole Rock
B. = biotite

their new dates. They presented four new dates of glauconite—17.5 ± 0.9 m.y. and 16.3 ± 1 m.y. from Cristotel, Rumania, and 17.6 ± 1.4 m.y. and 17.1 ± 1.5 m.y. from Briglez, Rumania.

It must be emphasized that all new dates of Ottnangian and Egerian glauconites as well as Obradovich's date for the Eggenburgian do not agree with radiometric ages obtained from volcanic rocks. The discordance between volcanic rock and glauconite radiometric dates is remarkable particularly for Ottnangian, Karpathian, and some lower Badenian volcanic rocks. At the moment it seems that it would be better if the glauconite ages were omitted in the constructed radiometric time scale.

The Ottnangian volcanic rocks were still dated only by one, but repeated radiometric age 8/AV-24—21.9 m.y.—rhyolite tuff from Kalonda, Slovakia (Vass et al, 1971, p. 322; Vass and Bagdasarjan *in* Papp et al, 1973, p. 37). The biostratigraphic age of the rock is not supported by fauna reliable for interregional correlation and the possibility of its being Eggenburgian in age is not excluded.

Calcareous nannoplankton of the Ottnangian in Austria indicates the correlation with the upper part of NN3 and lower part of NN4 standard nannoplankton zones (Martini and Müller, 1975, p. 121-123). On the other hand the age of 21.9 m.y. (Ryan et al, 1974), in the paleomagnetic and radiometric World Neogene time scale corresponds to NN2 standard nannoplankton zone (Fig. 3).

Several reasons for this disagreement may be taken into consideration, but in any case we need more radiometric ages with better biostratigraphic control to solve the problem. At present we must consider the age of 21.9 m.y. as the maximum age of the Ottnangian.

Karpathian

Only the lower part of the Karpathian was radiometrically dated. The lower Karpathian is a time interval before the *Globigerinoides sicanus* first appearance (base of N8 standard planktonic zone). In South Slovakia the lower Karpathian contains calcareous nannoplankton of NN4 (upper part) including *Helicopontosphaera ampliapertura* (Lehotayova, personal communication, 1975). Three radiometric ages come from the lower Karpathian of southern Slovakia and northern Hungary:

9/AV2: 20.7 ± 1.3 m.y.   glassy tuff from Klenany, Slovakia (Vass et al, 1971, p. 322)
48/M 1a: 20.7 ± 3.2 m.y.   tuff from Magyarkut, Hungary
49/M 5: 19.4 ± 1.3 m.y.   dacite between villages Nagyirtas and Ipolytarnocz, Hungary (Hamor et al, in press).

From the biostratigraphic evidences it could be concluded that the lower Karpathian is older than N8 and contemporaneous with NN4 zone.

The comparison with the Neogene radiometric time scale shows that three mentioned ages correspond to the lower Miocene. But the NN4 zone in the Ryan et al (1974) scale is younger (approximately 16.5 to 18.5 m.y.; Fig. 3). It must be stressed that the existing ages of the lower Karpathian are charged with relatively large errors. Perhaps two older ages with larger errors could be considered as the maximum ages for the lower Karpathian and the actual age is closer to the youngest radiometric age of 19.4 m.y. The upper Karpathian has not been radiometrically dated.

Badenian

The base of the Badenian, according to the biostratigraphic evidence, corresponds to the base of the Langhian; it means the Karpathian-Badenian boundary is inside the N8 standard planktonic zone. The radiometric ages concerning the lower Badenian are as follows:

10/AV-12: 16.45 ± 1.5 m.y. and 11/AV-13: 15.3 ± 0.7 m.y.   andesites from Hrusov, Slovakia (Vass et al, 1971, p. 323)
12/3/891: 15.6 ± 0.5 m.y.   rhyolite tuff from Novoselica, Transcarpathia, USSR
13/49/1: 15.2 ± 0.5 m.y.   rhyodacite tuff from Neresnica, Transcarpathia, USSR (Bagdasarjan and Danilovich, 1968; Table 1)
42/AVR-1: 16.8 ± 0.75 m.y., 43/AVR-2: 16.0 ± 0.3 m.y.   basalts from

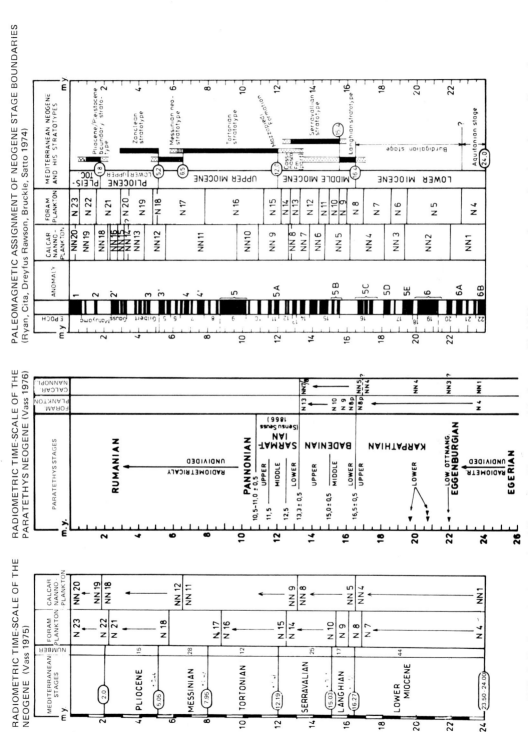

FIG. 3.–Radiometric time scale of Paratethys Neogene and both recent World Neogene time scales, one based purely on radiometric ages (Vass, 1975), the second based on paleomagnetic assignment of Neogene stage boundaries (Ryan et al, 1974).

Weitendorf, Austria

44/AVR-3: 16.3 ± 0.9 m.y., 45/AVR-4: 15.5 ± 0.1 m.y.   andesites from Klause, Austria, (Steininger and Bagdasarjan, in press; Bagdasarjan et al, in preparation)

47/STW 12B: 15.2 ± 0.9 m.y.   basalt from Weitendorf, Austria (Lippolt et al, 1975)

All sequences from which the dated rocks come are biostratigraphically a little higher than the base of the Badenian because *Orbulina suturalis* is always present (*Orbulina* datum equals the base of N9 standard planktonic zone).

The majority of the existing radiometric lower Badenian ages range from 16.45 to 15.2 m.y. We suggest 16.5 ± 0.5 m.y. as the radiometric age of the base of the Badenian. The datum agrees well with the base of the Langhian in both recent World Neogene time scales of Ryan et al (1974) and Vass (1975).

Beginning from the middle Badenian (Wieliczkian) the Paratethys Neogene is free of the plankton and nannoplankton forms which are typical for the standard zonation.[4] The biostratigraphic correlation with the Neogene regions outside the Paratethys area is not possible. The only possible way of correlating is based on radiogeochronology. The Wieliczkian has not been radiometrically dated.

The upper Badenian (Kosovian) was dated by five radiometric ages[5]:

14/AV-28: 15.2 ± 2 m.y.   rhyolite tuff from borehole Zipov-1, 950 to 995 m, near the village of Zipov, Slovakia

15/AV-27: 15.0 ± 0.8 m.y.   andesite from the borehole Zatin-1, 1709 to 1714 m, near the village of Zatin, Slovakia

17/AV-24: 14.4 ± 2 m.y.   rhyodacite from borehole Zamutov-1, 597 to 600 m, near the village of Zamutov, Slovakia

18/AV-30: 13.9 ± 0.3 m.y.   rhyolite tuff from Kralovce, Slovakia (Bagdasarjan et al, 1971, p. 88)

16   : 15.3 ± 2 m.y.   rhyolite from Mala Bara, Slovakia (Tsonj and Slavik, 1971, p. 215).

Regarding the listed ages, the radiometric age of the lower limit of the upper Badenian is estimated to be 15 m.y. ± 0.5 m.y. The top of the upper Badenian regarding not only the listed ages of the Badenian but the radiometric ages of the lower Sarmatian is estimated to be 13.3 ± 0.5 m.y.

The time span of the upper Badenian from 13.3 to 15.0, ± 0.5 m.y. corresponds in the World Neogene radiometric time scale (Ryan et al, 1974; Vass, 1975) to the lower Serravallian and to N11-N12 and(or) NN5 p-NN7 p of standard planktonic and nannoplanktonic zones.

Sarmatian

The lower Sarmatian (Sarmatian *sensu* Seuss, 1866) is dated by four ages but two dated samples come from the same site:

19/AV-31: 13.6 ± 1 m.y.   andesite fiom Ruskov, Slovakia (Bagdasarjan et al, 1971). Repeated dating (19a) has given an age of 12.45 ± 0.05 m.y. (Slavik et al, 1976)

21/AV-4: 13.1 ± 2.1 m.y.   andesite tuff from Nizna Mysla, Slovakia

20/AV-18: 13.0 ± 0.3 m.y.   rhyolite tuff from Sazdice, Slovakia (Vass et al, 1971, p. 323)

Two ages come from andesite dikes which intrude sediments that are lower Sarmatian in age: 9.5 ± 0.5 m.y. dike from Vyshkovo, and 10.5 ± 1 m.y. dike from Velikaja Dobronj, Transcarpathia, USSR (Bagdasarjan and Danilovich, 1968; Table 1). Of course, both ages give information, that the lower Sarmatian must be older than 9.5

---

[4] Recently Kehotsyov (oral commun.) has found the Wieliczkian and Kosovian to correspond to nannoplanktonic zones NN6-NN7.

[5] In Vass and Slavik (1975, p. 134) and in Vass et al (1975, p. 294) six radiometric ages for the upper Badenian are mentioned. The sixth age is an age of the tuff from Buchtino of 15.7 ± 0.5 m.y., Transcarpathia, USSR (originally published by Bagdasarjan and Danilovich, 1968; Tables 1 and 2). The biostratigraphic control is rather poor and indirect and the age itself is relatively old, closer to the lower than to the upper Badenian. Because of it, the age is omitted here.

to 10.5 m.y., but do not tell anything about the upper boundary of the lower Sarmatian.

Eight radiometric ages concern the lower - middle Sarmatian:

22/14 to 29/22: 12.7 ± 2.1 m.y.; 13.0 ± 1.3 m.y.; 11.3 ± 2 m.y.; 11.9 ± 1.3 m.y.; 12.7 ± 1.3 m.y.; 12.6 ± 1.8 m.y. 12.4 ± 1.4 m.y.; 12.7 ± 1.2 m.y. adularised and silicified tuffs from Beregovo, Kosino, Kvasno (Bojko et al, 1970, p. 223, 224; Table 1).

Nine other radiometric ages concern the same biostratigraphic time interval but the actual biostratigraphic age is closer to the uppermost part of the lower Sarmatian than to the middle Sarmatian[6]:

30/AV-34: 11.4 ± 0.5 m.y.   perlitic rhyolite from Vinicky, Slovakia

30a: 11.1 ± 0.8 m.y.   obsidian from the same locality, the age determination was made by fission track (Bagdasarjan et al, 1971, p. 90, 92; Repcok, 1975)

31/6: 12.1 ± 0.75 m.y.   perlitic rhyolite from the borehole KO-1, village Komarovce, Slovakia (Bagdasarjan et al, 1968, p. 423-425)

32/4: 11.9 ± 1.2 m.y.   obsidian from Dlinnaja Hill

33/12: 12.3 ± 1.6 m.y.   rhyolite from Sarok Hill, both localities in Transcarpathia, (Bojko et al, 1970, p. 227; Table 1)

34/1004: 12.4 ± 1.0 m.y.   perlitic rhyolite from Pelikan Hill near Muzievo, Transcarpathia (Bagdasarjan and Danilovich, 1968, p. 16, 17; Table 1)

35/AAD-24: 12.2 ± 0 m.y.   andesite from Tuhrina, Slovakia (Slavik et al, in press)

50/23: 12.2 ± 0.1 m.y.   andesite from Lesicek, Slovakia

51/24: 12.15 ± 0.05 m.y.   andesite between village Zehna and Tuhrina, Slovakia (both ages in Slavik et al, 1976)

The lower - middle Sarmatian boundary, regarding all ages concerning the lower and middle Sarmatian, was estimated to be 12.5 m.y. The middle - upper Sarmatian is dated by the next three radiometric ages:

36/AV-35: 11.2 ± 0.25 m.y.   andesite from quarry Lancoska near Vinne, Slovakia (Bagdasarjan et al, 1971, p. 89, 92)

36a: 12.1 ± 0.15 m.y.   andesite from the same quarry was dated by Kreuzer (Slavik et al, 1976). It seems the last age is closer to the actual age of Lancoska andesite

52/3: 12.05 ± 0.05 m.y.   andesite from Vinne, Slovakia (Slavik et al, 1976)

The middle - upper Sarmatian boundary was estimated to be approximately 11.5 ± 0.5 m.y.

As previously mentioned, the lower boundary of the Sarmatian is radiometrically calibrated to 13.3 m.y. The upper boundary is about 10.5 to 11 ± 0.5 m.y. The time span of 10.5 to 13.3 m.y. corresponds in the World Neogene time scales (mentioned previously) to the upper Serravallian and lower Tortonian, that is to N13 p-N16 p and NN7 p-NN10 and(or) the lowermost part of NN11 of the standard planktonic and nannoplanktonic zones.

Pannonian

A continental analogue of the Meotian (the upper part of the lower Pannonian to the middle Pannonian) was dated by three samples 41, 53 and 54 from the Eastern Paratethys:

41: 10.6 ± 1.5 m.y.

53: 10.1 ± 1.5 m.y.

[6]The radiometric age of 13.9 m.y., according to the biostratigraphic control of the dated rock and its petrographic analogy with ignimbrites and perlites of the Sarmatian, is correlated with the lower - middle Sarmatian and is mentioned among the basic ages in the last papers dealing with the Paratethys Neogene radiometric time scale. (Vass and Slavik, 1975, age No. 18, p. 133; Vass et al, 1975, p. 294). It is an age of perlitic rhyolite from Ardo-Hill near Beregovo, Transcarpathia (Afanas'yev et al, 1963, p. 8).

The radiometric age in comparison with other ages of the lower and the middle Sarmatian is relatively old and the origin of the age determination is rather ancient, carried out by an ancient technique of K-Ar dating. Because of this, we consider the age as insufficiently reliable, omitting it in our new time scale.

54: 8.3 ± 1 m.y.   rhyodacites and dacites from Saro, Shabaneli, Toloshi, Georgia, USSR (Gabunia and Rubinstein, 1977).

The Pannonian younger than zone C was dated by four radiometric ages but all of the second order reliability:

37/AAD-25: 9.3 ± 0.2 m.y.

38/AAD-26: 8.7 ± 0.7 m.y.   andesites from Vihorlat Mountains, Slovakia (Slavik et al, in press)

39/765: 9.1 ± 0.5 m.y.   andesite from Sinjak Hill

40/252/106: 7.6 ± 0.3 m.y.   andesite from Bolschaja Benganj Hill, both samples come from Transcarpathia (Bagdasarjan and Danilovich, 1968; Table 1)

The lower boundary of the Pannonian is approximately 10.5 to 11 m.y. The upper radiometric limit of the Pannonian is still unknown. Perhaps the four last radiometric ages are close to the top of the Pannonian but it is not proved, and because of it, we could not define the Pannonian time span. According to existing evidence, it may be concluded that at least the lower Pannonian (if not all Pannonian) corresponds to the upper Tortonian as it was radiometrically calibrated in the Vass (1975) and Ryan et al (1974) Neogene time scales.

Pliocene

The Pontian, Dacian, and Rumanian were not radiometrically dated. Only the continental analogue of the uppermost Neogene—the Akchagylian was dated in the Eastern Paratethys by two radiometric ages:

57/18: 0.95 ± 0.3 m.y.   andesite-basalt from Deshbash

58/30: 2.26 ± 0.6 m.y.   andesite-basalt from Trialeti (Rubinstein et al, 1972, p. 14-15).

The latter is perhaps closer to the actual age of Akchagylian. In this case the Akchagylian corresponds to the upper Pliocene of the previously mentioned Neogene radiometric time scales.

Conclusion

Egerian and Eggenburgian, the oldest Paratethys Neogene stages, are dated only by the radiometric ages of glauconites. The only one existing Ottnangian volcanic rock age (21.9 m.y.) and three ages of Karpathian (between 19.4 and 20.7 m.y.) are considered as the maximum age of the Ottnangian and the Karpathian; the actual age of the Karpathian is probably close to the youngest radiometric age, 19.4 m.y.

The base of the Badenian is estimated to be 16.5 ± 0.5 m.y.; the base of the upper Badenian (Kosovian) 15 ± 0.5 m.y.; the base of the Sarmatian to 13.3 m.y.; and the base of the Pannonian 10.5 to 11 m.y.

The radiometric calculation of bases of the Pontian, Dacian, and Rumanian has not been established, because reliable dates with biostratigraphic control are missing. Two ages coming from the top of the Pliocene are of 0.95 and 2.26 m.y. Probably the older one is closer to the actual age of the Pliocene-Pleistocene boundary.

The submitted radiometrical time scale of the Paratethys Neogene represents already its third revision. Still it cannot be considered final. New datings, particularly for the lower and upper Miocene and the Pliocene are inevitable. On the other hand we do not expect the eventual new datings of the middle Miocene to bring any important changes in the respective part of the time scale.

Appendix: Data for Samples Cited in Report

The number of the sample is a fraction in which the numerator is the order number of the collection and the denominator is the number given by the authors of the dating in the original paper.

Samples No. 1, 2/G229, 3/G469

A. Name of Stratigraphic Unit: Egerian.

B. Geographic location of sample: Wind brick kiln, Eger, Hungary.

C. Stratigraphic position of sample: The samples come from 5th horizon of the Wind brick kiln (Baldi, 1973, Fig. 34, p. 80-82), corresponding to the *Lepidocyclina* limestone and the glauconitic marl of the Marjas by Novai, where the *Miogypsina septentrionalis* was found (Baldi et al, 1961; Baldi, 1973, p. 86).

D. Petrologic designation of sample: Glauconite from the glauconitic sandstone.

    1.   Method used for analysis: K-Ar.

    2a.   Age and analytical uncertainty: No. 1, 30.5 m.y.; No. 2, 20.8 ± 1.3 m.y.; No. 3, 19.6 ± 1.5 m.y.

    3.   Decay constant used: $\lambda_k = 0.584 \cdot 10^{-10} y^{-1}$; $\lambda_\beta = 4.72 \cdot 10^{-10} y^{-1}$

    4.   Evaluation of reliability: biostratigraphic control is reliable. There is a large difference between the first dating and the other two. It seems the glauconite does not give reliable radiometric data. The radiometric ages of the samples 2 and 3 are too young in comparison with their biostratigraphic correlation (latest Oligocene) and with the radiometric age of Miocene-Oligocene boundary suggested by Turner (1970), Berggren (1969a, b, c, d, 1972a, b) Berggren and Van Couvering (1974), and Theyer and Hammond (1974).

    8.   Reference: Dating 30.5 m.y. was reported by Odin et al (1970). Datings 20.8 m.y. and 19.6 m.y. were realized by Odin and Hunziker *in* Inst. für Mineralogie und Petrographie Bern, published by Odin and Hunziker, 1975.

Sample No. 4/PTS82

A. Name of Stratigraphic Unit: Eggenburgian.

B. Geographic location of sample: Borehole FZ 6, 45.0-47.7 m, Bad Hall, Austria.

C. Stratigraphic position of sample: Subjacent and overjacent fauna (macro and micro, the last is represented mainly by benthonic Foraminifera) corresponds to "sand-shalerreiche Zone" Eggenburgian in age (Steininger et al, 1971, p. 44).

D. Petrologic designation of sample: Glauconite from the glauconitic sandstone.

    1.   Method used for analysis: K-Ar.

    2a.   Age and analytical uncertainty: 25 m.y.

    3.   Decay constant used: $\lambda_k = 0.584 \cdot 10^{-10} y^{-1}$; $\lambda_\beta = 0.72 \cdot 10^{-10} y^{-1}$

    4.   Evaluation of reliability: Fauna giving the biostratigraphic age is reliable inside of the Central Paratethys region, but it is not reliable for an interregional correlation so far. The radiometric dating seems unreliable because of a large difference between this sample and sample 5 coming from the same biostratigraphic horizon (see sample No. 5). The author of dating himself doubts its reliability (Evernden and Evernden, 1970). According to Odin and Hunziker (1975) the content of potassium in the sample shows the inconvenient mineralogical quality of the glauconite.

    6.   Analyst: Evernden.

    8.   Reference: Evernden et al (1961).

Sample No. 5/K-A1309

A. Name of Stratigraphic Unit: Eggenburgian.

B. Geographic location of sample: Borehole 25.7 - 28.7 m, Bad Hall, Austria.

C. Stratigraphic position of sample: The same as for sample No. 4.

D. Petrologic designation of sample: Glauconite from the glauconitic sandstone.

    1.   Method used for analysis: K-Ar.

    2a.   Age and analytical uncertainty: 18.8 m.y.

    3.   Decay constant used: $\lambda_k = 0.584 \cdot 10^{-10} y^{-1}$; $\lambda_\beta = 0.72 \cdot 10^{-10} y^{-1}$

    4.   Evaluation of reliability: Fauna giving the biostratigraphic age is reliable inside the Central Paratethys region, but it is not reliable for an interregional correlation so far. The radiometric age seems to be too young in comparison with the radiometric ages of the biostratigraphically younger volcanic rocks (see samples No. 8, 9).

6.   Analyst: Obradovich.

8.   Reference: Obradovich (1964).

Sample No. 6/447

A. Name of Stratigraphic Unit: lower Ottnangian.

B. Geographic location of sample: Cristotel, Rumania.

C. Stratigraphic position of sample: The glauconitic horizon from which the sample comes is inside of the Chechis member (Popescu, personal communication to Odin). The Chechis member is above the Corus member (Eggenburgian and Chechis member contains *Globigerina langhiana, G. dehiscens*. According to the correlation tables made by Senes et al, 1975 the Chechis member represents Ottnangian.

D. Petrologic designation of sample: Glauconite from glauconitic horizon.

1.   Method used for analysis: K-Ar.

2a. Age and analytical uncertainty: 17.5 ± 0.9 m.y.[+] and 16.6 ± 1 m.y.[++]

3.   Decay constant used: $\lambda_k = 0.584 \cdot 10^{-10} y^{-1}$; $\lambda_\beta = 0.72 \cdot 10^{-10} y^{-1}$

4.   Evaluation of reliability: Fauna giving the biostratigraphic age is reliable inside of the Central, eventually Eastern Paratethys region. The radiometric age seems to be too young in comparison with the radiometric ages of the biostratigraphically younger volcanic rocks (see sample No. 8, 9).

6.   Analyst: Odin[+] and Hunziker.[++]

7.   Institution: B.R.G.M.[+] and Institut für Mineralogie und Petrographie Bern.[++]

8.   Reference: Odin and Hunziker (1975).

Sample No. 7/422 and 7a/G470

A. Name of Stratigraphic Unit: lower Ottnangian.

B. Geographic location of sample: Briglez, Rumania.

C. Stratigraphic position of sample: The glauconitic marl from which the sample comes is just above the Corus member which contains Eggenburgian molluscan assemblage of Loibersdorf type (Chlamys gigas, etc). The glauconitic marl contains Foraminifera and among them *Uvigerina posthantkeni, U. parviformis* (Steininger, written communication, 1974).

D. Petrologic designation of sample: Glauconite from the glauconitic marl.

1.   Method used for analysis: K-Ar.

2a. Age and analytical uncertainty: 22.6 ± 1.2 m.y.[+], repeated analysis 17.6 ± 1.4 m.y.[++]

3.   Decay constant used: $\lambda_k = 0.584 \cdot 10^{-10} y^{-1}$; $\lambda_\beta = 0.72 \cdot 10^{-10} y^{-1}$

4.   Evaluation of reliability: Fauna giving the biostratigraphic age is reliable inside of the Central and(or) Eastern Paratethys region. The radiometric age of 22.6 m.y. seems to be close to the real age, but the repeated analysis of 17.6 m.y. seems too young in comparison with the radiometric age of the biostratigraphically younger volcanic rocks (see sample No. 8, 9).

6.   Analyst: Odin and Hunziker.

7.   Institution: B.R.G.M.[+], Institut für Mineralogie und Petrographie Bern.[++]

8.   Reference: Odin and Hunziker (1975).

Sample No. 8/AV-24

A. Name of Stratigraphic Unit: lower Ottnangian.

B. Geographic location of sample: Kalonda, district Lucenec, Slovakia (Czechoslovakia).

C. Stratigraphic position of sample: The dated rock comes from a continental formation containing the flora. Under the continental formation there are marine deposits with mollusca of Eggenburgian. Above the formation there are deposits with upper Ottnangian-lower Karpathian fauna (Vass et al, 1971, p. 322). Karpathian of South Slovakia (without *Globigerinoides sicanus*) corresponds to the upper part of NN4 calcareous nannoplankton standard zone with *Helicopontosphaera ampliaperta*. (Lehotayova, personal communication, 1975).

D. Petrologic designation of sample: Ryolite tuff—whole rock was dated.

1. Method used for analysis: K-Ar.
2a. Age and analytical uncertainty: 21.9 ± 0.0 m.y.
3. Decay constant used: $\lambda_k = 0.584 \cdot 10^{-10} y^{-1}$; $\lambda_\beta = 0.72 \cdot 10^{-10} y^{-1}$
4. Evaluation of reliability: The possible biostratigraphic age is between Eggenburgian and Karpathian, but the upper Eggenburgian is not excluded. The fauna supporting and interregional correlation is missing.
6. Analyst: Bagdasarjan.
7. Institution: Armenian Academy of Science, Erivan, UdSSR.
8. Reference: Vass et al (1971), p. 322, Table 1; Vass and Bagdasarjan *in* Papp et al (1973), p. 37.

Sample No. 9/AV-2

A. Name of Stratigraphic Unit: Karpathian.
B. Geographic location of sample: Klenany, district Levice, Slovakia (Czechoslovakia).
C. Stratigraphic position of sample: The dated rock is inside a marine sequence containing the assemblage of Mollusca and Foraminifera, but the plankton is not typical. The *Globigerinoides sicanus* is not present (Vass et al, 1971, p. 322). The dated rock must be older than the upper part of N8 zone.
D. Petrologic designation of sample: Volcanic acid glassy ash—whole rock was dated.
1. Method used for analysis: K-Ar.
2a. Age and analytical uncertainty: 20.7 ± 1.3 m.y.
3. Decay constant used: $\lambda_k = 0.584 \cdot 10^{-1} y^{-1}$; $\lambda_\beta = 0.72 \cdot 10^{-10} y^{-1}$
4. Evaluation of reliability: Fauna giving the biostratigraphic age is reliable inside of the Central Paratethys region but absence of the *Globigerinoides sicanus*, and its presence in overjacent beds gives certain possibility to correlate the dating with Blow's planktonic Foraminifera zones.
6. Analyst: Bagdasarjan.
7. Institution: Armenian Academy of Science, Erivan, UdSSR.
8. Reference: Vass et al (1971), p. 322.

Sample No. 10/AV-12, 11/AV-13

A. Name of Stratigraphic Unit: lower Badenian.
B. Geographic location of sample: Hrusov, district Lucenec, Slovakia (Czechoslovakia).
C. Stratigraphic position of sample: The dated andesites are the extrusive bodies which together with their volcanoclastic equivalents are part of a volcanic sedimentary formation containing the fauna of lower Badenian including *Orbulina suturalis* (first appearance in whole South Slovakian Neogene sequence; Vass et al, 1971, p. 323). "The calcareous nannoplankton (including *Sphenolithus heteromorphus*) indicates the upper part of NN5 standard nannoplankton zone." (Lehotayova, personal communication).
D. Petrologic designation of sample: Andesites, whole rocks were dated.
1. Method used for analysis: K-Ar.
2a. Age and analytical uncertainty: 16.45 ± 1.5 m.y. and 15.3 ± 0.7 m.y.
3. Decay constant used: $\lambda_k = 0.584 \cdot 10^{-10} y^{-1}$; $\lambda_\beta = 0.72 \cdot 10^{-10} y^{-1}$
4. Evaluation of reliability: The presence of the *Orbulina suturalis* as the first appearance in the Neogene sequence gives the possibility of correlating with the N9 zone of Blow but it seems the *Orbulina universa* is present too.
6. Analyst: Bagdasarjan.
7. Institution: Armenian Academy of Science, Erivan, UdSSR.
8. Reference: Vass et al (1971), p. 323.
9. Additional comments: Originally six radiometric analyses were realized on sample 10/AV-12 but the published mean age was calculated only from three analyses (18.5 ± 0.9 m.y.); after recalculation with constant $\lambda_k = 0.584 \cdot 10^{-10} y^{-1}$ (17.8 ± 0.9 m.y.).

Sample No. 12/3/891

A. Name of Stratigraphic Unit: lower Badenian.
B. Geographic location of sample: Novoselica, Transcarpathia, UdSSR.

C. Stratigraphic position of sample: The dated tuff is inside a marine sequence containing the fauna of lower Badenian: *Candorbulina universa (Orbulina suturalis* according to Pishvanova, 1974, p. 775), *Globigerinoides trilobus, Gl.-des sicanus, Praeorbulina transitoria, Turborbulina mayeri* according to Pishvanova, (Mrs. Danilovich, written communication, 1973).
D. Petrologic designation of sample: Rhyolite tuff—whole rock was dated.
   1. Method used for analysis: K-Ar.
   2a. Age and analytical uncertainty: 15.6 ± 0.5 m.y.
   3. Decay constant used: $\lambda_k = 0.584 \cdot 10^{-1} y^{-1}$; $\lambda_\beta = 0.72 \cdot 10^{-1} y^{-1}$
   4. Evaluation of reliability: The planktonic Foraminifera accompanying the dated tuff allow correlation with zones N9 of Blow.
   6. Analyst: Bagdasarjan.
   7. Institution: Armenian Academy of Science, Erivan, UdSSR.
   8. Reference: Bagdasarjan and Danilovich (1968); Table 1.

Sample No. 13/49/1

A. Name of Stratigraphic Unit: lower Badenian.
B. Geographic location of sample: Neresnica, Transcarpathia, UdSSR.
C. Stratigraphic position of sample: The dated tuff is inside a marine sequence containing the fauna of lower Badenian: *Candorbulina universa (Orbulina suturalis* according to Pishvanova, 1974, p. 775), *Globigerinoides trilobus, Gl.-des sicanus, Praeorbulina transitoria, Turborbulina mayeri* according to Pishvanova, (Mrs. Danilovich, written communication, 1973).
D. Petrologic designation of sample: Rhyodacite tuff—whole rock was dated.
   1. Method used for analysis: K-Ar.
   2a. Age and analytical uncertainty: 15.2 ± 0.5 m.y.
   3. Decay constant used: $\lambda_k = 0.584 \cdot 10^{-1} y^{-1}$; $\lambda_\beta = 0.72 \cdot 10^{-1} y^{-1}$
   4. Evaluation of reliability: The planktonic Foraminifera accompanying the dated tuff allow correlation with zones N9 of Blow.
   6. Analyst: Bagdasarjan.
   7. Institution: Armenian Academy of Science, Erivan, UdSSR.
   8. Reference: Bagdasarjan and Danilovich (1968); Table 1.

Sample No. 14/AV-28

A. Name of Stratigraphic Unit: middle-upper Badenian.
B. Geographic location of sample: Zipov, borehole Zipov-1, 950-995 m, district Trebisov, Slovakia (Czechoslovakia).
C. Stratigraphic position of sample: Dated rock is in the middle of a sedimentary sequence, under the dated rock there is microfaunal assemblage of the middle Badenian and above the assemblage of upper Badenian. (Bagdasarjan et al, 1971, p. 88).
D. Petrologic designation of sample: Rhyolite tuff—whole rock was dated.
   1. Method used for analysis: K-Ar.
   2a. Age and analytical uncertainty: 15.2 ± 2.0 m.y.
   3. Decay constant used: $\lambda_k = 0.584 \cdot 10^{-10} y^{-1}$; $\lambda_\beta = 0.72 \cdot 10^{-10} y^{-1}$
   4. Evaluation of reliability: The biostratigraphic age of dated tuff is reliable inside the Central Paratethys region. According to the preliminary correlation based on planktonic Foraminifera the middle Badenian corresponds to the N13 zone of Blow (Lehotayova, 1971), and to NN5p - NN6 nannoplanktonic zone.
   6. Analyst: Bagdasarjan.
   7. Institution: Armenian Academy of Science, Erivan, UdSSR.
   8. Reference: Bagdasarjan et al (1971), p. 88.

Sample No. 15/AV-27

A. Name of Stratigraphic Unit: upper Badenian.
B. Geographic location of sample: Zatin, borehole Zatin-1, 1709-1714 m, district Trebisov, Slovakia (Czechoslovakia).
C. Stratigraphic position of sample: Andesite lava flows intercalated and buried by

marine sediments containing the upper Badenian Foraminifera assemblage (Bagdasarjan et al, 1971, p. 88).
D. Petrologic designation of sample: Andesite—whole rock was dated.
1. Method used for analysis: K-Ar.
2a. Age and analytical uncertainty: 15.0 ± 0.8 m.y.
3. Decay constant used: $\lambda_k = 0.584 \cdot 10^{-10} y^{-1}$; $\lambda_\beta = 0.72 \cdot 10^{-10} y^{-1}$
4. Evaluation of reliability: The biostratigraphic age of the dated rock is reliable inside the Central Paratethys region. According to the preliminary correlation based on planktonic Foraminifera the upper Badenian corresponds to or is younger than the N13 zone of Blow, and corresponds to NN6-NN7 (NN8?) nannoplanktonic zone.
6. Analyst: Bagdasarjan.
7. Institution: Armenian Academy of Science, Erivan, UdSSR.
8. Reference: Bagdasarjan et al (1971), p. 88.

Sample No. 16

A. Name of Stratigraphic Unit: middle and/or upper Badenian.
B. Geographic location of sample: Mala Bara, district Trebisov, Slovakia (Czechoslovakia).
C. Stratigraphic position of sample: The dated rock comes from a rhyolite body in the vicinity of which the similar rhyolite fragments were found in the marine beds containing the upper Badenian fauna (Tsonj and Slavik, 1971, p. 215).
D. Petrologic designation of sample: Rhyolite—whole rock was dated.
1. Method used for analysis: K-Ar.
2a. Age and analytical uncertainty: 15.3 ± 2.0 m.y.
3. Decay constant used: $\lambda_k = 0.584 \cdot 10^{-10} y^{-1}$; $\lambda_\beta = 0.72 \cdot 10^{-10} y^{-1}$
4. Evaluation of reliability: The biostratigraphic age of the dated rock is reliable inside the Central Paratethys region. According to the preliminary correlation based on planktonic Foraminifera the upper Badenian corresponds to or is younger than the N13 zone of Blow, and corresponds to NN6-NN7 (NN8?) nannoplanktonic zone.
6. Analyst: Tsonj.
8. Reference: Tsonj and Slavik (1971), p. 215.

Sample No. 17/AV-29

A. Name of Stratigraphic Unit: upper Badenian (uppermost part).
B. Geographic location of sample: Zamutov, borehole Zamutov-1, (597-600 m), district Vranov, Slovakia, (Czechoslovakia).
C. Stratigraphic position of sample: The dated rock is a part of Kolcov Formation which is uppermost Badenian to Lower Sarmatian in age (Bagdasarjan et al, 1971, p. 88).
D. Petrologic designation of sample: Rhyolite—whole rock was dated.
1. Method used for analysis: K-Ar.
2a. Age and analytical uncertainty: 14.4 ± 2.0 m.y.
3. Decay constant used: $\lambda_k = 0.584 \cdot 10^{-10} y^{-1}$; $\lambda_\beta = 0.72 \cdot 10^{-10} y^{-1}$
4. Evaluation of reliability: The biostratigraphic correlation is reliable inside the Central Paratethys region.
6. Analyst: Bagdasarjan.
7. Institution: Armenian Academy of Science, Erivan, UdSSR.
8. Reference: Bagdasarjan et al (1971), p. 88.

Sample No. 18/AV-30

A. Name of Stratigraphic Unit: upper Badenian (uppermost part).
B. Geographic location of sample: Kralovce, district Kosice, Slovakia (Czechoslovakia).
C. Stratigraphic position of sample: The dated rock comes from a freshwater sedimentary sequence on the top of the Badenian and under the Sarmatian. It is an equivalent of Kolcov Formation (Bagdasarjan et al, 1971, p. 88).
D. Petrologic designation of sample: Rhyolite tuff—whole rock was dated.

1. Method used for analysis: K-Ar.
2a. Age and analytical uncertainty: 13.9 ± 0.3 m.y.
3. Decay constant used: $\lambda_k = 0.584 \cdot 10^{-10} y^{-1}$; $\lambda_\beta = 0.72 \cdot 10^{-10} y^{-1}$
4. Evaluation of reliability: The biostratigraphic correlation is reliable inside the Central Paratethys region.
6. Analyst: Bagdasarjan.
7. Institution: Armenian Academy of Science, Erivan, UdSSR.
8. Reference: Bagdasarjan et al (1971), p. 88.

Sample No. 19/AV-31

A. Name of Stratigraphic Unit: lower Sarmatian.
B. Geographic location of sample: Ruskov, quarry NE of the village, district Kosice, Slovakia (Czechoslovakia).
C. Stratigraphic position of sample: The dated lava flow is buried by the marls containing the fauna of lower Sarmatian
D. Petrologic designation of sample: Andesite lava flow—whole rock was dated.
1. Method used for analysis: K-Ar.
2a. Age and analytical uncertainty: 13.6 ± 1.0 m.y.
3. Decay constant used: $\lambda_k = 0.584 \cdot 10^{-10} y^{-1}$; $\lambda_\beta = 0.72 \cdot 10^{-10} y^{-1}$
4. Evaluation of reliability: The biostratigraphic correlation is reliable inside of the Central and Eastern Paratethys regions.
6. Analyst: Bagdasarjan.
7. Institution: Armenian Academy of Science, Erivan, UdSSR.
8. Reference: Bagdasarjan et al (1971), p. 89.
9. Additional comments: The andesite from the same quarry was dated by Kreuzer (Hannover). The age 19a is of 12.45 ± 0.05 m.y.

Sample No. 20/AV-18

A. Name of Stratigraphic Unit: lower Sarmatian.
B. Geographic location of sample: Nizna Mysla, quarry on the southern periphery of the village, district Kosice, Slovakia, (Czechoslovakia).
C. Stratigraphic position of sample: The dated rock is inside the sequence containing the fauna of lower Sarmatian (Svagrovsky, 1962, Lehotayova in Pesl et al, 1968).
D. Petrologic designation of sample: Andesite tuff—whole rock was dated.
1. Method used for analysis: K-Ar.
2a. Age and analytical uncertainty: 13.0 ± 0.3 m.y.
3. Decay constant used: $\lambda_k = 0.584 \cdot 10^{-10} y^{-1}$; $\lambda_\beta = 0.72 \cdot 10^{-10} y^{-1}$
4. Evaluation of reliability: The biostratigraphic correlation is reliable inside of the Central and Eastern Paratethys regions.
6. Analyst: Bagdasarjan.
7. Institution: Armenian Academy of Science, Erivan, UdSSR.
8. Reference: Vass et al (1971), p. 323.

Sample No. 21/AV-4

A. Name of Stratigraphic Unit: lower Sarmatian.
B. Geographic location of sample: Sazdice, district Levice, abandoned quarry west of the village, Slovakia, (Czechoslovakia).
C. Stratigraphic position of sample: The dated tuff is lying on the marine upper Badenian and is covered by lower Sarmatian (brackish water) rocks (Vass et al, 1971, p. 324). Both Badenian and Sarmatian contain characteristic faunal assemblage.
D. Petrologic designation of sample: Rhyolite tuff—whole rock was dated.
1. Method used for analysis: K-Ar.
2a. Age and analytical uncertainty: 13.1 ± 1.1 m.y.
3. Decay constant used: $\lambda_k = 0.584 \cdot 10^{-10} y^{-1}$; $\lambda_\beta = 0.72 \cdot 10^{-10} y^{-1}$
4. Evaluation of reliability: The biostratigraphic correlation is reliable inside the Paratethys region.
6. Analyst: Bagdasarjan.

7.   Institution: Armenian Academy of Science, Erivan, UdSSR.
8.   Reference: Vass et al (1971), p. 324.

Samples No. 22/14, 23/15, 24/16, 25/17, 26/18, 27/19, 28/21, and 29/22

A. Name of Stratigraphic Unit: lower and middle Sarmatian.
B. Geographic location of sample: Beregovo (mine field), Kosino, and boreholes in the vicinity of Kvasovo, Transcarpathia, UdSSR.
C. Stratigraphic position of sample: The dated rocks are covered by the Almash Formation (Fishkin *in* Bojko et al, 1970, p. 225). The Almash Formation corresponds to upper Sarmatian (*sensu* Seuss, 1866). It means the zone with *Porosononion subgranosum* equals Protelphidium subgranosum (Vialov et al, *in* Semenenko et al, 1966, p. 251 and Fig. 41).
D. Petrologic designation of sample: Adularised and silicified tuffs.
  1.   Method used for analysis: K-Ar.
  2a. Age and analytical uncertainty: 12.7 ± 2.1; 13.0 ± 1.3; 11.3 ± 2.0; 11.9 ± 1.3; 12.7 ± 1.3; 12.6 ± 1.8; 12.4 ± 1.4; 12.7 ± 1.2.
  3.   Decay constant used: $\lambda_k = 0.584 \cdot 10^{-10} y^{-1}$; $\lambda_\beta = 0.72 \cdot 10^{-10} y^{-1}$
  4.   Evaluation of reliability: The biostratigraphic correlation is reliable inside the Central Paratethys region.
  8.   Reference: all dates were reported by Bojko et al (1970), p. 223-224.

Sample No. 30/AV-34

A. Name of Stratigraphic Unit: lower Sarmatian (upper part) middle Sarmatian.
B. Geographic location of sample: Vinicky, district Trebisov, Slovakia, (Czechoslovakia).
C. Stratigraphic position of sample: The dated rock and similar rocks in the vicinity of the village of Vinicky rest on the lower Sarmatian marine brackish sediments containing the typical fauna assemblage (Ivan, 1964, p. 145).
D. Petrologic designation of sample: Perlitic rhyolite.
  1.   Method used for analysis: K-Ar.
  2a. Age and analytical uncertainty: 11.4 ± 0.5 m.y.
  3.   Decay constant used: $\lambda_k = 0.584 \cdot 10^{-10} y^{-1}$; $\lambda_\beta = 0.72 \cdot 10^{-10} y^{-1}$
  4.   Evaluation of reliability: The biostratigraphic correlation is reliable inside the Paratethys region.
  6.   Analyst: Bagdasarjan.
  7.   Institution: Armenian Academy of Science, Erivan, UdSSR.
  8.   Reference: Bagdasarjan et al (1971), p. 90, 92.
  9.   Constants used (isochron method): fission track = 11.1 ± 0.8 m.y. (Repcok, 1975) using: $\lambda_f = 6.85 \cdot 10^{-17} y^{-1}$; $\lambda_a = 1.54 \cdot 10^{-10} y^{-1}$; and $\sigma = 5.82 \cdot 10^{-24} \cdot cm^2$. $U^{235}/U^{238} = 7.26 \cdot 10^{-3}$.

Sample No. 31/6

A. Name of Stratigraphic Unit: lower - middle Sarmatian.
B. Geographic location of sample: Borehole KO-1 Komarovce, district Kosice, Slovakia (Czechoslovakia).
C. Stratigraphic position of sample: The dated rock is petrographically similar to sample No. 30 and both samples come from the same region. Other nonbiostratigraphic data support the mentioned age (Vass, 1969; Pulec and Vass, 1969).
D. Petrologic designation of sample: Perlitic rhyolite.
  1.   Method used for analysis: K-Ar.
  2a. Age and analytical uncertainty: 12.10 ± 0.75 m.y.
  3.   Decay constant used: $\lambda_k = 0.584 \cdot 10^{-10} y^{-1}$; $\lambda_\beta = 0.72 \cdot 10^{-10} y^{-1}$
  4.   Evaluation of reliability: The direct biostratigraphic data are missing.
  6.   Analyst: Bagdasarjan.
  7.   Institution: Armenian Academy of Science, Erivan, UdSSR.
  8.   Reference: Bagdasarjan et al (1968), p. 423-425.

Sample No. 32/4, 33/12

A. Name of Stratigraphic Unit: lower - middle Sarmatian.

B. Geographic location of sample: Dlinnaja Hill, Sarok Hill (Beregov Hills), Transcarpathia, UdSSR.

C. Stratigraphic position of sample: The dated rocks are petrographically similar to sample No. 30. They are close (by petrography, and also by stratigraphic position), to upper tuffs of Dobrotovo Formation (Fishkin *in* Bojko et al, 1970, p. 224), which correspond to lower Sarmatian zone with *Cibicides badenensis* (Vialov et al, *in* Semenenko et al, 1966).

D. Petrologic designation of sample: (a) Obsidian from rhyolite, (b) rhyolite tuff with perlitic glass.

  1. Method used for analysis: K-Ar.

  2a. Age and analytical uncertainty: (a) 11.9 ± 1.2 m.y.; (b) 12.3 ± 1.6 m.y.

  3. Decay constant used: $\lambda_k = 0.584 \cdot 10^{-10} y^{-1}$; $\lambda_\beta = 0.72 \cdot 10^{-10} y^{-1}$

  4. Evaluation of reliability: The biostratigraphic correlation with lower - middle Sarmatian is reliable inside the Paratethys region.

  8. Reference: Radiometric ages were reported by Bojko et al (1970), p. 223, datings No. 4 and 12.

Sample No. 34/1004

A. Name of Stratigraphic Unit: lower - middle Sarmatian.

B. Geographic location of sample: Pelikan Hill, near Muzievo, Transcarpathia, UdSSR.

C. Stratigraphic position of sample: The dated rock is petrographically similar to samples 30, 32, 33. The dated rock is also in some relation with marine brackish beds containing *Cardium transcarpaticum = C. gleichenbergense* (L. G. Danilovich, written communication, 1973). *Cardium* mentioned is known from lower and middle Sarmatian of Paratethys.

D. Petrologic designation of sample: Perlitic rhyolite.

  1. Method used for analysis: K-Ar.

  2a. Age and analytical uncertainty: 12.4 ± 1.0 m.y.

  3. Decay constant used: $\lambda_k = 0.584 \cdot 10^{-10} y^{-1}$; $\lambda_\beta = 0.72 \cdot 10^{-10} y^{-1}$

  4. Evaluation of reliability: The relation between dated rock and fauna-bearing beds is rather obscure.

  6. Analyst: Bagdasarjan.

  7. Institution: Armenian Academy of Science, Erivan, UdSSR.

  8. Reference: Bagdasarjan and Danilovich (1968), p. 16, 17; Table 1.

Sample No. 35/AAD-24

A. Name of Stratigraphic Unit: lower - middle Sarmatian.

B. Geographic location of sample: SW of Tuhrina, district Presov, Slovakia (Czechoslovakia).

C. Stratigraphic position of sample: The dated rock rests on Rankovce tuffs, lower Sarmatian in age (Jiricek *in* Cvercko, 1968, p. 48) and are partially covered by volcanic complex of Brestov-Abramovce which is indirectly correlated with middle - upper Sarmatian (Slavik and Tozser, 1973, p. 32).

D. Petrologic designation of sample: Andesite—whole rock was dated.

  1. Method used for analysis: K-Ar.

  2a. Age and analytical uncertainty: 12.2 ± 0.0 m.y.

  3. Decay constant used: $\lambda_k = 0.584 \cdot 10^{-10} y^{-1}$; $\lambda_\beta = 0.72 \cdot 10^{-10} y^{-1}$

  4. Evaluation of reliability: The biostratigraphic age of underlying rocks is well defined, but the direct biostratigraphic evidence concerning the age of the overjacent complex is missing.

  6. Analyst: Bagdasarjan.

  7. Institution: Armenian Academy of Science, Erivan, UdSSR.

  8. Reference: Slavik et al (1976).

Sample No. 36/AV-35

A. Name of stratigraphic unit: middle - upper Sarmatian.

B. Geographic location of sample: Vinne, quarry Lancoska, near Medvedova Hill, district Michalovce, Slovakia (Czechoslovakia).

C. Stratigraphic position of sample: The dated rock is a member of the Vinne-Zavadka volcanic formation, the pyroclastic equivalents of which rest on the *Elphidium hauerinum* zone (middle Sarmatian, Seuss, 1866) and is covered by the *Protelphidium subgranosum* zone.

D. Petrologic designation of sample: Andesite—whole rock was dated.

   1.   Method used for analysis: K-Ar.

   2.   Age and analytical uncertainty: 11.20 ± 0.25 m.y.

   3.   Decay constant used: $\lambda_k = 0.584 \cdot 10^{-10} y^{-1}$; $\lambda_\beta = 0.72 \cdot 10^{-10} y^{-1}$

   4.   Evaluation of reliability: The biostratigraphic correlation seems to be reliable. But the result of one of the three control datings was high (17 m.y.) and was not included in the mean of 11.20 m.y. If the extremely high value is calculated, the mean value is higher (13.5 ± 2.3 m.y.).

   6.   Analyst: Bagdasarjan.

   7.   Institution: Armenian Academy of Science, Erivan, UdSSR.

   8.   Reference: Bagdasarjan et al (1971), p. 89, 92.

   9.   Additional comments: The andesite from the same quarry was dated by Kreuzer *in* Hannover. The age (36a) is 12.1 ± 0.15 m.y.

Samples No. 37/AAD-25, 38/AAD-26

A. Name of Stratigraphic Unit: Pannonian, zone C, or younger.

B. Geographic location of sample: Vihorlat Hill in Vihorlat Mountains, Slovakia (Czechoslovakia).

C. Stratigraphic position of sample: The dated rocks represent the upper complex of the uppermost volcanic formation in the volcanic mountains Vihorlat-Poprichny. The Valaskovce Formation, which rests on a lacustrine formation, correlated by ostracodes (according to Jiricek, 1972) with Pannonian C.

D. Petrologic designation of sample: Andesites—whole rocks were dated.

   1.   Method used for analysis: K-Ar.

   2a. Age and analytical uncertainty: 8.7 ± 0.2 m.y. and 9.2 ± 0.7 m.y.

   3.   Decay constant used: $\lambda_k = 0.584 \cdot 10^{-10} y^{-1}$; $\lambda_\beta = 0.72 \cdot 10^{-10} y^{-1}$

   4.   Evaluation of reliability: The reliability of biostratigraphic correlation is of the second order.

   6.   Analyst: Bagdasarjan.

   7.   Institution: Armenian Academy of Science, Erivan, UdSSR.

   8.   Reference: Slavik et al (1976).

   9.   Relations to other lines of chronostratigraphic evidence (e.g., paleontologic, paleomagnetic, etc)—the dated andesites are of reverse remanent paleomagnetization.

Samples No. 39/765, 40/252/106

A. Name of Stratigraphic Unit: Pannonian, zone C or younger.

B. Geographic location of sample: Sinjak Hill, Bolchaja Beganj Hill, Transcarpathia, UdSSR.

C. Stratigraphic position of sample: The dated rocks have the same stratigraphic position in the volcanic sequence as the samples 37 and 38. The reverse remanent paleomagnetization (Mikhailova et al, 1974) is the same as samples 37, 38.

D. Petrologic designation of samples: Andesites—whole rocks were dated.

   1.   Method used for analysis: K-Ar.

   2a. Age and analytical uncertainty: 9.1 ± 0.5 m.y.; 8.0 ± 0.3 m.y.

   3.   Decay constant used: $\lambda_k = 0.584 \cdot 10^{-10} y^{-1}$; $\lambda_\beta = 0.72 \cdot 10^{-10} y^{-1}$

   4.   Evaluation of reliability: The biostratigraphic evidence of the age is missing.

   6.   Analyst: Bagdasarjan.

   7.   Institution: Armenian Academy of Science, Erivan, UdSSR.

   8.   Reference: Bagdasarjan and Danilovich (1968), Table 1.

   9.   Relation to other lines of chronostratigraphic evidence (e.g., paleontologic,

paleomagnetic, etc)—the dated andesites are of reverse remanent paleomagnetization.

Samples No. 41, 53, 54

A. Name of stratigraphic unit: Meotian (upper Miocene)
B. Geographic location of sample: No. 41 Saro, Georgia, USSR; No. 53 Shabanebeli, Georgia, USSR; No. 54 Toloshi, Georgia, USSR.
C. Stratigraphic position of sample: The beds containing the remnants of *Hipparion garedzikum* (a species of lower Meotian mammals) underlie the dated volcanics.
D. Petrologic designation of sample: 41, rhyodacite; 53, dacite; 54, rhyodacite.
   1. Method used for analysis: K-Ar.
   2a. Age and analytical uncertainty: 41, 10.6 ± 1.5 m.y.; 53, 10.1 ± 1.5 m.y.; 54, 08.3 ± 1.0 m.y.
   3. Decay constant used: $\lambda_k = 0.584 \cdot 10^{-10} y^{-1}$; $\lambda_\beta = 0.72 \cdot 10^{-10} y^{-1}$
   4. Evaluation of reliability: The relations with marine stratigraphy are not perfectly done. The Meotian corresponds mostly to the lower Pannonian in the Central Paratethys.
   6. Analyst: Rubinstein.
   7. Institution: Georgian Academy of Science, Tbilisi, USSR.
   8. Reference: Gabunia and Rubinstein (1977).

Samples No. 42/AVR-1, 43/AVR-2

A. Name of stratigraphic unit: lower Badenian.
B. Geographic location of sample: Weitendorf, quarry, lower stratum, 42/AVR-1; upper stratum, 43/AVR-2 (Steyermark, Austria).
C. Stratigraphic position of sample: The planktonic forams from the subjacent beds correspond to the upper part of the N8 zone of Blow because of the presence of *Praeorbulina* (first appearance); forams from the overjacent beds correspond to N9 zone of Blow; *Orbulina suturalis* is present (Steininger, personal communication, 1974).
D. Petrologic designation of sample: Basalts—whole rocks were dated.
   1. Method used for analysis: K-Ar.
   2. Age and analytical uncertainty: 42/AVR-1, 16.8 ± 0.75 m.y.; 43/AVR-2, 16.0 ± 0.3 m.y.
   3. Decay constant used: $\lambda_k = 0.584 \cdot 10^{-10} y^{-1}$; $\lambda_\beta = 0.72 \cdot 10^{-10} y^{-1}$
   4. Evaluation of reliability: The biostratigraphic control is perfect.
   6. Analyst: Bagdasarjan.
   7. Institution: Armenian Academy of Science, Erivan, UdSSR.
   8. Reference: Steininger, Rogl, and Martini (1976); Steininger, and Bagdasarjan (in press).
   9. Additional comments: From the same locality, Lippolt et al (1975) published a whole rock radiometric age of 15.2 ± 0.9.

Samples No. 44/AVR-3, 45/AVR-4

A. Name of stratigraphic unit: lower Badenian.
B. Geographic location of sample: Klause bei Gleichenberg, quarry lower stratum, 44/AVR-3, upper stratum, 45/AVR-4 (Steyermark, Austria).
C. Stratigraphic position of sample: The same as for samples 43/AVR-2, 42/AVR-1.
D. Petrologic designation of sample: Andesites—whole rocks were dated.
   1. Method used for analysis: K-Ar.
   2. Age and analytical uncertainty: 44/AVR-3, 16.3 ± 0.9 m.y.; 45/AVR-4; 15.5 ± 0.1 m.y.
   4. Evaluation of reliability: The biostratigraphic control is perfect.
   6. Analyst: Bagdasarjan.
   7. Institution: Armenian Academy of Science, Erivan, UdSSR.
   8. Reference: Steininger, Rogl, and Martini (1976); Steininger and Bagdasarjan (in press).

Sample No. 46/7

A. Name of stratigraphic unit: upper Sarmatian - lower Pannonian.

B. Geographic location of sample: Quarry near the village Kremnicka, district Ziar/ Hron, Slovakia (Czechoslovakia).

C. Stratigraphic position of sample: The dated rock is a rhyolite extrusive body. The limnic sediments in the vicinity of the extrusive body bear the bentonitic rhyolite tuffs, possible explosive equivalents of the same rhyolite volcanic activity. The biostratigraphic age of the sediments according to the pollen spectra is upper Sarmatian - lower Pannonian (Ciesarik and Planderova, 1965).

D. Petrologic designation of the sample: Rhyolite—whole rock was dated.

    1.   Method used for analysis: K-Ar.

    2.   Age and analytical uncertainty: 10.7 ± 0.3 m.y.

    3.   Decay constant used: $\lambda_k = 0.584 \cdot 10^{-10} y^{-1}$; $\lambda_\beta = 0.72 \cdot 10^{-10} y^{-1}$

    4.   Evaluation of reliability: The biostratigraphic age mentioned above is proved indirectly and only by pollen spectra.

    6.   Analyst: Bagdasarjan.

    7.   Institution: Armenian Academy of Science, Erivan, UdSSR.

    8.   Reference: Bagdasarjan et al (1968), p. 421, 423.

Sample No. 47/STW 12 B

A. Name of stratigraphic unit: middle Miocene; lower Badenian (Base of Badenian equals base of Langhian).

B. Geographic location of sample: Weitendorf quarry, Steyermark, Austria.

C. Stratigraphic position of sample: The planktonic forams from the subjacent beds correspond to the upper part of the N8 zone of Blow because of the presence of *Praeorbulina* (first appearance); forams from the overjacent beds correspond to N9 zone of Blow; *Orbulina suturalis* is present (Steininger, personal communication, 1974).

D. Petrologic designation of sample: Basalts—whole rocks were dated.

    1.   Method used for analysis: K-Ar.

    2a. Age and analytical uncertainty: 15.2 ± 0.9 m.y.

    4.   Evaluation of reliability: First order reliability; the biostratigraphic control is perfect.

    6.   Analyst: Lippolt.

    8.   Reference: Lippolt et al (1975).

    9.   Additional comment: Relation to other lines of chronostratigraphic evidence— paleomagnetic measurements show that the dated basalt comes from the same paleomagnetic interval as sub and overjacent marine sediments containing the fauna (Steininger, personal communication, 1975).

Samples No. 48/M 1a, 49/M 5

A. Name of stratigraphic unit: lower Miocene (Karpathian).

B. Geographic location of sample: 48/M 1a, Magyarkut, Northern Hungary; 49/M 5, Road cut between Kisinocz and Nagyirtas, Borzsony Mountains, Hungary.

C. Stratigraphic position of sample: Both dated rocks are inside the marine sequence containing the assemblage of Mollusca and Foraminifera of Karpathian age but the plankton is not typical. *Globigerinoides sicanus* is missing and appears in overjacent Badenian deposits (it is not the first appearance). The same sequence in nearby Ipel basin (southern Slovakia) corresponds to the upper part of NN4 calcareous nannoplankton standard zone with *Helicopontosphaera ampliaperta* (Lehotayova, personal communication, 1975; see comments to sample 9/AV-2).

D. Petrologic designation of sample: 48/M 1a, biotite from tuff; 49/M 5, dacite lava flow; the biotite was dated.

    1.   Method used for analysis: K-Ar.

    2a. Age and analytical uncertainty: 48/M 1a, 20.7 ± 3.2 m.y.; 49/M 5, 19.4 ± 1.3 m.y.

    3.   Decay constant used: $\lambda_k = 0.584 \cdot 10^{-10} y^{-1}$; $\lambda_\beta = 0.72 \cdot 10^{-10} y^{-1}$

    4.   Evaluation of reliability: Second order reliability because the most important

biostratigraphic dates do not come from the dated sequence (calcareous nanno-plankton). Also the error of 48/M 1a is too high.
6. Analyst: Balogh.
7. Institution: Institute of Nuclear Research, Hungarian Acad. of Science, Debrecen, Hungary.
8. Reference: Hamor et al. (in press).
9. Additional comments: The third age obtained from a tuff very probably of the same age but without the biostratigraphic proof (quarry in Tar, Matra Mountains) is of 19.0 ± 1.5 m.y.

Sample No. 50/23

A. Name of stratigraphic unit: lower - middle Sarmatian.
B. Geographic location of sample: Outcrop near the village Lesicek, district Presov, Slovakia (Czechoslovakia).
C. Stratigraphic position of sample: See comments concerning the stratigraphic position of the sample 35/AAD-24.
D. Petrologic designation of sample: Andesite—whole rock was dated.
  1. Method used for analysis: K-Ar.
  2a. Age and analytical uncertainty: 12.2 ± 0.1 m.y.
  4. Evaluation of reliability: First order reliability. See comments concerning the evaluation of reliability of sample 35/AAD-24.
  6. Analyst: Kreuzer (head of laboratory).
  7. Institution: Bundesanstalt f. Bodenforschung, Hannover, GFR.
  8. Reference: Not yet published.

Sample No. 51/24

A. Name of stratigraphic unit: lower - middle Sarmatian.
B. Geographic location of sample: Road cut between villages Zehna and Tuhrina, district Presov, Slovakia (Czechoslovakia).
C. Stratigraphic position of sample: See comments concerning the stratigraphic reliability of sample 35/AAD-24.
D. Petrologic designation of sample: Andesite—whole rock was dated.
  1. Method used for analysis: K-Ar.
  2a. Age and analytical uncertainty: 12.15 ± 0.05 m.y.
  4. Evaluation of reliability: The biostratigraphic age of underlying rocks is well defined, but the direct biostratigraphic evidence concerning the age of the overjacent complex is missing.
  6. Analyst: Kreuzer (head of laboratory).
  7. Institution: Bundesanstalt f. Bodenforschung, Hannover, GFR.
  8. Reference: Not published.

Sample No. 52/3

A. Name of stratigraphic unit: middle - upper Sarmatian.
B. Geographic location of sample: Quarry on the east border of the village Vinne, district Michalovce, Slovakia (Czechoslovakia).
C. Stratigraphic position of sample: The dated rock is a member of the Vinne-Zavadka volcanic formation, the pyroclastic equivalents of which rest on the *Elphidium hauerinum* zone (middle Sarmatian, Seuss, 1866) and is covered by the *Protelphidium subgranosum* zone.
D. Petrologic designation of sample: Andesite—the whole rock was dated.
  1. Method used for analysis: K-Ar.
  2. Age and analytical uncertainty: 12.05 ± 0.05 m.y.
  3. Decay constant used: $\lambda = 5.32 \cdot 10^{-10} y^{-1}$; R = 0.123.
  4. Evaluation of reliability: The biostratigraphic correlation seems to be reliable but biostratigraphic control is indirect; second order reliability.
  6. Analyst: Kreuzer (head of laboratory).
  7. Institution: Bundesanstalt f. Bodenforschung, Hannover, GFR.
  8. Reference: Not yet published.

Samples No. 55/Ko-10, 56/Ko-11

A. Name of stratigraphic unit: lower Sarmatian - lower Pontian.

B. Geographic location of sample: 55/Ko-10—Quarry Rohy near Pstrusa; 56/Ko-11—Stara Huta. Both localities are in Middle Slovakia (Czechoslovakia).

C. Stratigraphic position of sample: Both samples come from a stratovolcanic complex. The stratovolcanic complex rests on sediments containing the microflora (pollen spectrum) of lower Sarmatian. The marginal parts of the complex are covered by sediments containing the microflora of uppermost Pannonian to lower Pontian.

D. Petrologic designation of sample: Amphibole - pyroxene andesites (lavas)—whole rock was dated.

1. Method used for analysis: K-Ar.

2a. Age and analytical uncertainty: 55/Ko-10, 11.7 ± 0.4 m.y.; 56/Ko-11, $11.4 \pm 0.$ m.y.

3. Decay constant used: $\lambda_k = 0.557 \cdot 10^{-10} y^{-1}$; $\lambda_\beta = 0.72 \cdot 10^{-10} y^{-1}$

4. Evaluation of reliability: The biostratigraphic control is rather large; second order reliability.

6. Analyst: Bagdasarjan.

7. Institution: Armenian Academy of Science, Erivan UdSSR.

8. Reference: Bagdasarjan, Konecny, Lexa (in preparation).

9. Additional comments: radiometric ages recalculated by decay constant: $\lambda_k = 0.584 \cdot 10^{-10} y^{-1}$ is for Ko-10: 11.1 ± 0.4 m.y. Ko-11: 10.8 ± 0.0 m.y.

Samples No. 57/18, 58/30

A. Name of stratigraphic unit: Akchagylian - upper Pliocene.

B. Geographic location of sample: 57/18, Dashbash; 58/30, Trialeti.

C. Stratigraphic position of sample: The dated rocks come from a lava complex covered by two volcanic complexs. One of them contains in its lower part the mammals of upper Pliocene, another is lying on lacustrine beds containing the Akchagylian mammals.

D. Petrologic designation of sample: Andesite-basalts—whole rocks were dated.

1. Method used for analysis: K-Ar.

2a. Age and analytical uncertainty: 57/18, 0.95 ± 0.30 m.y.; 58/30, 2.26 ± 0.60 m.y.

3. Decay constant used: $\lambda_k = 0.584 \cdot 10^{-10} y^{-1}$; $\lambda_\beta = 0.72 \cdot 10^{-10} y^{-1}$

4. Evaluation of reliability: The biostratigraphic control is from continental mammals and the continental biochronology is reliable.

6. Analyst: Rubinstein.

7. Institution: Georgian Academy of Science, Tbilisi, UdSSR.

8. Reference: Rubinstein et al (1972), p. 14-16.

References Cited

Afanas'yev, G. D., et al, 1963, Materialy k obosnovaniu vozrasta rubezej mezdu nekotorymi geologiceskimi sistemami i epochami: Izv. Akad. Nauk SSSR, Ser. Geol., v. 11, p. 7-31.

Bagdasarjan, G. P., and L. G. Danilovich, 1968, Novye dannye ob absoljutnom vozraste vulkaniceskich obrazovanij Zakarpatja: Akad. Nauk SSSR, Izv. Ser. Geol., v. 8, p. 15-23.

—— D. Vass and V. Konecny, 1968, Results of Absolute age determination of Neogene rocks in central and eastern Slovakia: Geol. Zb. (Geologica Carpath.), v. 19, no. 2, p. 419-425.

—— J. Slavik and D. Vass, 1971, Chronostratigraficky a biostratigraficky vek niektorych vyznamnych neovulkanitov vychodneho Slovenska: Geol. Prace, Zpr., v. 55, p. 87-96.

Baldi, T., 1973, Moluscs fauna of the Hungarian upper Oligocene (Egerian): Budapest Akad. Kiado, 511 p.

—— T. Kecskemeti and M. R. Nyiro, 1961, A kati es akvitani emelet kerdese a karpat-medenceben Eger kornyeki uj adatok alapjan: Foeldt. Koezl., v. 91, no. 3.

Berggren, W. A., 1969a, Biostratigraphy: Cenozoic foraminiferal faunas, in M. Ewing et al, eds., Initial Repts. Deep Sea Drilling Project, v. I: Washington, D.C., U.S. Govt., p. 584-607.

—— 1969b, Cenozoic chronostratigraphy, planktonic foraminiferal zonation and the radiometric time-scale: Nature, v. 224, no. 5224, p. 1072-1075.

—— 1969c, Rates of evolution in some Cenozoic planktonic foraminifera: Micropaleontology, v. 15, no. 3, p. 351-365.

—— 1969d, Micropaleontologic investigations of Red Sea cores—summation and synthesis of results, *in* E. T. Degens and D. A. Ross, eds., Hot brines and Recent heavy metal deposits in the Red Sea: New York, Springer-Verlag, p. 329-335.

—— 1971a, Tertiary boundaries and correlations, *in* B. M. Funnell and W. R. Riedel, eds., The Micropaleontology of Oceans: Cambridge, Cambridge University Press, p. 693-809.

—— 1971b, Neogene chronostratigraphy, planktonic foraminiferal zonation and the radiometric time-scale: Hungary Geol. Soc. Bull., v. 101, p. 162-169.

—— 1971c, Multiple phylogenetic zonations of the Cenozoic based on planktonic Foraminifera, *in* A. Farinacci ed., 2nd Planktonic Conf. (Rome) Proc., p. 41-56.

—— 1972a, A Cenozoic time-scale—some implications for regional geology and paleobiogeography: Lethaia, v. 5, p. 195-215.

—— 1972b, Late Pliocene—Pleistocene glaciation, *in* A. S. Laughton, W. A. Berggren et al, Preliminary reports of the Deep Sea Drilling Project: U.S. Govt., v. 12, p. 953-963.

—— 1973, The Pliocene time-scale; calibration of planktonic foraminiferal and calcareous nannoplankton zones: Nature, v. 243, no. 5407, p. 391-397.

—— 1974, Biostratigraphy and biochronology of the late Miocene (Tortonian and Messinian), *in* C. W. Drooger, ed., Symposium on Messinian Events in the Mediterranean, Utrecht, Amsterdam, North-Holland, p. 10-20.

—— and J. A. Van Couvering, 1974, The late Neogene: Amsterdam, Elsevier, 216 p.

Bojko, R. K., et al, 1970, Absolutnaja geochronologia glavnejshich kompleksov Ukrainskich Karpat: Akad. Nauk SSSR Kom. Opred. Absol. Vozrasta Geol. Form. Tr., p. 202-226.

Ciesarik, M., and E. Planderova, 1965, Geologicka pozicia limnokvarcitov loziska: Stara Kremnicka Geol. Prace, Zpr., v. 35, p. 87-98.

Cvercko, J., 1968, Vysledky geologickeho vyskumu v neogene Kosicko-presovskej kotliny so zretelom k problemom zivic: Manuskript, Geofond (Bratislava).

Evernden, J. F., and R. K. S. Evernden, 1970, The Cenozoic time scale: Geol. Soc. America Spec. Paper 124, p. 71-90.

—— et al, 1961, On the evaluation of glauconite and illite for dating sedimentary rocks by the potassium-argon method: Geochim. et Cosmochim. Acta, v. 23, p. 78-99.

Gabunia, L. K., and M. M. Rubinstein, 1977, Ob absolutrom vozraste gipparion iz: Saro Geol. Zbor. (Geologica Carpath.), v. 28, no. 1, p. 7-11.

Hamor, G., K. Balogh, and L. Ravasz Baranyai, in press, Az eszak-magyarovszagi harmadidoszaki formaciok radioaktiv Kora.

Ivan, L., 1964, Geologicka pozicia perlitov v juz. casti Zemplinskych vrchov: Geol. Prace Zpr., no. 2, p. 143-145.

Jiricek, R., 1972, Problem hranice sarmat/panon ve videnske, podunajske a vychodoslovenske panvi: Mineral. Slovaca, v. 4, no. 14, p. 39-81.

Laskarev, V., 1924, Sur les equivalents du Sarmatien superieur en Serbie: Rec. Trav. M.I. Cvijic (Belgrade), p. 1-5.

Lehotayova, R. H., 1971, Foraminifery a vapnita nanoflora badenienu jv. casti Podunajskej niziny: Manuskript, Geofond (Bratislava).

Lippolt, J., I. Baranyi, and W. Todt, 1975, Das kalium-argon alter des basaltes vom Lavant-Tal in Karnten: Der Auschlus, Jg. (Heidelberg), v. 26, no. 6, p. 218-242.

Martini, E., and C. Muller, 1975, Calcareous nannoplankton and silicoflagellates from the type Ottnangian and equivalent strata in Austria (lower Miocene): 6th Cong. Reg. Comm. Mediterranean Neogene Stratigraphy Proc., p. 121-124.

Mikhailova, N. P., A. M. Glevasskaja, and V. M. Sykora, 1974, Paleomagnetizm vulkanogennych porod i rekonstrukcia paleomagnitnogo polja neogena: Izd. Nauk Dumka, Kiev, 252 p.

Obradovich, J. D., 1964, Problem in the use of glauconite and related minerals for radioactivity dating: PhD thesis, Univ. of California, Berkeley.

Odin, G. S., and J. C. Hunziker, 1975, Donnees nouvelles sur l'age nummerique des glauconies Oligo-Miocenes de la Paratethys: Geol. Prace, Zpr., v. 63, p. 141-148.

—— et al, 1970, Geochronologie de niveaux glauconieux paleogenes d Allemagne du nord: Soc. Geol. France Compte Rendu, no. 6, p. 220-221.

Papp, A., et al, 1973, Chronostratigraphie und Neostratotypen Miozan der zentralen Paratethys, v. 3, Ottnangien: Slov. Akad. Vied Bratislava, p. 1-841.

Pesl, V., J. Salaj, and D. Vass, 1968, The flysh and Klippen belts, Neogene basin: 23rd Int. Geol. Cong. (Prague) Ex. 6 AC Guide, 140 p.

Pishvanova, L. S., 1974, Le Tortonien inferieur de l'Ukraine et ses analogues en Europe occidental: (France) Bur. Recherches Geol. et Minieres Mem., v. 78, p. 775-783.

Pulec, M., and D. Vass, 1969, Les textures et les structures des tufs soudes du Neogene superieur de la Slovaquie orientale: Geol. Zb. (Geologica Carpath.), v. 20, p. 65-80.

Repcok, I., 1975, Priprava metodik na datovanie mladych formacii: Manuskript, Archiv GUDS (Bratislava).

Rubinstein, M. M., and L. K. Gabunia, 1972, Nekotorye voprosy geochronologii Kainozoia: Akad.

Nauk SSSR Izv. Ser. Geol., no. 3, p. 3-8.

—— et al, 1972, Datirovanie nekotorych verkhne neogenovych i chetvertichnykh effuzivov Zakavkazia: Akad. Nauk SSSR, Izv. Ser. Geol., no. 4, p. 13-16.

Ryan, W. B. F., et al, 1974, A paleomagnetic assignment of Neogene stage boundaries and the development of isochronous datum planes between the Mediterranean, the Pacific and Indian Oceans: Riv. Italiana Paleontologia e Stratigraphia, v. 80, p. 631-688.

Selli, R., 1970, Report on absolute age: Gior. Geologia, Ser. 2a, v. 35, no. 1, p. 51-59.

Semenenko, N. P., et al, 1966, Geologia SSSR, Karpaty: Moscow, Fad Nedra, v. 48, no. 1, 540 p.

Senes, J., et al, 1975, Regional stages of the Central Paratethys Neogene and the definition of their lower boundaries: 6th Cong. Reg. Comm. Mediterranean Neogene Stratigraphy Proc., p. 259-265.

Seuss, E., 1866, Untersuchungen uber den charakter der osterreichischen Tertiarablagerungen; II. Uber die Bedeutung der sogenannten "brakischen Stufe" oder "Cerithienschichten". Sitzungsber: Akad. Wiss. Wien Math-Naturw. Kl., v. 54, p. 218-357.

Slavik, J., and J. Tozser 1973, Geological structure of the Presovske pohorie Mts. and its relation to the boundary of the West and East Carpathians: Geol. Zb. (Geologica Carpath.), v. 24, no. 1, p. 23-52.

—— et al, 1976, Radiometric age and paleomagnetism Vihorlat and Presovske pohorie volcanic mountains: (in Russian) Mineral. Slovaca, v. 8, no. 4, p. 319-334.

Steininger, F., and G. P. Bagdasarjan, in press, Neue radiometrische Alter mittelmoizaner Vulkanite der Steiermark (Osterreich) ihre biostratigraphische Korrelation und ihre mogliche Stellung innerhalb der palaomagnetischen Zeitskale: Verh. Geol. Bundesanst (Wien).

—— F. Rogl, and E. Martini, 1976, Current Oligocene-Miocene biostratigraphic concept of the Central Paratethys: Newsletter Stratigraphy, v. 4, p. 174-202.

—— et al, 1971, Die definitionden Zeiteinheit M. (Eggenburgien) in Chronostratigraphie und neostratotypen Miozan der zentralen Paratethys, v. 2: Bratislava, Slov. Akad. Vied, (Geologica Carpath.), 827 p.

Svagrovsky, J., 1962, Geologia uzemia medzi Torysour a Olsavou na vychodnom Slovensku: Geol. Prace, Zpr., v. 63, p. 185-192.

—— 1964, Zur Torton-Sarmat Grenze im ostslowakischen Neogen. Geol.: Geol. Zb. (Geologica Carpath.), v. 15, no. 1, p. 79-84.

Theyer, F., and S. R. Hammond, 1974, Cenozoic magnetic time scale in deep-sea cores: completion of the Neogene: Geology, v. 2, no. 10, p. 487-492.

Tsonj, O. V., and J. Slavik, 1971, Vek ryolitov zemplinskeho ostrova: Geol. Prace, Zpr., v. 55, p. 215-216.

Turner, D. L., 1970, Potassium argon dating of Pacific Coast Miocene foraminiferal stages: Geol. Soc. America Spec. Paper 124, p. 91-129.

Vass, D., 1969, Vek a petrograficke zlozenie neogennej vyplne komarovskej depresie: Zbornik vychodoslov. muzea (Kosice), ser. A., v. 7, p. 87-95.

—— 1975, Report of the working group on radiometric age and paleomagnetism: 6th Reg. Comm. Mediterranean Neogene Stratigraphy Proc., p. 103-117.

—— and J. Slavik, 1975, The radiometric calibration of Paratethys Neogene: Geol. Prace, Zpr., v. 63, p. 131-139.

—— G. P. Bagdasarjan, and V. Konecny, 1971, Determination of the absolute age of the West Carpathians Miocene: Foeldt. Koezl, v. 101, no. 2-3, p. 321-327.

—— J. Slavik, and G. P. Bagdasarjan, 1975, Radiometric time scale for Neogene of Paratethys (to August 1, 1974): 6th Cong. Reg. Comm. Mediterranean Neogene Stratigraphy Proc., p. 259-297.

On Dating of the Paleogene[1]

M. RUBINSTEIN[2] and L. GABUNIA[2]

Abstract Differences in data reported by various authors raise questions on the absolute dating of the main Paleogene boundaries. Ignoring established geochronometric and biostratigraphic principles only complicates the task of correlating stratigraphic units and establishing the precision of the absolute geochronologic scale. Absolute ages of younger, radiologically dated samples provides a lesser absolute error but greater relative error. It is suggested that age determinations on effusives are closer to true values (or the error is minimized), whereas interpretation on glauconites is more complicated because characteristically high argon loss in authigenic minerals seems to yield younger trending dates. With new data becoming available on continental rock units, biostratigraphic and geochronometric correlation with coeval marine facies appears to be increasingly important.

Discussion

After the publication of the two most frequently cited absolute geochronologic scales of the Phanerozoic—the one by the commission for absolute age determination of geologic formations of the Earth Science Division (ESD) of the Academy of Sciences of the USSR (Afanas'yev and Rubinstein, 1964) and the one by the London Symposium dedicated to A. Holmes (Harland et al, 1964)—a number of articles appeared giving greater precision and detail for parts of the scales. Particular attention has been given to the Cenozoic part of the scale (Evernden et al, 1964; Evernden and James, 1964; Gabunia and Rubinstein, 1964, 1965, 1968; Rubinstein, 1967; Berggren, 1969, 1972; Rubinstein and Gabunia, 1972).

Differences in data reported by the various authors raise questions on the absolute dating of the main Paleogene boundaries. Ignoring well-known geochronometric and biostratigraphic principles unnecessarily complicates the difficult task of correlating stratigraphic units and establishing the precision of the absolute geochronologic scale.

The younger the absolute age of a radiologically dated sample the less the absolute error and the greater the relative error. For example, tentative calculations (Rubinstein, 1967) show that the most probable error for Carboniferous age determinations (300 to 360 m.y.) by K-Ar dating (mainly used for the construction of the Phanerozoic geochronologic scale) are within 10 to 15 m.y., whereas for the Paleogene (25 to 65 m.y.) the error is between 1.5 and 2.5 m.y.

It is natural that requirements for the accuracy of stratigraphic positions for Cenozoic critical points must be higher than for the Mesozoic, and Mesozoic points higher than for Paleozoic ones.

Most authors strive to use well-stratified rocks, particularly effusive rocks, and glauconite-bearing rocks for geochronometric measurements. In geochronometric respects, even effusive and glauconitic rocks have their disadvantages and do not yield equivalent results.

Effusive rocks (both whole-rock analyses or individual-mineral analyses) commonly yield age figures close to true values or minimized to variable extent as a result of the partial loss of radiogenic argon. The argon loss is especially great in volcanic glasses. As a rule, excess age figures for effusive units are uncommon (the exception is pyroclastics which may contain an admixture of older clastogenic elements; this exception causes

---

[1] Manuscript received, October 11, 1976.
[2] Academy of Science of Georgian S.S.R., 380004 Tbilisi, Georgian, S.S.R., USSR.

no particular misunderstanding). The loss of argon depends on the thermal history of the rocks, their degree of crystallinity, their chemical composition, and other factors (Rubinstein, 1967; Evernden and Evernden, 1970). Thus, if several analytically reliable argon age dates are available for effusive rocks from a certain stratigraphic level, the oldest values come nearer to true values.

Interpretation of age dates from glauconites is more complicated. The study of authigenic minerals for argon dating and their characteristically high argon loss leads commonly to variable young dates although some dates apparently correspond to the true age. Argon dates older than the true age are not frequent. The graph[3] (Fig. 1) constructed on the basis of 390 Phanerozoic glauconite dates illustrates these relations clearly (Rubinstein and Polevaya, 1974). Rubinstein (1967) showed that such variable results may be explained by the presence of admixtures of older clastogenic hydromicas yielding older argon dates.

Unless objective mineralogic data establishes the reliability of glauconites for argon dating, we must avoid far-reaching conclusions or recognize the conditional nature of these dates. We shall have to return to this question later. In summary, argon determination errors are made up of analytic errors and errors caused by the loss of radiogenic argon from the sample, or rarely by a surplus of argon. The estimation of the excess argon in a sample contributing to the total error is a difficult problem. The probability is that more conclusive geochronologic determinations will come from effusive rocks than from glauconites[4].

Complications bound up with the use of radiologic data for correlation purposes are not caused by geochronometric errors alone. Stratigraphic errors are of no less importance. Radiologically dated samples cannot always be taken directly from the horizons containing fossil animals or plants but, if they could, paleontologic determination of the stratigraphic level would be as essential as the radiometric age determination. The significance of the stratigraphic error sharply increases when comparing sections far removed from one another, especially in intercontinental correlation or differing lithofacies. For example, it was shown long ago (Gabunia and Rubinstein, 1964, 1965; Evernden and James, 1964; Evernden et al, 1964) that the lower Pliocene in the stratigraphic sequence of North American continental deposits (Clarendon) corresponds to the upper Miocene of Europe (Sarmatian). Of primary importance for the absolute age of the upper boundary of the Paleogene, drawn at the base of the Aquitanian stage according to most modern investigators, is the geochronometric data for the Paratethys (Rubinstein and Gabunia, 1972; Vass et al, 1975). In the Paratethys almost every horizon of a widespread stratigraphic sequence of deposits from uppermost Oligocene to Pannonian, inclusive, has an argon date, predominantly from effusives but partly from glauconites. Perfect conformity of age figures[5] obtained for rhyolite tuff of the Ottnangian (M_2)–21.9 (23.0) m.y. and vitroclastic tuff of the lower Carpathian (M_3)–20.7 (21.7) ± 1.3 m.y. on the one hand and the glauconite of the Ottnangian (M_2)–22 (23.7) ± 1.2 m.y. on the other is striking. However, considerably lesser age dates for three glauconites of the Eggenburgian (M_1) ranging from 1 $\_$.3 to 17.6 (17.1 to 18.5) m.y. are reported by Odin et al (1975).

In view of the high quality of investigations carried out by Bagdasarjan, we disagree with Odin (1973) who attributed the higher age dates for the effusives of the Paratethys, in comparison with the glauconites of the same stratigraphic position, to selection of data and method of measurement. According to Vass et al (1975) in a report to the 6th Congress on the Mediterranean Neogene, repeated measurements of the argon age for samples of effusive rocks carried out by other laboratories (Lippolt, private commun.) completely confirm the data published earlier by Vass et al (1971). This age confirmation suggests that the age dates obtained for the Ottnangian and lower Car-

---

[3] Age values on the graph are given for the constant: $\lambda_K = 0.557 \cdot 10^{-10}$ yr^{-1}.

[4] For the same reason, it would be a serious error when constructing a geochronologic scale to ignore radiologically reliable data obtained from micas in intersecting magmatic rock bodies for which only one stratigraphic limit—upper or lower—is known precisely.

[5] Ages calculated by the constant $\lambda_K = 0.584$ stand first; those calculated by the constant $\lambda_K = 0.557 \ 10^{-10}$ yr^{-1} accepted in the USSR are enclosed in brackets.

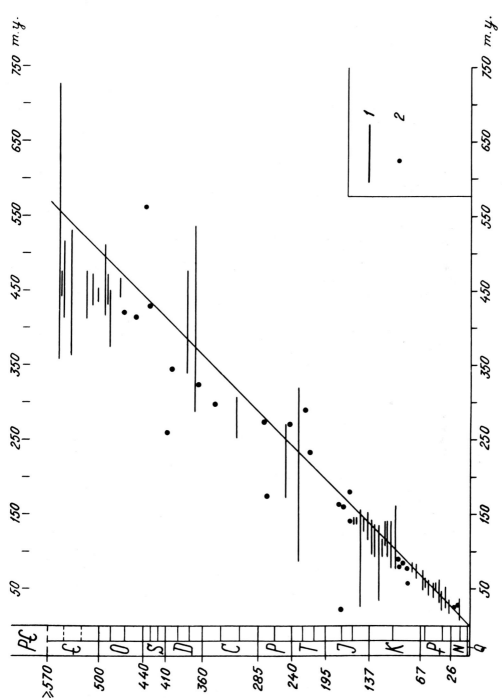

FIG. 1—Argon age of Phanerozoic glauconites showing: (1) limits of variation of time-figures, and (2) individual measurements (Rubinstein and Polevaya, 1974).

pathian effusives are close to the true dates and the corresponding glauconites studied by Odin and others evidently were rejuvenated, especially as the same argon dates (almost) were reported earlier (Vass et al, 1970) for the effusives of a significantly higher stratigraphic level—the Badenian (M_3): 15.1 to 17.2 (15.9 to 18.1) m.y. Absolute age dating is even more contradictory in the case of the Eggenburgian (M_1) and Egerian $O-M_1$ mentioned previously for which we have no direct argon dating of effusives, whereas previous data for glauconites—25.0 (26.2) m.y. (M_1) and 30.5 (32.0) m.y. ($O-M_1$) are in question (Obradovich, 1964; Odin et al, 1975) and far lesser ones, 18.8 (19.7) m.y. (M_1) and 20.8 to 19.6 (21.8 to 20.6) m.y. ($O-M_1$), are suggested. The latter values, in our opinion, are less subject to criticism for the simple reason that they are closer to the argon age of the Ottnangian effusives, and to the obviously understated age date of 22.5 (23.5) m.y. for the Oligocene-Miocene boundary in the European sense, suggested by Berggren's scale (1972). The age data, 33.5 (35.2) ± 2.4 m.y., reported recently by the Hungarian investigators (Baldi et al, 1975) for glauconites from Kiskel clays underlying the Egerian and apparently equivalent to the upper Rupelian, are in satisfactory agreement with former argon dates for the Egerian glauconite, 30.5 (32.0) m.y., and the Eggenburgian glauconite, 25.0 (26.2) m.y.

The foregoing data led to the conclusion that argon age dates for the main Miocene horizons of the Paratethys, obtained for effusives (and consequently some dates for glauconites conforming to them) are close to the true ones.

Berggren's age for the Oligocene-Miocene boundary in his Cenozoic geochronologic scale (Berggren, 1972) was based exclusively on glauconites, using data from French investigators published for the Paris basin (Bonhomme et al, 1968; Bødelle et al, 1969), London-Hampshire basin (Odin et al, 1969b), Belgian basin (Odin et al, 1969a), and North German basin (Odin et al, 1970). Inasmuch as the argon age dates for five samples of the upper Oligocene-Chattian glauconites from the Belgian and North German basins range between 28.8 to 31.2 (30.2 to 32.8) m.y.—on the average 30.0 (31.5) m.y., Berggren dated the Rupelian-Chattian boundary at 31 to 32 (32.7 to 33.6) m.y., assuming that in the North German basin a part of this period corresponds to a prolonged gap (5 to 6 m.y.).[6] A still longer gap on the order of 9 m.y. is postulated in the Aquitaine basin between the Stampian and the uppermost Chattian (Fig. 2), which is overlain by the stratotype sequence of the Aquitanian. Berggren correlated the base of the latter with the lower boundary of Blow's zone N4 (the *Globigerinoides* level, which places the Oligocene-Miocene boundary at 22.5 (23.6) m.y. He based this on Turner's argon dates (Turner, 1970) for foraminiferal stages of the Pacific coast of the United States, according to which the Zemorrian-Saucesian boundary is estimated at 22.5 (23.5 m.y.).

In Berggren's discussion there are weak points, including his ignoring of some geochronometric dates received for European sequences. It is necessary to remember that Selli and Tongiorgi (1967) proposed the date of 23.3 (24.5) m.y. for glass from the lower Miocene volcanic ash of Italy (1 m above the level of the first appearance of *Globigerinoides*). As mentioned previously, volcanic glass usually yields low argon ages; therefore the 23.3 (24.5) m.y. date should be considered minimal. The date of 25 (26.2) m.y. reported by Evernden et al (1961) for glauconites from the "Burdigalian" of the Vienna basin (Eggenburgian) has been repeatedly subjected to criticism for the precision of its stratigraphic position. The dated sample was taken above the *Globigerinoides trilobus* level (Bandy and Ingle, 1970) and though this date is questioned (Evernden and Evernden, 1970) it should not be totally disregarded. The European data for the present do not permit the dating of the Oligocene-Miocene boundary by an age date less than that accepted by us earlier as 25 (26) m.y. (Rubinstein and Gabunia, 1972).

From such a point of view there is no more necessity for the assumption of hardly probable gaps in the Oligocene of the North German and Aquitaine basins, the unjustifiedly long total duration of the Oligocene shortens (nearly to equal the duration of the entire Eocene in Berggren's scale), corresponding stratigraphic subdivisions of

---

[6] The time interval corresponding to the entire middle Eocene according to Berggren's scale.

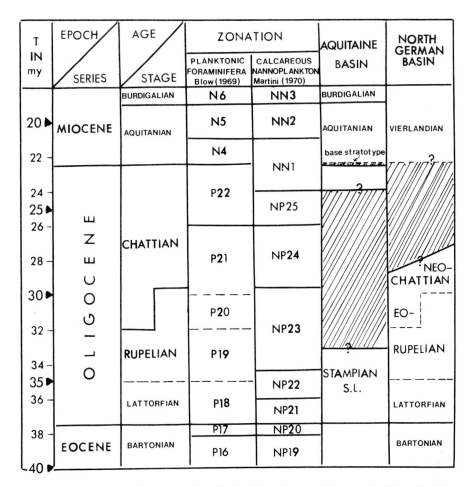

T IN my	EPOCH / SERIES	AGE / STAGE	ZONATION		AQUITAINE BASIN	NORTH GERMAN BASIN
			PLANKTONIC FORAMINIFERA Blow (1969)	CALCAREOUS NANNOPLANKTON Martini (1970)		
		BURDIGALIAN	N6	NN3	BURDIGALIAN	
20 ▶	MIOCENE	AQUITANIAN	N5	NN2	AQUITANIAN	VIERLANDIAN
22			N4	NN1	base stratotype	?
24			P22	NP25	?	
25 ▶	O L I G O C E N E	CHATTIAN				
26						
28			P21	NP24		NEO-CHATTIAN
30 ▶			P20	NP23		EO-
32						
34		RUPELIAN	P19		?	RUPELIAN
35 ▶				NP22	STAMPIAN S.L.	
36		LATTORFIAN	P18	NP21		LATTORFIAN
38		BARTONIAN	P17	NP20		BARTONIAN
40 ▶	EOCENE		P16	NP19		

FIG. 2—Suggested late Paleogene chronologic relations between time-stratigraphic units in Aquitaine and North German basins (Berggren, 1972).

the European continental and marine deposits become more easily correlable on the basis of the absolute age, and absolute dates for the effusive elements of the Paratethys no longer look anomalous.

We may now consider the Oligocene-Eocene boundary, dated by Berggren as 37.5 (39.4) m.y. on the basis of the argon age determination of only one glauconite from the uppermost Eocene of Belgium (Odin et al, 1969a). In the presence of repeatedly published geochronometric data for the post-middle Eocene - pre-Oligocene intrusive and effusive rocks of Georgia and Armenia (Rubinstein, 1961; Bagdasarian and Gukasyan, 1964; Rubinstein et al, 1973) it is possible to date the mentioned boundary 35 to 36 (37 to 38) m.y. The question does not seem to be debatable, at least at present.

Determining the probable age of the boundaries of the major Eocene subdivisions, in previous publications (Gabunia and Rubinstein, 1965, 1968; Rubinstein and Gabunia, 1972) we accepted the following figures with a considerable degree of question: late - middle Eocene, 43 (45) m.y.; middle - early Eocene, 52 (55) m.y.; and early Eocene - Paleocene, 57 (60) m.y. Berggren dates the first of these boundaries 43 (45) m.y. like us, whereas for the second and third ones he accepts considerably lesser values, accordingly 49 (51.4) m.y. and 53.5 (56.2) m.y., proceeding from the data for glauconites of western Europe, published in the French works cited previously.

It should be noted that the greatest argon age dates on gross samples of the middle Eocene alkaline and subalkaline effusive rocks of the western part of the Adjara-Thrialethi[7] (Rubinstein et al, 1973) vary within 40.8 to 47.5 m.y. which is in good accord with the previously published argon dating results of the Armenian middle Eocene effusives (Bagdasarjan and Gukasyan, 1964) of 41.8 to 45.6 (44 to 48) m.y., and do not contradict the same dating of the upper-middle Eocene boundary. However, we cannot consider justified the reduction in age of the middle-early Eocene and early Eocene - Paleocene boundaries by 3 to 3.5 m.y. as suggested by Berggren. We account for it by Berggren's tendency, so frequently manifested, to understate boundary ages between separate stratigraphic subdivisions. Ignoring of the argon age-determination results for the four glauconites from the Evekoro Formation (southeastern Nigeria) taken in the vicinity of the level corresponding to the lower Eocene - Paleocene boundary (Adegoke et al, 1970) is extremely significant. Noting that the age figures for these glauconites vary within 53.0 to 56.1 (55.6 to 58.9) ± 2.7 m.y., with a mean value of 54.4 (57.1) m.y., the author nevertheless leaves the age of the boundary —itself equal to 53.5 (56.2) m.y. Finally, as to the Paleogene-Cretaceous boundary, which is dated by an overwhelming majority of investigators as 64 to 65 (67 to 68) m.y., it is virtually only a semblance of "good luck," because there is no unity of opinion regarding the stratigraphic position of this boundary, whether it passes above or under the Danian. What are the relations between the Montian and Danian on the one hand and the Puerco and "Maastrichtian" of the North American scale, on the other? Until the new data come along we prefer to hole our previous opinion (Rubinstein and Gabunia, 1972) interpreting the age value for the Cretaceous-Paleogene boundary of 65 (68) m.y. as a synchronization of boundaries (Danian-Paleocene in Eurasia and "Maastrichtian"-Puerco in North America).

We concerned ourselves exclusively with the material of argon dating of the Paleogene deposits represented in the marine facies. In light of this, a critical revision of previous biostratigraphic and geochronometric correlations with coeval continental units seems highly important, especially as there are new data for the latter (Gabunia et al, 1975). We hope to return to this problem in the near future.

References Cited

Afanas'yev, G. D., and M. M. Rubinstein, 1964, Explanatory note to the geochronological scale in the absolute chronological system, *in* Absolute age of geologic formations—Papers presented by Soviet geologists at 22d International Geological Congress (Delhi, India): Moscow, Nauka, p. 289-324.

Bagdasarjan, G. P., and R. Kb. Gukasyan, 1964, Investigations on the working out of geologic markers for the absolute time scale (on the basis of data from the Armenian SSR), *in* Absolute age of geologic formations—Papers presented by Soviet geologists at 22d International Geological Congress (Delhi, India): Moscow, Nauka.

—— D. Vass, V. Konecny, 1968, Results of absolute age determination of Neogene rocks in central and eastern Slovakia: Geol. Zb. (Slov. Akad. Vied.), v. 19, no. 2.

Baldi, T., et al, 1975, On the radiometric age and the biostratigraphic position of the Kiscell Clay in Hungary: 6th Cong. Reg. Comm. Mediterranean Neogene Stratigraphy Proc.

Bandy, O. L., and J. C. Ingle, Jr., 1970, Neogene planktonic events and radiometric scale, California—radiometric dating and paleontologic zonation: Geol. Soc. America Spec. Paper 124, p. 131-172.

Berggren, W. A., 1969, Cenozoic chronostratigraphy, planktonic foraminifera zonation and the radiometric time scale: Nature, v. 224, p. 1072-1075.

—— 1972, A Cenozoic time-scale—some implication for regional geology and paleogeography: Lethaia, v. 5, p. 195-215.

Bodelle, J., C. Lay, and G. S. Odin, 1969, Determination d'age par la methode geochronologique "potassium-argon" de glauconies du Bassin de Paris: Acad. Sci. Comptes Rendus, v. 268, p. 1474-1477.

Blow, W. H., 1969, Late middle Eocene to Recent planktonic foraminiferal biostratigraphy, *in* P. Bronnimann and H. H. Renz, eds., Int. Conf. Planktonic Microfossils, Geneva (1967), Proc., p. 199-421.

---

[7] They most probably have undergone some argon rejuvenation.

Bonhomme, M., G. S. Odin, and C. Pomerol, 1968, Age de formation glauconieuses de l'Albien et de l'Eocene du Bassin de Paris: Rech. Geol. Minieres Mem. 58, p. 339-346.

Dessauvagie, T. F., O. S. Adegoke, and C. A. Kogbe, 1971, Radioactive age determination of glauconite from the Ewekoro type locality, Nigeria (abs.), *in* 1st Conference on African geology: London, Commonwealth Geol. Liason Office, p. 3-4.

Evernden, J. F., and R. K. S. Evernden, 1970, The Cenozoic time scale, *in* Radiometric dating and paleontologic zonation: Geol. Soc. America Spec. Paper 124, p. 71-90.

—— and G. T. James, 1964, Potassium-argon dates and the Tertiary floras of North America: Am. Jour. Sci., v. 262, p. 945-974.

—— et al, 1961, On the evaluation of glauconite and illite for dating sedimentary rocks by the potassium–argon method: Geochim. et Cosmochim. Acta, v. 23, p. N1-2.

—— et al, 1964, Potassium-argon dates and the Cenozoic mammalian chronology of North America: Am. Jour. Sci., v. 262, p. 145-198.

Gabunia, L. K., and M. M. Rubinstein, 1964, On the question of the correlation of the Neogene and late Paleogene deposits of the Old and New Worlds (on the basis of data on fossil mammals and absolute age), *in* A. L. Tsagareli, ed., A collection of papers on aspects of the geology of Georgia, USSR, compiled for the 22d International Geological Congress (Delhi, India): Akad. Nauk Gruzin, Tiflis (in Russian, with English summary).

—— —— 1965, Biostratigraphic correlation of the Cenozoic deposits of Eurasia and North America in the light of absolute geochronological data: Izv. Geol. Ob-va Gruzii, v. 4, no. 1.

—— —— 1968, On the correlation of the Cenozoic deposits of Eurasia and North America based on the fossil mammals and absolute age data: 23d Int. Geol. Cong. (Prague) Proc., sec. 10, p. 9-17.

—— Y. V. Devyatkin, and M. M. Rubinstein, 1975, Data on the absolute age of the Cenozoic continental formations of Asia and their biostratigraphic significance: Akad. Nauk SSSR Dokl., v. 225, no. 4.

Harland, W. B., A. G. Smith, and B. Wilcock, eds., 1964, The Phanerozoic time scale, a symposium: Geol. Soc. London Quart. Jour., v. 120, supp., 458 p.

Martini, E., 1971, Standard Tertiary and Quaternary calcareous nannoplankton zonation: Int. Conf. Planktonic Microfossils, Rome (1970), Proc., p. 739-785.

Obradovich, J., 1964, Problems in the use of glauconite and related minerals for radioactivity dating: PhD thesis, Univ. California, Berkeley, 160 p.

Odin, G. S., 1973, Sur datations radiometriques du Miocene de la Parathethys: Soc. Geol. France Comptes Rendu, v. I, p. 33-34.

—— J. C. Hunziker, and C. R. Zorenz, 1975, L'age radiometrique du Miocene inferieur en Europe Occidentale et Centrale: Geol. Rundsch., v. 64, p. 570-572.

—— et al, 1969a, Geochronologie de niveaux glauconieux tertiaires du bassin de Belgique (methode potassium-argon): Soc. Geol. France Comptes Rendu, v. 6, p. 198-200.

—— et al, 1969b, Geochronologie de niveau glauconieux tertiaires des bassins de Londes et du Hampshire (methode potassium-argon): Soc. Geol. France Comptes Rendu, v. 8, p. 309-310.

—— et al, 1970, Geochronologie de niveau glauconieux paleogenes d'Allemagne du Nord: Soc. Geol. France Comptes Rendu, v. 6, p. 220-221.

Rubinstein, M. M., 1961, Some critical points of the post-cryptozoic geological time-scale geochr. of rock systems: New York Acad. Sci. Annals, v. 91, p. 364-368.

—— 1967, The argon method applied to some problems of regional geology: Akad. Nauk Gruz. SSR Geol. Inst. Tr. n.s., no. 11, 239 p.

—— and L. K. Gabunia, 1972, Some problems of Cenozoic geochronology: Int. Geol. Rev., v. 15, p. 664-668.

—— N. I. Polevaya, 1974, Geochronologic scale of the Phanerozoic, *in* N. I. Polevaya, ed., Fanerozoy: Izd. Nedra, Leningrad Otd., v. 2, p. 304-314.

—— et al, 1973, On the argon age of some magmatic formations of the Adjara-Thrialethian folded system: Akad. Nauk Gruz. SSR Geol. Inst. Tr. n.s., no. 38.

Selli, R., and E. Tongiorgi, 1967, Report of the working group: absolute age: 4th Cong. Reg. Mediterranean Neogene Stratigraphy (Bologna).

Turner, D. L., 1970, Potassium-argon dating of Pacific Coast Miocene foraminiferal stages–radiometric dating and paleontologic zonation: Geol. Soc. America Spec. Paper 124, p. 91-130.

Vass, D., G. Bagdasarjan, and V. Konecny, 1970, Absolute age of several stages of West-Carpathian Miocene: Geol. Prace Zpr., v. 51.

—— —— —— 1971, Determination of the absolute age of the West Carpathian Miocene: Foeldt. Koezl., v. 101, p. 321-327.

—— J. Slavik, and G. P. Bagdasarjan, 1975, Radiometric time scale for Neogene of Parathethys (as of August 1, 1974): 6th Cong. Reg. Comm. Mediterranean Neogene Stratigraphy (Bratislava) Proc.

A New Paleogene Numerical Time Scale[1]

J. HARDENBOL[2] and W. A. BERGGREN[3]

Abstract New data and a reinterpretation of existing data are the bases for this emended Paleogene numerical time scale. Biostratigraphy, radiometric dating, stratotypes, and paleomagnetic
stratigraphy have been independently evaluated and subsequently integrated into a single numerical
time scale framework.

In the three Paleogene series, Oligocene, Eocene, and Paleocene, we have recognized eight
commonly used stages. The corresponding age-units vary in duration from 3 to 8 million years
(average 5 Ma). As a result of the improvement in establishing the relationships between the
stratotypes of these stages and plankton biostratigraphy, the position of the stages within the series
has been modified from earlier integrated chronostratigraphic schemes.

Stratotypes (type sections) represent the formal basis for relating rock and time. However, the
stratotypes of the western European Paleogene stages were not originally established on the basis
of planktonic microfossils, and it was therefore not possible to recognize these stages worldwide.
Subsequently, independent plankton zonations were established that could not be properly related
to the stratotypes of the standard European chronostratigraphic units. Nevertheless, tentative relationships between these planktonic zonations and the relative chronostratigraphic units have been
suggested and have been followed for practical reasons by nearly all plankton biostratigraphers.

Calcareous nannoplankton have recently provided an indirect means of relating the stratotypes
of the northwestern European Paleogene stages to the planktonic foraminiferal biostratigraphy and
have revealed discrepancies between these two stratigraphic systems. The most significant of these
is in the Eocene, where the type section of the traditional late Eocene Bartonian contains planktonic microfossils generally considered indicative of middle Eocene age according to most biostratigraphers. This situation is corrected by placing the middle-late Eocene boundary between the
Priabonian and the Bartonian. The alternative solution, moving planktonic foraminiferal zones P13
and P14 from middle to late Eocene, would cause far greater confusion in the worldwide correlation framework.

An evaluation of published radiometric dates provided us with the following age ranges for the
Paleogene geochronologic units:

Oligocene: 24 to 37 Ma   Late: 24 to 32 Ma (Chattian)
                                 Early: 32 to 37 Ma (Rupelian)

[1] Manuscript received, January 10, 1977.

[2] Exxon Production Research Company, Houston, Texas 77001.

[3] Woods Hole Oceanographic Institution, Woods Hole, Massachusetts 02543.

We are grateful to A. Salvador, Exxon USA, and R. C. Tjalsma, Woods Hole, for reading the
manuscript and suggesting many improvements. We thank T. C. Foster for the special care in preparing the illustrations and D. O. Smith and J. Fewox for the editing and preparation of the manuscript.

Research by W. A. Berggren supported by grant OCE76-21274 from the Submarine Geology
and Geophysics Branch, Oceanography Section of the Division of Ocean Sciences, National Science
Foundation. This is Woods Hole Oceanographic Institution contribution No. 3889.

Article Identification Number: 0149-1377/78/SG06-0016/$03.00/0

Editor's Note: The writers prefer the usage of Ma (which equals megayear, or one million years)
to the abbreviation m.y. which has been used through most of this volume.

Eocene: 37 to 53.5 Ma   Late: 37 to 40 Ma (Priabonian)
                                 Middle: 40 to 49 Ma (Bartonian 40-44 Ma)
                                              (Lutetian: 44-49 Ma)
                                 Early: 49 to 53.5 Ma (Ypresian

Paleocene: 53.5 to 65 Ma   Late: 53.5 to 60 Ma (Thanetian)
                                 Early: 60 to 65 Ma (Danian)

Introduction

The attention that numerical time scales have received in recent years is the result of an increased economic and scientific interest in the rates at which geologic processes such as sedimentation, subsidence, and seafloor spreading take place. Quantification in geology requires reliable, detailed, numerical time scales that are the result of an integration of stage stratotype information, biostratigraphy, paleomagnetic stratigraphy, and radiometric dating.

The history of the Paleogene numerical time scale dates back to 1959, when Holmes first published numerical ages for epoch boundaries. Subsequent improvements by Kulp (1961), Evernden et al (1961), Funnell (1964), Evernden and Curtis (1965), and Odin (1973) added detail by subdividing the Paleogene epochs (Fig. 1). Berggren (1969a, b) published a more detailed time scale, integrated with biostratigraphy and, where possible, with standard stage stratotypes. Further improvements were made by Berggren in 1971 and 1972 (Fig. 2). The Paleogene time scale proposed by Odin in 1975 differs considerably from his 1973 scale, as a result of changes in his measurement techniques.

The changes proposed in this paper are based on new biostratigraphic and radiometric data and reinterpretation of the relative positioning between stage stratotypes and the plankton-based biostratigraphic framework.

HISTORY OF PALEOGENE TIME SCALE

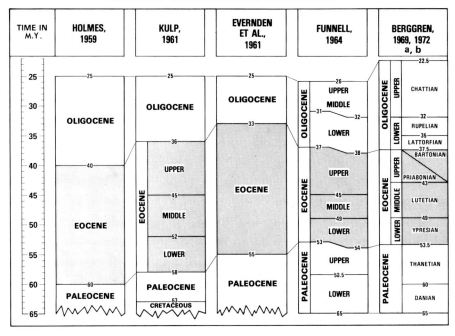

FIG. 1—History of Paleogene time scale.

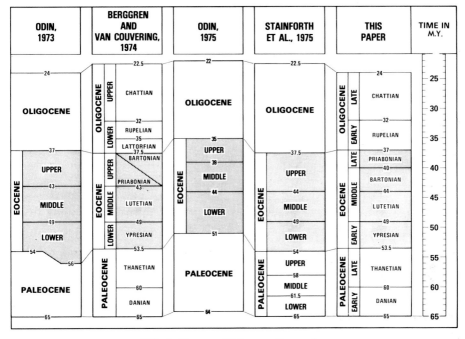

FIG. 2–History of Paleogene time scale.

Stage Stratotypes

Stage stratotypes have an important reference function in stratigraphy. They represent the formal basis for relating rock and time. Let us therefore begin with a closer look at the Paleogene stage stratotypes (Fig. 3). The majority of Paleogene stages have been defined in basins of France, England, Belgium, Germany, and Denmark.

Throughout the Paleogene the western European continental margin subsided slowly. Marine transgressions during periods of high sea-level stands brought open-marine sediments far onto the continental margin (Vail et al, 1974, 1977). These marine units are separated from each other by nonmarine deposits or even by unconformities that were formed during periods when sea level was standing lower. That most open-marine units were preserved during these periods of low sea level is the result of the continuing subsidence. Neogene tilting and erosion have provided us with access to all those units. Unfortunately, a continuous section is not available. We are faced with a multitude of relatively small outcrops scattered throughout the western European basins. Facies changes further complicated the correlation and relative positioning of the Paleogene stratotypes. Nevertheless, for basinwide correlation some of these type sections could be successfully used as a reference. For worldwide correlation the reference function of most western European stratotypes was initially not so successful, because of their lack of continuity and the nature of their fossil content.

The recommendation by the International Subcommission on Stratigraphic Classification to define stages by boundary stratotypes cannot be readily carried out for the Paleogene stages in the western European basins. The Paleogene stages in western Europe that are the most promising for long-distance correlation are those defined in sediments deposited in the most open-marine conditions. However, those open-marine units represent natural eustatic cycles and are separated by marginal marine or nonmarine deposits or even unconformities. The definition of boundary stratotypes between the marine units would be impractical because of the limited chance of worldwide recognition of boundaries defined in marginal marine and nonmarine sediments. However, definition of boundary stratotypes within the marine units would place the Paleogene boundaries within the natural cycles and thus within the traditional litho-

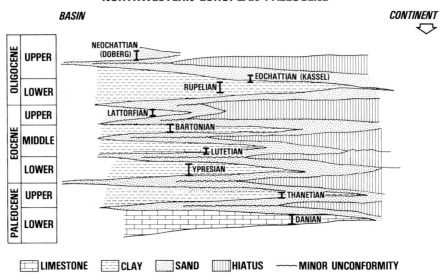

FIG. 3–Schematic reconstruction of depositional history resulting from major eustatic sea-level changes in the Paleogene. Northwestern European stage stratotypes are defined in the marine units and represent periods of high sea-level stands (Vail et al, 1974).

logic units on which European stratigraphy is based.

Boundary stratotypes can be used in basins along compressional continental margins where marine units do not necessarily coincide with natural eustatic cycles because of erratic vertical movement. The possibility of piecing together a complete marine section may be real in such a basin (Mediterranean Neogene). However, in basins situated in tensional continental margins it would be quite impossible to piece a complete marine section together in surface outcrops.

Biostratigraphic Framework

With the development of planktonic foraminiferal biostratigraphy, it became obvious that planktonic microfossils were much better suited for detailed zonation and long-range correlation, at least for open-marine sediments in low- to middle-latitude areas. With the addition of calcareous nannofossils and radiolarian biostratigraphy, the area of successful application of plankton biostratigraphy was extended into higher latitude areas and also into shallower and very deep water deposits. The Deep Sea Drilling Project (DSDP) provided abundant core material in which all planktonic microfossil groups could be studied and their zonations correlated and integrated into a sophisticated biostratigraphic framework. With this plankton-based biostratigraphic framework as reference, other stratigraphic entities that lack the continuity of the plankton accumulation, such as stage type sections, magnetic polarity events, and zonations based on other fossil groups, can be placed in their most plausible relative position. Recognizing that stage type sections represents the formal basis for relating rock and time; we do need the continuum of the plankton-based biostratigraphic framework to place the stage type sections in their proper relative position. Included in this biostratigraphic framework are the planktonic foraminiferal zonal schemes of Bolli (1957a, b, c, 1966) with modifications by Stainforth et al (1975).

Another widely used zonal scheme is the one proposed by Blow in 1969, complemented by Berggren (1969b) for the earlier Paleogene. For the calcareous nannoplankton, zonations by Bramlette and Wilcoxon (1967), Hay et al (1967), Baumann and Roth (1969), Martini (1970), Roth (1970), Gartner (1971), Muller (1974), and Bukry

(1973, 1975) have been considered. The radiolarian zonation is based on Riedel and Sanfilippo (1970, 1971) with modifications by Moore (1971) and Foreman (1973).

An important step in establishing a biostratigraphic framework based on three fossil groups is the relative positioning of the respective zonations. Records of the DSDP provided most of the necessary documentation for this correlation. Ambiguities in the relative positioning of the more recent radiolarian zonation could be caused by discontinuities of cores or preservation problems with one or more of the fossil groups. The correlation between planktonic Foraminifera and calcareous nannofossil zonal schemes proposed by Berggren (1972) for the Paleogene has been largely confirmed with minor modifications for zones NP20 and NP14 (Fig. 4). The top of NP20 is defined by the extinction of *Discoaster saipanensis*; this extinction takes place prior to the extinction of *Turborotalia cerroazulensis* (Ellis, 1975; Ujiie, 1975). *D. saipanensis* is also absent from the top samples of Shubuta Hill in Mississippi, whereas *T. cerroazulensis* is still present. We therefore placed the top of NP20 within the lower part of P17.

The *Discoaster sublodoensis* zone (NP14) has been recognized in the Lutetian by Bouché (1962) and Hay et al (1967). We therefore placed NP14 in the lower part of the *Hantkenina aragonensis* zone (=P10). However, in the Possagno section of Italy, NP14 is correlated with P9 (Luterbacher, 1975; Proto Decima et al, 1975), but this may be caused by preservation problems with *Hantkenina* in these deep-water deposits.

Integration of radiolarian zones with planktonic foraminiferal and calcareous nannoplankton zones is, in the Paleogene, not yet as well established as in the Neogene. The correlation accepted here is based mainly on the co-occurrences with other planktonic fossil groups in the DSDP records (Goll, 1972; Foreman, 1973; Riedel and Sanfilippo, 1973; Sanfilippo and Riedel, 1974). Correlations in the early Eocene and late Paleocene are after Caro et al (1975).

Stratotypes and Biostratigraphic Framework, Relations

The original low-latitude planktonic foraminiferal zonations (Bolli, 1957a, b, c; Blow, 1969) could not be properly related to the western European stratotypes that are located in higher latitudes and consist mostly of shallow or moderately deep-water deposits, which lack the diversified planktonic foraminiferal assemblages of the low-latitude deeper water deposits, However, for practical reasons early planktonic foram workers established a tentative relation between their zones and relative geologic time. This tentative relation has been followed by nearly all plankton biostratigraphers.

Recently, calcareous nannoplankton have provided an indirect means of relating some European Paleogene stratotypes to the planktonic foraminiferal biostratigraphy (Roth, 1970; Roth et al, 1971; Martini and Müller, 1971). These studies have led to the conclusion that discrepancies exist between the standard time scale based on stratotypes and the tentative relation based on planktonic Foraminifera. The most significant difference is in the Eocene, where the type section of the traditional late Eocene Bartonian Stage contains planktonic microfossils generally considered indicative of middle Eocene age by most biostratigraphers. To correct this situation, we propose to place the middle-late Eocene boundary between the Priabonian and the Bartonian. An alternative solution would be to move planktonic foraminiferal zones P13 and P14 from the middle to the late Eocene. The latter solution would create far greater confusion in the worldwide correlation framework.

Paleocene

Two distinct marine lithostratigraphic units can be recognized in the European Paleocene—a lower carbonate unit in which the Danian Stage is defined, and an upper clastic unit in which the Thanetian Stage is defined.

The Danian Stage defined in Denmark has few equivalent stages in Europe. Apart from the Montian in Belgium, which is in part equivalent with the type Danian, no other stages of equivalent age have been proposed. However, sediments correlated with the Danian are widespread, especially in central and eastern Europe. In a comprehensive study concerning the Danian and its equivalents, Hansen (1970) defined the Danian ". . .as the time interval between the rocks found above the Maastrichtian White Chalk exposed on Stevns Klint up to and including the time of deposition of the

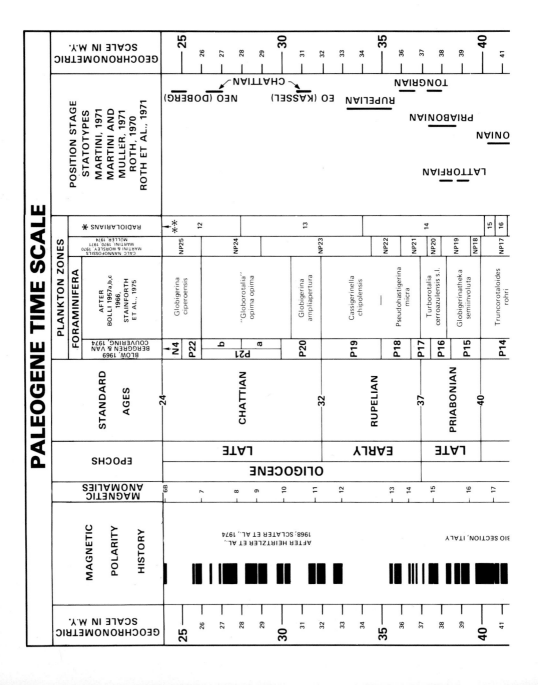

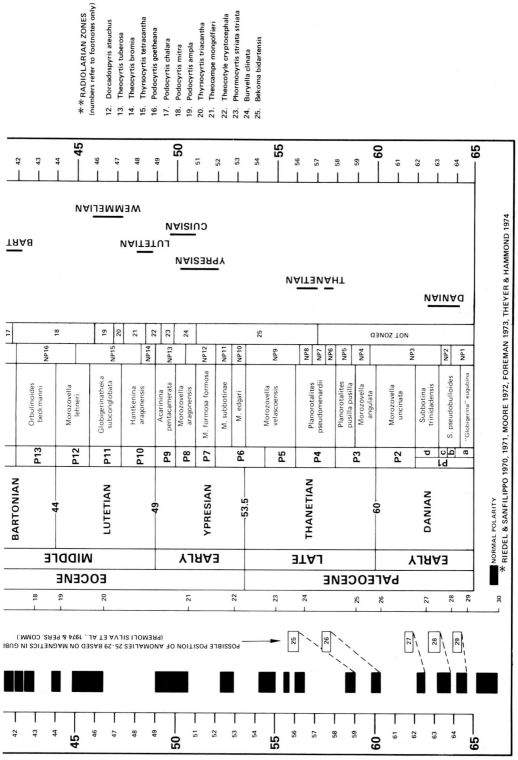

FIG. 4—Improved Paleogene time scale with approximate relative position of stage stratotypes.

rocks found below the Selandian basal conglomerate exposed at the locality Hvallose, in Jyland." This definition would include part of planktonic foraminiferal zone Pa and all of zone P2. We extended the Danian unit to include all of planktonic foraminiferal zones P1 and P2, corresponding with calcareous nannoplankton zones NP1-NP3 (part). Martini (1971) and Perch-Nielsen (1972) reported these nannoplankton zones NP1-NP3 from Danian outcrops in the type area.

The upper clastic unit in which the Thanetian is defined is widespread in western Europe. Equivalents are known from Denmark (type Selandian) and from Belgium (type Landenian and type Heersian). The "Sables de Bracheux" represent this unit in the Paris basin. The biostratigraphic position is well established; calcareous nannoplankton zone NP8 has been reported from the Thanetian (Martini, 1971). This zone can be correlated with planktonic foraminiferal zone P4 (Bramlette and Sullivan, 1961). El-Naggar (1967a, b) reported zone P3 from the type Heersian in Belgium. The Thanetian chronostratigraphic unit has been extended to include sediments identified biostratigraphically as planktonic foraminiferal zones P3-P6 (lower part), corresponding to calcareous nannoplankton zones NP4-NP9 (Berggren, 1971, 1972).

Eocene

Traditionally, three chronostratigraphic units have been recognized in the Eocene of western Europe. The lower unit includes the Ypresian Stage, defined in Belgium in shallow to moderately deep-water sand and clay deposits. The middle unit, in which the Lutetian Stage is defined, consists of mostly shallow-water carbonate deposits in the Paris basin. The upper unit, or Bartonian Stage, is defined in shallow to moderately deep-water sands and clays in southern England. The Priabonian Stage, defined in northern Italy, has long been considered a Mediterranean equivalent of the northern European Bartonian.

Lithostratigraphic units in other western European basins equivalent to the Ypresian in its type area are the London Clay in England, Rosnaes Clay in Denmark, and Eocene 3 in Germany. The sandstones on which the Cuisian Stage is based in the Paris basin are believed to be equivalent to the upper part of the Ypresian. Calcareous nannoplankton belonging to zones NP11 and NP12 have been identified from the type area of the Ypresian in Belgium (Hay and Mohler, 1967; Martini, 1971). Zone NP12 can be correlated with planktonic foraminiferal zone P8 (Hay et al, 1967). The same calcareous nannoplankton zones NP11 and NP12 have been reported from the Rosnaes Clay in Denmark and the Eocene 3 in Germany. The well-documented position of the type Ypresian relative to the biostratigraphic framework makes it a useful stage to serve as a reference for the lower Eocene. The Ypresian chronostratigraphic unit has been extended here to include sediments identified by biostratigraphic means as planktonic foraminiferal zones P6 (upper part only) - P9, or calcareous nannoplankton zones NP10-NP13 (Berggren, 1971, 1972).

Lithostratigraphic units in other western European basins equivalent to the Lutetian in its type area include part of the Bracklesham beds in England and the Brussels sands in Belgium. Calcareous nannoplankton typical for zone NP14 have been reconnized in the Lutetian type area by Bouche (1962) and by Hay et al (1967). A tentative correlation with planktonic foraminiferal zone P10 is made by Bukry and Kennedy (1969). The Lutetian chronostratigraphic unit is often extended to include all planktonic foraminiferal zones traditionally placed in the middle Eocene (P10-P14). However, the identification of calcareous nannoplankton zone NP17 in the Barton beds in England (Martini and Ritzkowski, 1970; Martini, 1971), which is believed to be equivalent with P14 (Berggren, 1972), or with P13 as suggested by Roth et al (1971), would place the Bartonian within the Lutetian chronostratigraphic unit. Since there is no indication of an overlap between the Lutetian s.s. and the Bartonian s.s., we suggest a restriction of the Lutetian chronostratigraphic unit to include sediments identified with biostratigraphy as planktonic foraminiferal zones P10-P12, corresponding to calcareous nannoplankton zones NP14, NP15, and NP16 (lower part).

The Bartonian Stage has long been considered a northern European equivalent of the Mediterranean Priabonian Stage. The absence of *Isthmolithus recurvus* from the type Bartonian suggests that the Bartonian is older than NP19; this excludes an equivalence with the type Priabonian. The planktonic microfossils in the Bartonian type area (NP17 and P14-P13) are by general usage considered indicative of a middle Eocene age. We therefore believe it appropriate to place the Bartonian in the middle Eocene and extend the chronostratigraphic unit to include planktonic foraminiferal zones P13 and P14, which correspond to calcareous nannoplankton zones NP16 (upper part) and NP17.

The Priabonian type section contains planktonic Foraminifera indicative of zones P16 and P17 (part) (Hardenbol, 1968) and calcareous nannoplankton zones NP19 (Martini, 1971) and NP19 and NP20 (Roth et al, 1971). The Priabonian s.l. was considered at the Eocene Colloquium in 1968 to include planktonic foraminiferal zones P15-P17.

Oligocene

The Oligocene in northwestern Europe is represented by two distinct lithostratigraphic units. The upper, predominantly shallow marine sandy unit (or upper Oligocene) contains the Chattian type section, whereas the lower, deeper marine clayey unit (or lower Oligocene) includes the typical Rupelian. A third sedimentary unit contained in the original definition of the Oligocene, and including the Lattorfian and the lower part of the Tongrian type sections, is very likely equivalent with the late Eocene or Priabonian. Although this unit has repeatedly been placed in the early Oligocene nannoplankton zone NP21 by Martini (1969), Martini and Ritzkowski (1968, 1969), Roth (1970), and Martini and Müller (1971), others such as Marks and Van Vessem (1971), Cavelier (1972, 1976), and Cavelier and Pomerol (1976) provide evidence for a much older position, around P15-P16, for the Lattorfian deposits (neotype Lattorfian) near Helmstedt. The calcareous nannofossils in the Silberberg beds near Helmstedt (neotype Lattorfian, Martini and Ritzkowski, 1968) cannot be older than NP19 and not younger than NP21. The absence of *Discoaster saipanensis* and *D. barbadiensis* does not provide sufficient evidence for placement in zone NP21. As mentioned earlier, the extinction of *D. saipanensis* occurs within the late Eocene before the extinction of *Turborotalia cerroazulensis*.

The Oligocene-Miocene boundary (defining the top of the Paleogene) is placed between the Chattian and the Aquitanian Stages. Studies of calcareous nannoplankton in the Chattian (Neochattian) type section at Doberg near Bünde in Germany have tentatively identified zones NP24 and NP25, although the zonal markers for these zones were not found (Martini, 1971; Benedek and Müller, 1974, 1976; Martini and Müller, 1975). Nannozone NP25 correlates with the *Globigerina ciperoensis* zone, which in turn is correlative (in part) with zones P22 and lower part of N4 (Blow, 1969). Although the initial appearance of *Globigerinoides primordius* has been shown to occur in upper Oligocene pre-Aquitanian levels (Anglada, 1971a, b; Scott, 1972; Alvinerie et al, 1973; Theyer and Hammond, 1974b; Lamb and Stainforth, 1976; Van Couvering and Berggren, 1976), the *Globigerinoides* datum is of general biostratigraphic utility in recognizing (i.e., correlating) the approximate position of the Oligocene-Miocene boundary. The base of the Chattian depends on the position of the "Kasseler Meeressand" or Eochattian near Kassel in Germany. If these sands are the regressive phase of the "Rupel" transgression, they could be of nearly the same age as the Boom clay in Belgium. However, Ritzkowski (1967) described a considerable unconformity between these Kassel sandstones and the "Rupel" clay in Germany. Roth (1970) studied nannoplankton from the equivalents of the Kassel sandstones near Glimmerode and found them to be of the same age as the Boom clay in Belgium. However, this could have been caused by reworking. Martini (1971) placed samples from the Boom Clay in Belgium in nannofossil zone NP23, but the next higher zone NP24 could not be identified.

We arbitrarily placed the boundary between the Chattian and Rupelian Stages at the top of P19, which corresponds approximately with the middle of NP23. This boundary position may have the best chance of falling between the Rupelian and Chattian Stages as defined in their respective type areas. Since only two main sedimentary units make up the Oligocene in western Europe, an informal subdivision into lower and upper Oligocene seems to be the most logical solution—the lower Oligocene including the Rupelian, and the upper Oligocene, the Chattian Stage.

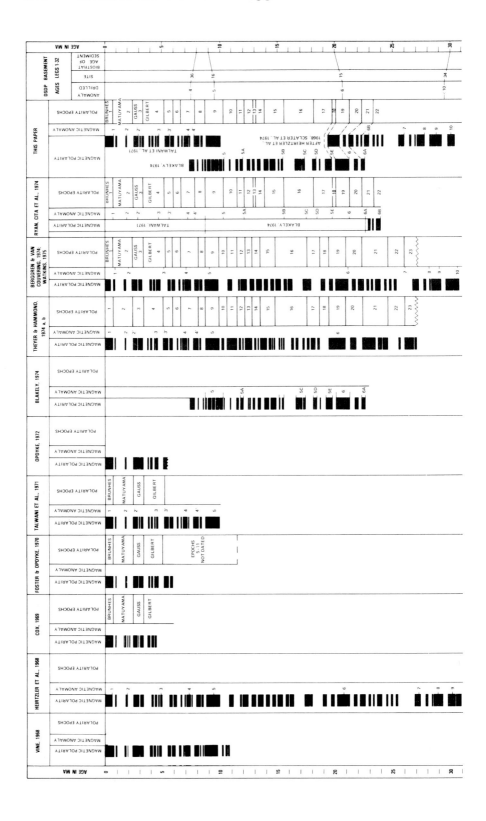

await more DSDP results. However, we have recalculated the reversal scale on the basis of the spreading rate suggested by Sclater et al (1974; Table 1). The improvement is considerable, although discrepancies between apparent anomaly ages and "biostratigraphic ages" still exist (Fig. 5), especially for the interval between anomalies 20 and 26, which is, according to our time scale, between 45 Ma and 60 Ma. Similar discrepancies were noted by Premoli Silva et al (1974) in a land-based section at Gubbio in the central Apennines of Italy. Anomalies 25 to 29 could be recognized in a sequence that also contained planktonic Foraminifera. If we use the numerical ages for the biostratigraphic zones proposed in this paper, anomaly 26 would be 57 Ma which is the same age obtained for anomaly 26 at site 213 of the DSDP (Table 2).

The reasons for these discrepancies could be that the spreading rate in the Paleogene is not constant or that the numerical ages on the biostratigraphic framework scale are too young. As long as we do not know what the problem is, there is no justification to abandon the constant spreading rate model for a more complicated one (Sclater et al, 1974).

Table 2. Oceanic basement ages from selected DSDP sites. Sites were selected only where basement was reached and sediments overlying basement could be reliably dated using biostratigraphy.

Magnetic Anomaly	DSDP Site	Foraminifera Zone	Nannoplankton Zone	Age (Ma) This Paper
10 (younger side)	34	P20	--	30
13	17	P19/20	NP23 (low)	33.5
13	32	--	NP22	35.5
14/15	14	P17	NP20	38
20	221	--	NP17	42
21 (younger side)	19	P10	NP15	48
24 (older side)	39	P6	NP12	52
26 (younger side)	213	P4	NP7	57
28/29	245	P1 c-d	NP3	63
30	20C	UC 16	--	66

Tarling and Mitchell (1976) proposed a new magnetic reversal scale that does not show the discrepancies between the apparent ages of anomalies and the "biostratigraphic ages." They essentially adjusted spreading rates to a recently proposed time scale by Odin (1975). Where a single linear correction could not achieve this, they proposed three linear corrections for different parts of the reversal scale. These parts are from anomaly 6 to 13, from anomaly 13 to 24, and older anomalies. Unfortunately, the uncertainties surrounding potassium-argon (K-Ar) dating of glauconites are in this manner introduced in paleomagnetic stratigraphy. Berggren et al (in press) strongly disagreed with this approach by Tarling and Mitchell (1976), because of the noncritical acceptance of Odin's 1975 Paleogene time scale. Berggren et al (in press) have contested Odin's 1975 time scale for reasons such as sample preparation techniques, but also because Odin's new dates do not agree with K-Ar determinations from North American Gulf Coast glauconites (Ghosh, 1972), nor with K-Ar dates on basalts in North American mammalian sequences, biostratigraphically correlated with their marine equivalents.

Radiometric Ages

Radiometric rate processes provide us with the unique opportunity to add numerical ages to our framework of stages and biostratigraphic zones. We have selected only those radiometric ages from recent literature that can be reliably correlated with either the plankton zones or the stage type sections. Because the biostratigraphic framework and the stage stratotypes discussed in this study are defined in marine sediments, most of the ages selected are K-Ar dates based on glauconites, with only a few based on basalts or bentonites (Fig. 6).

Radiometric ages available from glauconites of the western European basins are those from Bonhomme et al (1968) and Bodelle et al (1969) for the Paris basin; from

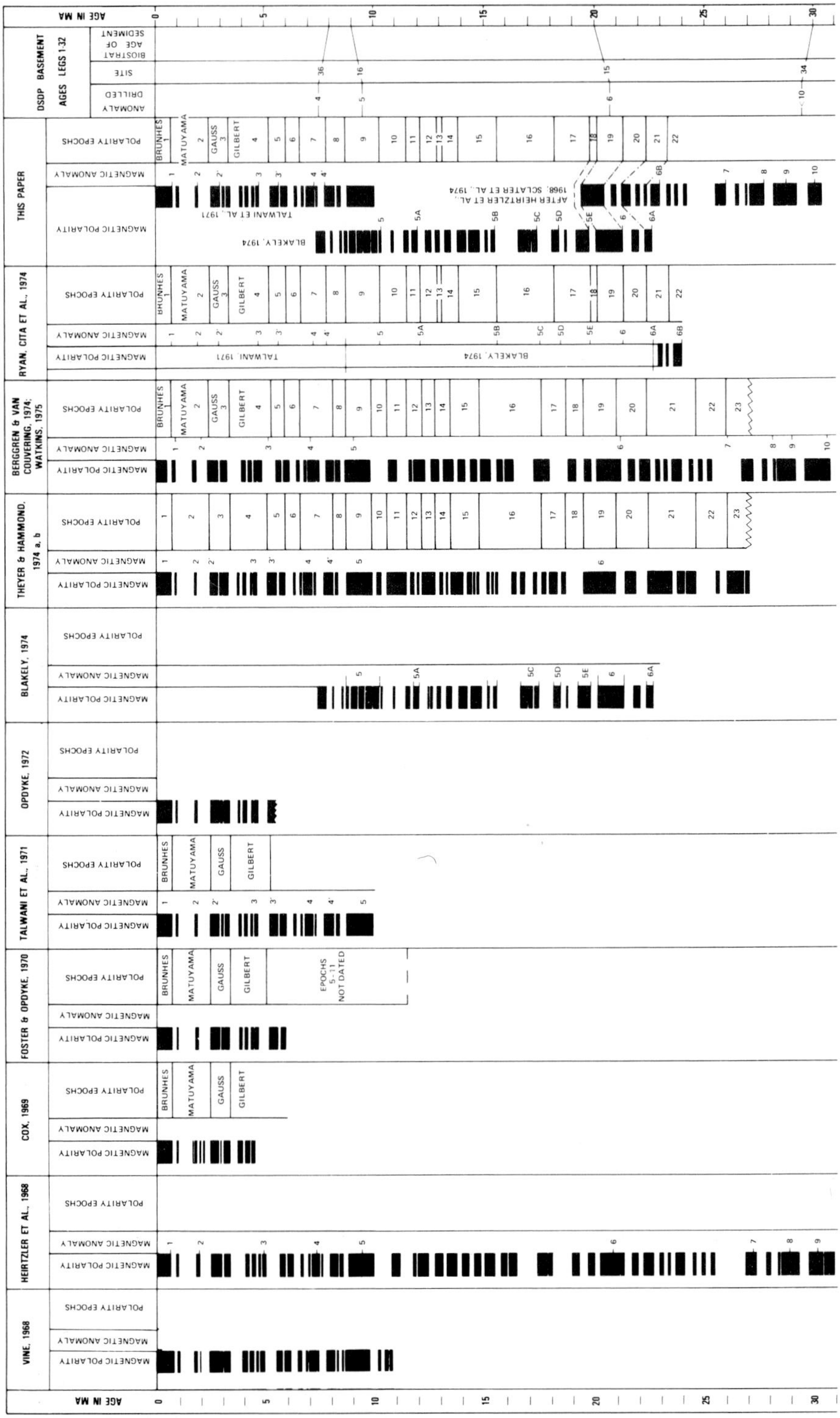
AGE IN MA
DSDP BASEMENT AGES LEGS 1-32
BIOSTRAT AGE OF SEDIMENT
SITE
ANOMALY DRILLED
THIS PAPER
POLARITY EPOCHS
MAGNETIC ANOMALY
MAGNETIC POLARITY
TALWANI ET AL., 1971
AFTER HEIRTZLER ET AL., 1968; SCLATER ET AL., 1974
BLAKELY, 1974
RYAN, CITA ET AL., 1974
BERGGREN & VAN COUVERING, 1974; WATKINS, 1975
THEYER & HAMMOND, 1974 a, b
BLAKELY, 1974
OPDYKE, 1972
TALWANI ET AL., 1971
FOSTER & OPDYKE, 1970
EPOCHS 5-11 NOT DATED
COX, 1969
HEIRTZLER ET AL., 1968
VINE, 1968
BRUNHES
MATUYAMA
GAUSS
GILBERT

lence with the type Priabonian. The planktonic microfossils in the Bartonian type area (NP17 and P14-P13) are by general usage considered indicative of a middle Eocene age. We therefore believe it appropriate to place the Bartonian in the middle Eocene and extend the chronostratigraphic unit to include planktonic foraminiferal zones P13 and P14, which correspond to calcareous nannoplankton zones NP16 (upper part) and NP17.

The Priabonian type section contains planktonic Foraminifera indicative of zones P16 and P17 (part) (Hardenbol, 1968) and calcareous nannoplankton zones NP19 (Martini, 1971) and NP19 and NP20 (Roth et al, 1971). The Priabonian s.l. was considered at the Eocene Colloquium in 1968 to include planktonic foraminiferal zones P15-P17.

Oligocene

The Oligocene in northwestern Europe is represented by two distinct lithostratigraphic units. The upper, predominantly shallow marine sandy unit (or upper Oligocene) contains the Chattian type section, whereas the lower, deeper marine clayey unit (or lower Oligocene) includes the typical Rupelian. A third sedimentary unit contained in the original definition of the Oligocene, and including the Lattorfian and the lower part of the Tongrian type sections, is very likely equivalent with the late Eocene or Priabonian. Although this third unit has repeatedly been placed in the early Oligocene nannoplankton zone NP21 by Martini (1969), Martini and Ritzkowski (1968, 1969), Roth (1970), and Martini and Müller (1971), others such as Marks and Van Vessem (1971), Cavelier (1972, 1976), and Cavelier and Pomerol (1976) provide evidence for a much older position, around P15-P16, for the Lattorfian deposits (neotype Lattorfian) near Helmstedt. The calcareous nannofossils in the Silberberg beds near Helmstedt (neotype Lattorfian, Martini and Ritzkowski, 1968) cannot be older than NP19 and not younger than NP21. The absence of *Discoaster saipanensis* and *D. barbadiensis* does not provide sufficient evidence for placement in zone NP21. As mentioned earlier, the extinction of *D. saipanensis* occurs within the late Eocene before the extinction of *Turborotalia cerroazulensis*.

The Oligocene-Miocene boundary (defining the top of the Paleogene) is placed between the Chattian and the Aquitanian Stages. Studies of calcareous nannoplankton in the Chattian (Neochattian) type section at Doberg near Bünde in Germany have tentatively identified zones NP24 and NP25, although the zonal markers for these zones were not found (Martini, 1971; Benedek and Müller, 1974, 1976; Martini and Müller, 1975). Nannozone NP25 correlates with the *Globigerina ciperoensis* zone, which in turn is correlative (in part) with zones P22 and lower part of N4 (Blow, 1969). Although the initial appearance of *Globigerinoides primordius* has been shown to occur in upper Oligocene pre-Aquitanian levels (Anglada, 1971a, b; Scott, 1972; Alvinerie et al, 1973; Theyer and Hammond, 1974b; Lamb and Stainforth, 1976; Van Couvering and Berggren, 1976), the *Globigerinoides* datum is of general biostratigraphic utility in recognizing (i.e., correlating) the approximate position of the Oligocene-Miocene boundary. The base of the Chattian depends on the position of the "Kasseler Meeressand" or Eochattian near Kassel in Germany. If these sands are the regressive phase of the "Rupel" transgression, they could be of nearly the same age as the Boom clay in Belgium. However, Ritzkowski (1967) described a considerable unconformity between these Kassel sandstones and the "Rupel" clay in Germany. Roth (1970) studied nannoplankton from the equivalents of the Kassel sandstones near Glimmerode and found them to be of the same age as the Boom clay in Belgium. However, this could have been caused by reworking. Martini (1971) placed samples from the Boom Clay in Belgium in nannofossil zone NP23, but the next higher zone NP24 could not be identified.

We arbitrarily placed the boundary between the Chattian and Rupelian Stages at the top of P19, which corresponds approximately with the middle of NP23. This boundary position may have the best chance of falling between the Rupelian and Chattian Stages as defined in their respective type areas. Since only two main sedimentary units make up the Oligocene in western Europe, an informal subdivision into lower and upper Oligocene seems to be the most logical solution—the lower Oligocene including the Rupelian, and the upper Oligocene, the Chattian Stage.

Paleomagnetic Reversal Scale

The contributions of paleomagnetic stratigraphy to the Paleogene time scale are modest in comparison with the achievements in the Neogene (Theyer and Hammond, 1974a, b; Ryan et al, 1974). In the Paleogene the process of calibrating seafloor anomalies with magnetic epoch and fossiliferous sediments has just begun (Premoli Silva et al, 1974). The paleomagnetic reversal scale most often cited for the Paleogene is the one proposed by Heirtzler et al (1968) based on a profile (V-20) in the south Atlantic. Heirtzler et al (1968) assumed a constant spreading rate, which was calculated at 1.9 cm/yr for the youngest part of the profile and extrapolated beyond anomaly 32 in the Cretaceous. Sclater et al (1974) showed, on the basis of the DSDP records (through leg 22), that for the anomalies actually drilled, the biostratigraphic age (Berggren, 1972) was considerably younger than could be expected from the Heirtzler et al (1968) reversal scale. They concluded that Heirtzler's spreading rate was too low, and proposed a slightly higher spreading rate of 2.04 cm/yr between anomalies 5 and 30 at 10 Ma and 66 Ma,   respectively. Sclater et al (1974) declined to update the magnetic reversal scale on the basis of their findings. Such an update, they stated, should

Table 1. Numerical ages for normal polarity zones from 10 to 65 m.y. (after Heirtzler et al, 1968), recalculated for a constant spreading rate of 2.04 cm/yr (Sclater et al, 1974), computed with the equation $t_n' = 0.92 (t_n - 10.0) + 10.0$.

10.71 – 11.05	24.03 – 24.20	38.66 – 38.94
11.58 – 11.70	25.51 – 25.62	39.00 – 39.40
11.78 – 12.24	25.69 – 25.98	39.70 – 40.60
12.50 – 12.84	26.40 – 26.59	40.67 – 40.88
13.03 – 13.41	26.88 – 26.96	40.95 – 41.29
13.64 – 13.94	27.04 – 27.78	41.47 – 41.91
14.15 – 14.43	28.20 – 28.79	41.99 – 42.42
14.58 – 15.01	28.84 – 29.26	42.49 – 42.93
15.25 – 15.52	29.78 – 30.09	43.82 – 44.28
15.55 – 15.90	30.15 – 30.40	44.88 – 46.41
16.74 – 17.18	31.31 – 31.67	49.02 – 50.63
17.20 – 17.38	31.72 – 32.14	52.25 – 52.93
18.20 – 18.52	32.56 – 33.00	54.20 – 55.02
18.85 – 19.16	35.40 – 35.59	55.48 – 55.71
19.37 – 20.41	35.66 – 36.00	56.01 – 56.49
20.72 – 20.96	36.39 – 36.47	58.53 – 59.02
21.20 – 21.63	36.52 – 36.61	59.81 – 60.25
21.87 – 22.03	36.71 – 36.78	62.12 – 62.53
22.23 – 22.33	37.07 – 37.11	63.15 – 63.83
22.54 – 22.94	37.39 – 37.60	64.13 – 64.68
23.26 – 23.42	37.63 – 37.83	65.14 – 66.23
23.63 – 23.77	38.25 – 38.49	

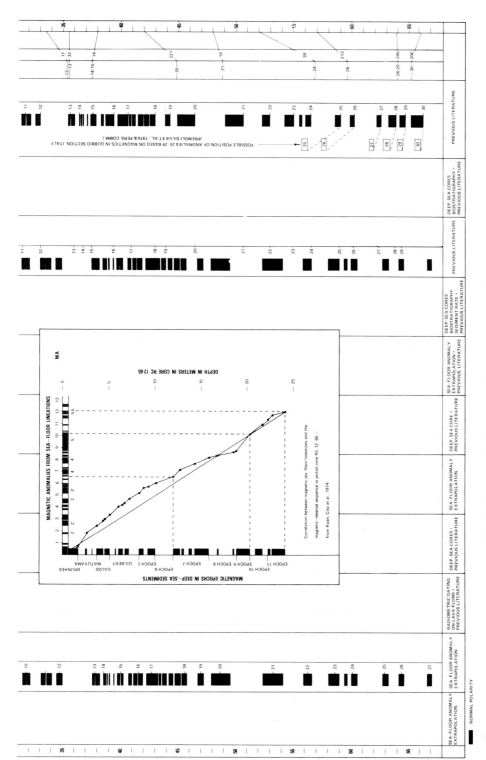

FIG. 5 – Historical development of Cenozoic geomagnetic reversal scale.

EPOCHS

OLIGOCENE — LATE | EARLY | LATE | MIDDLE
EOCENE

AGE: CHATTIAN | RUPELIAN | PRIABONIAN | BARTONIAN

BIOSTRAT. ZONES: NP21 | P15 | P15

EUROPEAN LOCALITIES

- Lower Miocene Austria (78) — 25.0 MA
- G. Astrup, Chattian A (7) — 28.8 MA
- G. Doberg, Chattian C (6) — 29.5 MA
- G. Doberg, Chattian B (5) — 30.1 MA
- B. Voort sands (17) — 30.6 MA
- G. Doberg, Chattian A (4) — 31.2 MA
- G. Lehrte, corehole (168) — 35.5 MA
- G. Helmstedt, Gehlberg Fm. (172) — 36.5 MA
- G. Helmstedt, Silberberg Fm. (171) — 36.6 MA
- G. Helmstedt, Gehlberg Fm. (173) — 36.9 MA
- B. Neerrepen sands, Tongrian (16) — 37.5 MA
- G. Lehrte, corehole (169) — 38.5 MA
- G. Lehrte, corehole (170) — 38.7 MA
- G. Helmstedt, Gehlberg Fm. (175) — 38.7 MA
- G. Helmstedt, Gehlberg Fm. (174) — 39.0 MA
- F. Mont Cassel - "Asse clay" (33) — 41.8 MA
- G. B. Barton beds (25) — 42.0 MA
- F. Meaux, Lower Lutetian II (32) — 43.0 MA
- G. B. Studley wood, overlying beds with N. prestwichianus (23) — 43.1 MA
- G. Lehrte, green sands (3) — 43.1 MA
- G. B. Barton beds (24) — 43.6 MA
- G. B. Bracklesham beds (22) — 44.0 MA
- B. Asse clay (15) — 44.0 MA
- G. B. "Bartonian" (14) — 44.4 MA
- F. Fosses, "Lower Lutetian" II (31) — 44.7 MA

37.1 MA — DSDP (35)
38.2 MA — DSDP (34)

NORTH AMERICAN LOCALITIES

- (96) "Zemorrian" — 24.3 MA
- (66) Vicksburg, Catahoula Fm — 35.7 MA
- (64) Jackson, Pachuta Mbr — 36.7 MA
- (65) Jackson, Shubuta Mbr — 37.0 MA
- (61) Jackson, Moodys Branch Fm — 37.3 MA
- (63) Jackson, Yazoo Fm — 38.1 MA
- (62) Jackson, Yazoo Fm — 38.5 MA
- (59) Claiborne, Gosport Fm — 38.5 MA
- (72) Jackson, Moodys Branch Fm — 39.0 MA
- (52) Claiborne, Cook Mt. Fm — 40.6 MA
- (60) Claiborne, Gosport Fm — 40.8 MA
- (57) Claiborne, Cook Mt. Fm — 42.7 MA
- (74) Basal Kreyenhagen — 43.0 MA
- (48) Claiborne, Zilpha Fm — 43.6 MA

BIOSTRAT. ZONES: P18/19 | P16 | P17 | P14/15 | P15 | P15 | P13/14 | P13/14 | P12

GULF COAST FORMATIONS: CHICKASAWHAY | VICKSBURG | JACKSON | CLAIBORNE

AGE IN M.Y.: 25 | 30 | 35 | 40 | 45

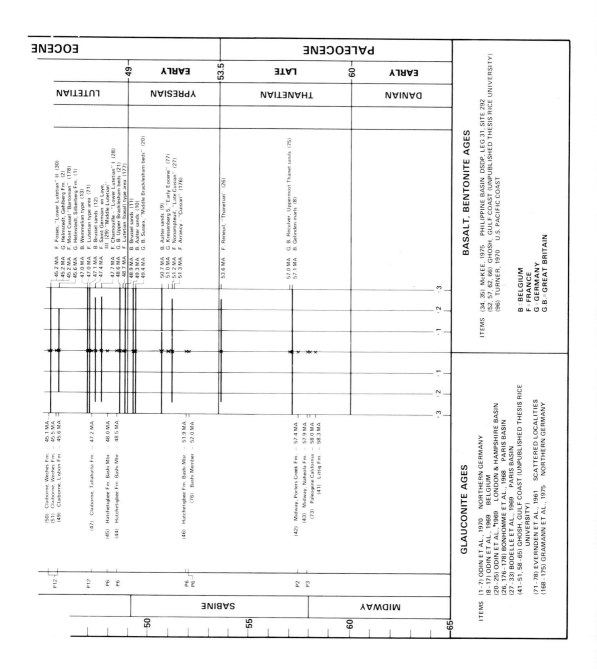

FIG. 6—Radiometric ages from selected North American, European, and DSDP localities.

Odin et al (1969) for the Belgium and London-Hampshire basins, respectively; and from Odin et al (1970) and Gramann et al (1975) for northern Germany. Most ages from North American localities are from Ghosh (1972). Other ages included are: Evernden et al (1961), from scattered localities in Europe and the U.S.; McKee (1975), two dates on a basalt in the Philippine sea; and Turner (1970), one date on a California basalt.

Ages obtained from glauconites in the western European basins generally lack detailed biostratigraphic information. However, several of the samples are from type sections of Paleogene stages, and many others are from lithologic units that can be reliably correlated with those stages. Glauconite ages from North American localities in the Gulf Coast area can be more successfully correlated with biostratigraphic zones.

No attempt has been made in this paper to evaluate the overall quality of the published ages, or to compare the measurement techniques used for the European and American ages. Quality of glauconites, possible reworking, potassium content, atmospheric argon, argon loss, and decay rates, have not been seriously considered. Figure 6 lists the Paleogene radiometric ages obtained from published sources. The majority of the selected ages concern Eocene deposits, and only a few are available for the Paleocene and Oligocene. Berggren (1972) included most of the glauconite ages from the western European basins in his revision of the Cenozoic time scale. It is therefore obvious that the radiometric ages listed in Figure 6 can hardly form a basis for important changes in Berggren's 1972 time scale. However, the availability of Gulf Coast ages allows for a number of interesting observations.

When we compare ages obtained from the Jackson and Claiborne Groups in the United States Gulf Coast area (Ghosh, 1972) with ages obtained from the Bartonian and Lutetian type areas (Fig. 7), the following conclusions can be made: (1) most samples from the Jackson Group are younger than those from the Claiborne group; (2) samples from the type area of the Bartonian are generally younger than those from the type area of the Lutetian; (3) Bartonian and Lutetian are age equivalent with the Claiborne Group and are older than the Jackson Group; (4) the samples from the Jackson Group contain upper Eocene planktonic foraminifers and can be correlated with the type Priabonian; and (5) samples from the Claiborne Group contain middle Eocene planktonic foraminifers, whereas Bartonian as well as Lutetian samples contain middle Eocene calcareous nannofossils.

On the basis of these observations, we conclude that the radiometric age for the Eocene-Oligocene boundary is close to 37.0 Ma (Item 65, Fig. 4, Shubuta marl, close to the extinction of *Turborotalia cerroazulensis*). The traditional age of 37.5 Ma is based on a sample from the Neerrepen sandstones in Belgium (Item 16), with a much more uncertain stratigraphic position.

For the middle-late Eocene boundary two options are open: (1) 40 Ma which is at the commonly used biostratigraphic boundary between zones P15 and P14—the Bartonian would then be in the middle Eocene; and (2) 44 Ma, which is between the Bartonian and the Lutetian. Biostratigraphic zones P14 and P13 would be in the upper Eocene, and for American stratigraphers part of the Claiborne would be in the upper Eocene. These changes in the age relationships of the widely used biostratigraphic framework could cause considerable confusion and may be strongly contested.

For practical reasons we have opted for the first solution. Placement of the Bartonian in the middle Eocene would be much more readily accepted and less confusing.

Odin (1975) obtained an age of 39 Ma for the middle-late Eocene boundary at the results because of different preparation techniques. His original ages were believed to be too old because of equipment insensitivity to atmospheric argon. However, his new ages are considered too young because of argon loss due to high bakeout temperatures (Obradovich, personal communication).

Odin (1975) obtained an age of 39 m.y. for the middle/late Eocene boundary at the base of the Bartonian. This age seems to correspond well with the results obtained by Ghosh (1972) on the base of the Jackson. However, as explained above, the base of the Bartonian is biostratigraphically much older than the base of the Jackson.

Odin (1975) revised the age of the early-middle Eocene boundary at the base of the Lutetian to 44 Ma; this is 5 Ma younger than the original age. We cannot accept

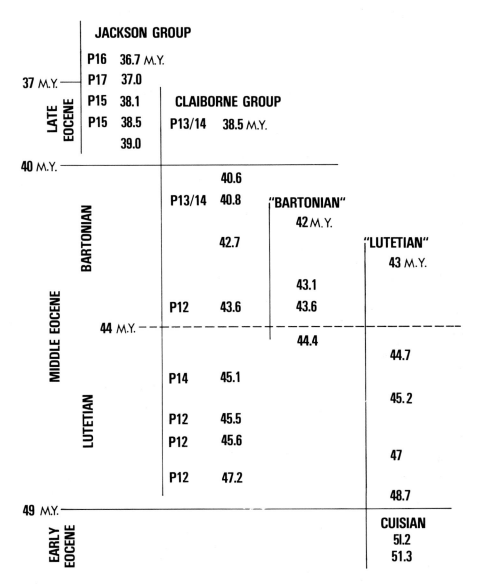

FIG. 7–Comparison of radiometric dates from samples of the Jackson and Claiborne Groups from the United States Gulf Coast with dates from samples attributed to the Bartonian, Lutetian, and Cuisian Stages in western Europe (see also Fig. 6).

these new ages on stratigraphic grounds. Ghosh (1972) dated samples from the Claiborne Group, tentatively correlated with planktonic foraminiferal zone P12, at 45 to 47 Ma, which corresponds very well with an early-middle Eocene boundary at 49 Ma at the base of the Lutetian, which is the base of zone P10.

Berggren et al (in press) disagreed with Odin's 1975 time scale on similar stratigraphic grounds. Using intercontinental vertebrate biostratigraphy, they noted that reliable North American biotite-sanidine dates are much older than Odin's (1975) new glauconite dates. An important reference level is the early-middle Eocene boundary. The boundaries between the European marine Ypresian-Lutetian and the North American continental Wasatchian-Bridgerian can be shown to be biostratigraphically correlative (Berggren et al, in press). Potassium-argon (biotite-sanidine) dates on the latter boundary center on 49 Ma and provide supportive evidence for the 49 Ma estimate adopted by Berggren (1969a, b, 1971, 1972) for the early-middle Eocene boundary. A number of additional dates for various post-Wasatchian rocks are older than 44 Ma, the upper limit of Odin's (1975) lower Eocene. Published dates on the Wiggins Formation in northwestern Wyoming, a rock unit that overlies early Uintan faunas in the type section of the Teepee Trail Formation, range in age from about 43 to 47 Ma with the majority lying between 46 and 47 Ma. They are all younger (by reason of superposition) than the early Uintan faunas in the type section of the Teepee Trail Formation. This means that early Uintan (post-Bridgerian D and therefore post-Bridgerian) time had begun by about 47 Ma at the latest. Therefore, Odin's (1975) estimate of 44 Ma for the early-middle Eocene boundary is seen to be younger than dated levels well within the middle Eocene according to their biostratigraphic correlation (Berggren et al, in press).

Berggren et al (in press) concluded that geochronology should proceed in close correlation with biostratigraphy. Available radiometric dates on marine glauconites in the lowermost Eocene and Paleocene do not warrant any changes in the numerical ages proposed by Berggren (1969a, b, 1971, 1972) for the early Paleogene.

The radiometric ages reported by Odin et al (1970) on samples from Chattian localities at Astrup and Doberg appeared somewhat old in relation to the reported biostratigraphic position. Martini and Müller (1975) placed those localities in calcareous nannoplankton zones NP25 and NP24, although the zonal markers for these zones were not found.

Because neither biostratigraphic nor radiometric evidence is conclusive, we maintain the early-late Oligocene boundary at 32 Ma (Berggren, 1969b, 1971, 1972).

The numerical age for the Oligocene-Miocene boundary is placed at 24 Ma, which is the same date used by Ryan et al (1974) for the base of the Neogene. The 24 Ma estimate is based on Turner's 1970 dates on basalts near the Saucesian-Zemorrian boundary of California, which is somewhat below the planktonic foraminiferal zone boundary N4/N5 (Lamb and Hickernell, 1972; Hornaday, 1972; Van Couvering and Berggren, 1976). The Oligocene-Miocene boundary lies between this level and the *Globigerinoides* datum, which is known to be of latest Oligocene pre-Aquitanian age. Turner (1970) bracketed the boundary between the San Emigdio volcanics, of about 22.5 Ma,   and the Iversen and Santa Cruz basalts, dated as 22 ± 1.2, 24.3 ± 0.3, and 23 ± 0.7. Lamb (personal communication) placed the boundary near the Iversen basalt at 23 to 24 Ma,   because of the presence of hundreds of feet of Miocene (Saucesian) section below the San Emigdio basalt. Theyer and Hammond (1974b) arrived at essentially the same conclusion by placing the boundary within polarity Epoch 21 at about 23 to 24 Ma,   close to the *Lychnocanoma elongata* datum.

The changes in the Paleogene numerical time scale proposed here are only in part the result of new radiometric data (Fig. 8). New information concerning the relative position between stage type sections and the biostratigraphic framework contributed significantly to this general revision.

References Cited

Alvinerie, J., et al, 1973, A propos de la limite oligo-miocene; resultats preliminaires d'une recherche collective sur les gisements d'Escornebeou (Saint Geours de Maremme, Landes, Aquitaine meridionale). Presence de *Globigerinoides* dans les faunes de l'Oligocene superieur: Soc. Geol. France Compte Rendu, v. 15, no. 3-4, p. 75-76.

Anglada, R., 1971a, Sur la position du datum a *Globigerinoides* (Foraminiferida) la zone N4 (Blow 1967) et la limite oligo-miocene en Mediterranee: Acad. Sci. Comptes Rendus, v. 272, p. 1067-1070.

—— 1971b, Sur la limite Aquitanien-Burdigalien, sa place dans l'echelle des Foraminiferes planctoniques et sa signification dans le Sud-Est de la France: Acad. Sci. Comptes Rendus, v. 272, p. 1948-1950.

Baumann, P., and P. H. Roth, 1969, Zonierung des Obereozans und Oligozans des Monte Cagnero (Zentralapennin) mit planktonischen Foraminiferen und Nannoplankton: Eclogae Geol. Helvetiae, v. 62, no. 1, p. 303-323.

von Benedek, P. N., and C. Muller, 1974, Nannoplankton-Phytoplankton Korrelation im Mittel

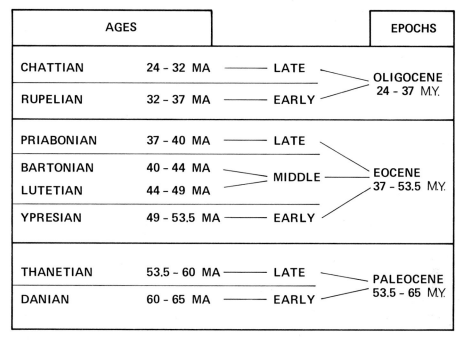

AGES			EPOCHS
CHATTIAN	24 – 32 MA	—— LATE	OLIGOCENE
RUPELIAN	32 – 37 MA	—— EARLY	24 – 37 M.Y.
PRIABONIAN	37 – 40 MA	—— LATE	EOCENE
BARTONIAN	40 – 44 MA	MIDDLE	37 – 53.5 M.Y.
LUTETIAN	44 – 49 MA		
YPRESIAN	49 – 53.5 MA	—— EARLY	
THANETIAN	53.5 – 60 MA	—— LATE	PALEOCENE
DANIAN	60 – 65 MA	—— EARLY	53.5 – 65 M.Y.

FIG. 8–Proposed Paleogene numerical time scale.

und Ober-Oligozan von NW-Deutschland: Neues Jahrb. Geologie u. Palaontologie Monatsh., no. 7, p. 385-397.

—— —— 1976, Die Grenze Unter/Mittel Oligozan am Doberg bei Bunde/Westfalen, I. Phyto. und Nannoplankton: Neues Jahrb. Geologie u. Palaontologie Monatsh., p. 129-144.

Berggren, W. A., 1969a, Rates of evolution in some Cenozoic planktonic foraminifera: Micropaleontology, v. 15, p. 351-365.

—— 1969b, Cenozoic chronostratigraphy, planktonic foraminiferal zonation and the radiometric time-scale: Nature, v. 224, p. 1072-1075.

—— 1971, Tertiary boundaries and correlations, in B. F. Funnel and W. R. Riedel eds., The Micropaleontology of the oceans: Cambridge, Cambridge University Press, p. 693-809.

—— 1972, A Cenozoic time-scale; some implications for regional geology and paleobiogeography: Lethaia, v. 5, p. 195-215.

—— and J. A. Van Couvering, 1974, Biostratigraphy, geochronology and paleoclimatology of the last 15 million years in marine and continental sequences: Palaeogeography, Palaeoclimatology, Palaeoecology, v. 16, p. 1-216.

—— et al, in press, Revised Paleogene polarity time scale.

Blakely, R. J., 1974, Geomagnetic reversals and crustal spreading rates during the Miocene: Jour. Geophys. Research, v. 79, p. 2979-2985.

Blow, W. H., 1969, Late middle Eocene to Recent planktonic foraminiferal biostratigraphy: 1st Int. Conf. Planktonic Microfossils Proc., p. 199-421.

Bodelle, C., C. Lay, and G. S. Odin, 1969, Determination d'age par la methode geochronologie "potassium-argon" de glauconies du Bassin de Paris: Acad. Sci. Comptes Rendus, v. 268, p. 1474-1477.

Bolli, H. M., 1957a, The genera *Globigerina* and *Globorotalia* in the Paleocene - lower Eocene Lizard Springs Formation of Trinidad, B.W.I.: U.S. Natl. Mus. Bull., v. 215, p. 61-81.

—— 1957b, Planktonic Foraminifera from the Oligocene-Miocene Cipero and Lengua Formations of Trinidad, B.W.I.: U.S. Natl. Mus. Bull., v. 215, p. 97-123.

—— 1957c, Planktonic Foraminifera from the Eocene Navet and San Fernando formations of Trinidad, B.W.I.: U.S. Natl. Mus. Bull., v. 215, p. 155-172.

—— 1966, Zonation of Cretaceous to Pliocene marine sediments based on planktonic Foraminifera: Assoc. Venezolana Geologia, Mineria y Petroleo Bol. Inf., v. 9, p. 3-32.

Bonhomme, M., G. S. Odin, and C. Pomerol, 1968, Age de formations glauconieuses de l'Albien et de l'Eocene du Bassin de Paris: (France) Bur. Recherches Geol. et Minieres Mem., v. 58, p. 339-345.

Bouche, P. M., 1962, Nannofossiles calcaires du Lutetien du bassin de Paris: Rev. Micropaleontologie, v. 5, p. 75-103.

Bramlette, M. N., and F. R. Sullivan, 1961, Coccolithophorids and related nannoplankton of the early Tertiary in California: Micropaleontology, v. 7, p. 129-188.

—— and J. A. Wilcoxon, 1967, Middle Tertiary calcareous nannoplankton of the Cipero section, Trinidad, W.I.: Tulane Studies Geology, v. 5, no. 3, p. 93-130.

Bukry, D., 1973, Low-latitude coccolith biostratigraphic zonation, in N. T. Edgar et al, eds., Initial reports of the Deep Sea Drilling Project: Washington, D.C., U.S. Govt., v. 15, p. 685-703.

—— 1975, Coccolith and silicoflagellate stratigraphy, northwestern Pacific ocean leg 32, in R. L. Larson et al, eds., Initial reports of the Deep Sea Drilling Project: Washington, D.C., U.S. Govt., v. 32, p. 677-701.

—— and M. P. Kennedy, 1969, Cretaceous and Eocene coccoliths at San Diego, California: California Div. Mines and Geology Spec. Rept. 100, p. 33-43.

Caro, Y., et al, 1975, Zonations a l'aide de microfossiles pelagiques du Paleocene superieur et de l'Eocene inferieur: Soc. Geol. France Bull., ser. 7, v. 17, no. 2, p. 125-147.

Cavelier, C., 1972, L'age priabonien superieur de la zone a *Ericsonia subdisticha* en Italie et l'attribution des Latdorf Schichten a l'Eocene superieur: (France) Bur. Recherches Geol. et Minieres Bull., ser. 2, v. 1, no. 4, p. 115-123.

—— 1976, La limite Eocene-Oligocene en Europe Occidentale: PhD thesis, Universite P. et M. Curie, Paris, p. 1-354.

—— and C. Pomerol, 1976, Les rapports entre le Bartonien et le Priabonien. Incidence sur la position de la limite Eocene moyen-Eocene superieur: Soc. Geol. France Compte Rendu, v. 2, p. 49-51.

Cox, A., 1969, Geomagnetic reversals: Science, v. 163, no. 3864, p. 237-244.

Ellis, H. C., 1975, Calcareous nannofossil biostratigraphy, leg 31, DSDP, in D. E. Karig et al, eds., Initial reports of the Deep Sea Drilling Project: Washington, D.C., U.S. Govt., v. 31, p. 655-676.

El-Naggar, Z. R., 1967a, On the so-called Danian s.l. or Dano-Montian of authors: Dansk Geol. Foren. Medd., v. 17, p. 103-111.

—— 1967b, Remarques sur les divisions du Paleocene; resultats d'etude dans les localites types en Europe occidentale: Rev. Micropaleontologie, v. 10, p. 215-216.

Evernden, J. F., and G. H. Curtis, 1965, Potassium-argon dating of late Cenozoic rocks in East Africa and Italy: Current Anthropology, v. 6, p. 343-385.

—— et al, 1961, On the evaluation of glauconite and illite for dating sedimentary rock by the potassium-argon method: Geochim. et Cosmochim. Acta, v. 23, p. 78-99.

Foreman, H. P., 1973, Radiolaria of leg 10 with systematics and ranges for the families *Amphipyndacidae, Artostrobiidae,* and *Theoperidae,* in J. L. Worzel et al, eds., Initial reports of the Deep Sea Drilling Project: Washington, D.C., U.S. Govt., v. 10, p. 407-474.

Foster, J. H., and N. D. Opdyke, 1970, Upper Miocene to Recent magnetic stratigraphy in deep-sea sediments: Jour. Geophys. Research, v. 75, no. 23, p. 4465-4473.

Funnell, B. F., 1964, The Tertiary Period, in W. B. Harland, A. G. Smith and B. Wilcock, eds., The Phanerozoic Time-Scale—a symposium: Geol. Soc. London Quart. Jour., v. 120 supp., p. 179-191.

Gartner, S., Jr., 1971, Calcareous nannofossils from the JOIDES Blake plateau cores; and revision of Paleogene nannofossil zonation: Tulane Studies Geology, v. 8, no. 3, p. 101-121.

Ghosh, P. K., 1972, Use of bentonites and glauconites in potassium-40/argon-40 dating in Gulf Coast stratigraphy: PhD Thesis, Rice Univ.

Goll, R. M., 1972, Leg 9, synthesis, Radiolaria, in J. D. Hays et al, eds., Initial reports of the Deep Sea Drilling Project: Washington, D.C., U.S. Govt., v. 9, p. 947-1058.

Gramann, F., et al, 1975, K-Ar ages of Eocene to Oligocene glauconitic sands from Helmstedt and Lehrte (northwestern Germany): Newsletter Stratigraphy, v. 4, no. 2, p. 71-86.

Hansen, H. J., 1970, Danian Foraminifera from Nugssuaq, West Greenland: Gronlands Geol. Undersogelse Bull., no. 93, p. 1-132.

Hardenbol, J., 1968, The Priabonian type section: (France) Bur. Recherches Geol. et Minieres Mem., no. 58, p. 629-635.

Hay, W. W., and H. P. Mohler, 1967, Calcareous nannoplankton from early Tertiary rocks at Pont Labau, France, and Paleocene-early Eocene correlations: Jour. Paleontology, v. 41, p. 1505-1541.

—— et al, 1967, Calcareous nannoplankton zonation of the Cenozoic of the Gulf Coast and Carribbean-Antillean area and transoceanic correlation: Gulf Coast Assoc. Geol. Socs. Trans., v. 17, p. 428-480.

Heirtzler, J. R., et al, 1968, Marine magnetic anomalies, geomagnetic field reversals, and motions of the ocean floor and continents: Jour. Geophys. Research, v. 73, no. 6, p. 2119-2136.

Holmes, A., 1959, A revised geological time-scale: Edinburgh Geol. Soc. Trans., v. 17, p. 183-216.

Hornaday, G. R., 1972, Oligocene smaller Foraminifera associated with an occurrence of *Mio-*

gypsina in California: Jour. Foraminiferal Research, v. 2, no. 1, p. 35-46.

Kulp, J. L., 1961, Geologic time scale: Science, v. 133, p. 1105-1114.

Lamb, J. L., and R. L. Hickernell, 1972, The late Eocene to early Miocene passage in California, *in* E. H. Stinemeyer and C. C. Church, eds., Pacific coast Miocene biostratigraphic symposium: Proc., p. 63-76.

—— and R. M. Stainforth, 1976, Unreliability of *Globigerinoides* datum: AAPG Bull., v. 60, no. 9, p. 1564-1569.

Luterbacher, H. P., 1975, Planktonic Foraminifera of the Paleocene and early Eocene, Possagno section: Schweizer. Paleont. Abh., v. 97, p. 57-67.

Marks, P., and E. J. Van Vessem, 1971, Foraminifera from the Silberberg Formation (lower Oligocene) at Silberberg, near Helmstedt (Germany): Paleontol. Zh., v. 45, p. 53-68.

Martini, E., 1969, Nannoplankton aus dem Latdorf (locus typicus and weltweite parallellisierungen im oberen Eozan und unteren Oligozan: Senckenbergiana Lethaea, v. 50, no. 2-3, p. 117-159.

—— 1970, Standard Paleogene calcareous nannoplankton zonation: Nature, v. 226, no. 5245, p. 560-561.

—— 1971, Standard Tertiary and Quaternary calcareous nannoplankton zonation: 2nd Int. Conf. Planktonic Microfossils Proc., p. 739-785.

—— 1973, Nannoplankton-Massenvorkommen in den Mittleren Pechelbronner Schichten (unter-Oligozaen): Oberrhein. Geol. Abh., v. 22, no. 3, p. 1-12.

—— and C. Muller, 1971, Das marine altertiar in Deutschland und seine Einordnung in die Standard Nannoplankton Zonen: Erdol u. Kohle, v. 24, p. 381-384.

—— —— 1975, Calcareous nannoplankton from the type Chattian (upper-Oligocene): 6th Cong. Reg. Comm. Mediterranean Neogene Stratigraphy Proc., p. 37-41.

—— and S. Ritzkowski, 1968, Was ist das "Unter-Oligocan?: Akad. Wiss. Gottingen, Nachr., Math.-Phys. Kl., no. 13, p. 231-250.

—— —— 1969, Die Grenze Eozaen/Oligozaen in der typus-Region des Unteroligozaens (Helmstedt - Egeln - Latdorf): (France) Bur. Recherches Geol. et Minieres Mem., v. 69, p. 233-237.

—— —— 1970, Stratigraphische Stellung der Obereozanen Sande von Mandrikovka (Ukraine) und Parallelisierungs - Moglichkeiten mit Hilfe des fossilen Nannoplanktons: Newsletter Stratigraphy, v. 1, no. 2, p. 49-60.

—— and T. Worsley, 1970, Standard Neogene calcareous nannoplankton zonation: Nature, v. 225, no. 5229, p. 289-290.

McKee, E. H., 1975, K-Ar ages of deep-sea basalts, Benham Rise, West Philippine basin, leg 31, DSDP, *in* D. E. Karig et al, eds., Initial reports of the Deep Sea Drilling Project: Washington, D.C., U.S. Govt., v. 31, p. 599-600.

Moore, T., 1971, Radiolaria, *in* J. I. Tracey, Jr., et al, eds., Initial reports of the Deep Sea Drilling Project: Washington, D.C., U.S. Govt., v. 8, p. 727-775.

Muller, C., 1974, Calcareous nannoplankton, leg. 25, (western Indian Ocean), *in* E. W. Simpson et al, eds., Initial reports of the Deep Sea Drilling Project: Washington, D.C., U.S. Govt., v. 25, p. 579-633.

Odin, G. S., 1973, Resultats de datations radiometriques dans les series sedimentaires du Tertiaire de l'Europe occidentale: Rev. Geographie Phys. et Geologie, Dynam., v. 15, no. 3, p. 317-330.

—— 1975, Recherches sedimentologiques et geochimiques sur la genese des glauconies actuelles et anciennes; application a la revision de l'echelle chronostratigraphique par l'analyse isotopique des formations sedimentaries d'Europe occidentale (du Jurrassique superieur au Miocene inferieur): These, Université de Paris, 250 p.

—— et al, 1969a, Geochronologie de niveaux glauconieux tertiaires des bassins de Londres et du Hampshire (methode potassium-argon): Soc. Geol. France Compte Rendu, v. 8, p. 309-310.

—— et al, 1969b, Geochronologie de niveaux glauconieux tertiaires du bassin de Belgique (methode potassium-argon): Soc. Geol. France Compte Rendu, v. 6, p. 198-200.

—— et al, 1970, Geochronologie de niveaux glauconieux paleogenes d'Allemagne du Nord (methode potassium-argon). Resultats preliminaires: Soc. Geol. France Compte Rendu, v. 6, p. 220-221.

Opdyke, N. D., 1972, Paleomagnetism of deep sea cores: Rev. Geophys. Space Phys., v. 10, no. 1, p. 213-249.

Perch-Nielsen, K., 1972, Les nannofossiles calcaires de la limite Cretace-Tertiaire: (France) Bur. Recherches Geol. et Minieres Mem., no. 77, p. 181-188.

Premoli Silva, I., G. Napoleone, and A. G. Fisscher, 1974, Risultati preliminari sulla stratigrafia paleomagnetica della scaglia Cretaceo-Paleocenica della sezione di Gubbio (Appennino centrale): Soc. Geol. Italiana Boll., v. 93, p. 647-659.

Proto Decima, F., P. H. Roth, and L. Todesco, 1975, Nannoplancton calcareo del Paleocene e dell-Eocene della sezione di Possagno: Schweizer. Palaont. Abh., v. 97, p. 35-55.

Riedel, W. R., and A. Sanfilippo, 1970, Radiolaria, leg 4, Deep Sea Drilling Project, *in* R. G. Bader et al, eds., Initial reports of the Deep Sea Drilling Project: Washington, D.C., U.S. Govt., v. 4, p. 503-575.

—— —— 1971, Cenozoic Radiolaria from the western tropical Pacific, leg 7, *in* E. L. Winterer et al, eds., Initial reports of the Deep Sea Drilling Project: Washington, D.C., U.S. Govt., v. 7, p. 1529-1672.

—— —— 1973, Cenozoic Radiolaria from the Caribbean, DSDP, leg 15, *in* N. T. Edgar et al, eds., Initial reports of the Deep Sea Drilling Project: Washington, D.C., U.S. Govt., v. 15, p. 705-751.

Ritzkowski, S., 1967, Mittel-Oligozan, Ober-Oligozan und die Grenze Rupel/Chatt im nordlichen Hessen: Neues Jahrb. Geologie u. Palaontologie Abh., v. 127, no. 3, p. 293-336.

Roth, H. P., 1970, Oligocene calcareous nannoplankton biostratigraphy: Eclogae Geol. Helvetiae, v. 63, no. 3, p. 799-881.

—— P. Baumann, and V. Bertolino, 1971, Late Eocene-Oligocene calcareous nannoplankton from central and northern Italy: 2nd Int. Conf. Planktonic Microfossils Proc., p. 1069-1097.

Ryan, W. B. F., et al, 1974, A paleomagnetic assignment of Neogene stage boundaries and the development of isochronous datum planes between the Mediterranean, the Pacific and Indian oceans in order to investigate the response of the world ocean to the mediterranean salinity crisis: Riv. Italiana Paleontologia, v. 80, no. 4, p. 631-688.

Sanfilippo, A., and W. R. Riedel, 1974, Radiolaria from the west-central Indian Ocean and Arabian sea, leg 24, *in* D. L. Fisher et al, eds., Initial reports of the Deep Sea Drilling Project: Washington, D.C., U.S. Govt., v. 24, p. 997-1035.

Sclater, J. G., et al, 1974, Comparison of the magnetic and biostratigraphic time scales since the Late Cretaceous, *in* C. C. von der Borch et al, eds., Initial reports of the Deep Sea Drilling Project: Washington, D.C., U.S. Govt., v. 22, p. 381-386.

Scott, G. H., 1972, *Globigerinoides* from Escornebeou (France) and the basal Miocene *Globigerinoides* datum: New Zealand Jour. Geology and Geophysics, v. 15, no. 2, p. 287-295.

Stainforth, R. M., et al, 1975, Cenozoic planktonic foraminiferal zonation and characteristics of index forms: Kansas Univ. Paleont. Contr., a. 62, p. 1-425.

Talwani, M., C. C. Windisch, and M. G. Langseth, Jr., 1971, Reykjanes ridge crest, A detailed geophysical study: Jour. Geophys. Research, v. 76, no. 2, p. 473-517.

Tarling, D. H., and J. G. Mitchell, 1976, Revised Cenozoic polarity time scale: Geology, v. 4, no. 3, p. 133-136.

Theyer, F., and S. R. Hammond, 1974a, Paleomagnetic polarity sequence and radiolarian zones, Brunhes to polarity Epoch 20: Earth and Planetary Sci. Letters, v. 22, p. 307-319.

—— —— 1974b, Cenozoic magnetic time scale in deep sea cores: Completion of the Neogene: Geology, v. 2, no. 10, p. 487-492.

Turner, D. L., 1970, Potassium-argon dating of Pacific coast Miocene foraminiferal stages: Geol. Soc. America Spec. Paper 124, p. 91-129.

Ujiie, H., 1975, Planktonic foraminiferal biostratigraphy in the western Philippine Sea, leg 31 of the DSDP, *in* D. E. Karig et al, eds., Initial reports of the Deep Sea Drilling Project: Washington, D.C., U.S. Govt., v. 31, p. 677-691.

Vail, P. R., R. M. Mitchum, and S. Thompson, 1974, Eustatic cycles based on sequences with coastal onlap (abs.): Geol. Soc. America Abs. with Programs, v. 6, p. 993.

—— et al, 1977, Seismic stratigraphy and global changes of sea level, *in* C. E. Payton, ed., Seismic stratigraphy—applications to hydrocarbon exploration: AAPG Memoir 26, p. 49-213.

Van Couvering, J. A., and W. A. Berggren, 1976, Biostratigraphical basis of the Neogene time-scale *in* J. Hasel and E. Kaufmann, eds., New concepts in stratigraphy: Stroudsburg. Pa., Dowden, Hutchinson and Ross.

Vine, F. J., 1968, Magnetic anomalies associated with mid-ocean ridges, *in* R. A. Phinney, ed., The history of the earth's crust: Princeton, Princeton Univ. Press., p. 73-89.

Watkins, N. D., 1975, Correlating stratigraphic zones and magnetic polarities: Geotimes, v. 20, no. 6, p. 26-27.

Critical Review of Isotopic Dates in Relation to Paleogene Stratotypes[1]

CHARLES POMEROL[2]

Abstract: With respect to the uncertainties of the method, measurements on glauconites of the Paleogene of western Europe effected during the last decade provide isotopic dates within a margin of error of 5%, that is about 2 m.y. (Table 2). A discussion of the age of the main limits: Cretaceous-Paleocene (63 to 65 m.y.), Paleocene-Eocene (53 to 55 m.y.), Cuisian-Lutetian (47 to 49 m.y.), Lutetian-Bartonian (42 to 44 m.y.), Eocene-Oligocene (34 to 36 m.y.), Oligocene-Miocene (22 to 24 m.y.) is presented. The middle-late Eocene boundary may be placed either between the Lutetian and the Bartonian (42 to 44 m.y.) or between the Bartonian and the Priabonian (39 to 41 m.y.).

During the last decade several hundred datings have been made in Europe on Paleogene glauconites, principally from France, England, Belgium, Germany, and the U.S.S.R. Others, fewer in number, have been obtained in the Mediterranean region and in central Europe. Before a presentation of the results, the inherent uncertainties of the method are briefly discussed. These originate either in the specimen, in the mode of operation, or in the choice of physical "constants."

Choice of the Specimen

One must first ensure that there has been no reworking from older deposits. This is not always easy and certain apparently "recent" glauconites of the Spanish plateau have given an apparent age of 6 m.y. (Odin, 1975). In addition, the closing of the crystalline system could have taken place before the end of the time interval corresponding to the whole formation. In this case the measurement gives an age that is older than that of the end of sedimentation. Burial at a depth of more than 1,000 meters can alter the results, but this is not a factor for the sedimentary basins of western Europe. Another danger that must be taken into consideration is contamination by atmospheric argon, which can increase the apparent age.

Mode of Operation

The reliability of the results depends to a large degree on the choice and the preparation of the grains of glauconite. Grains must be unaltered and any physical action taken to purify the sample (with formic acid, hydrochloric acid, ammonia, barium chloride, or ultrasonics) must be moderate (see discussion in Gramann et al, 1975). Furthermore, it is not recommended that glauconites with potassium content less than 7% (as K_2O) be used and, lastly, verification of the crystallinity by X-rays is mandatory. As for the analysis apparatus, one supposes it to be accurate, as guaranteed by its constructor.

[1] Manuscript received, December 14, 1976.

[2] Laboratoire de Géologie des Bassins sédimentaires, Université Pierre et Marie Curie, Paris.

The author thanks W. A. Berggren, D. Curry, D. E. Russell, and D. Vass for helpful discussions, translation, and criticism of the manuscript.

Article Identification Number: 0149-1377/78/SG06-0017/$03.00/0

Choice of Physical "Constants"

The decay constant, or rather decay rate of the argon utilized in calculations by most workers of Western Europe is: $\lambda_e = 0.584 \times 10^{-10}$ yr^{-1}. In the USSR a slower rate is used (0.557×10^{-10} yr^{-1}), and the new constant favored by Armstrong (1974) is intermediate: 0.575×10^{-10} yr^{-1}. The adoption of the last of these for the Paleogene would result in older ages (by 2.5%) in the West, and younger ones (in the same proportion) for the East; the present discrepancy is on the order of 5%. This uncertainty is not negligible whereas, for the middle Eocene, it leads to differences of 2 m.y., or *the equivalent of at least a biozone*. All dates quoted in this paper are calculated on the basis of the standards used in western Europe ($\lambda_e = 0.584 \times 10^{-10}$ yr^{-1}) unless there is a statement to the contrary.

Moreover, for some workers (H. C. Dudley *in* M. Catacosinos, 1975) the decay rate is influenced by, among other factors, pressure, temperature, chemical state, and electric potential. The result of this is that the "constant" of decay is a "variable;" there are obviously repercussions due to this situation in the domain of isotopic chronology, which becomes less "absolute" than initially thought.

General Considerations

With these reservations, let us examine some Paleogene datings from western European basins. The first measurement was effected at Fosses (Val d'Oise), with an age of 47 m.y. by Evernden and colleagues, on the "glauconie grossière" from the early Lutetian. But it was at the Glasgow Symposium (Funnell, 1964) that the first coherent scale was presented, founded on measurements obtained from other parts of the world. The first general reviews of radiometric dating concerning western Europe were published by Berggren (1972), Pomerol (1972) and Odin (1975). In Odin's second scale (1975) most of the ages are 5 to 10% younger after a recalibration of the calculations based on analytical corrections (Table 1). Though this viewpoint has been accepted by Tarling and Mitchell (1976) without discussion, it is opposed by Berggren, et al (1976) for whom ". . .this revised time scale is premature at best, seriously misleading at worst." Moreover, a certain inconsistency must be noted in Odin's recalibration whereas the ages are younger by 1 m.y. (2%) at the Cretaceous-Paleocene limit (fn. 3; Fig. 1), by 5 m.y. at the limit between the early and the middle Eocene (that is, the base of the Lutetian, 3, Fig. 1) and even by 10 m.y. (or 32%) for the Egerian (Fig. 2). With respect to the scale of 1973 a lengthening in the duration of the early and middle Eocene is evident, accompanied by a shortening of the late Eocene s.l. (Bartonian + Priabonian). A distortion results that is inexplicable in relation to the duration of foraminiferal biozones: 12 m.y. for 7 biozones in the early and middle Eocene (1 biozone equals 1.7 m.y.) and 4 m.y. for 6 biozones in the late Eocene (1 biozone equals 0.67 m.y., or 2.6 times shorter). In fact it appears that the reduction in age by 10% proposed by Odin (1975) for ages in the middle Eocene should be regarded as exaggerated and not at all confirmed by datings obtained from volcanic rocks or by a comparison with biostratigraphic data.

The desire for prudence has led to the proposal here that each date limit be included in a time bracket of 2 m.y. (Table 2). The margin of uncertainty varies from 3% for the oldest age (65 m.y., Cretaceous-Paleocene limit) to 8% for the youngest (22 m.y.). But concerning the Oligocene-Miocene limit there still exists a great imprecision. An interval of 2 m.y. is of the same order as the difference between the ages calculated from the decay rates in Western Europe and those of the USSR (5%). The utilization of the rates envisaged by Armstrong (1974; ± 2.5% with respect to the extreme ages) gives dates situated within this interval. The dates here proposed are in good accordance with the different phases of the Eocene volcanic activity in the northern part of the Massif Central (Bellon, et al, 1974).

Cretaceous-Paleocene Limit

To the author's knowledge there exists no dating information taken directly from the Maestrichtian and the Danian at their type localities. The most recent dates obtained from the Maestrichtian in other regions of the globe are situated between 61 and 64 m.y. (Obradovich and Cobban, 1975; Van Hinte, 1976) up to 68.5 m.y. for the

Table 1. Paleogene geochronology—Comparison of Funnell (1964), Berggren (1972), Odin (1973, 1975), Rubinstein and Gabunia (1972; and Gabunia and Rubinstein, in press) and Cavelier and Pomerol (1976b) scales. Recent studies (Cavelier and Pomerol, 1976a) indicate that the Priabonian corresponds to the upper Bartonian s.l. (Ludian). The question of the middle Eocene - late Eocene boundary is open (see Table 2). Rubinstein and Gabunia column A shows dates published by the authors in 1972, 1975, and in press, using decay constant: $0.557 \times 10^{-10} \ a^{-1}$; column B shows dates recalculated with the decay constant used in Western Europe ($0.584 \times 10^{-10} \ a^{-1}$). These values are in good accordance with Cavelier and Pomerol (1976a, b) except for the early Eocene. This fact is probably due to a discrepancy in the biostratigraphic correlations (i.e., the Simferopolian generally considered by the Russian authors as middle Eocene begins in fact in the early Eocene and ends in the middle part of the Lutetian (see Pomerol, 1973, p. 136).

Epoch / Series			Time in Millions of years					
			Funnell 1964	Berggren 1972	Odin 1973	Odin 1975	Rubinstein Gabunia 1972 - 76 A B	Cavelier Pomerol 1976
Early Miocene			26	22.5	24	22	25 23.5	22 - 24
Oligocene			37 - 38	37.5	37	35	37.5 35.5	34 - 36
Eocene	Late	B/P						39 - 41
		L/B		44	43	39	45 43	42 - 44
	Middle			49	49	44	54 51	47 - 49
	Early		53 - 54	53.5	54	51	60 57	53 - 55
Paleocene			65	65	65	64	68 64.5	63 - 65
Late Cretaceous								

glauconites below the Cretaceous-Tertiary limit in North Carolina (Harris and Bottino, 1974). This is why the Cretaceous-Tertiary limit is provisionally retained here in the 63 to 65 m.y. interval, which is generally adopted by most authors, including Rubinstein and Gabunia (1972) after correction of the decay constant (Table 1).

Thanetian, Ilerdian and the Paleocene-Eocene Limit

Most workers today accept that the Paleocene-Eocene limit is situated between the Thanetian and the Sparnacian; the typically Eocene mammalian fauna of the latter is comparable to that of the North American Wasatchian (Russell, 1968). Unfortunately, the type Sparnacian is lagunal and contains no glauconite. However, its equivalence

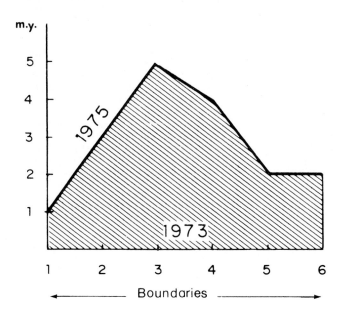

FIG. 1–Alteration in the dates proposed by Odin between 1973 and 1975. The new dates are from 1 m.y. to 5 m.y. younger in absolute value. The variation is greater in relative value (1.8 to 10%). **1** Cretaceous-Paleocene boundary; **2** Paleocene-Eocene b.; **3** early Eocene - middle Eocene b.; **4** Middle Eocene - late Eocene b.; **5** Eocene-Oligocene b.; **6** Oligo-Miocene b.

with at least the lower part of the Ilerdian marine stage has been established in Spain and in the south of France near the border of the Pyrenees (Colloque sur l'Ilerdien, 1974). As a result, the Paleocene limit is here placed near the base of zone P5 (*Morozovella velascoensis*) and within the zone NP9 (*Discoaster multiradiatus*).

Until now the only usable dates concern the Thanetian and its equivalents. In the type region (Herne Bay, Kent, England), Odin (1975) proposed two ages: 55.8 ± 3.5 and 53.3 ± 3.2. Other ages that are found to be younger, for the London Clay and the Landenian of Belgium, are included between 56 ± 2.2 and 47.8 ± 2.3. The Sable de Bracheux of the Paris basin (Butte de Reneuil, Aisne) has been dated at 53.6 ± 2.5 (Bonhomme et al, 1968). This age has been recalculated by Obradovich and increased to 58.5 and 59.2 m.y. (Berggren et al, in press). In the coastal plain of New Jersey and Maryland, the Vincentown Formation, base of the P5 zone of *Morozovella velascoensis* (late Thanetian) has yielded 56 m.y. (Owens and Sohl, 1973). As for the *Globorotalia subbotinae* zone (P6, Ilerdian), this has been dated by the same authors at 53.7 m.y. in the Manasquan Formation of New Jersey. All these dates suggest that the Paleocene-Eocene limit should be placed in the interval of 53 to 55 m.y. (Table 2).

Cuisian-Lutetian and Early - Middle Eocene Limit

The Sables de Cuise in the Paris basin have been dated at between 53.3 and 51.2 m.y. (Bodelle et al, 1969), but these dates seem too old because of an under-estimation of atmospheric argon (Odin, 1975).

The Lutetian and its English and Belgian equivalents contain the most often dated formations in Western Europe. In 1972, a dozen specimens were dated and the bracket for the early Lutetian was placed between 48.7 and 43 m.y. (Pomerol, 1972). Since then, numerous other measurements have been made. In considering the eight dates published by Odin (1975) that are middle Eocene, one obtains from the recalibrated

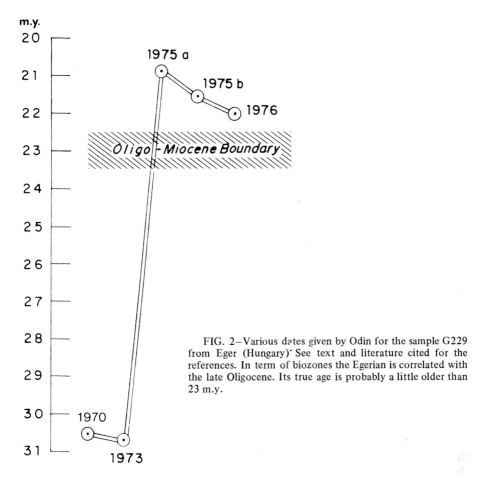

FIG. 2—Various dates given by Odin for the sample G229 from Eger (Hungary) See text and literature cited for the references. In term of biozones the Egerian is correlated with the late Oligocene. Its true age is probably a little older than 23 m.y.

values extremes of 48 to 37.1 m.y. (Table 3), or a mean of 42.5 m.y. The average age for the oldest dates is 45.8 and for the youngest, 41.6 m.y. In view of these results the author's choice for the stage limits (44 to 39 m.y.), equaling a displacement of 2 m.y. toward the younger end of the scale, is somewhat surprising.

If in this series one considers the four specimens closest to the base of the Lutetian (G941, G144, G435, and G128), the dates range from 47 to 42.5 m.y. with a mean of 45, reduced by Odin to 44 for no apparent reason. The same observation holds for the upper limit, comprised between 41.7 and 38.7 m.y. average—40.2, reduced to 39 m.y.[3]

Again, numerous K-Ar datings carried out in North America on volcanic formations in the terminal part of the Wasatchian, which corresponds to the terminal Ypresian (Russell and McKenna, 1975), range with remarkable consistancy between 49.3 and 49 m.y. (Berggren et al, in press).

The Wasatchian is overlain by the Bridgerian, which corresponds to the basal Lutetian and is situated within a range of 49 to 50 and 47 to 48 m.y. Now this continental stage of the base of the middle Eocene can be directly correlated with the Rose Canyon Shale (Ulatisian marine stage, see Pomerol, 1973). It belongs to the NP14 (*Discoaster sublodoensis*) biozone which, with the P10 (*Hantkenina aragonensis*) biozone,

---

[3] It seems that Odin later revised his dates as suggested here; a new scale presented at the Sydney Symposium gives 45 to 40 m.y. for the middle Eocene, but the late Eocene is consequently reduced to 3 m.y. (40 to 37 m.y.), which means 0.5 m.y. for each biozone of both Bartonian and Priabonian stages.

TABLE 2. STANDARD STRATIGRAPHIC SCALE PROPOSED FOR THE
PALEOGENE (AFTER CAVELIER AND POMEROL, 1976b).

Dates My	SERIES	STAGES			BIOZONES	
22 – 24	OLIGOCENE	Upper Oligocene (undefined stage) (1)			P22/N3	NP 25
					P21/N2	NP 24
12 My					P20/N1	
		STAMPIAN			P19	NP 23
					P18	NP 22
34 – 36						
	LATE EOCENE	PRIABONIAN			P17	NP 21
5 My					P16	NP 20
					P15	NP 19
39 – 41						NP 18
		BARTONIAN			P14	NP 17
3 My					P13	NP 16
42 – 44	MIDDLE EOCENE				P 12	
		LUTETIAN			P11	NP 15
5 My					P10	
47 – 49						NP 14
	EARLY EOCENE	YPRESIAN	CUISIAN		P 9	NP 13
6 My					P 8	
					P 7	NP 12
			ILERDIAN		P 6	NP 11
						NP 10
53 – 55					P 5	NP 9
	PALEOCENE	THANETIAN				NP 8
					P 4	NP 7
						NP 6
10 My					P 3	NP 5
						NP 4
		DANIAN				NP 3
					P 2	NP 2
					P 1	NP 1
63 – 65						

(1) Chattian (s.l.) P = Planktonic Foraminifera (Blow) NP = Nannoplankton (Martini)

TABLE 3. OLDER AND YOUNGER RECALIBRATED DATES FOR SAMPLES OF THE MIDDLE EOCENE OF WESTERN EUROPE (ODIN, 1975). SAMPLES G 144 (CARDITA BED OF THE BRACKLESHAM BEDS) and G 94 (1) (SANDS OF AELTRE) PROBABLY ARE JUST UNDER THE LOWER BOUNDARY OF THE LUTETIAN.

Samples	Older dates	Younger dates
G 150	41, 5	37, 1
G 104	41, 8	38, 2
G 437	46	41, 6
G 145	44, 8	41
G 435	46, 1	41, 7
G 144	48	43, 6
G 128	45, 6	42, 4
G 94 (1)	48	42, 6

characterizes the base of the middle Eocene (Berggren et al, in press). This is still another argument for keeping the early - middle Eocene limit in the 47 to 49 m.y. interval. To place it at 44 m.y. (Odin, 1975) or even 45 m.y. (Odin, in press) would lead to inconsistency in the correlation between the continental and marine formations of North America and the Caribbean.

Bartonian, Priabonian, and the Middle - Late Eocene Limit

Until now an equivalence between the Bartonian and the Priabonian has been generally admitted, these two stages representing the late Eocene—the former in Western Europe and the latter in the Mediterranean region. But, simulatenously in France, (Cavelier and Pomerol, 1976a) and in the United States (Berggren et al, in press), it has been demonstrated that the Priabonian corresponds only to the upper part of the Bartonian s.l. of the Nordic basins (Ludian). This is why Cavelier and Pomerol (1976b) have proposed the superposition of the two stages in a standard stratigraphic scale of the Paleogene with the corresponding biozones (Table 2). They left undetermined the middle - late Eocene limit, considering that the decision to place this limit either at the base of the Bartonian, as is habitual in the basins of northwest Europe, or at the base of the Priabonian, as is practiced in the Mediterranean area, should be made by an international working group.

Let us now consider the isotopic age of the base of the Bartonian. As early as 1969, as a result of a series of analyses effected on specimens from the lower levels of the Bartonian in the type locality, it was proposed to place the limit at 43 m.y. (Odin et al, 1969). Most of these specimens have been recalibrated by Odin (1975). In what follows, the first number is that of 1969, the second is that of 1975: base of the Barton Beds, Highcliffe near Barton on Sea 42 ± 2, 38.7; Alum Bay, Isle of Wight base of the Barton Beds 43.6 ± 1.8, 38 ± 1.8; Studley Wood, New Forest, Huntingbridge Bed 43.1 ± 2, 37.5 ± 1.8; Brook Bed, Bramshaw, New Forest 44.0 ± 2, 39.3 ± 2.2. It is possible that the first dates are a little too high because of contamination by atmospheric argon. But a reduction in age of 10% is certainly exaggerated and the real age is probably an intermediate one. The date obtained by Decarreau and Bellon (1976) on the illite of the Marinesian (middle Bartonian of the Paris basin), 41 ± 2 m.y., seems more credible.

New measurements at ± 4% (Elewaut et al, in press) give for the Barton Beds 38 m.y. (K-Ar) and 36.5 (Rb-Sr; the latter age is apparently low according to Pasteels, *in litteris*, who does not exclude a loss of radiogenic strontium) and for the Black Band of the Argile d'Asse (early Bartonian of Belgium) 38.1 to 40.1 m.y. (K-Ar, 6 specimens) and 39.5 to 43.5 m.y. (Rb-Sr, 5 specimens). The interval of 42 to 44 m.y. proposed here for the Lutetian-Bartonian boundary (Tables 1 and 2) agrees with the last measurements and is confirmed by previous datings of the Argile d'Asse published in 1969 (mean 45.2 m.y.) and recalculated by Obradovich (*in* Berggren et al, in press) 43.7 to 44 m.y.

If the base of the Bartonian is chosen as the limit between the middle and late Eocene, the duration of the late Eocene would then be 8 m.y. and would correspond to five and a half planktonic foraminiferal biozones (P12 pp to P17). From this point of view it is clear that the duration of the late Eocene which appears in Odin's new scale (4 m.y., Table 1) and even 3 m.y. (oral presentation at the Sydney Symposium, 1976) is obviously too short to contain five and a half foraminiferal biozones and six nannoplankton biozones (NP16 to NP21).

The other alternative is to choose the base of the Priabonian as the middle - late Eocene limit (Table 2). In this case, and until there is a possibility of dating the Italian Priabonian, it is estimated here that the middle - late Eocene limit is situated at about 40 m.y. (within the interval of 39 to 41 m.y.), a value that is in accordance with the ages obtained from Gehlberg Formation, in the NP16 to NP21 zones (late Bartonian-Priabonian) at Helmstedt in northwest Germany—39.3 ± 0.9 to 37.3 ± 0.9 (Kreuzer et al, 1973) and 39 ± 0.8 to 36.5 ± 0.7 (Gramann et al, 1975).

Eocene-Oligocene Limit

The Eocene-Oligocene limit is still difficult to establish because neither the late Bartonian s.l. - Priabonian, nor the Stampian-Rupelian has furnished a datable glauconite. In addition, the position of the limit itself is subject to controversy. Cavelier (1976) presented numerous arguments for placing it between the P17 and P18 zones (for planktonic Foraminifera, Blow, 1969) and between NP21 and NP22 (of Martini for nannoplankton, Table 2), that is to say, above the Lattorfian. Now the Silberberg Formation of Helmstedt (northwest Germany), belonging to the base of the NP21 zone, that is to say, to the Lattorfian, has been dated at 36.6 ± 0.7 m.y. by Gramann et al (1975), which could situate the Eocene-Oligocene limit at around 35 m.y. This figure corresponds to the recalculated value of Rubinstein and Gabunia (1972) and it is not in apparent disagreement with the date proposed by Berggren in 1972 (37.5 m.y.), because this author considered the Lattorfian at that time to belong to the Oligocene. The difference of 2.5 m.y. represents approximately the duration of this stage.

Oligocene-Miocene Limit

The age proposed by Kreuzer et al (1973) for the Oligocene-Miocene limit in northwest Germany is between 20 and 24 m.y., according to measurements effected on the Neo-Chattian (23.4 to 24.8 m.y.) and on the Hemmoorian early Miocene (19.2 to 20.4 m.y.). Two glauconites situated below the Oligocene-Miocene transition zone have been dated at 23.3 and 23.6 m.y. However, these results should be considered younger by 1% (Gramann et al, 1975).

A glauconite situated in the terminal part of the Aquitanian of the type region (bore hole of Soustons Landes) has been dated at 20.2 ± 1.3 m.y. by Odin (1975). A few measurements come from the Paratethys region, particularly from the locality of Eger (Hungary). By the presence of *Miogypsina septentrionalis, M. complanata, M. chormosensis, Chlamys decussata, C. delata, C. hertlei* and *Flabellipecten burdigalensis*, this bed of the Eger Formation belongs to the late Oligocene. It is overlain by marls, sandstone, and sands of Aquitanian age. The sands and sandstone of Eger, formerly attributed to the Chattian, are today designated by the name of Egerian, the oldest stage of the Paratethys, of which Eger is the type locality. In 1970, these sands (sample G229) were dated at 30.5 m.y. (Odin et al, 1970) a value that was slightly increased in 1973 (30.6 m.y.; Fig. 2). In 1973, still according to the same author in collaboration with Hunziker, this date was reduced to 20.8 ± 1.3 m.y., a decrease in age of 32%. This is the youngest age it has so far received. Since then, this Egerian has again been made older without apparent reason—21.5 ± 1.1 m.y. (Odin, 1975) and 22 ± 1.1 (Odin, in press).

Even in admitting that the first measurements (1970) were in error, the correction made for other datings effected in the same conditions, like those of the Gehlberg and Silberberg Formations at Helmstedt (northwest Germany), leads to a reduction in age on the order of 20%. This correction applied to the Eger beds would give an age of 24.7 m.y., which is probably close to reality (Vass, *in litteris*). Moreover, there exists a stratigraphic contradiction between the Oligocene-Miocene limit proposed at 22 m.y. (Odin, 1975) and the datings made by the same author of an Oligocene formation, that is older than this limit, at 20.8 or 21.5 m.y. It is regrettable that the publication, for the same specimen and by the same author of five different dates in six years with variations of as much as 32% of the initial age, tends to place in doubt the reliability of the method.

Taking also into account the results published by Ryan et al (1974) and by Vass et al (1975), it is probable that the Oligocene-Miocene limit is situated in the 22 to 24 m.y. interval, which includes the 22.5 m.y. age proposed by Berggren in 1972 and 23.5 m.y. (recalculated) of Gabunia and Rubinstein (1976). But the latest stage of the Oligocene is not correctly defined (Chattian, Bormidian, Egerian, ...) and the isotopic age of the Oligocene-Miocene limit will depend, to a certain degree, on the choice of this stage.

Conclusions

1. The isotopic age of some limits still depends on the choice by stratigraphers of the boundary stratotype. We need first to improve the stratigraphy. More precise isotopic dates for boundaries will follow.

2. The uncertainties of the isotopic datings lead us to propose that, provisionally, each date limit be included in an interval of 2 m.y. (Table 2). This suggestion is made to avoid a cluttering of the geochronologic literature with fluctuating dates (sometimes concerning the same specimen and the same analysis) which leads to scepticism about glauconite K-Ar dates. It would be desirable that authors resist the urge to immediately publish still uncertain new results and that specimens utilized for the datings be analysed in several laboratories. Publication should then not occur until after discussion between geochronologists, sedimentologists, stratigraphers, and paleomagneticians.

3. It is premature to adopt, without discussion, isotopic dates which are not in agreement with biostratigraphic correlations established between marine and continental formations previously dated by several laboratories.

4. In the present state of our knowledge it is obvious that, for Paleogene correlations, the precision of isotopic datings is inferior to that of classic biostratigraphic methods.

References Cited

Armstrong, R. L., 1974, Proposal for simultaneous adoptions of new U, Th, Rb and K decay constants for calculation of radiometric dates: Int. Mtg. Geochronology, Cosmochronology, and Isotope Geology (Paris) Abs. vol.

Bellon H., P. Y. Gillot, and P. Nativel, 1974, Eocene volcanic activity in Bourgogne, Charollais, Massif Central: Earth and Planetary Sci. Letters, v. 23, p. 53-58.

Berggren, W. A., 1972, A Cenozoic time scale—some implications for regional geology and paleobiogeography: Lethaia, v. 5, p. 195-215.

——— M. C. McKenna, and J. Hardenbol, in press, Revised Cenozoic polarity time scale.

Blow, W. H., 1969, Late middle Eocene to Recent planktonic foraminiferal biostratigraphy: 1st Int. Conf. Planktonic Microfossils Proc., p. 199-421.

Bodelle, C., C. Lay, and G. S. Odin, 1969, Determination d'age par la methode geochronologie "potassium-argon" de glauconites du Bassin de Paris: Acad. Sci. Comptes Rendus, v. 268, p. 1474-1477.

Bonhomme, M., G. Odin, and C. Pomerol, 1968, Age des formations glauconieuses de l'Abien et de l'Eocene du Bassin de Paris, *in* Colloque sur l'Eocene: (France) Bur. Recherches Geol. Minieres Mem. 58, p. 339-346.

Catacosinos, P. A., 1975, Do decay rates vary: Geotimes, v. 20, no. 4, p. 11.

Cavelier, C., 1976, La limite Eocene-Oligocene en Europe occidentale: These de Doc. Etat, Universite P. et M. Curie, Paris, 353 p.

—— and C. Pomerol, 1976a, Les rapports entre le Bartonien et le Priabonien, incidence sur la position de la limite Eocene moyen-Eocene superieur: Soc. Geol. France Comptes Rendu, p. 49-51.

—— —— 1976b, Proposition d'une echelle stratigraphique standard pour le Paleogene: Newsletter Stratigraphy, v. 5, no. 3.

Colloque Sur L'Ilerdien, 1974, Le contenu de l'Ilerdien et sa place dans le Paleogene: C. Pomerol, ed.: Soc. Geol. France Bull., v. 17, p. 123-223.

Decarreau, H., and H. Bellon, 1976, Resultats de datations radiometriques par la methode potassium-argon dans le Bartonien moyen du Bassin de Paris: Soc. Geol. France Bull., v. 7, no. 18, p. 769-772.

Elewaut, E., et al, in press, K-Ar and Rb-Sr radiometric ages on glauconites from calcareous nannoplankton zones NP14, NP15, NP16, NP21, in the North Sea Basins (contribution of IGCP projects no. 133 and 74/1/89.

Evernden, J. F., et al, 1961, On the evaluation of glauconite and illite for dating sedimentary rates by the K-Ar method: Geochim. Cosmochim. Acta, no. 23, p. 78-99.

Funnell, B. M., 1964, The Tertiary Period, *in* Phanerozoic Time Scale, a symposium: Geol. Soc. London Quart. Jour., v. 120, p. 179-191.

Gabunia, L. K., and M. Rubinstein, in press, A propos de l'age isotopique de la limite Paleogene-Neogene.

—— E. V. Deviatkin, and M. Rubinstein, 1975, Donnes sur l'age absolu des formations continentales cenozoiques d'Asie, et leur signification biostratigraphique: Akad. Nauk S.S.S.R., t. 225, no. 4, p. 895-898.

Gramann, F., et al, 1975, K-Ar ages of Eocene to Oligocene glauconitic sands from Helmstedt and Lehre (NW Germany). Newsletter Stratigraphy, v. 4, no. 2, p. 71-86.

Harris, W. B., and M. L. Bottino, 1974, Rb-Sr study of Cretaceous lobate glauconitic pellets, North Carolina: Geol. Soc. America Bull, v. 85, p. 1475-1478.

Holmes, A., 1964, *in* W. D. Harland, A. G. Smith, and B. Wilcock, eds., Phanerozoic time scale—a symposium: Jour. Geol. Soc. London Quart., v. 120, supp., p.

Kreuzer, H., et al, 1973, K-Ar dates of some glauconites of the NW German Tertiary Basin: Fortschr. Mineral, v. 50, no. 3, p. 94-95.

McKenna, M. C., et al, 1973, K-Ar recalibration of Eocene North American land-mammal "Ages" and European ages: Geol. Soc. America Abs. with Programs, v. 5, no. 7, p. 733.

Obradovich, J. D., and W. A. Cobban, 1975, A time scale for the late Cretaceous of the Western Interior of North America: Geol. Assoc. Canada Spec. Paper, no. 13, p. 31-54.

Odin, G. S., 1975, Les glauconies—constitution, formation, age: These Doct. Etat, Paris, 250 p.

—— in press, Isotopic dates for a Paleogene time scale: this volume.

—— and J. C. Hunziker, 1975, Donnees nouvelles sur l'age numerique des glauconies oligo-miocenes de la Paratethys (methode K-Ar): Geol. Prace, Zpr., v. 63, p. 141-148.

—— et al, 1969, Geochronologie de niveaux glauconieux tertiaires des bassins de Londres et de Hampshire (methode K-Ar): Soc. Geol. France Comptes Rendu, no. 8, p. 309-310.

—— et al, 1970, Geochronologie de niveaux glauconieux paleogenes d'Allemagne du Nord. Resultats preliminaires: Soc. Geol. France Comptes Rendu, no. 6, p. 220-221.

Owens, J. P., and N. Sohl, 1973, Glauconites from the New Jersey-Maryland coastal plain; their K-Ar ages and application in stratigraphic studies: Geol. Soc. America Bull., v. 84, p. 2811-2838.

Pomerol, C., 1972, Ages radiometriques dans le Tertiaire des bassins du Nord-Quest de l'Europe: Assoc. Geol. Bassin Paris, Bull. Int., no. 32, p. 3-6.

—— 1973, Stratigraphie et Paleogeographie—Ere Cenozoique (Tertiaire et Quaternaire): Paris, Doin, 269 p.

Rubinstein, M. M., and L. K. Gabunia, 1972, Certain aspects of Cenozoic geochronology: Akad. Nauk, SSSR, Izv., Ser. Geol., no. 3, p. 3-8.

Russell, D. E., 1968, Succession en Europe des faunes mammaliennes au debut du Tertiaire: France Bur. Recherches Geol. Minieres. Mem., no. 58, p. 291-296.

—— and M. C. McKenna, 1975, European fossil land mammals, radiometrically dated marine glauconites and Paleogene intercontinental correlations: 2nd Geol. Cong. Latino-American, (Caracas) Proc.

Ryan, W. B., et al, 1974, A Paleomagnetic assignment of Neogene Stage boundaries and the development of isochronous datum planes between the Mediterranean, the Pacific and Indian Oceans: Riv. Italian Paleontol., v. 80, no. 4, p. 631-688.

Tarling, D. H., and J. G. Mitchell, 1976, Revised Cenozoic polarity time scale: Geology, v. 4, no. 3, p. 133-136.

Van Hinte, J. E., 1976, A Cretaceous time scale: AAPG Bull., v. 60, p. 498-516.

Vass, D., J. Slavik, and G. P. Bagdasarjan, 1975, Radiometric time scale for Neogene of Paratethys: 6th Cong. Reg. Comm. Mediterranean Neogene Stratigraphy, (Bratislava) Proc., p. 293-297.

Isotopic Dates for a Paleogene Time Scale[1]

G. S. ODIN[2]

Abstract Numerous results obtained since 1964 are presented. A scheme of the reliability and accuracy of the different chronometers used for numerical dating is given. All the rocks and minerals are not of identical value. Some comparisons between Rb-Sr and K-Ar data are mentioned and comparisons between different chronometers are presented.

The probable numerical age of the boundaries between the stages are discussed, taking into account the new suggestions on the best position of these stratigraphic subjective boundaries.

The Paleogene has a duration of 42 million years—65 to 23 m.y. The Paleocene-Eocene boundary is at about 51 m.y. The appearance of *Pseudohastigerina* is a little more than 52 m.y. ago. The Lutetian stage is between 45.5 and 40 m.y. The base of the *Ericsonia subdisticha* zone is younger than 37 m.y. while the Priabonian-Rupelian boundary is between 34.5 and 36.5 m.y. All the apparent ages are calculated with the new standardized constants proposed during this symposium.

History

Radiometric investigations on Paleogene rocks were initially made between 1956 and 1964. Results were collected by Funnell in 1964. These dates were obtained primarily in the North American laboratories of Berkeley (Evernden and collaborators, K-Ar method) and M.I.T. (Hurley and collaborators, K-Ar and Rb-Sr methods).

Sedimentary and extrusive rocks have been utilized. Biostratigraphic control, upon which correlation with classical sections is based, is commonly poor or inadequately documented. Estimates of the geochronometric reliability remain very subjective and current scales must still be viewed as tentative.

The thesis by Obradovich (1964) represents the first systematic experimental investigation on glauconites as a geochronometric tool. The effects of different laboratory preparation techniques of glauconites on the eventual dating results and the ultimate reliability of glauconite as a geochronometer was discussed. It was shown that disparate ages older or younger than the age of "sedimentation," frequently can result owing to retention or loss of argon for various reasons.

Over the past decade additional information on glauconites has been obtained through the investigations of Obradovich, Odin and his colleagues from Bern, Brussels, and Strasbourg, and by Kreuzer. We are now in a position to select the most reliable data—in general, uncontaminated, potassium-rich minerals whose geochemical history is known. Recent age determinations appear more reliable than earlier ones (Odin, 1976, this volume).

[1] Manuscript received, January 6, 1977.

[2] Lab, de Géologie, Université P. & M. Curie, 75230, Paris, cédex 05.

I am indebted to D. Curry and W. A. Berggren who offered their competent advice concerning the bio- and litho-stratigraphic questions. However, I am solely responsible for the presented conclusions.

Article Identification Number: 0149-1377/78/SG06-0018/$03.00/0

Over the period 1964 to 1971, data which appeared allowed us to make reliable estimates on the base (Folinsbee et al, 1970) and top (Turner, 1970) of the Paleogene. An evaluation of the existing data base led to the formulation of a Cenozoic time scale by Berggren (1969; 1972). A compilation of isotopic dates on European formations has been made by Odin (1973). Modifications to existing scales within the Paleogene are rather minor, although it would appear that the Oligocene-Miocene boundary is somewhat younger than that suggested by Funnell (1964).

In North America, recent studies have been published on early Paleocene bentonites (Obradovich and Cobban, 1975), and Maastrichtian and Paleocene glauconites of the Atlantic Coastal Plain (Owens and Sohl, 1973). The thesis of Ghosh (1972) contains data on the Paleogene glauconite and bentonite minerals of the Gulf Coastal Plain (although biostratigraphic control generally is lacking). Paleogene data from eastern Europe have recently been compiled by Afanas'yev and his collaborators.

Since 1973, many new dates have been obtained in the Paleogene—Paleocene and Eocene (Odin, 1975) and latest Eocene or earliest Oligocene (Kreuzer et al, 1973; 1975). These recent data show the necessity of some modifications due to analytical errors in preceding works.

Up to the present time a large majority of Paleogene dates have been obtained by the K-Ar method. It would appear that problems concerning the reliability of some glauconite dates may be resolved by the Rb-Sr method whenever a new decay constant is adopted based on current work. Studies are underway with J. C. Hunziker at Bern, E. Keppens and P. Pasteels at Brussels, and new results have been proposed by Harris et al.

Choice of the Age Data

Numerical data will be discussed utilizing new standardized, recently proposed decay constants (NSC). Armstrong (1974) and McDougall (1974) proposed: $\lambda^{40}K = 0.575 \cdot 10^{-10}$ y^{-1}; $^{40}K/K = 1.18 \cdot 10^{-4}$ (atoms)[3] which give apparent ages older by 2.4% than the Phanerozoic time scale (Harland et al, 1964), and $\lambda^{87}Rb = 1.42 \cdot 10^{-11}$ y^{-1} which gives apparent ages younger by 2.2% than the 1.39 decay constant and older by 3.4% than the 1.47 decay constant. The aim of this work is not to support one or the other decay constant; I use the more probable after the recent data.

Various geochronometers are available for time-scale dating, namely plutonic rocks, extrusive rocks, and sedimentary rocks. In each of them only authigenic minerals are useful. As a general rule, total rock samples are considered equivalent or worse than bad minerals. A comparison of the qualities of various chronometers is given in Figure 1. Without taking into account the frequent exceptions it seems that biotite and sanidine of the bentonites (effusive rocks deposited and altered in the seas) are better than glauconites (a less instantaneous chronometer) which are better than poorly correlatable volcanic rocks, the last being better than minerals of the plutonic rocks. All these chronometers may be useful under specific conditions (choice of field, level, sample, and mineral).

It is known (Obradovich and Cobban, 1975) that biotite of the bentonites must be potassium rich and that sanidine and biotite of the same level must give the same apparent age. After our own research on glauconites (Odin, 1975) we know that only potassium-rich green grains can be used for a limitation of the geochemical uncertainty. Some authors used glass shards of volcanic rocks (Evernden and Evernden, 1970); they must be used with caution as with the total rocks.

After the choice of the sample and determination of its biostratigraphic age, a good isotopic measurement is done on carefully selected material. There is no other secret technique.

The last quality for good data is the number of measurements on the same horizon or in the neighborhood by the same (or preferably by different) authors or methods. In these cases, the geochemical uncertainty due to the chronometer (which is often greater than the analytical one) is greatly lowered.

---

[3] The new dates concerning the K-Ar method and proposed at Sydney during the IGC meeting— $\lambda^{40}K = 0.5811 \cdot 10^{-10} y^{-1}$; $^{40}K/K = 1.167 \cdot 10^{-4}$ —give the same results in our field of work.

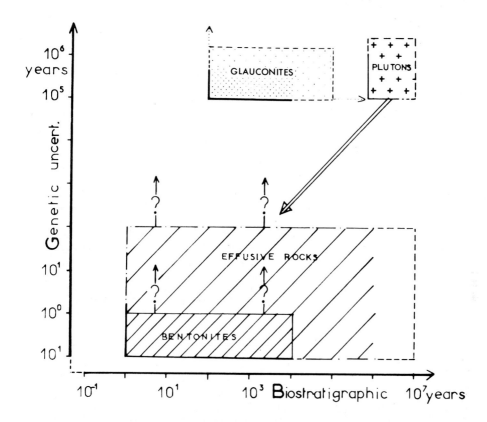

FIG. 1–Biostratigraphic and initial genetic uncertainties of the various chronometers used to build the radiometric time scale (separated minerals only are concerned). Original scheme shows (vertical axis) that time 0 of the glauconites and plutons is not well known because it is not instantaneously formed. On the contrary, genesis of bentonites and other effusive rocks is short–from a few months to tens of years. However it is not demonstrated that at the time 0 these effusive rocks do not contain inherited radiogenic isotopes. Data are lacking on recent bentonites whereas there are some studies on other effusive rocks (McDougall et al, 1969).

The horizontal axis shows that all chronometers are not instantaneously dated by biostratigraphic zonation. In rare cases effusive rocks are projected in seawater but in other cases they are never enclosed in marine or continental biozonation. On the contrary, glauconites are almost always introduced in a marine biozonation, rarely after 10^5 years.

The geochemical problems after burial are not discussed. We must say that they may be important in general for total rocks, glass shards (Evernden and Evernden, 1970), and micaceous minerals –biotite or glauconite with little potassium. Alteration by burial or pressure or temperature is important when K-Ar method is used while weathering is inauspicious during Rb-Sr dating.

In the following paragraphs I discuss each boundary based on published data selected according to the above criteria. It seems no longer useful to report old data obtained on biotites or glauconites with less than 4 or 5% potassium. For technical reasons the age data of Evernden et al (1961) are rarely used here (too old) while those published by Evernden et al (1964) and Obradovich (1964) are considered as better (Curtis, personal communication). In the same way only the apparent ages given (since 1973) by the author are utilized and the preceding ones abandoned (surely too old). Because the different existing time scales are based on the first works, one will understand that some changes are necessary today; one must use the more recent data (Baadsgaard et al, 1964; Ghosh, 1972).

It has been shown in different cases that the duration of biozones (Odin, 1976) and the rates of sedimentation (Lancelot, 1976) are not always regular even in deep sea basins. Then these methods of dating should not be used when only a minimum number of radiometric ages is available. In general these methods may give indications (but never more information) and the uncertainty of such extrapolations becomes too important when we are far from the reference points.

Considering these uncertainties, the general view is to give an analytical probability. I think that in most cases this is insufficient for the glauconites. I must say that the minimum geochemical uncertainty is at least in the order of a few 10^5 to 10^6 years due to the peculiar genesis. This uncertainty adds to the analytical one. In that way, for the glauconites younger than lower Miocene, the total relative uncertainty becomes very important, even if in a few cases dating of young glauconites has been done with success (Obradovich, 1968). For glauconites older than Eocene this relative uncertainty remains little. Data on recent bentonites are insufficient to give an evaluation of these questions, however it is known that recent volcanic rocks show excessive argon (McDougall et al, 1969).

Concerning the so-called low age given by the glauconites, we find in the literature (for potassium-rich glauconites of outcrops—the better case) many more examples of too-old ages than too-young. It is true that the more probable time of closing the chronometer glauconite is at the end of the genesis and the deposition of the overlying sediment. But the systematic choice of the older data given by the glauconites in the past time scales is another reason to obtain younger apparent ages today (Ghosh, 1972).

Comments on Paleogene Dates

The Mesozoic-Cenozoic boundary is essentially known from data from North America (Fig. 2). Considering as a better chronometer the high temperature minerals, the different works of Folinsbee gave us 65 m.y. After Obradovich and Cobban (1975) the boundary is rather near 66 m.y. Half of the data from glauconites agreed, especially the work of Harris and Bottino (1974), using the Rb-Sr method. However the data obtained on the Maastrichtian glauconites published by Owens and Sohl (1973) is taken by Obradovich as an example to show that some glauconites give too young and unreliable ages. Owens and Sohl, considering the good internal self consistance and lateral reproducibility of the used glauconites, proposed an age of 62.5 m.y. for the same boundary (Ghosh, 1972). The older evaluation of 65 to 66 m.y. is more probable at this time but a definitive solution will be accepted only when samples correlatable with marine faunas are dated.

Within the Paleocene, the Danian-Thanetian (that is Hornerstown-Vincentown or P3 biozone) boundary is dated by two "good" glauconites, near 57.5 m.y. in Owens and Sohl (1973). Tentatively, one can propose that the space time 58 to 60 m.y. includes the boundary.

In the upper Paleocene seven ages obtained by Obradovich (1964) or Odin et al (1976) on the sables de Bracheux, Thanet sands, and Landenian from Belgium, are included in the 53.5 to 55.0 time span. Bonhomme et al (1970) gave two ages—55.5 and 57 m.y. by the Rb-Sr method (NSC) on glauconites from the sables de Bracheux and the Thanet beds, and Harris et al (1976) proposed two ages near 55.5 m.y. and one 10% older for the Beaufort Formation of Carolina (Rb-Sr method on glauconites, Thanetian, biozone P4, in age).

The Paleocene-Eocene boundary placed at the first appearance of *Pseudohastigerina* seems fairly well dated by the glauconite of the Reading Beds below, and above by those from the London Clays, argiles de Varengeville (Odin et al, 1976), and Bashi member (Ghosh, 1972). An age near 51 m.y. is more probable than the older one given by Owens and Sohl after glauconites not really correlatable. A larger definition of the Ypresian including the Reading and Woolwich beds (part of the biozones P5 and NP 9- NP 10) rejects the limit at 52 m.y. and more.

Inside the Eocene there are good data on the boundaries of the Lutetian (Fig. 3). The base can be estimated at 45.5 ± 0.5 m.y. (Odin, 1976) and there is no biostratigraphic problem. Assuming that the upper limit is near the base of the biozone N.P. 16,

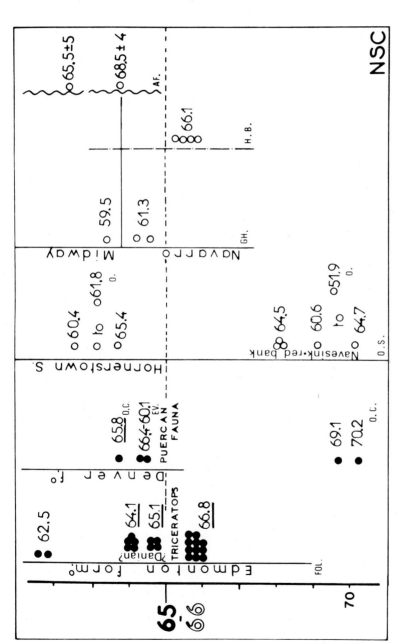

FIG. 2—Isotopic dates for the Mesozoic-Cenozoic boundary. Underlined ages, in megayears, seem more reliable. Most data must be understood with an uncertainty of ± 5% or less for recent or mean ages. Each black point is a sample of a high temperature mineral or rock and a white point is a date obtained on a selected glauconite (more than 5.8% K). Different studies by Folinsbee et al (1963, 1966, 1970) on tuffs and bentonites from the western plains of North America lead to a very well-documented age of 65 ± 0.5 m.y. (NSC) for the end of the Mesozoic Era (column FOL). This age slightly lower than that given in the Phanerozoic time scale (Funnel, 1964) is confirmed by ages from the works of Evernden et al (1964) and Obradovich and Cobban (1975) who, however, put the limit near 66 m.y.

Owens and Sohl (1973) published some ages from glauconites of New Jersey and Maryland after Obradovich (1964). In spite of a high potassium content and of a detailed study there is a problem in these results—the Maastrichtian and Danian glauconites give the same age. Owens and Sohl proposed an age of 62.5 m.y. for the boundary but Obradovich preferred to say that the glauconites in general are not a reliable chronometer (personnal communication). Data on glauconites from the Gulf Coastal province led to an age near the preceding one (Ghosh, 1972). Using the Rb-Sr method, Harris and Bottino (1974) proposed ages confirming the older age. Without giving many details Afans'yev (1970) reported two ages rather high for the Danian from SSSR.

All the ages are recalculated with the normalized decay constants.

Legend: Evernden, EV; Obradovich and Cobban, OC; Owens and Sohl, OS; Obradovich, O; Ghosh, GH; Afans'yev, AF.

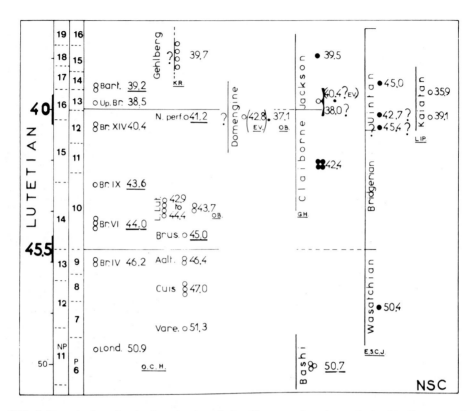

FIG. 3–Isotopic dates for the Lutetian boundaries. Many ages have been obtained in Europe on glauco-nites (white point for each sample) by Obradovich (1964), Gramann et al (1975), and Odin et al (1976). The sampled formations are London Clays, Lond.; argiles de Varengeville, Vare.; sables de Cuise, Cuis.; sables de Bruxelles, Brus.; glauconie grossière of the lower Lutetian, L. Lut.; Bracklesham beds of the Isle of Wight or the equivalents, Br.; latest Bracklesham beds, Up. Br.; limestones with *Nummulites perforatus* from Rumania, N. perf.; and lower Barton beds, Bart. The ages must be understood with an uncertainty of ± 5% or less and they are calculated with new constants. The notes of interrogations are either an uncertainty on the strati-graphical position of the formation or a measurement on a less reliable sample. The underlined ages are con-sidered as the most reliable.

The base of the Lutetian is older than 45 m.y. and younger than 46 m.y. The sands of Aalter (Aalt.) how-ever, are considered Ypresian near the boundary (Odin et al, 1972). The upper limit, considering the latest Bracklesham beds as already Bartonian, is near 40 m.y. Recent results obtained by the group at Brussels (Elewaut, Keppens, and Pasteels) show an age near 41 m.y. for the glauconites of the Bande Noire by both Rb-Sr and K-Ar methods.

Fairly good sequence of the ages measured is evidence of the reproducibility of glauconites in these sedi-mentary basins especially when we observe that ages are given by different laboratories—Berkeley, Berne, and Strasbourg on various samples.

I give as an example two ages from Evernden et al (1961) in parentheses. In both cases their ages are older than the more recent measurements by Obradovich (1964) or Ghosh (1972). This seems due to a problem in the potassium dosing in the old laboratory (Curtis, personal communication). Bentonite minerals from the Claiborne and Jackson Formations of the coastal province corroborate the age obtained on glauconites from Europe. Dates published by Evernden et al (1964) on biotites from the Uintan (Texas and Wyoming) are not well correlated. However one may note that they are clearly high, while the Wasatchian age is more in accord with the other dates.

I give the oldest dates on the Kaiatan glauconites from New Zealand by Lipson (1956). This stage was considered as upper Lutetian-lower Bartonian by Berggren (1972). Biozonations given are only tentative because little of the samples can be directly correlated with calcareous nannoplankton (NP zones of Martini, 1971) or planktonic Foraminifera (P zones of Blow, 1969). The Lutetian can be correlated with P zones 10-11-12 (or proparte) and the NP zones 14-15.

Author legend: Obradovich, OB; Gramann, KR; Odin et al, OCH; Evernden et al (1961), EV; Ghosh, GH; Evernden et al (1964), ESCJ; Lipson, LIP.

the age of 40 ± 0.5 m.y. is a good estimation after the glauconites of Europe. Some bentonites have been dated by Ghosh (1972) in the Gulf Coastal Province and show a good correlation between 39.5 and 42.4 m.y. The data from Evernden et al (1964) are less correlatable. The base of the Barton beds of England is clearly younger—39 m.y. (Odin et al, 1976) and younger than the sands of Asse and Wemmel from the Belgian basin studies under way by and with the group at Brussels who showed that Rb-Sr and K-Ar apparent ages of the "bande noire" are remarkably close.

Concerning the new interpretation of the upper Eocene, Hardenbol and Berggren (1976), one must say that there are no radiometric ages for the Priabonian in the mesogean basins. The duration of the Bartonian s.l. (including N.P. 16 to N.P. 20) is of the same order as the other six stages of the Paleogene while the new Bartonian s.s. (approximately N.P. 16 + N.P. 17) seems very short. However this is first of all a biostratigraphical question.

The boundary between Eocene and Oligocene presents two questions, a biostratigraphic one and a radiometric one. Let us admit that this boundary (Bartonian s.l. - Lattorfian) is under the *Ericsonia subdisticha* appearance, the base of which can be estimated at approximately 37 m.y. (NSC) from glauconites (Gramann et al, 1975), as in high temperature rocks (boundary Duchesnean-Chadronian dated by Evernden et al, 1964). One glauconite from the sables de Neerepen gave an apparent age of approximately 31 m.y., but this glauconite is potassium-poor while those of the Silberberg are richer. Curry had a very interesting observation (Odin et al, 1976) on the duration of biozones and I subscribe to it (Fig. 4). According to this observation, the base of the N.P. 21 is approximately 33 m.y. (the glauconites of the Silberberg-Schichten are probably inherited); the duration of biozones is much more regular with this estimation. The question cannot be solved today. While waiting for new information I still prefer the radiometric dates given by the best geochronometers.

Now the biostratigraphic question—the *Ericsonia* zone from Germany is assumed to be Priabonian by most authors (see Cavelier, 1972). Consequently the Eocene-Oligocene boundary (Priabonian-Rupelian) is younger than 37 m.y. I proposed 36 m.y. in Odin (1975); this boundary is probably between 34.5 and 36.5 m.y. which is a large uncertainty because the duration of the biozones is (locally) rather problematic. But we have to remember that the above considerations end at an age of approximately 31 m.y. The Rupelian-Chattian boundary cannot be estimated from direct measurements.

There are various studies concerning the Oligocene–Miocene boundary (Fig. 5). Turner (1970) gave us the best apparent ages on high temperature minerals and rocks while Kreuzer et al (1973) gave us a few details on corresponding glauconites from Germany. The literature also contains interesting discussions by Page and McDougall (1970), Berggren (1972), Ikebe (1973), and Odin et al (1975). As a conclusion, the Oligocene-Miocene boundary is not older than 24 m.y. and is probably very near 23 m.y.

An abstract of the results, presented as a selection here, is given in Figure 6. Taking into account the recent progress in the knowledge of the chronometers and of the decay constants, and the new data of the biostratigraphic studies, it seems that the ages given above are now not very far from reality at least for half of them. It is to be hoped that on this basis a standardization of the revised numerical time scale will be possible by removing the former evaluations often based on data obtained as long as 15 years ago.

References Cited

Abele, C., and R. W. Page, 1970, Stratigraphic isotopic ages of Tertiary basalts from Maud and Aircy's Inlet Victoria Australia: Royal Soc. Victoria Proc., v. 86, no. 2, p. 143-150.

Afanas'yev, G. D., 1970, Certain key data for the Phanerozoic time-scale: Eclogae Geol. Helvetiae, v. 63, no. 1, p. 1-7.

Armstrong, R. L., 1974, Proposal for simultaneous adoption of new U, Th, Rb, and K decay constants for calculation of radiometric dates: Int. Mtg. Geochronology, Cosmochronology, Isotope Geology (Paris), Abst.

Baadsgaard, H., et al, 1964, Limitations of radiometric dating, *in* Geochronology of Canada: Royal Soc. Canada Spec. Pub. 8, p. 20-38.

proposed age NSC					[33]		
66-65	52	45.5	40		37	23.5	6-5
1	9	13|14	16	NP20	NP21	25	
1	5	9|10	12|13	P17	P18	22	

number of zones

NP	8.5	4.5	2.5	4.5	5	12
P	4.5 (7.5)	4.5 (5.5)	3	5	5	15

duration

13.5	**6.5**	**5.5**	**3** [7]	**13.5** [9.5]	**18**
PALEOCENE	E O C E N E			OLIGOCENE	MIOCENE

mean duration of zones

NP	1.6	1.4	2.0	0.7	2.7	1.5
P	3.0(1.8)	1.4(1.2)	1.8	0.6	2.7	1.2

NP [1.6]   [1.9]

□ Base of Oligocene at 33   P [1.2]   [1.9]

FIG. 4—Duration of biozones from the Paleogene and the Eocene-Oligocene boundary (after Curry *in* Odin et al, 1976). The measured age of the base of the biozone NP 21 is 37 m.y. (with a great [unmeasurable] uncertainty in the history of the chronometer).

Assuming that correlations between ages and position of the other limits are accepted, we observe that the mean duration of the zones or subzones (in parentheses) is "too high" for the Oligocene and "too low" for the upper Eocene. All durations are in megayears. With the exception above noted the duration of the zones (or subzones for the Paleocene) is from 1.4 to 2.0 m.y. for the nannoplankton and from 1.2 to 1.8 for the planktonic Foraminifera.

This is very regular compared to the Upper Cretaceous for the planktonic Foraminifera (as for the Ammonites). The mean duration of the biozones in the stages is from 1 to 3.5 m.y. Considering this regularity, we can observe that an age of 33 m.y. for the base of NP 21 would eliminate the above mentioned exception. In this case, the Eocene is twice as long as the Oligocene.

This observation is a reason to confirm the probability of inheritance of the glauconite of the Silberberg Schichten. However the mean duration of the biozone is not and will never be a method of calculation of an age (only a wide evaluation) when the time space is so long (10 biozones and 17 m.y.). This is not the place to discuss the causes and effects of evolution, but it seems certain that analysis of such appearances and extinctions is not merely an exercise in probability.

It is not desirable in the writer's opinion to compromise by proposing the mean between 37 and 33. Such a solution is almost certain to be wrong as it is probable that one or the other of the alternative dates will prove to be correct. As a conclusion, and considering the probability of inheritance, let us say that the maximum age of the base of the biozone NP 21 is 37 m.y. The age of the boundary Eocene-Oligocene is another question of stratigraphy.

Berggren, W. A., 1969, Cenozoic chronostratigraphy, planktonic foraminiferal zonation and the radiometric time scale: Nature, v. 224, no. 5224, p. 1072-1075.

—— 1972, A Cenozoic time-scale—Some implications for regional geology and paleobiogeography: Lethaia, v. 5, p. 195-215.

Blow, W. H., 1969, Late middle Eocene to Recent planktonic foraminiferal biostratigraphy: 1st Int. Conf. Planktonic Microfossils Proc., p. 199-422.

Bonhomme, M., N. Clauer, and G. S. Odin, 1970, Resultats preliminaires de datation, rubidium-strontium sur des sediments glauconieux dans le Paleogene d'Angleterre: Alsace-Lorraine Serv. Carte Geol. Bull., v. 23, no. 3-4, p. 209-213.

Cavelier, C., 1972, L'age Priabonien superieur de la "zone a *Ericsonia subdisticha*" en Italie et l'attribut des Latdorfschichten allemands a l'Eocene superieur: (France) Bur. Recherches Geol. Minieres Bull., v. 4, no. 1, p. 15-24.

Evernden, J. F., and R. K. Evernden, 1970, The Cenozoic time scale: Geol. Soc. America Spec. Paper, 124, p. 71-90.

—— et al, 1961, On the evaluation of glauconite and illite for dating sedimentary rocks by the K-Ar method: Geochim. Cosmochim. Acta, v. 23, p. 78-99.

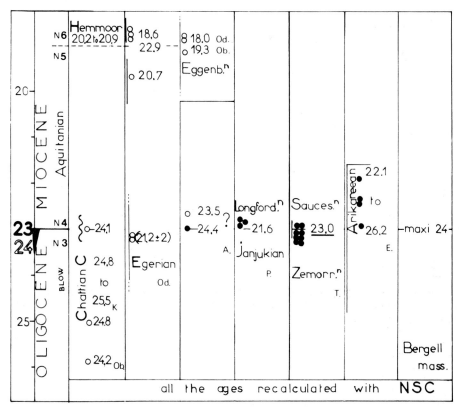

FIG. 5–Isotopic dates for the Oligocene-Miocene boundary. The papers used for this study are by Kreuzer et al (1973), Obradovich (1964), Page and McDougall (1970), Abele and Page (1970), Turner (1970), and Evernden et al (1964). The question of the Bergell Massiv is discussed by Odin et al (1975). The data of the high temperature minerals and rocks near the Zemorrian-Saucesian boundary are the best we know on the question. Since the studies of Turner (1967) in his thesis, the age of the end of the Oligocene is better known, and younger than in the Phanerozoic time scale. The dates of Evernden et al (1964) on the Arikareean (a mammalian stage of North America) are of the same order but the correlations are uncertain (Evernden and Evernden, 1970). The question is similar for the age of the Janjukian - Longfordian boundary from Australia.

The data by Afanas'yev (1970) are published without details and we cannot know their reliability. The age of the Egerian given in parentheses is not very useful as the potassium content of these glauconites is low and the geochemical history is unknown. The other ages obtained on glauconites from mesogean basins seem better. The data by Kreuzer et al (1973) gave us another example of a good accord between high and low temperature minerals.

The conclusion of this review is that the studied boundary is not far from 23 m.y. and surely more recent than 24 m.y.

Legend: Kreuzer et al, K; Obradovich, Ob; Page and McDougall and Abele and Page, P; Turner, T; Evernden et al, E; Afanas'yev, A.

—— et al, 1964, Potassium-argon dates and the Cenozoic mammalian chronology of North America: Am. Jour. Sci., v. 262, p. 145-198.

Folinsbee, R. E., H. Baadsgaard, and G. L. Cumming, 1963, Dating of volcanic ash beds (bentonites) by the K-Ar method: Nuclear Geophysics, Nuclear Sci. Ser. Rept., 38, p. 70-82.

—— —— —— 1970, Geochronology of the Cretaceous-Tertiary boundary of the Western plains of North America: Eclogae Geol. Helvetiae, v. 63, no. 1, p. 91.

—— et al, 1966, Late Cretaceous radiometric dates for the Cypress hills of western Canada: Canadian Soc. Petroleum Geologists, p. 162-174.

Funnell, B. M., 1964, The Tertiary period in The Phanerozoic time-scale—A Symposium: Geol. Soc. London, Quart. Jour., 120 Supp., p. 179-191.

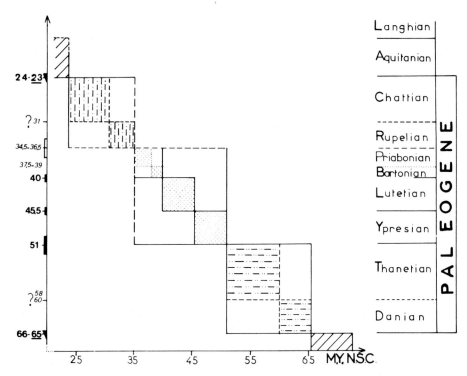

FIG. 6—Proposed time scale for the Paleogene. The stages used are those proposed by Hardenbol and Berggren (1976) and Van Couvering (1976). We can be "sure" only of the ages given by bold black numbers. Others are either extrapolated (note of interrogation) or very inaccurate (Priabonian boundaries).

Ghosh, P. K., 1972, Use of bentonites and glauconites in potassium 40-argon 40 dating in Gulf coast stratigraphy: PhD Thesis, Rice Univ., 136 p.

Gramann, F., et al, 1975, K-Ar ages of Eocene to Oligocene glauconitic sands from Helmstedt and Lehrte (N.W. Germany): Newsletter Stratigraphy, v. 4, p. 71-86.

Hardenbol, J., and W. A. Berggren, 1976, A new paleogene numerical time scale: 25th Int. Geol. Cong. (Sydney) Abs. 106-6, (also, this volume).

Harland, W. B., A. G. Smith, and B. Wilcock, 1964, The Phanerozoic time scale—A symposium: Geol. Soc. London Quart. Jour., v. 120, supp., 458 p.

Harris, W. B., and M. L. Bottino, 1974, Rb-Sr study of Cretaceous Lobate glauconite pellets: Geol. Soc. America Bull., no. 85, p. 1475-1478.

—— G. R. Baum, and M. L. Bottino, 1976, The Paleocene Beaufort formation, North-Carolina, microfauna and radiometric ages: Geol. Soc. America Abs. with Programs, v. 8, no. 2, p. 191.

Hurley, P. M., et al, 1960, Reliability of glauconite for age measurement by K-Ar and Rb-Sr methods: AAPG Bull., v. 44, no. 11, p. 1793-1808.

Ikebe, N., 1973, Neogene biostratigraphy and radiometric time scale: Osaka City Univ. Jour. Geosciences, v. 16, no. 4, p. 51-67.

Kreuzer, H., et al, 1973, K/Ar dates of some glauconites of the northwest German Tertiary basin: Fortschr. Mineral., v. 50, no. 3, p. 94-95.

—— and I. Wendt, 1977, Fourth annual report on the works of the I.G.C.P. project *Calibration of stratigraphical methods*: Bull. Liais. Inf. (IGCP) Project 133, v. 3, p. 14-15.

Lancelot, Y., 1976, Comparaison et evolution des bassins Atlantique Nord et Pacifique: These Doct. Etat. Paris.

Lipson, J., 1956, K-A dating of sediments: Geochim. Cosmochim. Acta, v. 10, no. 11.

Martini, E., 1971, Standard Tertiary and Quaternary calcareous nannoplankton zonation: 2nd Int. Conf. Planktonic Microfossils Proc., v. 2, no. 2, p. 739-777.

McDougall, I., 1974, The present status of the decay constants: *in* Report of activities of the I.U.G.S. subcommission of Geochronology: Int. Mtg. Geochronology, Cosmochronology, and Isotope Geology (Paris) Proc., p. 1-5.

—— M. A. Polach, and J. J. Stipp, 1969, Excess radiogenic argon in young subaerial basalts from the Auckland volcanic field N.Z.: Geochim. Cosmochim. Acta, v. 33, p. 1485-1520.

Obradovich, J., 1964, Problems in the use of glauconite and related minerals for radioactivity dating: PhD Thesis, Univ. California, Berkeley, 160 p.

—— 1968, The potential use of glauconite for late Cenozoic geochronology, *in* Means of correlation of Quaternary succession: 7th Cong. Int. Assoc. Quaternary Res., Proc., v. 8, no. 7, p. 267-279.

—— and W. A. Cobban, 1975, A time scale for the late Cretaceous of the Western interior of North America: Geol. Assoc. Canada, Spec. Paper 13, p. 31-54.

Odin, G. S., 1973, Resultats de datations radiometriques dans les series sedimentaires du Tertiaire de l' Europe occidentale: Rev. Geographie Phys. Geologie Dynam., v. 15, no. 3, p. 317-330.

—— 1975, De glauconiarum, constitutione, origine, aetateque: These Doct. Etat, Paris, 280 p.

—— 1976, Results of dating Cretaceous and Paleogene sediments of Europe: 25th Int. Geol. Cong. (Sydney), Abs. 106-6, (also, this volume).

—— D. Curry, and J. C. Hunziker, in press, Radiometric dating by glauconites from N.W. Europe and the time-scale of the Paleogene: Geol. Soc. London Quart. Jour.

—— J. C. Hunziker, and C. R. Lorenz, 1975, Age radiometrique du Miocene inferieur en Europe occidentale et centrale: Geol. Rundschau, v. 64, no. 2, p. 570-592.

—— et al, 1972, Le sondage du Mont-Cassel: Assoc. Geol. Bassin Paris Bull. Inf., no. 32, p. 21-52.

Owens, J. P., and N. F. Sohl, 1973, Glauconites from New Jersey - Maryland Coastal plain, their K-Ar ages and application in stratigraphic studies: Geol. Soc. America Bull., v. 84, p. 2811-2838.

Page, R. W., and I. McDougall, 1970, Potassium-argon dating of the Tertiary f.1-2 stage in New Guinea and its bearing on the geological time-scale: Am. Jour. Sci., v. 269, p. 321-342.

Turner, D. L., 1967, Potassium-argon dates concerning the Tertiary foraminiferal time scale: PhD thesis, Univ. California, Berkeley, 99 p.

—— 1970, Potassium argon dating of Pacific Coast Miocene Foraminiferal stages: Geol. Soc. America Spec. Paper, 124, p. 91-129.

Van Couvering, J. A., 1976, Calibration and correlation of Neogene biochronology: 25th Int. Geol. Cong. (Sydney) Abs. 106-6.

Cretaceous Time Scale from North America[1]

MARVIN A. LANPHERE[2] and DAVID L. JONES[2]

Abstract The Upper Cretaceous strata of the Western Interior of North America contain a rich molluscan fauna and abundant intercalated bentonites. This combination has permitted the development of a detailed time scale for the Late Cretaceous. Volcanic rocks are rare in the Lower Cretaceous of North America. Several isolated time-scale points have been reported for Lower Cretaceous rocks, but the Early Cretaceous time scale is not well defined.

Nearly all of the available isotopic ages are K-Ar mineral ages. Data have been recalculated using decay constants for K^{40} based on the refined activity data for K^{40} and the isotopic abundance of K^{40}. The best current estimates for system and series boundaries are: Cretaceous-Tertiary boundary—65 to 66 m.y.; Lower Cretaceous-Upper Cretaceous boundary—96 m.y.; Jurassic-Cretaceous boundary—138 ± 3 m.y.

Introduction

A successful geologic time scale must be based on detailed and precise biostratigraphic and geochronologic evidence. This criterion is good for the Late Cretaceous of North America but not nearly so good for the Early Cretaceous. In the Western Interior of North America, deposition of Upper Cretaceous rocks under marine conditions was nearly continuous and incorporated a rich and varied molluscan fauna. In addition, abundant intercalated altered volcanic ash beds (bentonites) are present. Bentonites are valuable marker beds; they represent isochronous layers and commonly contain minerals such as biotite and sanidine whose ages can be measured using isotopic techniques. Volcanic rocks are rare in the Lower Cretaceous of North America, and only isolated time-scale points have been reported.

Obradovich and Cobban (1975) proposed a detailed time scale for the Late Cretaceous of the Western Interior of North America. This scale is based on their K-Ar data on 15 ammonite zones in the United States and on K-Ar ages from adjacent areas in Canada reported by R. T. Folinsbee and his colleagues (1961, 1963, 1966).

Precise correlations of the Cretaceous of North America and the type strata of the Cretaceous in Europe are not well established although studies now in progress on inocerami and ammonites will greatly improve the situation. The type strata in Europe do not contain interbedded volcanic rocks but instead are characterized by abundant glauconite. Many isotopic ages have been measured on these glauconites; unfortunately, glauconite has not proved to be a reliable mineral for isotopic dating. Thus, it appears likely that a reliable physical time scale cannot be constructed directly for the type strata of the Cretaceous.

Objectives of this paper are to review the current status of the Cretaceous time scale for North America, to present new data on some Early Cretaceous time-scale points, and to identify problems (particularly with boundaries) that remain to be solved.

New measurements of the decay constants of radioactive isotopes commonly used in isotopic dating have been completed recently or are in progress. An international

[1] Manuscript received, November 27, 1976.
[2] U.S. Geological Survey, Menlo Park, California 94025.

We thank J. D. Obradovich for general discussion of the Late Cretaceous time scale for North America and for specific comments which improved this manuscript. A. L. Berry, B. M. Myers, and J. C. Von Essen measured the new K-Ar ages presented herein.

effort is underway to consider a general change to refined decay constants. All data used to construct a Cretaceous time scale for North America are K-Ar mineral ages. Refined decay constants for K^{40} were proposed by Beckinsale and Gale (1969), and these refined constants are being used in several laboratories.

More recently, the isotopic abundance of K^{40} has been redetermined by Garner et al (1975). In this paper, all K-Ar ages have been calculated or recalculated using decay constants based on the refined activity data for K^{40} proposed by Beckinsale and Gale (1969) and the isotopic abundance of K^{40} measured by Garner et al (1975).

Cretaceous-Tertiary Boundary

Near the end of the Cretaceous Period the epicontinental sea that covered large parts of the Western Interior of the United States and Canada retreated so that the youngest marine strata are Maastrichtian in age. The youngest marine rocks in Canada contain the *Baculites grandis* Zone of early Maastrichtian age (Obradovich and Cobban, 1975). In the Western Interior of the United States the youngest marine strata contain *Discoscaphites cheyennesis* and are no younger than early late Maastrichtian. The top of the Cretaceous sequence in the Western Interior consists of widespread nonmarine deposits that contain a terrestrial fauna correlated by characteristic reptiles. The genus *Triceratops* March occurs widely, and it is common to refer to the nonmarine sequence as the "Triceratops beds." Jeletzky (1960) reviewed the Upper Cretaceous biostratigraphy of the Western Interior of North America and concluded that the *Triceratops* beds are late Maastrichtian and correspond to part or all of the *Belemnitella junior* and *Belemnella casimirovensis* zones of northern Europe.

The Maastrichtian-Tertiary boundary in North America is drawn just above the latest appearance of the Triceratopsian fauna in nonmarine units. The Kneehills Tuff Zone of the upper member of the Edmonton Formation of Alberta is a widespread zone of volcanic ash and bentonitic clay that occurs just below the *Triceratops* beds. Some 60 to 120 meters above the Kneehills Tuff Zone, bentonites interbedded with coal seams mark the top of the *Triceratops* beds. A number of K-Ar ages on biotite and sanidine from the bentonites that bracket the *Triceratops* beds have been reported by Shafiqullah et al (1966) and Folinsbee et al (1966). The biotite ages commonly are younger than the ages for coexisting sanidine, probably an effect of weathering, and the sanidine ages provide the better data. No analytical data were reported by Shafiquallah et al (1966), but the data reported by Folinsbee et al (1966) apparently are for the same samples. The results for sanidine yield pooled ages of 66.9 ± 0.9 m.y. for the Kneehills Tuff Zone and 64.8 ± 1.2 m.y. for the post-*Triceratops* bentonites. The plus-or-minus values are the standard deviations of the pooled ages.

Some data for early Tertiary rocks assist in bracketing the age of the Cretaceous-Tertiary boundary. A recalculated age of 66.4 m.y. is obtained from data reported by Evernden et al (1964) for plagioclase from a pumice bed in the Denver Formation near Golden, Colorado. The pumice bed is located approximately 11 m (stratigraphically) above the Cretaceous-Tertiary boundary and is at the same level as mammal bones of Puercan age. G. Izett and J. Obradovich (*in* Obradovich and Cobban, 1975) measured an age of 65.9 m.y. for biotite from a pumice located approximately 11 m (stratigraphically) above the site collected by Evernden et al (1964).

There are slight discrepancies between the ages bracketing the Cretaceous-Tertiary boundary, but this may well reflect interlaboratory analytical bias. It appears that the best current age for the Cretaceous-Tertiary boundary in North America is 65 to 66 m.y.

Late Cretaceous Time Scales

Cobban and Reeside (1952) proposed the first complete biostratigraphy for Cretaceous rocks in the Western Interior of the United States. Their zonation is based chiefly on ammonites, and subsequent refined zonation proposed by Cobban relies entirely on ammonites. Cobban's proposed zonation includes 60 ammonite zones in the Late Cretaceous. Jeletzky (1968) proposed a comparable biostratigraphy for Cretaceous rocks in the Western Interior of Canada with zonation based on several types of mollusks. Although the two biostratigraphic systems are somewhat different because they are based on different faunas, some faunal assemblages are found in both the United States and Canada making correlations possible.

Obradovich and Cobban (1975) reviewed in detail the Late Cretaceous biostratigraphy of the Western Interior of North America and proposed a time scale based on available data plus a large number of newly measured mineral ages. Their time scale is shown in Figure 1 compared with time scales proposed previously for the Late Cretaceous of North America by Bandy (1967) and Kauffman (1970). Van Hinte (1976) proposed a time scale for the entire Cretaceous (not restricted to North America) that also is shown in Figure 1. The time scales of Kauffman (1970) and Obradovich and Cobban (1975) utilized a molluscan zonation whereas Bandy (1967) and Van Hinte (1976) utilized a planktonic foraminiferal zonation. Boundaries on all of the time scales have been recalculated using the refined decay constants for K^{40}. It is not our intention to discuss in detail the placement of stage boundaries within the Late Cretaceous; the interested reader is referred to Obradovich and Cobban (1975) for an excellent critical review. Instead, we will indicate those stage boundaries whose placement is a matter of disagreement on the basis of biostratigraphic evidence and also indicate those situations where additional radiometric ages could help define stage boundaries more precisely.

Campanian-Maastrichtian Boundary

There is general disagreement among biostratigraphers about placement of the Campanian-Maestrichtian boundary. Cobban placed the boundary between the *Baculites cureatus* and *B. reesidei* Zones (Obradovich and Cobban, 1975) whereas Jeletzky (1968) placed the boundary three zones higher at a position equivalent to between the *Baculites eliasi* and *B. baculus* Zones. Pessagno (1967, 1969), utilizing planktonic foraminifers, placed the Campanian-Maastrichtian boundary at a position equivalent to between the *Didymoceras nebrascense* and *D. stevensoni* Zones (Obradovich and Cobban, 1975). Olsson (1964), also on the basis of foraminifers, placed the boundary at Cobban's *Baculites scotti* Zone (Obradovich and Cobban, 1975). The various proposed boundaries differ by 10 ammonite zones in Cobban's classification. A number of K-Ar ages have been measured on bentonites in zones near the Campanian-Maastrichtian boundary. These results give the following ages for the different proposed boundaries: Pessagno and Olsson, 74 to 75 m.y.; Obradovich and Cobban, 72 to 73 m.y.; Jeletzky, approximately 71 m.y.

Santonian-Campanian Boundary

A biotite from a bentonite in the *Desmoscaphites bassleri* Zone, the uppermost ammonite zone in the Santonian, yielded an age of 84.5 m.y. (Obradovich and Cobban, 1975). No radiometric ages are available from the older four ammonite zones in the Campanian, but Obradovich and Cobban (1975) measured ages on biotite from three different bentonites from *Baculites* sp. (weak flank ribs) and *B. obtusus* Zones that average about 80 m.y. Obradovich and Cobban (1975) suggested that the boundary be placed at about 84 m.y., but that more data on the lower part of the Campanian would be desirable.

Coniacian-Santonian Boundary

No mineral ages have been reported that bear directly on the age of the Coniacian-Santonian boundary. Obradovich and Cobban (1975) assumed that the Coniacian Stage lasted approximately 1 m.y. because it encompasses no more than two ammonite zones. On this basis an age of about 88 m.y. can be extrapolated for the Coniacian-Santonian boundary from the age of a bentonite, described later, collected near the base of the Coniacian.

Turonian-Coniacian Boundary

There has been some dispute about whether the zone of *Scaphites preventricosus* and *Inoceramus deformus* should be placed at the top of the Turonian or the base of the Coniacian. Obradovich and Cobban (1975) measured the age of a biotite from near the base of *S. preventricosus* Zone. Their age of 88.9 m.y. suggests that 89 m.y. is a reasonable estimate for the age of the Turonian-Coniacian boundary.

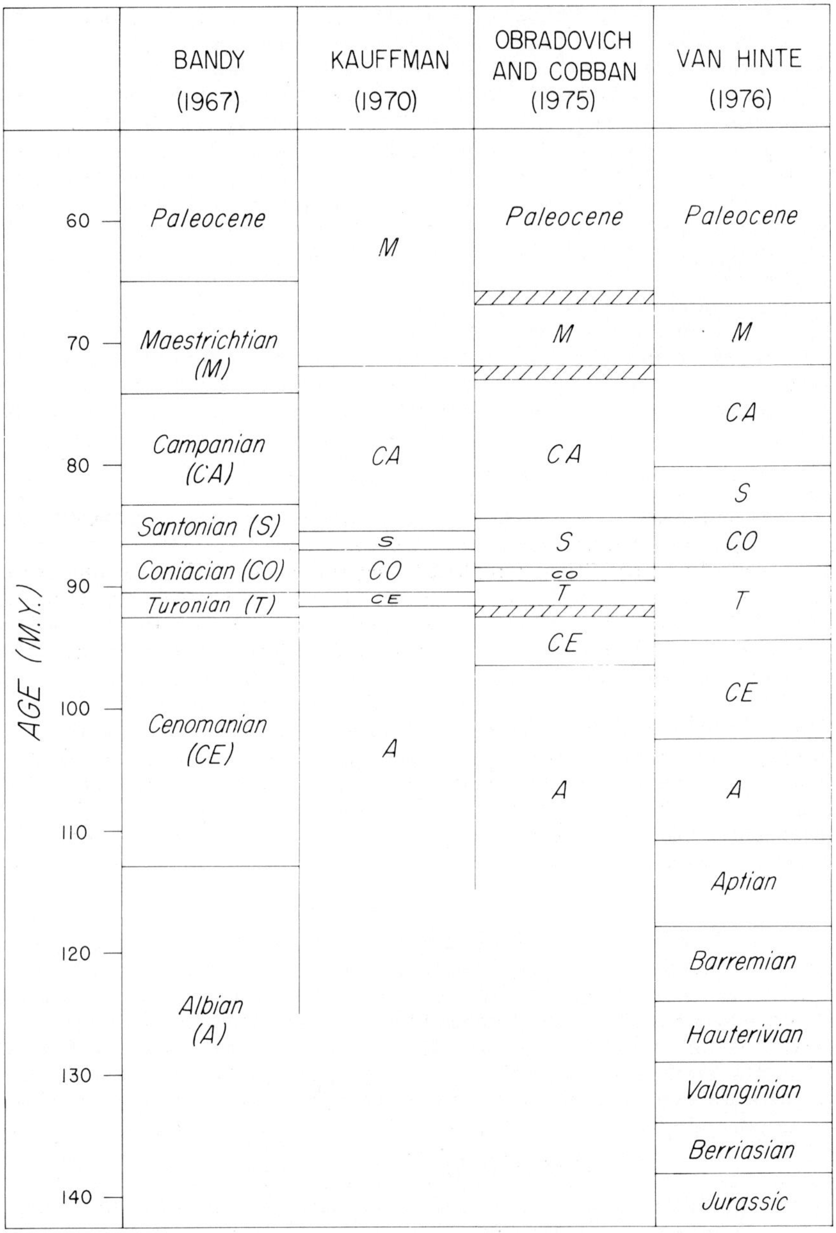

FIG. 1–Comparison of time scales for Late Cretaceous of North America proposed by Bandy (1967), Kauffman (1970), and Obradovich and Cobban (1975). A time scale for the entire Cretaceous proposed by Van Hinte (1976) is also shown. Diagonal hachures in the scale proposed by Obradovich and Cobban represent time interval in which they think a specific boundary may fall.

Cenomanian-Turonian Boundary

There seems general agreement that the boundary between the Cenomanian and Turonian Stages should be placed between the *Sciponoceras gracile* and *Inoceramus labiatus* Zones according to Cobban's zonation. A bentonite from the *I. labiatus* Zone, the oldest zone in the Turonian, yielded an age of 91.1 m.y.; biotites from bentonites from the late Cenomanian *Dunvegenoceras pondi* Zone and the middle Cenomanian *Acanthoceras amphibolum* Zone gave ages of 93.5 m.y. and 94.4 m.y., respectively (Obradovich and Cobban, 1975). M. A. Lanphere and I. L. Tailleur (unpublished data) measured ages on five bentonites from the *I. labiatus* Zone in northern Alaska that average 91.9 m.y. These data indicate that the Cenomanian-Turonian boundary is in the range 91 to 92 m.y.

Albian-Cenomanian Boundary

There has been some controversy about the age assignment of the ammonite *Neogastroplites.* Most workers have assigned *Neogastroplites* a late Albian age, although Reeside and Cobban (1960) noted that the genus cound be Cenomanian. However, Owen (1973), in an evaluation of Albian ammonite faunal provinces, pointed out that in the Kamchatka and Anadyr regions of Siberia (Pergament, 1969), rocks of the *Neogastroplites* spp. Zone are overlain by rocks containing *Stoliczkaia*, a genus of nearly worldwide occurrence in late Albian sediments. Thus, it seems certain that *Neogastroplites* is Albian and not Cenomanian in age. Obradovich and Cobban (1975) determined identical ages of 97.6 m.y. on biotite and sanidine in the *Neogastroplites cornutus* Zone, the fourth oldest of five chronologic species of *Neogastroplites* (Reeside and Cobban, 1960). Folinsbee et al (1963) reported K-Ar ages for three of the *Neogastroplites* Zones. All of the data are consistent with an age of about 96 m.y. for the Albian-Cenomanian boundary.

Early Cretaceous Data Points

Only four reasonably good time-scale points are available for the Early Cretaceous of North America. Two are based on volcanic rocks intercalated with sediments, but for each example only duplicate K-Ar measurements on a single mineral have been determined. The other two points are based on mineral ages for granitic intrusive rocks that provide a minimum age for one locality and a maximum age for the other on Lower Cretaceous fossiliferous strata.

Punta Cabra, Baja California, Mexico

The Alisitos Formation, a eugeosynclinal assemblage of volcanic and sedimentary rocks, is intruded by plutons of diorite, quartz diorite, and granodiorite near Punta Cabra on the Pacific Ocean side of the peninsula of Baja California. Silver et al (1969) noted that the Alisitos Formation is Aptian and possibly early Albian in age although in the vicinity of the intrusive rocks they studied, only Aptian fossils have been found. Zircons from the intrusive rocks yielded U-Pb ages of 115 ± 2 m.y. (Silver et al, 1969) using the decay constants for uranium measured by Kovarik and Adams (1955) and Fleming et al (1952). Complete documentation of the biostratigraphic and isotopic evidence from the Punta Cabra area has not been published. However, an age of 114 ± 2 m.y. for the zircon can be recalculated using the new decay constants for uranium measured by Jaffey et al (1971). Thus, the available biostratigraphic and isotopic data can be interpreted as indicating a minimum age of 114 m.y. for the Aptian or possibly some part of the early Albian.

Hudson Hope, British Columbia, Canada

Folinsbee et al (1961, 1963) reported duplicate ages of 118 and 122 m.y. (recalculated using the refined decay constants for K^{40}) on biotite from a bentonite near Hudson Hope, British Columbia. The bentonite occurs in what Folinsbee et al (1961) referred to as the Harmon Shale. This, apparently, is the Harmon Member, Peace River Formation, Fort St. John Group of the Peace River Plains of Alberta (Alberta Study Group, 1954). At Hudson Hope in the Rocky Mountain foothills, the sequence of shales containing the dated bentonite is the Hulcross Member, Commotion Formation, Fort St. John Group (Stott, 1968). The Hulcross and Harmon Members are lithologically similar, and the Hulcross in outcrop can be correlated with the Harmon in subsurface using electric log data (Stott, 1968).

The *Gastroplites* fauna occurs in both the Hulcross Member and the overlying Boulder Creek Member of the Commotion Formation (Stott, 1968). The Zone of *Gastroplites* is considered by Jeletzky (1968) to be late middle Albian in age. The Gates Member of the Commotion Formation, which underlies the Hulcross Member, contains the *Beudanticeras affine* fauna of early Albian age (Stott, 1975). Thus, the Hulcross Member definitely seems to be middle Albian in age.

Potassium-argon ages were measured on only a single sample of biotite from the Hulcross Member, and a more detailed study of this unit is desirable. In addition, the biotite from the Hulcross bentonite is highly leached as evidenced by its low potassium content of approximately 2.5%. Obradovich and Cobban (1975) compiled data showing the variation of age with potassium content for altered biotite. They concluded that weathered or altered biotites containing less than about 4% potassium may give ages older or younger than the true age though these biotites may also yield the true age. This potential problem with low-potassium biotites is an additional reason for more work on the Hudson Hope locality.

Shasta Bally Batholith, California

The Shasta Bally batholith is located west of Redding, California, along the southern boundary of the Klamath Mountains province of northern California and southern Oregon. A thick section of Lower Cretaceous clastic rocks lies unconformably on the south margin of the batholith. These rocks are part of a transgressive sequence that is of Valanginian age to the west and of Turonian age to the east (Jones and Irwin, 1971). The basal strata of this sequence are called the Rector Conglomerate Member of the Budden Canyon Formation by Murphy et al (1964). Fossils are not known from these strata lying directly on granitic rocks of Shasta Bally batholith, but from equivalent beds 15 km to the west near Begum, middle Hauterivian fossils (*Hollisites* cf. *H. dichotomous* Imlay) were found within 24 m of the basal contact. Upper Hauterivian or lower Barremian fossils (zone of *Hertleinites aquila*) occur near Ono in mudstone beds overlying the Rector that in turn unconformably overlies granitic rocks of Shasta Bally batholith (Murphy et al, 1969).

The age of the *Hertleinites aquila* fauna is in some dispute. Although traditionally considered to be of late Hauterivian age (Imlay, 1960), Casey (1964) has argued for an early Barremian age. Murphy (1975, p. 13) showed *H. aquila* and other associated ammonites spanning the Hauterivian-Barremian boundary. Because all known specimens of *Hertleinites aquila* occur above the Rector Conglomerate Member, it seems most probable that the basal beds are at least late Hauterivian in age.

Potassium-argon ages for coexisting biotite and hornblende from two samples of granodiorite and quartz diorite from Shasta Bally batholith were reported by Lanphere et al (1968), and duplicate measurements on the same mineral separates and U-Pb measurements on zircon from another sample have been measured by M. A. Lanphere, D. L. Jones, and L. T. Silver (unpublished data). The coexisting biotite and hornblende from each sample yielded slightly discordant ages, the biotite from each sample being older than the coexisting hornblende. The four ages were combined, weighting each age by the inverse of its variance, to yield a pooled age of 133.6 or approximately 134 m.y. for the two samples. The preferred interpretation of the zircon data yields an age of 136 ± 2 m.y.

The biostratigraphy indicates that isotopic ages on Shasta Bally batholith will yield a maximum or older limit on the age of the Hauterivian stage of the Lower Cretaceous. The K-Ar and U-Pb data establish a maximum age of about 136 m.y. for the Hauterivian.

Central Alaska

In the Hughes quadrangle of central Alaska a sequence of Late Jurassic and Early Cretaceous andesitic volcanic rocks and subordinate sedimentary rocks has been mapped by Patton and Miller (1966). A K-Ar age of 137 ± 5 m.y. on biotite from a lithic tuff in this sequence was reported by Patton and Miller (1966). Nearby and at nearly the same stratigraphic horizon *Buchia* cf. *B. keyserlingi* (Lahausen) of late Valanginian age has been found. A duplicate measurement has been made on the biotite and all of the analytical data are given in Table 1. The mean age is weighted by the inverse of the variance of the individual measurements. These data indicate an age of about 136 m.y. for the late Valanginian.

The same qualifications raised about the bentonite at Hudson Hope apply to this Alaskan tuff. That is, the Alaskan point is based on only a single mineral age and the biotite has a low potassium content of about 2.2%.

Table 1. K-Ar Ages and Analytical Data for Biotite from a Lithic Tuff (63APa-177), Hughes Quadrangle, Alaska

K_2O (wt. percent)	$^{40}Ar_{rad}$ (mol/gm)	$\dfrac{^{40}Ar_{rad}}{^{40}Ar_{total}}$	Calculated Age* (millions of years)
2.63, 2.64	5.425 x 10^{-10}	0.72	137.6 ± 3.0 ⎫
	5.278	0.14	134.0 ± 4.6 ⎭ 136.5 ± 2.7

*Weighted mean age and calculated standard deviation. $\lambda_\epsilon = 0.572 \cdot 10^{-10}$ yr^{-1}, $\lambda_\epsilon' = 8.78 \cdot 10^{-13}$ yr^{-1}, $\lambda_\beta = 4.963 \cdot 10^{-10}$ yr^{-1}, $^{40}K/K_{total} = 1.167 \cdot 10^{-4}$

Jurassic-Cretaceous Boundary

No (really definitive) new data on the age of the Jurassic-Cretaceous boundary has been reported since Lambert (1971) reviewed the situation. He concluded that there were no reliable ages available for the Late Jurassic or Early Cretaceous. Lambert suggested both that the boundary may fall in the interval 125 to 145 m.y. and that there is no sound reason for taking the mean figure. In previous compilations of the geologic time scale, ages assigned to Jurassic-Cretaceous boundary have included 135 ± 5 m.y. (Holmes, 1959), 135 m.y (Kulp, 1961), and 136 m.y. (Harland et al, 1964). At present, there is no basis to revise the age of about 135 m.y. (138 m.y. recalculated using the refined decay constants) used in previous time scales. The data for the Shasta Bally, California, and central Alaska localities are consistent with previous suggestions, and for purposes of discussion an age of 138 ± 3 m.y. is assigned to the Jurassic-Cretaceous boundary.

Discussion

Our estimate of the best time scale currently available for the Cretaceous of North America is given in Figure 2. This time scale incorporates both the Late Cretaceous proposed by Obradovich and Cobban (1975) and the four Early Cretaceous points discussed previously. The boundaries proposed by Obradovich and Cobban (1975) have been recalculated using the refined decay constants for K^{40}. The three Early Cretaceous points based on K-Ar data and the single point based on U-Pb data have also been recalculated using the refined decay constants. Boundaries in the Early Cretaceous (Fig. 2) are arbitrary estimates based on these four points.

Obradovich and Cobban (1975) noted that the time scale they proposed was a provincial one tied directly to the ammonite zonation of the Western Interior of North America. However, Jeletzky (1968) pointed out that the Western Interior of Canada and the United States as well as northern Eurasia were part of the boreal province throughout the Cretaceous. This resulted in close faunal affinity between North America and the type localities of Cretaceous rocks in Europe. Although the Cretaceous environment was dominated by epicontinental seas, faunal links occurred at various times that permit correlations to be made between the different faunal zona-

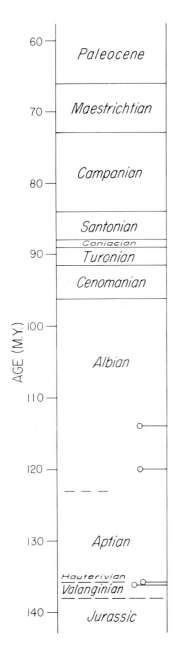

FIG. 2–Proposed time scale for Cretaceous of North America. Late Cretaceous time scale is that of Obradovich and Cobban (1975) recalculated using decay constants for K^{40} based on the refined activity data for K^{40} proposed by Beckinsale and Gale (1969) and the isotopic abundance of K^{40} measured by Garner et al (1975). Circles are the four Early Cretaceous time-scale points described in the text. Boundaries in the Early Cretaceous are arbitrary estimates based on these data points. No data are available relative to the age of the Barremian and the Berriasian and these stages are not shown in the proposed time scale.

tions in Europe and North America. Glauconite (the only datable mineral in the type sections) generally is not considered reliable material for isotopic age measurement; however, the results on glauconite from the type sections in Europe presented by G. S. Odin at the 25th International Geological Congress in Sydney in 1976 were quite encouraging. His results for glauconite from Upper Cretaceous rocks agree well with ages on biotite and sanidine from bentonites in North America, but ages for the glauconites from Early Cretaceous rocks are too young. The inherent problems with dating glauconites suggest that it probably will not be possible to establish directly an adequate time scale for the type-Cretaceous in Europe. Instead, the time scale for the

Cretaceous will have to be based on faunal correlations between the type section and localities elsewhere on which isotopic ages have been measured.

The time scale for the Late Cretaceous of North America seems reasonably well defined. There is controversy about placement of some of the stage boundaries and some of the boundaries are not bracketed by mineral ages as closely as is desirable. There undoubtedly will be some refinements as additional data are obtained, but for most purposes the Late Cretaceous time scale is well established.

The time scale for the Early Cretaceous of North America on the other hand is not well established. None of the ages of the stage boundaries are fixed by mineral ages. J. D. Obradovich and W. A. Cobban currently are studying bentonites in Lower Cretaceous rocks, and their results will provide fine structure for the upper part of the Albian, at least. Volcanic rocks are uncommon in the Lower Cretaceous strata of North America. In addition, the distribution of marine rocks of Neocomian and Aptian age is much more restricted than the distribution of Albian and younger rocks. Because of these factors, it probably never will be possible to work out as detailed a time scale for the Early Cretaceous as has been done for the Late Cretaceous.

References Cited

Alberta Study Group, 1954, Lower Cretaceous of the Peace River region, *in* L. M. Clark, ed., Western Canada sedimentary basin: AAPG Rutherford Mem. Vol., p. 268-278.

Bandy, O. L., 1967, Cretaceous planktonic foraminiferal zonation: Micropaleontology, v. 13, p. 1-31.

Beckinsale, R. D., and N. H. Gale, 1969, A reappraisal of the decay constants and branching ratio of K^{40}: Earth and Planetary Sci. Letters, v. 6, p. 289-294.

Casey, R., 1964, The Cretaceous period, *in* W. B. Harland, A. G. Smith, and B. Wilcock, eds., The Phanerozoic time scale: Geol. Soc. London Quart. Jour., v. 120 supp., p. 193-202.

Cobban, W. A., and J. B. Reeside, Jr., 1952, Correlation of the Cretaceous formations of the western interior of the United States: Geol. Soc. America Bull., v. 63, p. 1011-1044.

Evernden, J. F. et al, 1964, Potassium-argon dates and the Cenozoic mammalian chronology of North America: Am. Jour. Sci., v. 262, p. 145-198.

Fleming, E. H., Jr., A. Ghiorso, and B. B. Cunningham, 1952, The specific alpha-activities and half-lives of U^{234}, U^{235}, and U^{236}: Phys. Rev., v. 88, p. 642-652.

Folinsbee, R. E., H. Baadsgaard, and G. L. Cumming, 1963, Dating of volcanic ash beds (bentonites) by the K-Ar method: Natl. Acad. Sci.-Natl. Research Council Pub. 1075, p. 70-82.

—— —— and J. L. Lipson, 1961, Potassium-argon dates of Upper Cretaceous ash falls, Alberta, Canada, *in* J. L. Kulp, ed., Geochronology of rock systems: New York Acad. Sci. Annals, v. 91, art. 2, p. 352-363.

—— et al, 1966, Late Cretaceous radiometric dates from the Cypress Hills of western Canada: Alberta Soc. Petroleum Geologists, 15th Ann. Conf. Guidebook, pt. I, p. 162-174.

Garner, E. L. et al, 1975, Absolute isotopic abundance ratios and the atomic weight of a reference sample of potassium: Natl. Bur. Standards (USA) Jour. Research, v. 79A, p. 713-725.

Harland, W. B., A. G. Smith, and B. Wilcock, eds., 1964, The Phanerozoic time scale—a symposium: Geol. Soc. London Quart. Jour., v. 120 supp., p. 260-262.

Holmes, A., 1959, A revised geological time-scale: Edinburgh Geol. Soc. Trans., v. 17, p. 183-216.

Imlay, R. W., 1960, Ammonites of Early Cretaceous age (Valanginian and Hauterivian) from the Pacific Coast States: U.S. Geol. Survey Prof. Paper 334-F, p. 167-228.

Jaffey, A. H. et al, 1971, Precision measurements of half-lives and specific activities of U^{235} and U^{238}: Phys. Rev. C, v. 4, p. 1889-1906.

Jeletzky, J. A., 1960, Youngest marine rocks in western interior of North America and the age of the *Triceratops* beds; with remarks on comparable dinosaur-bearing beds outside North America: 21st Int. Geol. Cong. Proc. Report, pt. 5, p. 25-40.

—— 1968, Macrofossil zones of the marine Cretaceous of the western interior of Canada and their correlation with the zones and stages of Europe and the western interior of the United States: Canada Geol. Survey Paper 67-72, 66 p.

Jones, D. L., and W. P. Irwin, 1971, Structural implications of an offset Early Cretaceous shoreline in northern California: Geol. Soc. America Bull., v. 82, p. 815-822.

Kauffman, E. G., 1970, Population systematics, radiometrics, and zonation—A new biostratigraphy: North American Paleont. Conv. Proc., pt. F, p. 612-666.

Kovarik, A. F., and N. I. Adams, Jr., 1955, Redetermination of the disintegration constant of U^{238}: Phys. Rev., v. 98, p. 46.

Kulp, J. L., 1961, Geologic time-scale: Science, v. 133, p. 1105-1114.

Lambert, R. St. J., 1971, The pre-Pleistocene Phanerozoic time-scale – a review: Geol. Soc. London Spec. Pub. 5, p. 9-31.

Lanphere, M. A., W. P. Irwin, and P. E. Hotz, 1968, Isotopic age of the Nevadan orogeny and older plutonic and metamorphic events in the Klamath Mountains, California: Geol. Soc. America Bull., v. 79, p. 1027-1052.

Murphy, M. A., 1975, Paleontology and stratigraphy of the Lower Chickabally Mudstone (Barremian-Aptian) in the Ono quadrangle, northern California: California Univ. Pubs. Geol. Sci., v. 113, 52 p., 13 plates.

——— G. L. Peterson, and P. U. Rodda, 1964, Revision of Cretaceous lithostratigraphic nomenclature, northwest Sacramento Valley, California: AAPG Bull., v. 48, p. 496-502.

——— P. U. Rodda, and D. M. Morton, 1969, Geology of the Ono quadrangle, Shasta and Tehama Counties, California: California Div. Mines and Geol. Bull. 192, 28 p.

Obradovich, J. D., and W. A. Cobban, 1975, A time-scale for the Late Cretaceous of the western interior of North America, *in* W. G. E. Caldwell, ed., Cretaceous system in the western interior of North America: Geol. Assoc. Canada Spec. Paper 13, p. 31-54.

Olsson, R. K., 1964, Late Cretaceous planktonic Foraminifera from New Jersey and Delaware: Micropaleontology, v. 10, p. 157-188.

Owen, H. G., 1973, Ammonite faunal provinces in the Middle and Upper Albian and their paleogeographical significance, *in* R. Casey and P. F. Rawson, eds., The Boreal Lower Cretaceous: Geol. Jour. Spec. Issue no. 5, p. 145-154.

Patton, W. W., Jr., and T. P. Miller, 1966, Regional geologic map of the Hughes quadrangle, Alaska: U.S. Geol. Survey Misc. Geol. Inv. Map I-459.

Pergament, M. A., 1969, Zonal Cretaceous differentiation in north-east Asia correlated with the American and European scales: Akad. Nauk SSSR Izv. Ser. Geol. no. 4, p. 106-119.

Pessagno, E. A., Jr., 1967, Upper Cretaceous planktonic Foraminifera from the western Gulf Coast Plain: Palaeontogr. Amer., v. 5, p. 245-445.

——— 1969, Upper Cretaceous stratigraphy of the western Gulf Coast area of Mexico, Texas, and Arkansas: Geol. Soc. America Mem. 111, 139 p.

Reeside, J. B., Jr., and W. A. Cobban, 1960, Studies of the Mowry Shale (Cretaceous) and contemporary formations in the United States and Canada: U.S. Geol. Survey Prof. Paper 355, 126 p.

Shafiqullah, M. et al, 1966, Geochronology of Cretaceous-Tertiary boundary, Alberta, Canada: 22nd Int. Geol. Cong. Proc. Report, pt. 3, p. 1-20.

Silver, L. T., C. R. Allen, and F. G. Stehli, 1969, Geological and geochronological observations on a portion of the Peninsular Range batholith in northwestern Baja California, Mexico (abs.): Geol. Soc. America Spec. Paper 121, p. 279-280.

Stott, D. F., 1968, Lower Cretaceous Bullhead and Fort St. John Groups between Smoky and Peace Rivers, Rocky Mountain foothills, Alberta and British Columbia: Canada Geol. Survey Bull. 152, 279 p.

——— 1975, The Cretaceous System in northeastern British Columbia, *in* W. G. E. Caldwell, ed., Cretaceous System in the western interior of North America: Geol. Assoc. Canada Spec. Paper No. 13, p. 441-467.

Van Hinte, J. E., 1976, A Cretaceous time scale: AAPG Bull., v. 60, p. 498-516.

Van Hinte's article has been
reprinted from the April 1976
American Association of Petroleum Geologists Bulletin
Volume 60, Number 4

A Cretaceous Time Scale[1]

J. E. VAN HINTE[2]
Houston, Texas 77001

Abstract An emendation of published Cretaceous time scales is proposed on the basis of presently available age data. The standard geochronologic subdivision is combined with a linear numeric scale, biostratigraphic framework, and geomagnetic-reversal time scale. The biostratigraphic framework includes standard Tethyan and boreal pelagic macrofossil zonations and planktonic as well as benthic microfossil zones and biohorizons. Basement ages at some Deep Sea Drilling Project sites are evaluated and related to the geomagnetic-reversal scale.

INTRODUCTION

Integration of marine and terrestrial geophysical, paleontologic, paleomagnetic, and other historical geologic data and application of new quantitative techniques important in stratigraphic exploration necessitates an increasingly refined numeric geologic time scale. This paper is an attempt to combine a Cretaceous numeric time scale with biostratigraphic schemes and the geomagnetic-reversal scale.

In 1964 the Geological Society of London published the Phanerozoic time scale (PTS), which it emended in 1971 (Harland and Francis, 1971). Other scales have been published since 1964 (see Fig. 1), but the original PTS scale still is the most widely used for the Cretaceous. Casey (1964, p. 199), who covered the Cretaceous for the PTS, clearly stated that the scale was "set up purely as a basis for discussion in the absence of a more positive scale of calibration." In other words, the scale was considered to be "no more than a simple working hypothesis" (Internat. Union Geol. Sci., 1967, p. 378). The PTS 1964 scale assumed that all Cretaceous stages represented equal time intervals (6 Ma each; 1 Ma = 1 megayear = 1 million years).

Using the PTS during the past few years as a working model, the writer somewhat changed it (and will do so further) either when new relevant data became available or for pragmatic reasons when inconsistencies in the results of its application suggested such change. An early version of this scale was published in 1972 with a geomagnetic-reversal scale and a planktonic foraminiferal scheme (van Hinte, 1972). A later unpublished version was used successfully by Vail et al (1974), by Berggren et al (1975), and in unpublished reports. A wider distribution of its present emended

form (Fig. 2) seems useful in anticipation of establishment of 'a more authoritative scale. For further general considerations, see van Hinte (1976).

RADIOMETRIC DATA

Published Cretaceous radiometric age determinations have been compiled by the Geological Society of London (1964; Harland and Francis, 1971) in a numbered-item list and generally are referred to as "PTS items." To obtain an overview of the published information, the writer plotted the radiometric age of selected PTS items, and of data published after 1971, against a relative geologic time scale (Figs. 3, 4). In these figures, the numeric dates are presented without their analytic error (about 2 to 5 percent), but their possible relative-age span is given by a vertical line.

The relative-age span was determined from the published information in accordance with the biostratigraphic scheme of Figure 2. This scheme enables one to plot directly most of the later European dates, and very few adjustments had to be made in the stratigraphic ages of the PTS items. For the North American data, the writer followed Obradovich and Cobban (1975) by drawing the boundary of the Campanian/Maastrichtian[3] at the base of the *Baculites reesidei* Zone, the Santonian/Campanian boundary between the *Desmoscaphites bassleri* Zone and the *Scaphites*

[1]Manuscript received, August 29, 1975; accepted, October 10, 1975.

[2]Exxon Production Research Co.
I thank W. A. Berggren (Woods Hole) for continuously stimulating me to improve and expand on earlier published and unpublished versions of "a Cretaceous time scale," and am grateful to Exxon Production Research Company's management for giving the opportunity to complete this work. R. M. Jeffords and D. O. Smith (both of Houston, Texas) critically read the manuscript, and their corrections and suggestions are highly appreciated. This compilation would have been impossible without the support of EPRCo.'s library staff.

[3]The author's spelling of Maastrichtian is contrary to the spelling used by the *Bulletin* (Maestrichtian) but has been retained at the author's request. His request to use "planktic" in lieu of "planktonic" as the adjectival derivative of "plankton" was not honored.

Article Identification Number: 0149-1377/78/SG06-0021/$03.00/0

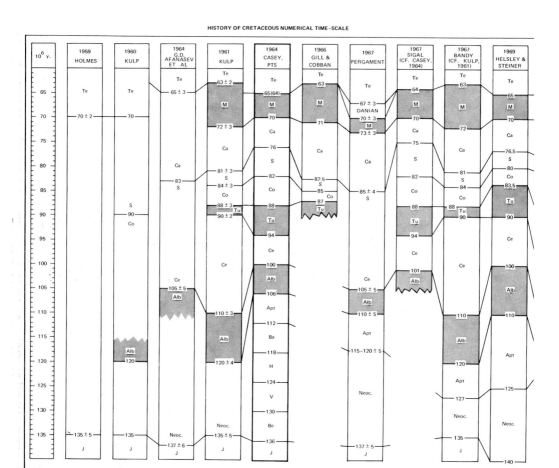

FIG. 1—Different numeric time scales proposed since 1959 for the Cretaceous. No attempt has been made to convert numbers of early and Russian scales to 1964 PTS standards. Maastrichtian, Turonian, and Albian ages are shaded for easy comparison. Gill and Cobban (1966), Pergament (1967), Kauffman (1970), and Obradovich and Cobban (in press) presented their scale with a molluscan biostratigraphic scheme; Bandy (1967), Sigal (1967), and van Hinte (1972) scales were accompanied by planktonic foraminiferal zonation.

Alb, Albian; *Apt*, Aptian; *Ba*, Barremian; *Be*, Berriasian; *Ca*, Campanian; *Ce*, Cenomanian; *Co*, Coniacian; *H*, Hauterivian; *J*, Jurassic; *M*, Maastrichtian; *Neoc*, Neocomian; *S*, Santonian; *Te*, Tertiary; *Tu*, Turonian; *V*, Valanginian.

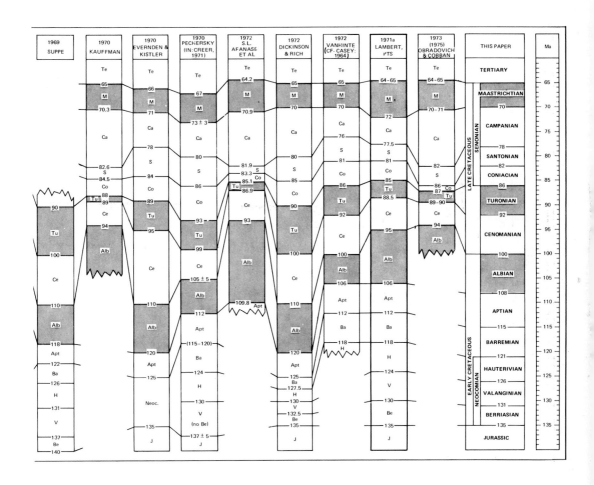

hippocrepis Zone, the Coniacian/Santonian boundary below the *Inoceramus undulataplicatus* Zone, the Cenomanian/Turonian boundary below the *Inoceramus labiatus* Zone, and the Albian/Cenomanian boundary above the *Neogastroplites* Zones (see also Owen, 1973). The Turonian/Coniacian boundary, however, is drawn above the *Inoceramus deformis* Zone (cf. Seitz, 1956; Ernst, 1970).

Owens and Sohl's (1973) glauconite dates from the New Jersey Monmouth Group were plotted as Maastrichtian and those from the Matawan Group as Campanian (Olsson, 1964, found *Globotruncana gansseri* and *G. calcarata* in these units, respectively), and the radiometric ages were averaged per formation.

The PTS item 365 Bearpaw Formation data were grouped as in Figure 3 of Folinsbee et al (1965), whose ages also were followed (365 a-h); the Kneehills Tuff sanidine dates were averaged per locality (365 i, j, l), the biotite date being kept separate (365 k). Caldwell (1968) and North and Caldwell (1970) reviewed the relative-age assignment of the Bearpaw Formation.

Lambert (1971a) suggested rounding of pre-Tertiary major boundary ages to 5 Ma. The present scale follows his −65 and −135? Ma for the ages of the end and the beginning of the Cretaceous. With respect to the Cretaceous-Jurassic boundary, Lambert stated somewhat pessimistically: "At the very best the boundary may fall in the range 125 to 145 m.y. and there is no sound reason for taking the mean figure as the most likely." The radiometric age of the end of the Cretaceous is known more accurately and in all likelihood could not be as far from −65 Ma as the round figures −60 or −70 Ma; the correct age, though, very well could be −64 rather than the −65 Ma generally used (Berggren, 1972) and here retained.

More controversial is the third starting point of our scale: −100 Ma for the Lower/Upper Cretaceous (= Albian/Cenomanian) boundary. Age assignments for this boundary in literature range from −93 to −110 Ma (Fig. 1). Bentonite (Obradovich and Cobban, 1975) and glauconite (Juignet et al, 1975) dates obtained recently suggest that the Albian indeed may have ended 5 or 6 Ma later, that is, at −95 or −94 Ma. Other dates (PTS items 202, 226, 319, 336; van Hinte et al, 1975), however, indicate that the early Cenomanian may be considerably older than −95 Ma, and we decided not to change Casey's 1964 estimate of −100 Ma (cf. Rubinshteyn, 1967, p. 113-121).

With the three main points of the scale fixed at conservative figures, somewhat more flexibility seems justified in diverging from the PTS "equal stage" principle with respect to other stage boundaries. Radiometric dates (Figs. 3, 4) strongly indicate that the Campanian/Maastrichtian boundary should be drawn at −69 or −70 Ma. The Santonian/Campanian and Coniacian/Santonian boundaries rather clearly are at −78 and −82 Ma. The Turonian/Coniacian boundary can be drawn at −86 Ma between the oldest Coniacian glauconite date and the youngest Turonian bentonite date. Acceptance of these boundaries gives the Coniacian and Santonian equal duration, both being shorter than other Late Cretaceous ages, and has the advantage that equal duration can be assigned to Coniacian, Santonian, and Campanian ammonite zones. Most authors place the Cenomanian/Turonian boundary at −90 Ma or slightly younger, but some have it at −94 or −95 Ma and few at −99 or −100 Ma (Fig. 1). In the absence of accurate data, a boundary age of −92 Ma seems reasonable—especially because this gives the same duration to ammonite zones in both the Turonian and the Senonian.

Radiometric dates for the Lower Cretaceous are scarce, and nearly all are based on glauconites which become less reliable with increasing age of the section. Two Albian dates (PTS items 203, 217) led some authors (Bandy, 1967; Suppe, 1969; Dickinson and Rich, 1972; Baldwin et al, 1974) to accept an "old Albian" (see Fig. 1), whereas Casey (1964) and others placed more weight on a younger (−99 Ma) date on item 217 and took into account that the Albian age for item 203 is a youngest stratigraphic age. The Albian and Aptian beginnings are drawn in this work near the oldest glauconite ages at −108 and −115 Ma, respectively.

Having no dates available to help determine the Hauterivian/Barremian boundary, with the Hauterivian likely being younger than 127 Ma (item 75), the Valanginian being older than 124 Ma (Suppe, 1973, 105-124 Ma old ages for metamorphic Hellhole-Williams rocks with Tithonian-Valanginian *Buchia*), and the Berriasian possibly older than 131 Ma (items 177, 215, 328), the writer drew the boundaries at −121, −126, and −131 Ma. Thus, the present model assumes equal duration for the Valanginian and Hauterivian, both being shorter than the Barremian and longer than the Berriasian. However, this assumption is based largely on intuition and on reasons of practical utility.

The feasibility of the proposed model may be judged from Figure 5, which summarizes the available chronometric information and illustrates its position with respect to the preferred scale and an "equal stage" scale.

BIOSTRATIGRAPHIC SCHEMES

The French colloquia on the Upper and Lower Cretaceous set the stage for relating macrofossil biostratigraphy to the type sections of stages. The conclusions of the colloquia (Dalbiez, 1959; Anon., 1963) form the basis for the Cretaceous ammonite zonation of Figure 2. Subsequent work has refined this "standard" and established its correlation with the boreal faunal succession (Fig. 2, authors listed at base of column).

The planktonic foraminiferal scheme is after van Hinte (1972) and is, with the alterations (Figs. 6-8) briefly discussed later, firmly related now to the stratotypes of stages through first-order and second-order (macrofossils, benthic foraminifers) correlation. The zones have been given letter-number designations for easy reference.

Recent work on stratotypes and on correlations between benthic and planktonic foraminiferal zones, between foraminiferal and macrofossil zones, and between boreal and Tethyan biostratigraphic subdivisions provided a better basis for relating the planktonic foraminiferal scheme to the Santonian and Coniacian stratotypes and stage boundaries (i.e., Ernst, 1970; Scheibnerova, 1972; Séronie-Vivien, 1972; micropaleontologic literature listed under Figure 6; see also discussion in Barr, 1972, p. 6).

The top of the *Rotalipora cushmani* Zone now is drawn to coincide with the Cenomanian/Turonian boundary for reasons illustrated in Figure 7, a conclusion that had been reached earlier by other authors for the same or for different reasons (cf. Bolli, 1966; Pessagno, 1969, p. 22).

Wider geographic recognition of Moullade's (1966) Albian-Aptian foraminiferal ranges and biostratigraphy and repeated confirmation of the phylogenetic interpretation suggest that a generalization of his scheme is justified. The derivation of the present zonation for this part of the section is illustrated in Figure 8.

Van Hinte (1972) selected planktonic foraminiferal biohorizons that are recognized widely and

most likely have time significance; in Figure 2 some Senonian biohorizons were added after Herm (1965), Premoli Silva and Bolli (1973), and Smith and Pessagno (1973).

The present biostratigraphic scheme includes a neritic and littoral benthic foraminiferal phylozonation for accurate dating of Senonian shallower water sediments in which pelagic markers are lacking (Fig. 6). In Figure 2, other widely distributed benthic markers of proved time significance were added for the same purpose.

Allemann et al (1971) and Thierstein (1973) related their respective calpionellid and Early Cretaceous nannofossil zonations to macrofossil zones and/or to foraminiferal zones and to type sections of stages. The stratigraphic ages of their boundaries and biohorizons, therefore, are reliable. Those of the Late Cretaceous nannofossil scheme and of the radiolarian zonation are less well established and, therefore, had to be drawn as broken lines in Figure 2.

GEOMAGNETIC-REVERSAL SCALE

Since 1968, various geomagnetic-reversal time scales have been published for the Cretaceous (Fig. 9), some based on measurements made on land, others on marine-anomaly data. The scale presented here combines several of the earlier proposals. The subdivision of the Beringov interval is after the Heirztler et al (1968) scale as revised by Sclater et al (1974). The Serra Geral interval is recalibrated after Larson and Pitman (1972) using − 145 and − 125 Ma as calibration points instead of − 155 and − 120 Ma (see van Hinte, 1976). Green and Brecher (1974), Green et al (1974), and Jarrard (1974) recorded three brief reversed zones in upper Albian sediments (Proto Decima, 1974) at DSDP Site 263 which are included in the conclusion column as "Site 263 Mixed" Zone. Keating et al (1975) mentioned that the Akoh Formation recently has been dated radiometrically as 92-104 Ma old, which suggests that the Akoh reversals are within the Site 263 Mixed Zone. But because no further reference to

←

FIG. 2—Numeric Cretaceous time scale proposed in this paper. Younger part of geomagnetic-reversal scale is after Sclater et al (1974), older part is after Larson and Pitman (1972, recalibrated see text), and middle zones are after McElhinny and Burek (1971), Green and Brecher (1974), and Jarrard (1974). Geomagnetic interval and zonal names are after McElhinny and Burek (1971) and Pechersky and Khramov (1973). Foraminiferal scheme is newly compiled (see text); other biostratigraphic data are after literature listed at bottom of each column in figure. Foraminiferal literature listed refers to benthic biohorizons. Note that first and last occurrences of *Globotruncana* and *Bolivinoides* species do not refer to typological species (see text). Most planktonic foraminiferal zones are phylozones, and they as well as selected biohorizons can be expected to have good geochronologic reliability.

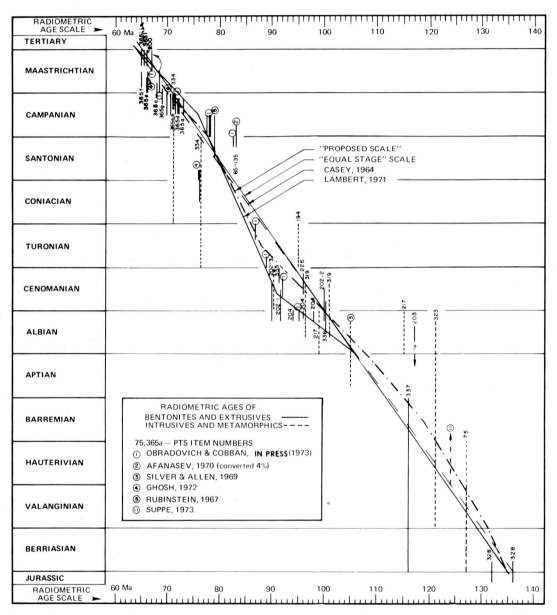

FIG. 3—Plot of radiometric age against relative age for Cretaceous dates on minerals other than glauconite. Data after literature listed on figure. PTS items are those selected by Casey (1964) plus items 334, 335, and 336 and minus items 198 and 199.

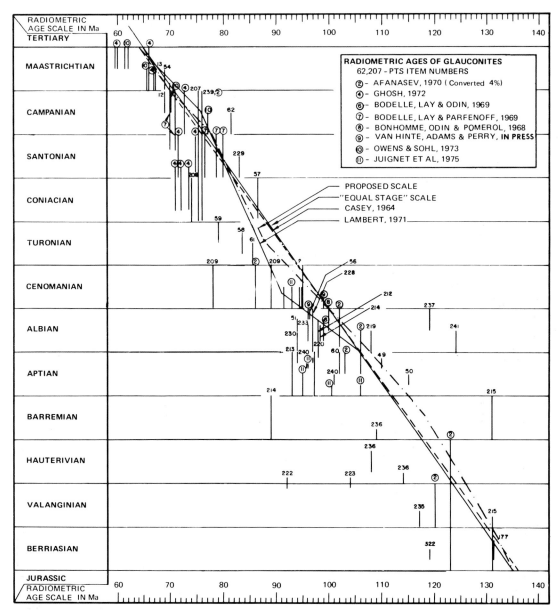

FIG. 4—Plot of radiometric age against relative age for Cretaceous glauconites. Data after literature listed on figure. PTS items are those selected by Casey (1964) plus items 51, 59, 207, 213, 228, 233, and 239.

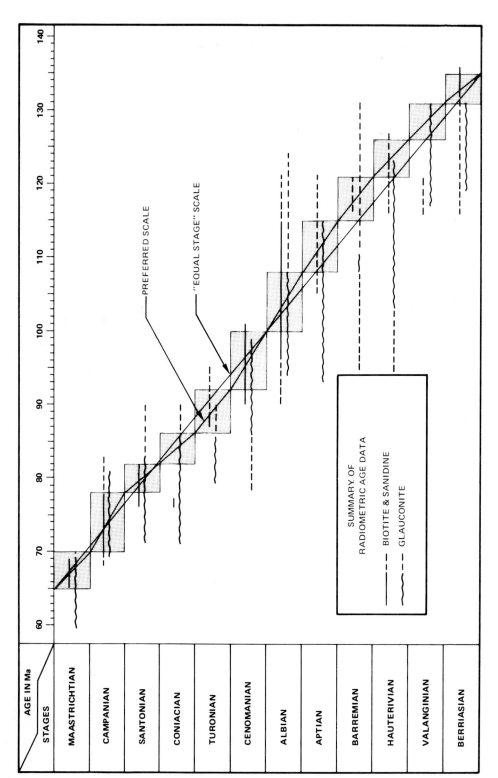

FIG. 5—Summary of Figures 3 and 4. Figure illustrates relation of published radiometric data with preferred scale (shaded) and compares line representing equal duration of stages with line representing preferred scale.

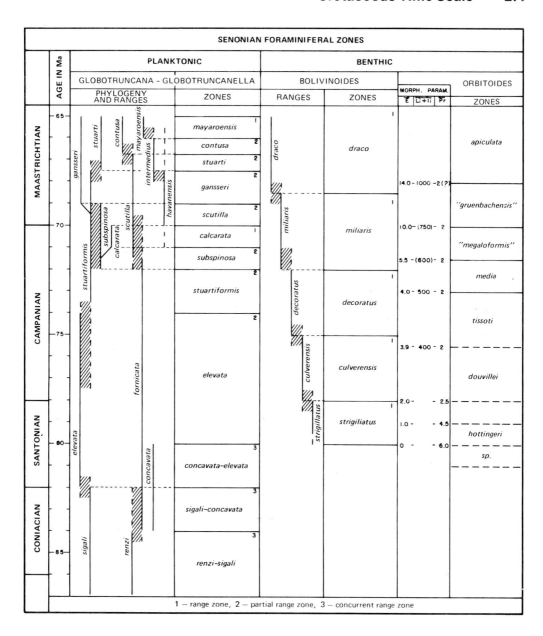

FIG. 6—Senonian foraminiferal zonal scheme. Multiple and single phylozones mainly after Papp (1956), Hilter-mann (1963), van Hinte (1963-1969), and Barr (1962-1972). Correlation with stratotypes and macrofossil zones after just mentioned authors, after older literature referred to in van Hinte (1972), and after Reiss (1962), Donze et al (1970), Kuhri (1970), Salaj and Bajanik (1972), Séronie-Vivien (1972), van Gorsel (1973a, b), and Colin (1974). Vertical lines represent ranges of typologic "species," interval of presence of two successive and of intermediate morphotypes is shaded. Nomenclature of parts of *Orbitoides* lineage is uncertain because quantitative descriptions are not available for all taxa. "Morph. param." stands for morphologic parameters; *E* is number of epiauxiliary chambers, *Li + li* is measure for size of embryo, *Pr* is budding number of progressive chamber; for definitions see van Hinte (1966a, b).

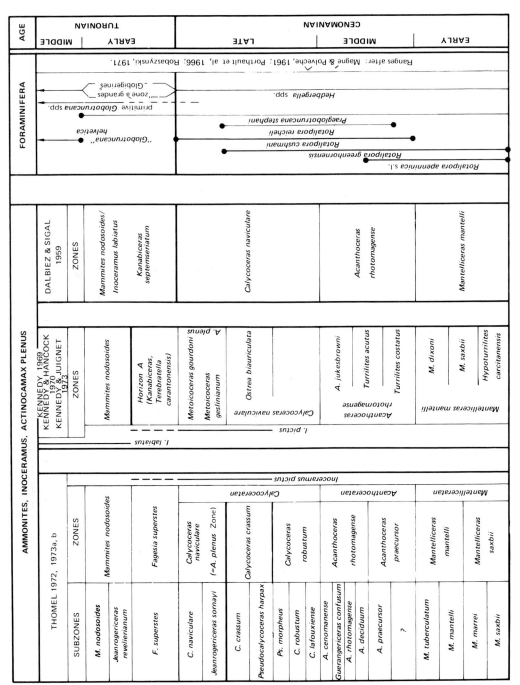

Fig. 7—Cenomanian-Turonian macrofossil zonations, some planktonic foraminiferal ranges, and different opinions on position of Cenomanian/Turonian boundary (heavy line). This paper follows resolutions of Colloque sur le Cretace Superieur Francais (Dalbiez, 1959) which are in accordance with Muller and Schenck (1943) in including *Actinocamax plenus* Zone in Cenomanian. No linear vertical scale.

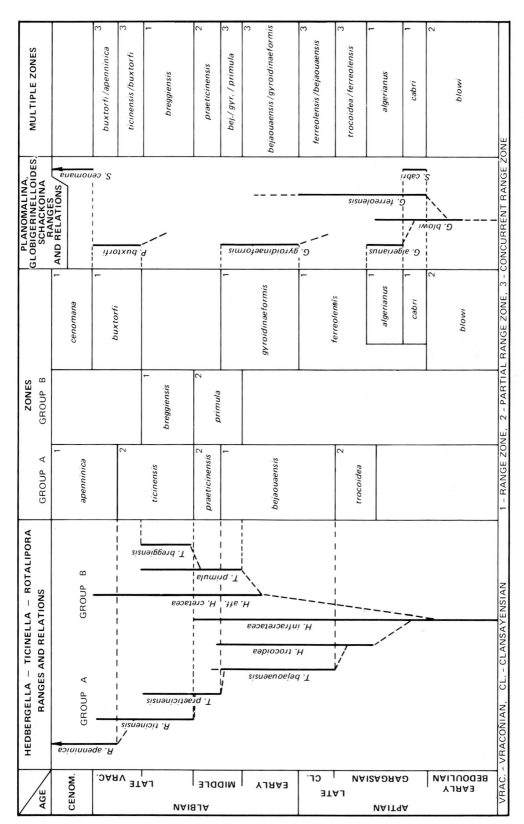

FIG. 8—Planktonic foraminiferal zonation of Aptian-Albian. Interpretation of phylogenetic relations largely after Risch (1971), ranges after Sigal and Moullade (*in* Anon., 1963), Sigal (1966), Hermes (1969), Moullade (1966), Cotillon (1971), Risch (1971), Kuhri (1971, 1972), and Salaj and Bajanik (1972). No linear vertical scale.

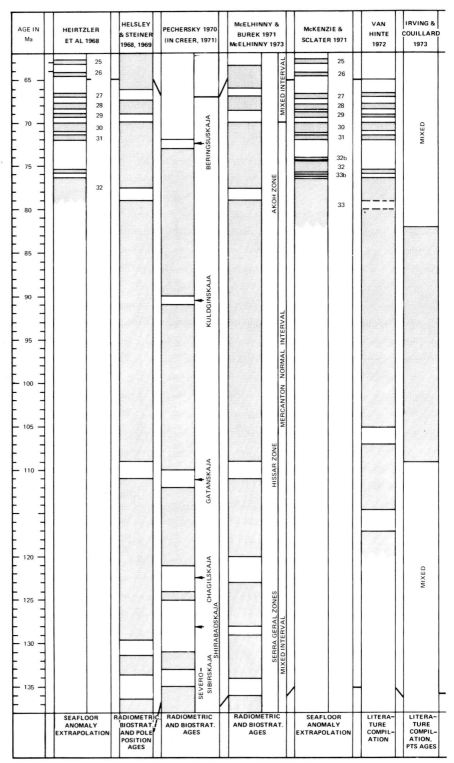

FIG. 9—Cretaceous geomagnetic-reversal scales proposed by different authors. Linear vertical scale for each column follows original authors. No attempt was made to convert radiometric ages to PTS standards or to reinterpret biostratigraphic ages except for *This Paper* column. *Site 263 Mixed Zone* refers to mixed polarity found by Green and Brecher (1974) and Jarrard (1974) in cores from DSDP Site 263. See also caption of Figure 2. Cretaceous basement ages at DSDP sites are discussed in text.

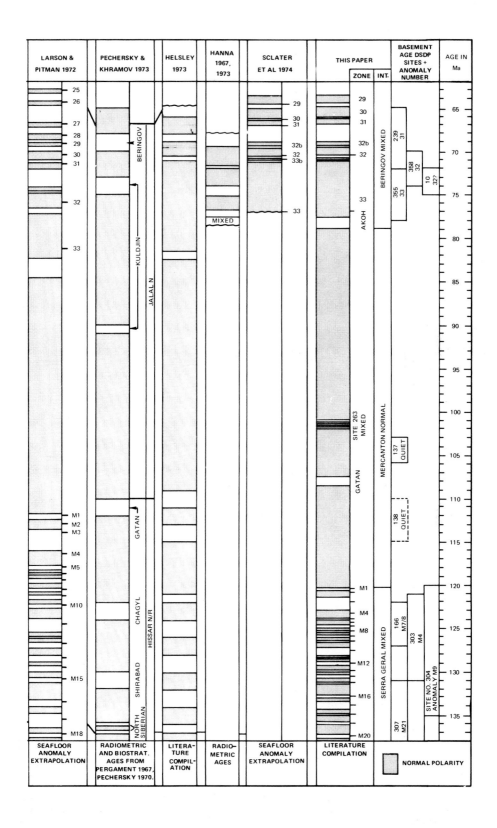

this new radiometric date is given, we provisionally leave the Akoh Reversed Zone at -79 to -77.5 Ma, following Helsley and Steiner (1969) and McElhinny and Burek (1971). The Gatan Reversed Zone is placed near -108 Ma, as it is assigned to the Albian-Aptian boundary by Pechersky and Khramov (1973). Jarrard (1974) suggested that he may have recorded this zone in sediments of DSDP Site 260.

Watkins and Cambray (1971) studied Campanian-Santonian dikes from Jamaica and found that "a normal geomagnetic field dominates (but not completely so)" and that their reversely magnetized dikes might be Maastrichtian or Santonian or "we may have detected the first Campanian reversed polarities." The second alternative would confirm the extrapolation of marine data by Heirtzler et al (1968). Irving and Couillard (1973) updated Helsley and Steiner's (1969) and McElhinny and Burek's (1971) compilation of Cretaceous radiometrically or paleontologically dated geomagnetic field data. They confirmed the long Cretaceous normal interval by consistently finding normal polarity for the PTS interval -109 to -82 Ma (middle Aptian-end of Coniacian), the PTS intervals -82 to -65 Ma and -160 to -109 Ma being of mixed polarity. Later paleomagnetic work by Hanna (1973) on the radiometrically dated Boulder batholith and by Keating et al (1975) on biostratigraphically dated sediments confirmed the mixed-polarity nature for post-Santonian time. A different result was obtained by Shive and Frerichs (1974), who found no reversed polarity in their study of paleomagnetism in 350 samples of the micropaleontologically dated Coniacian-early Maastrichtian Niobrara Formation in Wyoming. This is no surprise for the lower Coniacian/early Santonian part of the formation, *Globotruncana concavata* (= *G. c. primitiva*?), considered early Santonian by Shive and Frerichs (1974) and Frerichs and Adams (1973), but it poses a problem for the upper part unless the late Campanian/early Maastrichtian age assignment is not correct and the entire Niobrara Formation is Coniacian/Santonian as had been suggested by Cobban and Reeside (1962). The latter may be the case, for microfaunal evidence for a younger age is very weak. Furthermore, the Niobrara Formation is older than the Elkhorn Mountains volcanic rocks and the Boulder batholith (Klepper et al, 1957), both of which represent times of mixed polarity (Hanna, 1967, 1973, respectively); the younger Boulder batholith has been dated radiometrically as 78-68 Ma old (Hanna, 1973). This would make the base of the Maastrichtian older than 78 Ma, an unlikely high age.

OCEANIC BASEMENT AGES

Holes of the Deep Sea Drilling Project that reach basement of known magnetic-anomaly number offer an opportunity to age-calibrate the marine geomagnetic-reversal scale, and to provide a check on the consistency of the stratigraphic-radiometric time scale being used. In the following discussion, stratigraphic basement ages are estimated and expressed in numbers of the proposed time scale. These estimates are plotted in Figure 9 for ready comparison with the magnetic-reversal time scale. One age (Site 166) was used to calibrate the magnetic-reversal scale (Larson and Pitman, 1972; van Hinte, 1976); the others function as checks on the age assignment for the calibration point and its consequences for the geomagnetic-reversal scale. The figure demonstrates that none of the basement ages found by the drill contradict the age as it would have been predicted from the magnetic-anomaly position of the sites. This suggests that the present scale is a useful work model.

The following DSDP holes bottomed in Cretaceous oceanic basement of known linear magnetic-anomaly position: 10, 137, 138, 166, 239, 303, 304, 355, and 358 (nearly so). The estimates for the age of basement at Sites 166, 303, 304, and 307 (Late Jurassic) were discussed in van Hinte (1976).

DSDP Site 10 was drilled near anomaly 32. Campanian sediment (*Globotruncana stuartiformis* Zone) overlies basaltic basement at -456 m (Peterson et al, 1970; Cita and Gartner, 1971; Smali, 1973). The *Globotruncana calcarata* Zone is well defined in the section above. This zone is 20 m thick (-400 to -420 m); assuming that its duration is 1 Ma (-71 to -70 Ma), the average rate of sediment accumulation is 2 cm/1,000 years. Further, assuming that this rate is constant, it follows that the age at the base of sediment is $71 + 1.8 = 72.8$ Ma old. If the *G. calcarata* Zone represents not 1 but 2 Ma, the average rate of sediment accumulation then would be 1 cm/1,000 years and the lowest sediment would be 75.6 Ma old. An upward extrapolation of the two rates would predict the top of the Cretaceous at -300 and -350 m, respectively. The Cretaceous top was not cored, but recovery of the *Globotruncanella mayaroensis* Zone in Core 10 (-291 to -299 m) indicates that the average rate of sediment accumulation was even slightly higher than 2 cm/1,000 years and supports the assumption of a 1-Ma duration for the *G. calcarata* Zone. Even incorporating the possibilities that the rate of sediment accumulation is not constant and that basement is somewhat older than the oldest sediment, it seems reasonable to estimate basement to be between 72 and 75 Ma old.

Basement age at Site 137 can be estimated from planktonic foraminiferal biohorizons given by Beckmann (1972): *Planomalina buxtorfi* (−100 Ma) in Core 14 at −341.7 m and *Ticinella (Biticinella) breggiensis* (−102 Ma) in Core 16 at −381.7 m. An extrapolation of the average rate of sediment accumulation derived from these data (2 cm/1,000 years) dates the base of sediment (−397 m) as 103 Ma old. A fair estimate for the age of basement seems to be −103 to −106 Ma.

Only a broad paleontologic age has been determined for the lowest sediment at Site 138. Radiolaria of Core 6 (−425 to −431 m) compare with those of Core 7 at Site 137, which is at the oldest late Cenomanian and probably closer to 90 than to 95 Ma old. Petrushevskaya and Kozlova (1972) dated the fauna as Cenomanian but mentioned that it could be Albian as well. One estimate for the age of the lowest sediment could be −92 to −105 Ma. But, is this close to basement age? The fact that at Site 138 sediment above the basalt was deposited below calcite-compensation depth might suggest that at time of deposition the location was too deep for the basalt to be "normal" basement and that the hole bottomed in a sill (Berger and von Rad, 1972, p. 878; above the carbonate-clay boundary, or CCB, Berger and Winterer, 1974). Sites 137 and 138 were drilled at comparable levels with regard to basement topography, and we may assume that basement at 138 was, as at 137, originally somewhere near 2,700 m depth (Berger and von Rad, 1972, Fig. 57, Site 137). A bold extrapolation of Berger and von Rad's Site 138 curve (1972, Fig. 57) to 2,700 m suggests that basement is in the order of 110-115 Ma old (Berger and von Rad use a different time scale but have, in Fig. 57, Core 6 at −92 Ma with which we would agree).

DSDP Site 239 was drilled on the older side of anomaly 31 (Schlich, 1974) and bottomed in basaltic basement overlain by less than a meter of Cretaceous sediment that in turn was covered by Paleocene. No Cretaceous planktonic Foraminifera were found, but the benthic fauna and the calcareous nannoflora suggest a Maastrichtian-latest Campanian age (not late Campanian exclusive of Maastrichtian as assumed by Schlich, 1974, and by Larson and Pitman, 1975). Although a late Maastrichtian basement age (−65 to −68 Ma) seems most likely, the writer plotted Site 239 as −65 to −72 Ma in Figure 9.

Sites 355 and 358 were drilled on anomalies 33 and 32, respectively, and sediment above basement was dated as early-middle Campanian and as late Campanian (Anon., 1975). Basement ages

for these sites can be plotted as −78 to −72 Ma and −72 to −70 Ma, respectively.

REFERENCES CITED

Afanasyev, G. C., 1970, Certain key data for the Phanerozoic time scale: Eclogae Geol. Helvetiae, v. 63, p. 1-7.
—— et al, 1964, The project of a revised geological time scale in absolute chronology: 22d Internat. Geol. Cong., India, Soviet Contrib. Geol., p. 287-324.
Afanasyev, S. L., et al, 1972, Duration of the ages of the Late Cretaceous: Moscow Soc. Naturalists Bull., Geol. Ser., v. 42, p. 136-137.
Allemann, F., R. Catalano, F. Fares, and J. Remane, 1971, Standard calpionellid zonation (upper Tithonian-Valanginian) of the western Mediterranean province: 2d Internat. Conf. Planktonic Microfossils, Rome 1970, Proc., v. 2, p. 1337-1340.
Anonymous, 1963, Conclusions du colloque de stratigraphie sur le Cretace Inferieur en France, Lyon, Septembre 1963: Soc. Geol. France Compte Rendu, no. 8, p. 292-296.
—— 1975, Leg 39 examines facies changes in South Atlantic: Geotimes, v. 20, no. 3, p. 26-28.
Baldwin, B., P. J. Coney, and W. R. Dickinson, 1974, Dilemma of a Cretaceous time scale and rates of seafloor spreading: Geology, v. 2, p. 267-270.
Bandy, O. L., 1967, Cretaceous planktonic foraminiferal zonation: Micropaleontology, v. 13, p. 1-31.
Barr, F. T., 1962, Upper Cretaceous planktonic Foraminifera from the Isle of Wight, England: Palaeontology, v. 4, p. 552-580.
—— 1966, The foraminiferal genus *Bolivinoides* from the Upper Cretaceous of the British Isles: Palaeontology, v. 9, p. 220-243.
—— 1970, The foraminiferal genus *Bolivinoides* from the Upper Cretaceous of Libya: Jour. Paleontology, v. 44, p. 642-654.
—— 1972, Cretaceous biostratigraphy and planktonic Foraminifera of Libya: Micropaleontology, v. 18, p. 1-46.
Bartenstein, H., 1974, *Lenticulina (Lenticulina) nodosa* (Reuss 1863) and its subspecies—worldwide index Foraminifera in the Lower Cretaceous: Eclogae Geol. Helvetiae, v. 67, p. 539-562.
—— and F. Bettenstaedt, 1962, Marine Unterkreide (Boreal und Tethys), *in* Leitfossilien der Mikropalaeontologie: Berlin, Borntraeger, p. 225-297.
Beckmann, J. P., 1972, The Foraminifera and some associated microfossils of Sites 135 to 144: Deep Sea Drilling Project Initial Repts., v. 14, p. 389-420.
Berger, W. H., and U. von Rad, 1972, Cretaceous and Cenozoic sediments from the Atlantic Ocean: Deep Sea Drilling Project Initial Repts., v. 14, p. 787-954.
—— and E. L. Winterer, 1974, Plate stratigraphy and the fluctuating carbonate line: Internat. Assoc. Sedimentol. Spec. Pub. 1, p. 11-45.
Berggren, W. A., 1972, A Cenozoic time scale—some implications for regional geology and paleobiogeography: Lethaia, v. 5, p. 195-215.
—— D. P. McKenzie, J. G. Sclater, and J. E. van Hinte, 1975, Worldwide correlation of Mesozoic magnetic anomalies and its implications, discussion:

Geol. Soc. America Bull., v. 86, p. 267-269.

Bodelle, J., C. Lay, and G. S. Odin, 1969, Determination d'age par la methode geochronologique "potassium-argon" de glauconies du bassin de Paris: Acad. Sci. Comptes Rendus, ser. D, v. 268, p. 1474-1477.

———— ———— and A. Parfenoff, 1969, Age des glauconies cretacees du sud-est de la France (2, Vallee de l'Esteron, Alpes-Maritimes; resultats preliminaires de la methode potassium-argon: Acad. Sci. Comptes Rendus, ser. D, v. 268, p. 1576-1579.

Bolli, H. M., 1966, Zonation of Cretaceous to Pliocene marine sediments based on planktonic Foraminifera: Asoc. Venezolana Geologia, Mineria y Petroleo Bol. Inf., v. 9, p. 3-32.

Bonhomme, M., G. S. Odin, and C. Pomerol, 1968, Age de formations glauconieuses de l'Albien et de l'Eocene de bassin de Paris: France, Bur. Recherches Geol. et Minieres Mem. 58, p. 339-346.

Bukry, D., 1974, Coccolith stratigraphy, offshore Western Australia, Deep-Sea Drilling Project leg 27: Deep Sea Drilling Project Initial Repts., v. 27, p. 623-630.

Caldwell, W. G. E., 1968, The Late Cretaceous Bearpaw Formation in the South Saskatchewan River Valley: Saskatchewan Research Council Geology Div. Rept. 8, 86 p.

Casey, R., 1961, The stratigraphical palaeontology of the Lower Greensand: Palaeontology, v. 3, p. 487-621.

———— 1964, The Cretaceous Period, in The Phanerozoic time-scale: Geol. Soc. London Quart. Jour., v. 120, supp., p. 193-202.

Cita, M. B., and S. Gartner, Jr., 1971, Deep sea Upper Cretaceous from the western North Atlantic: 2d Planktonic Conf., Rome 1970, Proc., v. 1, p. 287-320.

Cobban, W. A., and J. B. Reeside, 1962, Correlation of the Cretaceous formations of the Western Interior of the United States: Geol. Soc. America Bull., v. 63, p. 1011-1044.

Colin, J-P., 1974, Precisions sur le Campanien de Dordogne (region de Belves-Saint-Cyprien, Dordogne, S-O France): Newsletters Stratigraphy, v. 3, p. 139-151.

Cotillon, P., 1971, Le Cretace Inferieur de l'Arc subalpin de Castellane entre l'Asse et le Var stratigraphie et sedimentologie: Bur. Recherches Geol. et Minieres Mem. 68, 243 p.

Creer, K. M., 1971, Mesozoic palaeomagnetic reversal column: Nature, v. 233, p. 545-546.

Dalbiez, F., 1959, Correlations et resolutions: 84th Cong. Soc. Sav. Paris, Dijon, Sec. Sci., Comptes Rendus Colloque Surle Cretace Superieur Francais, p. 857-867.

Dickinson, W. R., and E. I. Rich, 1972, Petrologic intervals and petrofacies in the Great Valley sequence, Sacramento Valley, California: Geol. Soc. America Bull., v. 83, p. 3007-3024.

Donze, P., B. Porthault, G. Thomel, and O. de Villoutreys, 1970, Le Senonien Inferieur de Puget-Theniers (Alpes-Maritimes) et sa microfaune: Geobios, no. 3, p. 41-103.

Ernst, G., 1970, The stratigraphical value of the echinoids in the boreal Upper Cretaceous: Newsletters Stratigraphy, v. 1, p. 19-34.

Evernden, J. F., and R. W. Kistler, 1970, Chronology of emplacement of Mesozoic batholithic complexes in California and western Nevada: U.S. Geol. Survey Prof. Paper 623, 42 p.

Folinsbee, R. E., H. Baadsgaard, G. L. Cumming, J. Nascimbene, and M. Shafiqullah, 1965, Late Cretaceous radiometric dates from the Cypress Hills of western Canada, in Cypress Hills Plateau, Alberta and Saskatchewan: Alberta Soc. Petroleum Geology, 15th Field Conf. Guidebook, p. 162-174.

Frerichs, W. E., and P. R. Adams, 1973, Correlation of the Hilliard Formation with the Niobrara Formation: Wyoming Geol. Assoc. Guidebook, 25th Field Conf., p. 187-192.

Geological Society of London, 1964, Geological Society Phanerozoic time-scale 1964, in The Phanerozoic time-scale: Geol. Soc. London Quart. Jour., v. 120, supp., p. 260-262.

Ghosh, P. K., 1972, Use of bentonites and glauconites in potassium-40/argon-40 dating in Gulf Coast stratigraphy: PhD thesis, Rice Univ., 136 p.

Gill, J. R., and W. A. Cobban, 1966, The Red Bird section of the Upper Cretaceous Pierre Shale in Wyoming: U.S. Geol. Survey Prof. Paper 393-A, 69 p.

Green, K. E., and A. Brecher, 1974, Preliminary paleomagnetic results for sediments from Site 263, leg 27: Deep Sea Drilling Project Initial Repts., v. 27, p. 405-413.

———— and J. Sclater, 1974, Geomagnetic field reversals in the Cretaceous from DSDP sediments (abs.): EOS (Am. Geophys. Union Trans.), v. 55, p. 236.

Hanna, W. F., 1967, Paleomagnetism of Upper Cretaceous volcanic rocks of southwestern Montana: Jour. Geophys. Research, v. 72, p. 595-610.

———— 1973, Paleomagnetism of the Late Cretaceous Boulder batholith: Am. Jour. Sci., v. 273, p. 778-802.

Harland, W. B., and E. H. Francis, eds., 1971, The Phanerozoic time-scale; a supplement: Geol. Soc. London Spec. Pub. 5, 356 p.

Heirtzler, J. R., G. O. Dickson, E. M. Herron, W. C. Pitman, III, and X. LePichon, 1968, Marine magnetic anomalies, geomagnetic field reversals, and motions of the ocean floor and continents: Jour. Geophys. Research, v. 73, p. 2119-2136.

Helsley, C. E., 1973, Post-Paleozoic magnetic reversals (abs.): Geol. Soc. America Abs. with Programs, v. 5, p. 665 (hand-out figure "Comparison of Mesozoic reversal sequences").

———— and M. B. Steiner, 1968, Evidence for long periods of normal magnetic polarity in the Cretaceous Period (abs.): Geol. Soc. America Abs. with Programs, p. 133.

———— ———— 1969, Evidence for long intervals of normal polarity during the Cretaceous Period: Earth and Planetary Sci. Letters, v. 5, p. 325-332.

Herm, D., 1965, Mikropalaeontologisch-stratigraphische Untersuchungen im Kreideflysch zwischen Deva und Zumaya (Prov. Guipuzcoa, Nordspanien): Deutsch. Geol. Gesell. Zeitschr., v. 115 (1963), p. 277-348.

Hermes, J. J., 1969, Late Albian Foraminifera from the Subbetic of southern Spain: Geologie en Mijnbouw, v. 48, p. 35-66.

Hiltermann, H., 1963, Zur Entwicklung der Benthos-Foraminifere *Bolivinoides, in* Koenigswald et al, eds., Evolutionary trends in Foraminifera: New York, Elsevier Pub. Co., p. 198-223.

——— and W. Koch, 1962, Oberkreide des nördlichen Mitteleuropa, *in* Leitfossilien der Mikropalaeontologie: Berlin, Borntraeger, p. 299-338.

Holmes, A., 1959, A revised geological time-scale: Edinburgh Geol. Soc. Trans., v. 17, p. 183-216.

Irving, E., and R. W. Couillard, 1973, Cretaceous normal polarity interval: Nature, Phys. Sci., v. 244, p. 10-11.

International Union of Geological Sciences (IUGS), 1967, A comparative table of recently published geological time scales for the Phanerozoic time—explanatory notice: Norsk Geol. Tidsskr., v. 47, p. 375-380.

Jarrard, R. D., 1974, Paleomagnetism of some leg 27 sediment cores: Deep Sea Drilling Project Initial Repts., v. 27, p. 415-423.

Juignet, P., J. C. Hunziker, and G. S. Odin, 1975, Datation numerique du passage Albien-Cenomanien en Normandie; etude preliminaire par la methode a l'argon: Acad. Sci. Comptes Rendus, ser. D., v. 280, p. 379-382.

Kauffman, E. G., 1970, Population systematics, radiometrics and zonation—a new biostratigraphy: North Am. Paleont. Conv. Proc., p. 612-666.

Keating, B., C. E. Helsley, and E. A. Pessagno, 1975, Late Cretaceous reversal sequence: Geology, v. 3, p. 73-76.

Kemper, E., 1971, Zur Gliederung und Abgrenzung des norddeutschen Aptium mit Ammoniten: Geol. Jahrb., v. 89, p. 359-390.

——— 1973a, The Valanginian and Hauterivian Stages in northwest Germany, *in* The boreal Lower Cretaceous: Liverpool, England, Seel House Press, p. 327-344.

——— 1973b, The Aptian and Albian Stages in northwest Germany, *in* The boreal Lower Cretaceous: Liverpool, England, Seel House Press, p. 345-360.

——— P. F. Rawson, Fr. Schmid, and C. Spaeth, 1974, Die Megafauna der Kreide von Helgoland und ihre biostratigraphische Deutung: Newsletters Stratigraphy, v. 3, p. 121-137.

Kennedy, W. J., 1969, The correlation of the lower Chalk of south-east England: Geologists' Assoc. Proc., v. 80, p. 459-560.

——— 1970, A correlation of the uppermost Albian and the Cenomanian of southwest England: Geologists' Assoc. Proc., v. 81, p. 613-677.

——— and J. M. Hancock, 1970, Ammonites of the genus *Acanthoceras* from the Cenomanian of Rouen, France: Palaeontology, v. 13, p. 462-490.

——— and P. Juignet, 1973, Observations on the lithostratigraphy and ammonite succession across the Cenomanian-Turonian boundary in the environs of Le Mans (Sarthe, N.W. France): Newsletters Stratigraphy, v. 2, p. 189-202.

Klepper, M. R., R. A. Weeks, and E. T. Ruppell, 1957, Geology of the southern Elkhorn Mountains, Jefferson and Broadwater Counties, Montana: U.S. Geol. Survey Prof. Paper 292, 82 p.

Kuhri, B., 1970, Some observations on the type material of *Globotruncana elevata* (Brotzen) and *Globotruncana concavata* (Brotzen): Rev. Espanola Micropaleontologia, v. 2, p. 291-304.

——— 1971, Lower Cretaceous planktonic Foraminifera from the Miravetes, Argos and Represa Formations (S.E. Spain): Rev. Espanola Micropaleontologia, v. 3, p. 219-237.

——— 1972, Stratigraphy and micropaleontology of the Lower Cretaceous in the subbetic south of Cravaca (province of Murcia, S.E. Spain): Koninkl. Nederlandse Akad. Wetensch. Proc., ser. B, v. 75, p. 193-222.

Kulp, J. L., 1960, The geological time scale: 21st Internat. Geol. Cong., Copenhagen, Rept., pt. 3, p. 18-27.

——— 1961, Geologic time scale: Science, v. 133, p. 1105-1114.

Lambert, R. St. J., 1971a, The pre-Pleistocene Phanerozoic time-scale; a review, *in* The Phanerozoic time-scale; a supplement: Geol. Soc. London Spec. Pub. 5, p. 9-31.

——— 1971b, The pre-Pleistocene Phanerozoic time-scale; further data, *in* The Phanerozoic time-scale; a supplement: Geol. Soc. London Spec. Pub. 5, p. 33-34.

Larson, R. L., and W. C. Pitman, III, 1972, World-wide correlation of Mesozoic magnetic anomalies, and its implications: discussion: Geol. Soc. America Bull., v. 83, p. 3645-3661.

——— 1975, World-wide correlation of Mesozoic magnetic anomalies and its implications: reply: Geol. Soc. America Bull., v. 86, p. 270-272.

Magné, J., and J. Polvêche, 1961, Sur le niveau a *Actinocamax plenus* (Blainville) du Boulonnais: Soc. Geol. Nord Annales, v. 81, p. 47-62.

Maync, W., 1959, Foraminiferal key biozones in the Lower Cretaceous of the western hemisphere and the Tethys province, *in* El sistema Cretacio, v. 1: 20th Internat. Geol. Cong., Mexico, p. 85-111.

McElhinny, M. W., 1973, Palaeomagnetism and plate tectonics: Cambridge, England, Cambridge Univ. Press, 358 p.

——— and P. J. Burek, 1971, Mesozoic palaeomagnetic stratigraphy: Nature, v. 232, p. 98-102.

McKenzie, D., and J. G. Sclater, 1971, The evolution of the Indian Ocean since the Late Cretaceous: Royal Astron. Soc. Geophys. Jour., v. 25, p. 437-528.

Michael, E., 1966, Die Evolution der Gavelinelliden (Foram.) in der NW-deutschen Unterkreide: Senckenbergiana Lethaea, v. 47, p. 411-459.

Moullade, M., 1966, Etude stratigraphique et micropaleontologique du Cretace Inferieur de la "fosse vocontienne": Lyons Univ. Fac. Sci. Lab. Geol. Doc. 15, 369 p.

Muller, S. W., and H. G. Schenck, 1943, Standard of Cretaceous system: AAPG Bull., v. 27, p. 262-278.

North, B. R., and W. G. E. Caldwell, 1970, Foraminifera from the Late Cretaceous Bearpaw Formation in the south Saskatchewan River Valley: Saskatchewan Research Council Geology Div. Rept. 9, 117 p.

Obradovich, J. D., and W. A. Cobban, 1975, A time scale for the Late Cretaceous of the Western Interior of North America: Geol. Assoc. Canada Spec. Paper No. 13, p. 31-54.

Olsson, R. K., 1964, Late Cretaceous planktonic Foraminifera from New Jersey and Delaware: Micropaleontology, v. 10, p. 157-188.

Owen, H. G., 1973, Ammonite faunal provinces in the middle and upper Albian and their palaeogeographical significance, in The boreal Lower Cretaceous: Liverpool, England, Seel House Press, p. 145-154.

Owens, J. P., and N. F. Sohl, 1973, Glauconites from New Jersey-Maryland coastal plain; their K-Ar ages and application in stratigraphic studies: Geol. Soc. America Bull., v. 84, p. 2811-2838.

Papp, A., 1956, Die morphologisch-genetische Entwicklung von Orbitoiden und ihre stratigraphische Bedeutung im Senon: Palaont. Zeitschr., v. 30, p. 45-49.

Pechersky, D. M., and A. N. Khramov, 1973, Mesozoic palaeomagnetic scale of the U.S.S.R.: Nature, v. 244, p. 499-501.

Pergament, M. A., 1967, Stages in Inoceramus evolution in the light of absolute geochronology: Paleont. Jour. no. 1, p. 27-34 (transl.).

Pessagno, E. A., Jr., 1969, Upper Cretaceous stratigraphy of the western Gulf Coast area of Mexico, Texas and Arkansas: Geol. Soc. America Mem. 111, 139 p.

Peterson, M. N. A., N. T. Edgar, M. B. Cita, S. Gartner, R. Goll, C. Nigrini, and C. von der Borch, 1970, Site 10: Deep Sea Drilling Project Initial Repts., v. 2, 501 p.

Petrushevskaya, M. G., and G. E. Kozlova, 1972, Radiolaria; Leg 14, deep sea drilling project: Deep Sea Drilling Project Initial Repts., v. 14, p. 495-648.

Porthault, B., G. Thomel, and O. de Villoutreys, 1966, Etude biostratigraphique du Cenomanien du bassin superieur de l'Esteron (Alpes-Maritimes); le probleme de la limite Cenomanien-Turonian dans le sud-est de la France: Soc. Geol. France Bull., ser. 7, v. 8, p. 423-439.

Premoli Silva, I., and H. M. Bolli, 1973, Late Cretaceous to Eocene planktonic Foraminifera and stratigraphy of Leg 15 Sites in the Caribbean Sea: Deep Sea Drilling Project Initial Repts., v. 15, p. 499-547.

Proto Decima, F., 1974, Leg 27 calcareous nannoplankton: Deep Sea Drilling Project Initial Repts., v. 27, p. 589-621.

Reiss, Z., 1962, Stratigraphy of phosphate deposits in Israel: Israel Geol. Survey Bull. 34, p. 1-23.

Riedel, W. R., and A. Sanfilippo, 1975, Radiolaria from the southern Indian Ocean, DSDP leg 26: Deep Sea Drilling Project Initial Repts., v. 26, p. 771-813.

Risch, H., 1971, Stratigraphie der hoheren Unterkreide der bayerischen Kalkalpen mit Hilfe von Mikrofossilien: Palaeontographica, v. 138A, p. 1-80.

Robaszynski, F., 1971, Les foraminiferes pelagiques des "Dieves" cretacees aux abords du golfe de Mons (Belgique): Soc. Geol. Nord Annales, v. 41, p. 31-38.

Roth, P. H., 1973, Calcareous nannofossils Leg 17 Deep Sea Drilling Project: Deep Sea Drilling Project Initial Repts., v. 17, p. 695-796.

Rubinshteyn, M. M., 1967, The argon method as applied to some problems of regional geology: Akad. Nauk Gruzin. SSR, Inst. Geol. Trudy, n.s., no. 11, 239 p. (in Russian; English summ.).

Salaj, J., and S. Bajanik, 1972, Contribution a la stratigraphie du Cretace et du Paleogene de la region de l'Oued Zarga: Tunisia Serv. Geol. Notes, no. 38, p. 63-71.

Scheibnerova, V., 1972, Some new views on Cretaceous biostratigraphy based on the concept of foraminiferal biogeoprovinces: New South Wales Geol. Survey Rec., v. 14, p. 85-87.

Schlich, R., 1974, Seafloor spreading history and deep-sea drilling results in the Madagascar and Mascarene basins, western Indian Ocean: Deep Sea Drilling Project Initial Repts., v. 25, p. 663-678.

Sclater, J. G., R. D. Jarrard, B. McGowran, and S. Gartner, 1974, Comparison of the magnetic and biostratigraphic time scales since the Late Cretaceous: Deep Sea Drilling Project Initial Repts., v. 22, p. 381-386.

Seitz, O., 1956, Uber Ontogenie, Variabilitat und Biostratigraphie einiger Inoceramen: Palaont. Zeitschr., v. 30, Sonderh., p. 3-6.

Séronie-Vivien, M., 1972, Contribution a l'etude de Senonien en Aquitaine septentrionale, ses stratotypes: Coniacien, Santonien, Campanien: France, Centre Natl. Recherche Sci. Editions, 195 p.

Shive, P. N., and W. E. Frerichs, 1974, Paleomagnetism of the Niobrara Formation in Wyoming, Colorado, and Kansas: Jour. Geophys. Research, v. 79, p. 3001-3007.

Sigal, J., 1965, Etat des connaissances sur les foraminiferes du Cretace Inferieur: Bur. Recherches Geol. et Minieres Mem. 34, p. 489-502.

—— 1966, Contribution a une monographie des rosalines, pt. 1, le genre Ticinella Reichel, souche des rotalipores: Eclogae Geol. Helvetiae, v. 59, p. 185-217.

—— 1967, Essai sur l'etat actuel d'une zonation stratigraphique a l'aide des principales especes de rosalines (foraminiferes): Soc. Geol. France Compte Rendu, p. 48-50.

Silver, L. T., C. R. Allen, and F. G. Stehli, 1969, Geological and geochronological observations on a portion of the Peninsular Range batholith of northwestern Baja California, Mexico (abs.): Geol. Soc. America Abs. with Programs, p. 279-280.

Smali, M., 1973, Foraminiferi Campaniano-Maastrichtiani della dorsale Medio-Atlantica: Riv. Italiana Paleontologia e Stratigrafia, v. 79, p. 35-100.

Smith, C. C., and E. A. Pessagno, 1973, Planktonic Foraminifera and stratigraphy of the Corsicana Formation (Maestrichtian), north-central Texas: Cushman Found. Foram. Research Spec. Pub. 12, 68 p.

Sornay, J., 1968, Termes stratigraphique majeurs; Aptien, in Lexique stratigraphique international, v. 8: Paris, Union Internat. Sci. Geol., Comm. Stratigr., 109 p.

Suppe, J., 1969, Times of metamorphism in the Franciscan terrain of the northern coast ranges, California: Geol. Soc. America Bull., v. 80, p. 135-142.

—— 1973, Geology of the Leech Lake Mountain; Ball Mountain region, California: California Univ. Pubs. Geol. Sci., v. 107, 82 p.

Thierstein, H. R., 1973, Lower Cretaceous calcareous nannoplankton biostratigraphy: [Austria] Geol. Bundesanst. Abh., v. 29, 52 p.

Thieuloy, J.-P., 1973, The occurrence and distribution of boreal ammonites from the Neocomian of southeast France (Tethyan province), *in* The boreal Lower Cretaceous: Liverpool, England, Seel House Press, p. 289-302.

Thomel, G., 1972, Les Acanthoceratidae cenomaniens des chaines subalpines meridionales: Soc. Geol. France Mem. 116, 204 p.

————— 1973a, A propos de la limite entre les etages Cenomanien et Turonien: Acad. Sci. Comptes Rendus, ser. D, v. 277, p. 761-764.

————— 1973b, A propos de la zone a *Actinocamax plenus*: principe et application de la methodologie biostratigraphique: Nice Museum d'Histoire Nat. Annales, v. 1, supp. h.s., p. 1-28.

Troeger, K. A., 1968, Zur Bedeutung okologischer Faktoren fur die Leitfossilien des Obercenoman-Unterturon Zeitabschnitts in Mitteleuropa: Geologie, v. 17, p. 68-75.

Vail, P. R., R. M. Mitchum, and S. Thompson, 1974, Eustatic cycles based on sequences with coastal onlap (abs.): Geol. Soc. America Abs. with Programs, v. 6, p. 993.

Van Gorsel, J. T., 1973a, The type Campanian and the Campanian-Maastrichtian boundary in Europe: Geologie en Mijnbouw, v. 52, p. 141-146.

————— 1973b, *Helicorbitoides* from southern Sweden and the origin of the *Helicorbitoides-Lepidorbitoides* lineage: Koninkl. Nederlandse Akad. Wetensch. Proc., ser. B, v. 76, p. 273-286.

————— 1974, Some complex Upper Cretaceous rotaliid Foraminifera from the northern border of the Aquitaine basin (SW France): Koninkl. Nederlandse Akad. Wetensch. Proc., ser. B, v. 77, p. 319-339.

van Hinte, J.E., 1963, Zur Stratigraphie und Mikropalaontologie der Oberkreide und des Eozans des Krappfeldes (Karnten): Austria, Geol. Bundesanst. Jahrb., Sonderb. 8, 147 p.

————— 1965, The type Campanian and its planktonic Foraminifera: Koninkl. Nederlandse Akad. Wetensch. Proc., ser. B, v. 68, p. 8-28.

————— 1966a, *Orbitoides* from the Campanian type section: Koninkl. Nederlandse Akad. Wetensch. Proc., ser. B, v. 69, p. 79-110.

————— 1966b, *Orbitoides hottingeri* n. sp. from northern Spain: Koninkl. Nederlandse Akad. Wetensch. Proc., ser. B, v. 69, p. 387-402.

————— 1967, *Bolivinoides* from the Campanian type section: Koninkl. Nederlandse Akad. Wetensch. Proc., ser. B, v. 70, p. 254-263.

————— 1968, The Late Cretaceous larger foraminifer *Orbitoides douvillei* (Silvestri) at its type locality Belves, SW France: Koninkl. Nederlandse Akad. Wetensch. Proc., ser. B, v. 71, p. 359-372.

————— 1969a, A *Globotruncana* zonation of the Senonian subseries: 1st Internat. Conf. Planktonic Microfossils, Geneva 1967, Proc., v. 2, p. 257-266.

————— 1969b, The nature of biostratigraphic zones: 1st Internat. Conf. Planktonic Microfossils, Geneva 1967, Proc., v. 2, p. 267-272.

————— 1972, The Cretaceous time scale and planktonic-foraminiferal zones: Koninkl. Nederlandse Akad. Wetensch. Proc., ser. B, v. 75, p. 1-8.

————— 1976, A Jurassic time scale: AAPG Bull., v. 60, p. 489-497.

————— J. A. S. Adams, and D. Perry, 1975, K/Ar age of Lower-Upper Cretaceous boundary at Orphan Knoll (Labrador Sea): Canadian Jour. Earth Sci., v. 12, p. 1484-1491.

Vermeulen, J., 1974, Sur une biostratigraphie homophyletique basee sur la famille des Pulchellidae: Acad. Sci. Comptes Rendus, ser. D, v. 278, p. 2885-2887.

Watkins, N. D., and F. W. Cambray, 1971, Paleomagnetism of Cretaceous dikes from Jamaica: Royal Astron. Soc. Geophys. Jour., v. 22, p. 163-179.

Van Hinte's article has been
reprinted from the April 1976
American Association of Petroleum Geologists Bulletin
Volume 60, Number 4

A Jurassic Time Scale[1]

J. E. VAN HINTE[2]
Houston, Texas 77001

Abstract Application of geologic ages expressed in numeric time (millions of years) is a recent but rapidly expanding factor in petroleum exploration (rates of tectonic movements; basin models; geohistory analysis; prediction of reservoir quality, source potential, and HC maturation; correlation of geophysics with geologic events, etc.). This paper proposes an emendation of published versions of the Jurassic time scale. The standard geochronologic subdivision is combined with a linear-numeric scale, with a biostratigraphic scheme (ammonites, calcareous nannofossils), and with the geomagnetic-reversal time scale. To this purpose, basement ages at some Deep Sea Drilling Project sites are evaluated and related to the geomagnetic-reversal scale. The numeric scale certainly cannot be regarded as precisely and firmly determined throughout the Jurassic, but it seems adequate as a broad framework.

INTRODUCTION

A rather reliable and detailed numeric time scale has become available for the Tertiary (e.g., Berggren, 1972). This scale is based on many radiometric age determinations which tie key units of the biostratigraphic scheme to a linear time scale. Until recently, constructing such a continuous linear scale was difficult because many biohorizons ("biostratigraphic datums") could not be related closely to a radiometrically dated level. The breakthrough came with recognition of a continuous record of geomagnetic reversals recorded in a linear magnetic-anomaly pattern in oceanic sea floors and subsequent paleontologic dating by deep-sea drilling of oceanic basement at key anomalies. The assumption of constancy of sea-floor-spreading rate between the dated anomalies then allowed for a linear age interpretation of the pattern between these points and, with additional extrapolation beyond the calibration points, yielded a linear geomagnetic-reversal time scale for the entire Tertiary. Because geomagnetic reversals also can be recognized in sediments, the reversal time scale now makes possible the assignment of numeric ages to biohorizons, which are not radiometrically dated, by interrelating the magnetostratigraphy and biostratigraphy of sedimentary sections. As a result, it has become possible to place a linear time scale beside the biostratigraphic zonal scheme. The addition of new and better data to this feedback system continues to improve it. Increased reliability and resolution are being obtained now, for example, by studies on both biostratigraphy and magnetostratigraphy of deep-sea cores (e.g., Theyer and Hammond, 1974) and by intercalating an increasing number of radiometric dates for which the biostratigraphic position is known.

This intricate approach using a paleomagnetic time scale for linear extrapolation between paleontologically correlated, radiometrically dated calibration points only with uncertainty can begin to be applied to construct a Jurassic numeric time scale. Only very few useful Jurassic radiometric dates are available, and the biostratigraphic position of most of these is vague. Also, oceanic sea floor of that age with a well-defined magnetic-anomaly pattern is uncommon, and calibration points provided by deep-sea drilling are uncommon and of rather uncertain numeric age. Consequently, the Jurassic magnetic-reversal time scale is not well established. Availability of the reliable Tertiary time scale, however, led to development of stratigraphic techniques that also can be applied for older parts of the geologic column,

[1]Manuscript received, July 7, 1975; accepted, September 26, 1975.

[2]Exxon Production Research Company.
I am most grateful to R. M. Jeffords who critically read the manuscript and suggested many improvements.

Article Identification Number: 0149-1377/78/SG06-0020/$03.00/0

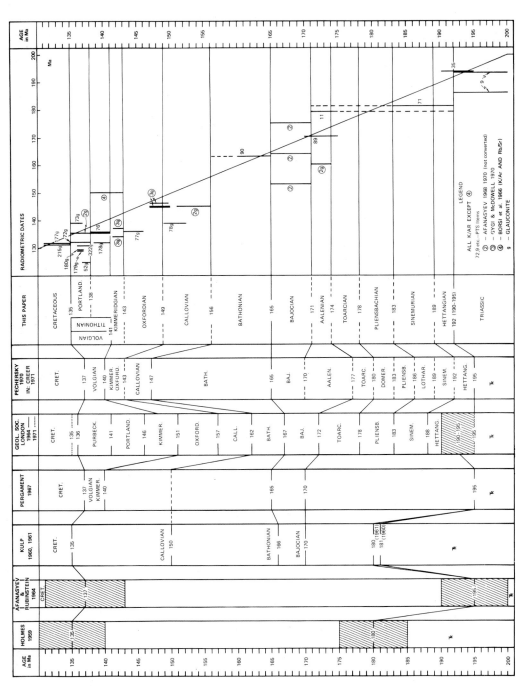

FIG. 1—History of Jurassic numeric time scale and comparison of radiometric dates with proposed scale. Russian dates have not been converted to accord with PTS constants.

be it with a wider margin of error. Thus, the present article provides a framework that serves the current needs and constitutes a basis for incorporation of additional information which will add needed precision for future requirements.

Our geochronologic terminology follows the recommendations of the International Geological Congress (1964), which discourage general usage of the term Purbeckian and support the use of the Aalenian as a stage covering the lower three ammonite zones of the Bajocian of others. Placement of the boundaries between Early, Middle, and Late Jurassic and between Lias, Dogger, and Malm also follows the IGC recommendations.

Calcareous nannofossil zonal boundaries are drawn in accordance with the definitions by Barnard and Hay (1974) following the species ranges given in their Figure 2 (which is not always the same as the boundaries in the third column of that figure). An exception had to be made for the *Stephanolithion speciosum* s.s. zone because the two lowest occurrences defining the zonal interval coincide on the range chart; for its top we used the lowest occurrence of *Diadozygus asymmetricus* instead of *Diazomatolithus lehmanni*, the lowest occurrence of *D. asymmetricus* also defining the base of the *D. lehmanni* zone.

PROPOSED JURASSIC TIME SCALE

During the past few years, the writer and associates successfully have used a Jurassic time scale that is an emended version of the scale proposed by the Geological Society of London, 1964 (the Phanerozoic time scale, in the following referred to as PTS). The PTS was "set up purely as a basis for discussion in the absence of a more positive scale of calibration" (Casey, 1964, p. 199) and is "no more than a simple working hypothesis" (Internat. Union Geol. Sci., 1967, p. 378). Incentive for emendation came when application of the PTS to exploration problems led locally to geologic inconsistencies and improbabilities. Some changes are based on new information, but others are pragmatic and intuitive; by necessity, the emended scale is no less a simple working model than the PTS. Nevertheless, it seems opportune to make the scale more widely available for practical use and for discussion that may test and improve it. This model has proved useful in oceanic studies (Berggren et al, 1975), has brought more "natural" proportions to the eustatic-cycle chart of Vail et al (1974), and, most significant, has yielded rates of sediment accumulation that compare well with those of reliably dated facies equivalents in the Tertiary.

The PTS proposed that the Jurassic covers the time-span −135 to −(190-195) Ma (megayears be-

fore present, one megayear = 1 million years), with − 165 Ma as its Bathonian approximate midpoint. Eleven Jurassic ages were recognized, and a linear scale was derived by (1) using the just-cited dates as calibration points and (2) assuming that the ages were of equal duration (5 Ma). Three exceptions to this assumption were includ-

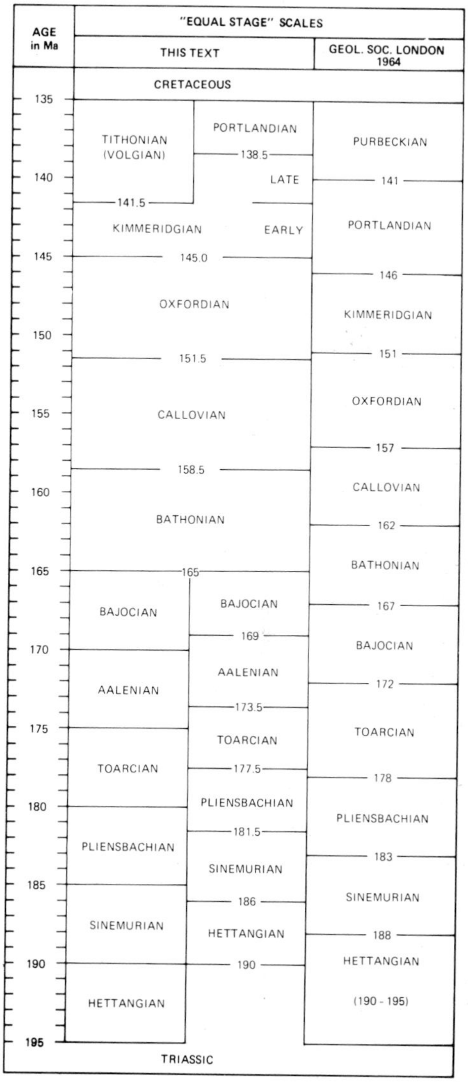

FIG. 2—Jurassic time scale proposed by Geological Society of London (1964) compared with scales derived on same assumptions but following chronostratigraphic subdivision proposed in Luxembourg in 1962 (Internat. Geol. Cong., 1964) and using −190 and −195 Ma as age of Jurassic-Triassic boundary.

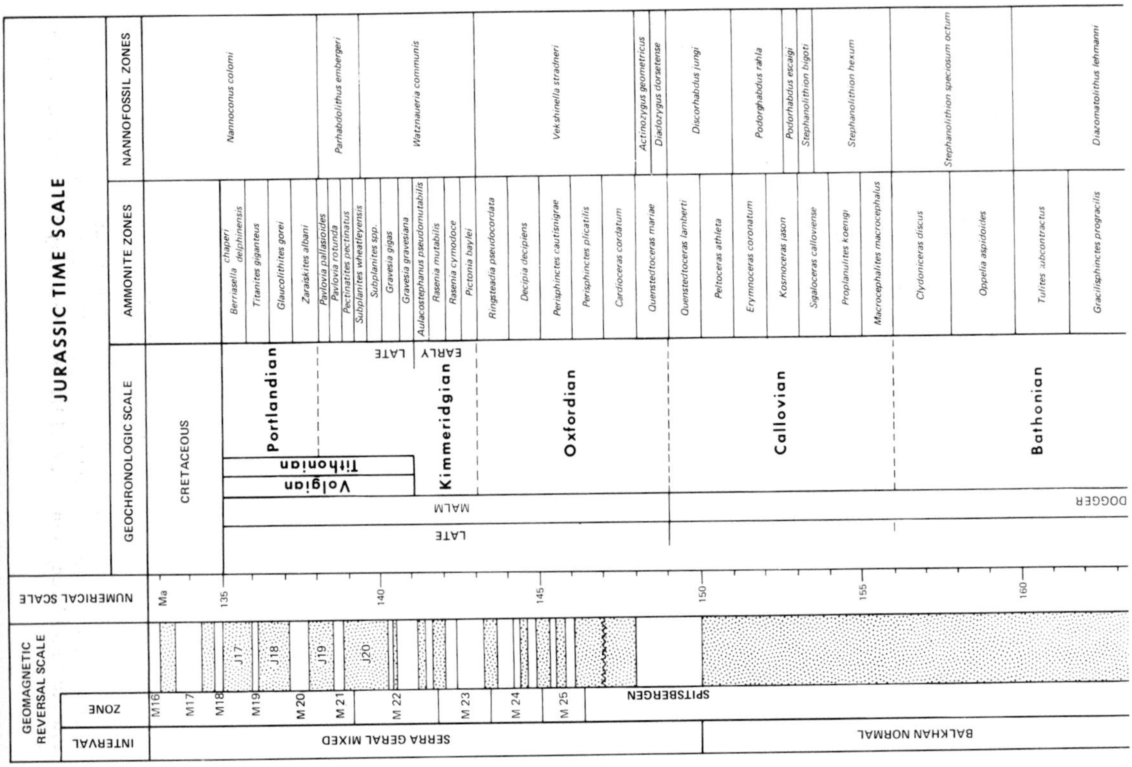

GEOMAGNETIC REVERSAL SCALE			NUMERICAL SCALE	JURASSIC TIME SCALE			
INTERVAL	ZONE			GEOCHRONOLOGIC SCALE		AMMONITE ZONES	NANNOFOSSIL ZONES
SERRA GERAL MIXED	M16		Ma	CRETACEOUS			
	M17						
	M18	J17	135	Volgian / Tithonian	Portlandian	Berriasella chaperi / delphinensis	Nannoconus colomi
	M19	J18				Titanites giganteus	
	M20					Glaucolithites gorei	
		J19				Zaraiskites albani	
	M21					Pavlovia pallasioides	Parhabdolithus embergeri
		J20				Pavlovia rotunda	
						Pectinatites pectinatus	
	M22		140			Subplanites wheatleyensis	
					LATE	Subplanites spp.	Watznaueria communis
						Gravesia gigas	
				Kimmeridgian		Gravesia gravesiana	
						Aulacostephanus pseudomutabilis	
	M23				EARLY	Rasenia mutabilis	
						Rasenia cymodoce	
						Pictonia baylei	
SPITSBERGEN	M24					Ringsteadia pseudocordata	Vekshinella stradneri
			145	Oxfordian		Decipia decipiens	
	M25					Perisphinctes cautisnigrae	
						Perisphinctes plicatilis	
						Cardioceras cordatum	
						Quenstedtoceras mariae	Actinozygus geometricus / Diadozygus dorsetense
						Quenstedtoceras lamberti	Discorhabdus jungi
			150			Peltoceras athleta	
BALKHAN NORMAL				Callovian		Erymnoceras coronatum	Podorghabdus rahla
						Kosmoceras jason	Podorhabdus escaigi
						Sigaloceras calloviense	Stephanolithion bigoti
						Proplanulites koenigi	Stephanolithion hexum
			155			Macrocephalites macrocephalus	
						Clydoniceras discus	Stephanolithion speciosum octum
				Bathonian		Oppelia aspidoides	
			160			Tulites subcontractus	
						Gracilisphinctes progracilis	Diazomatolithus lehmanni

(left geochronologic labels, read vertically: LATE, MALM, DOGGER)

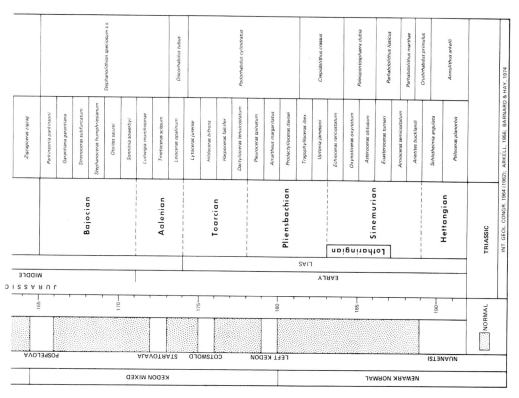

FIG. 3—Jurassic time scale proposed in this paper. Geomagnetic-reversal time scale is recalibrated after Larson and Pitman (1972) above wavy line at −147 Ma and after McElhinny and Burek (1971) and Pechersky and Khramov (1973) below.

ed (Oxfordian and Toarcian, 6 Ma; Hettangian, 4 Ma) for which no further reasons were given.

Howarth (1964), in an article which was the basis for the Jurassic part of the PTS, considered only 12 of all published radiometric dates of sufficient accuracy to be cited. These dates only broadly confirmed the proposed scale (Fig. 1).

Lambert (1971a, b) considered none of the glauconite ages cited by Howarth to be reliable and argued that the stratigraphic position of other radiometric dates is vague at best. His "most reliable" (1971a, p. 29) Jurassic radiometric age is 165 ± 4 Ma (K/Ar biotite) for the post-Bajocian Kelasury Granite. The second category of Lambert ("most reliable determinations without duplicate or supporting analyses," 1971a, p. 30) listed K/Ar biotite dates of −136 ±4 Ma for the post-early Kimmeridgian Horseshoe Bar diorite and −193? ±? for the Hotailuh batholith near the Jurassic-Triassic boundary. Lambert suggested rounding off to 5 Ma the ages of all major pre-Tertiary boundaries, proposing −135? Ma for the Cretaceous-Jurassic boundary but without deciding to −195? or −200? or −205? Ma for the Jurassic-Triassic boundary.

The present scale uses as basic points −135 Ma (Howarth, 1964; Lambert, 1971a, b) and −190 to −195 Ma (Howarth, 1964; Geol. Soc. London, 1964) boundaries and the 165 ±4 Ma (PTS, Item 90; Lambert, 1971a) mid-Jurassic age. Following the recommendations of the International Geological Congress (1964), the stage scheme of the PTS is altered in that Portlandian and Purbeckian are merged and Aalenian is added. Assuming equal-time representation of the stages, the Bajocian rather than the Bathonian becomes the middle stage of the Jurassic. The PTS conveniently could use the −165 ±4 Ma date (coincidentally the midpoint between −135 and −195) for the middle of their Middle Jurassic stage. Here it is placed as close to the middle as seems justified, the Bajocian-Bathonian boundary being drawn at −165 Ma (cf. Snelling, 1964).

A linear scale derived from the three calibration points in the PTS manner by assuming equal time (rounded to 0.5 Ma) for the stages would result in the scale of Figure 2. Another, equally hypothetical but perhaps slightly more meaningful, way to extrapolate a linear scale from calibration points is to assume time equivalence for biozones. Doing so for the 26 Arkell (1956) ammonite zones of the Jurassic older than 165 Ma gives each zone a duration of 1.15 Ma, if an age of 195 Ma is assumed for the base of the Jurassic and 0.96 Ma for a base 190 Ma old. Assignment of a duration of 1.0 Ma to each zone, therefore,

seemed a reasonable approximation, and the scale of Figure 3 was the result.

For the post-Bajocian part of the scale, a few recent radiometric dates and Russian opinions (Fig. 1) suggest partial divergence from the "equal stage" and from the "equal zone" approach. Rounding the beginning times of the Portlandian and Tithonian (Volgian) to the younger whole number gives −138 and −141 Ma, respectively. Zonal equivalence within most of the Kimmeridgian brings its beginning to −143 Ma. Although highly artificial, these boundary ages seem to be confirmed by the available radiometric ages.

On the basis of radiometric age determinations, Russian authors (Perchersky, in Creer, 1971) drew the Callovian/Bathonian boundary at −147 Ma so that the Bathonian has a very long time span (Fig. 1). Consequently, Bathonian biochronozones would last disproportionately long, and Bathonian sediment-accumulation rates would be exceedingly low. On the proposed scale we stayed within Gygi and McDowell's (1970) radiometric age range of −146 ±3 Ma for the lowermost Oxfordian and uppermost Callovian by simply assigning a 1-Ma duration to the Oxfordian and Callovian ammonite zones. The resulting Oxfordian/Callovian boundary is 149 Ma old, and the Callovian/Bathonian boundary is 156 Ma old.

The older part of the geomagnetic-reversal time scale of Figure 3 is after the terrestrial scale of McElhinny and Burek (1971) and Pechersky and Khramov (1973). Sallomy and Briden (1975) found reversed polarity in the upper Toarcian Cotswold sands and Rosedale iron ore, confirming Girdler's (1959) results which had been incorporated in the McElhinny and Burek (1971) scale as the Cotswold Zone (=? Nuanetsi Zone of Pechersky and Khramov, 1973). Probably the Bajocian Startovaja Reversed Zone of Pechersky (in Creer, 1971) and Pechersky and Khramov (1973) is the same as the Mateke Zone of McElhinny and Burek (1971) and possibly was recognized by Sallomy and Briden (1975) in their Cephalopod Bed. Further included are the Bathonian/Bajocian Pospelova and the Domerian Left-Kedon Zones of the just-mentioned Russian authors; the fact that these reversed-polarity zones have not been recognized outside the Soviet Union may well be because of the scarcity of paleomagnetic analyses for this part of the stratigraphic column. The Finish Mixed Zone of Pechersky and Khramov (1973) probably falls within the Nuanetsi Reversed Zone of McElhinny and Burek (1971).

The younger part of the geomagnetic-reversal time scale is after the marine scale of Larson and Pitman (1972; "J" anomaly numbers after Vogt et

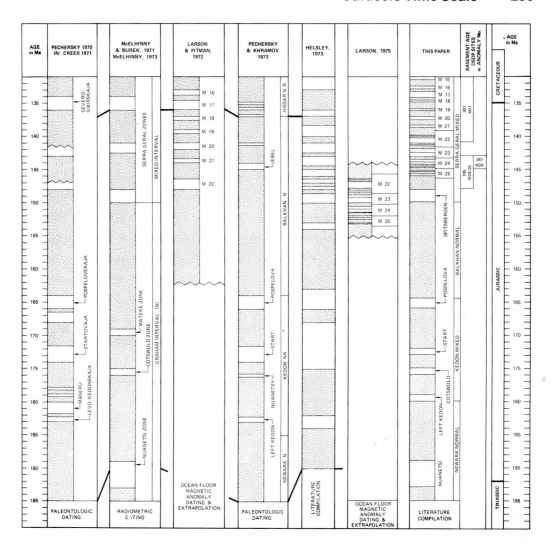

FIG. 4—Comparison of geomagnetic-reversal time scales published for Jurassic. Original Pechersky and Khramov (1973) scale is nonlinear; ages here have been estimated from Pechersky (1970, *in* Creer, 1971). Time intervals of normal field polarity are shaded. Russian radiometric ages have not been converted to accord with PTS constants.

al, 1971), but with 145 and 125 Ma instead of 155 and 120 Ma as the ages of the two calibration points used by Larson and Pitman—i.e., the age at the bases of Deep-Sea Drilling Project (DSDP) holes 105 and 166, respectively. The reasons for changing these estimates of basement age are given in the following.

OCEANIC BASEMENT AGES

Drawing the Cretaceous/Jurassic boundary (−135 Ma) at 550 m in DSDP hole 105 (calpion-

ellids) and placing the base of the Kimmeridgian (−143 Ma) at 603 m (ammonite fragment; Hollister et al, 1972) means that a section 53 m thick was deposited in 8 Ma, a rate of 0.65 cm/1,000 years. This rate compares well with sites 100 (0.75 cm/1,000 years) and 99 (0.60 cm/1,000 years). The base of the sediment lies 19 m lower than the base of the Kimmeridgian and, extrapolating the rate of sediment accumulation, is 3.1 Ma older. The base of the Kimmeridgian could be lower in hole 105 than just assumed, which would increase the rate of sediment accumulation and make the base of sediment younger. On the other hand, the dinoflagellate biostratigraphy suggests that it could be somewhat higher with the reversed effect. To give the base of sediment an age range of 143-148 Ma seems justified, hence the use here of − 145 Ma.

The age of the lowest sediment at DSDP Site 166 (cores 27, 28) is "definitely Neocomian, probably Late Hauterivian" (Douglas, 1973, p. 684). In the core description for the site, two incompatible ages were given: "Late Hauterivian (Forams, Nannos)" and "Late Albian-Cenomanian (Rads)" (Winterer et al, 1973, p. 126, 127). Accepting a Hauterivian age for the lower cores means that the base of sediment lies 18 m below the Hauterivian-Barremian boundary (− 121 Ma) and probably is considerably older because the rate of sediment accumulation for the "zeolitic nannofossil marl" deposited near calcite-compensation depth (CCD) most likely is low, and because basement is 700 m higher than "normal" (Winterer, 1973), which further suggests minimal sediment-accumulation rates. It seems reasonable to assume an age range of − 122 to − 127 Ma for basement at Site 166, and to use − 125 Ma for the calibration point.

The use of − 145 and − 125 Ma brings the ages found during DSDP Leg 32 for the Early Cretaceous and Jurassic anomalies M4, M9, and M21 in accord with the prediction on the scale (123, 126, and 138.5 Ma, respectively), whereas with − 120 and − 145 Ma as calibration points, two ages would be older than predicted (117.5 and 121 Ma, respectively). If − 155 and − 120 Ma are used as calibration points, the found ages for M4 and M9 are older than predicted, and the found age for anomaly M21 is younger than predicted (118.5, 121, and 142 Ma, respectively). Anomaly M4 was dated at DSDP Site 303 as Valanginian to Hauterivian, 131-121 Ma old; anomaly M9 was dated at DSDP Site 304 as Neocomian, probably Valanginian, 135-121 (131-126) Ma old; and anomaly M21 was dated at DSDP Site 307 as probably Tithonian or Berriasian, 141-131 Ma old (Anon., 1973).

Results obtained at DSDP Site 261 further confirm the usefulness of the scale given in this study. The section between 508 and 526 m is of Tithonian age (Proto Decima, 1974); using − 135 and − 141 Ma as boundary ages gives an average rate of sediment accumulation of 0.30 cm/1,000 years. Downward extrapolation of this rate suggests that the top of the Oxfordian (− 143 Ma) is to be expected at 532 m, which is precisely where the boundary has been drawn on paleontologic evidence (top of *Stephanolithion bigoti*). A basement

age of 143-145 Ma seems a reasonable estimate. Larson (1975) correlated the magnetic-anomaly pattern at site 261 with anomaly M24, the Atlantic DSDP Site 105 lying between M24 and M25. The preceding conclusions are summarized in Figure 4.

References Cited

Afanasyev, G. D., 1968, The geological time-scale in terms of absolute dating: 23d Internat. Geol. Cong., Czechoslovakia, Rept. Sec. 6, Proc., p. 33-43.

──── 1970, Certain key data for the Phanerozoic time-scale: Eclogae Geol. Helvetiae, v. 63, p. 1-7.

──── and M. M. Rubinstein, 1964, Geochronological scale of absolute age determinations based on data of the U.S.S.R. laboratories up to April 1964 with a consideration of foreign data: 22d Internat. Geol. Cong., Dokl. Sov. Geol., Problem 3, p. 287-324.

Ager, D. V., 1963, Jurassic stages: Nature, v. 198, p. 1045, 1046.

Anonymous, 1973, Leg 32; Deep Sea Drilling Project: Geotimes, v. 18, no. 12, p. 14-17.

Arkell, W. J., 1956, Jurassic geology of the world: New York, Hafner Pub. Co., 806 p.

Armstrong, R. L., and J. Besancon, 1970, A Triassic time-scale dilemma: K-Ar dating of Upper Triassic mafic igneous rocks, eastern U.S.A. and Canada and post-Upper Triassic plutons, western Idaho, U.S.A.: Eclogae Geol. Helvetiae, v. 63, p. 15-28.

Barnard, T., and W. W. Hay, 1974, On Jurassic coccoliths: a tentative zonation of the Jurassic of southern England and north France: Eclogae Geol. Helvetiae, v. 67, p. 563-585.

Berggren, W. A., 1972, A Cenozoic time-scale—some implications for regional geology and paleobiogeography: Lethaia, v. 5, p. 195-215.

──── D. P. McKenzie, J. G. Sclater, and J. E. van Hinte, 1975, World-wide correlation of Mesozoic magnetic anomalies and its implications: Discussion: Geol. Soc. America Bull., v. 86, p. 267-269.

Borsi, S., G. Ferrar, J. Mercier, and E. Tongiorgi, 1966, Age stratigraphique et radiometrique Jurassique superieur d'un granite des zones internes des Hellenides: Rev. Geographie Phys. et Geologie Dynam., ser. 2, v. 8, pt. 4, p. 279-287.

Casey, R., 1964, The Cretaceous Period, in The Phanerozoic time-scale: Geol. Soc. London Quart. Jour., v. 120, supp., p. 193-202.

Creer, K. M., 1971, Mesozoic palaeomagnetic reversal column: Nature, v. 233, p. 545, 546.

Dodson, M. H., D. C. Rex, R. Casey, and P. Allen, 1964, Glauconite dates from the Upper Jurassic and Lower Cretaceous, in The Phanerozoic time-scale: Geol. Soc. London Quart. Jour., v. 120, supp., p. 145-158.

Douglas, R. G., 1973, Planktonic foraminiferal biostratigraphy in the central north Pacific Ocean: Deep Sea Drilling Project Initial Repts., v. 17, p. 673-694.

Geological Society of London, 1964, Geological Society Phanerozoic time-scale 1964, in The Phanerozoic time-scale: Geol. Soc. London Quart. Jour., v. 120, supp., p. 260-262.

Girdler, R. W., 1959, A palaeomagnetic study of some Lower Jurassic rocks of N.W. Europe: Royal Astron. Soc. Geophys. Jour., v. 2, p. 353-363.

Gygi, R. A., and F. W. McDowell, 1970, Potassium-argon ages of glauconites from a biochronologically dated Upper Jurassic sequence of northern Switzerland: Eclogae Geol. Helvetiae, v. 63, p. 111-118.

Harland, W. B., and E. H. Francis, eds., 1971, The Phanerozoic time-scale—a supplement: Geol. Soc. London Spec. Pub. 5, 356 p.

Helsley, C. E., 1973, Post-Paleozoic magnetic reversals (abs.): Geol. Soc. America Abs. with Programs, v. 5, p. 665 (hand-out figure "Comparison of Mesozoic reversal sequences").

Hollister, C. D., J. I. Ewing, D. Habib, J. C. Hathaway, Y. Lancelot, H. Luterbacher, F. J. Paulus, C. W. Poag, J. A. Wilcoxon, and P. Worstell, 1972, Site 105—lower continental rise hills: Deep Sea Drilling Project Initial Repts., v. 11, p. 219-312.

Holmes, A., 1959, A revised geological time-scale: Edinburgh Geol. Soc. Trans., v. 17, p. 183-216.

Howarth, M. K., 1964, The Jurassic Period, in The Phanerozoic time-scale: Geol. Soc. London Quart. Jour., v. 120, supp., p. 203-205.

International Geological Congress, 1964, Recommendations, in P. L. Mauberge, ed., Colloque du Jurassique a Luxembourg, 1962 (Colloquium on the Jurassic, Luxembourg, 1962): Luxembourg, Inst. Grand-Ducal, Sect. Sci. Nat., Phys. Math., p. 84-88.

International Union of Geological Sciences (IUGS), 1967, A comparative table of recently published geological time scales for the Phanerozoic time—Explanatory notice: Norsk Geol. Tidsskr., v. 47, p. 375-380.

Kulp, J. L., 1960, The geological time scale: 21st Internat. Geol. Cong., Copenhagen, Rept., pt. 3, p. 18-27.

——— 1961, Geologic time scale: Science, v. 133, p. 1105-1114.

Lambert, R. St. J., 1971a, The pre-Pleistocene Phanerozoic time-scale; a review, in The Phanerozoic time-scale—a supplement: Geol. Soc. London Spec. Pub. 5, p. 9-31.

——— 1971b, The pre-Pleistocene Phanerozoic time-scale; further data, in The Phanerozoic time-scale—a supplement: Geol. Soc. London Spec. Pub. 5, p. 33-34.

Larson, R. L., 1975, Late Jurassic sea-floor spreading in the eastern Indian Ocean: Geology, v. 3, p. 69-71.

——— and W. C. Pitman, III, 1972, World-wide correlation of Mesozoic magnetic anomalies, and its implications: Geol. Soc. America Bull., v. 83, p. 3645-3661.

McElhinny, M. W., 1973, Palaeomagnetism and plate tectonics: Cambridge, England, Cambridge Univ. Press, 358 p.

——— and P. J. Burek, 1971, Mesozoic palaeomagnetic stratigraphy: Nature, v. 232, p. 98-102.

Pechersky, D. M., and A. N. Khramov, 1973, Mesozoic palaeomagnetic scale of the U.S.S.R.: Nature, v. 244, p. 499-501.

Pergament, M. A., 1967, Stages in *Inoceramus* evolution in the light of absolute geochronology: Paleont. Jour., no. 1, p. 27-34 (transl.).

Proto Decima, F., 1974, Leg 27 calcareous nannoplankton: Deep Sea Drilling Project Initial Repts., v. 27, p. 589-621.

Sallomy, J. T., and J. C. Briden, 1975, Paleomagnetic studies of Lower Jurassic rocks in England and Wales: Earth and Planetary Sci. Letters, v. 24, p. 369-376.

Snelling, N. J., 1964, A review of recent Phanerozoic time-scales, in The Phanerozoic time-scale: Geol. Soc. London Quart. Jour., v. 120, supp., p. 29-36.

Theyer, F., and S. R. Hammond, 1974, Paleomagnetic polarity sequence and radiolarian zones, Brunhes to polarity epoch 20: Earth and Planetary Sci. Letters, v. 22, p. 307-319.

Vail, P. R., R. M. Mitchum, and S. Thompson, 1974, Eustatic cycles based on sequences with coastal onlap (abs.): Geol. Soc. America Abs. with Programs, v. 6, p. 993.

Vogt, P. R., C. N. Anderson, and D. R. Bracey, 1971, Mesozoic magnetic anomalies, sea-floor spreading and geomagnetic reversals in the southwestern North Atlantic: Jour. Geophys. Research, v. 76, p. 4796-4823.

Winterer, E. L., 1973, Regional problems: Deep Sea Drilling Project Initial Repts., v. 17, p. 911-922.

——— J. I. Ewing, R. G. Douglas, R. D. Jarrard, Y. Lancelot, R. M. Moberly, T. C. Moore, P. H. Roth, and S. O. Schlanger, 1973, Site 166: Deep Sea Drilling Project Initial Repts., v. 17, p. 103-144.

Chronostratigraphy for the World Permian[1]

J. B. WATERHOUSE[2]

Abstract The marine Permian faunas of the world are subdivided into eight stages and nineteen substages: the Asselian Stage, with the Surenan, Uskalikian, and Kurmaian Substages; the Sakmarian Stage, with the Tastubian, Sterlitamakian, and Aktastinian Substages; the Baigendzinian Stage, with the Sarginian and Krasnoufimian Substages; the Kungurian Stage, with the Filippovian and Irenian Substages (the latter including the Nevolin, Elkin, and Ufimian horizons); the Kazanian Stage, with the Kalinovian and Sosnovian Substages; the Punjabian Stage, with the Kalabaghian and Chhidruan Substages; the Djulfian Stage, with the Urushtenian and Baisalian Substages, and the Dorashamian Stage with the Vedian and Ogbinan Substages and perhaps the Griesbachian Substage or a modification thereof. This offers the most extensive and refined scheme of correlation available for the Permian System, in which brachiopod correlations agree well with those proposed for the less abundant Permian ammonoids, and paleotropical Fusulinacea.

Short-lived normal events during the predominantly reversed paleomagnetic late Paleozoic interval offer prospects of a few well-defined horizons within the Permian, but are poorly dated and additional short-lived events may yet be recognized. Radiometric values, although showing considerable scatter, suggest that the length of the period should be extended from earlier estimates, and that it commenced about 300 m.y. B.P., and ended, as generally recognized, about 230 m.y. B.P. Perhaps the best support for these dates is provided by solar chronology, which postulates that the energy budget from the sun varied in rhythmic fashion, leading to major refrigerations every 30 million years or so.

Introduction

The Permian System was established on the basis of rocks and faunas of the Ural Mountains and Russian Platform by Murchison (1841), before Marcou (1859) proposed the word Dyas for equivalent rocks and faunas of Germany. The period has long been significant in our understanding of earth history. Its copper ores of Thuringia and Saxony led to the establishment of the mining school of Frieberg where Werner established geognosy. The peculiar distribution of its glacial deposits helped to suggest the theory of continental drift, and its strong and predominantly reversed magnetism has helped substantiate that theory. Its economic significance has always been great, with Permian rocks yielding oil, gas, coal, copper, gold, uranium, salts, and other minerals. Moreover the period ended with the greatest of catastrophes to all life, and this is naturally of current interest to our present doomsday philosophers.

Biochronology

Faunal Divisions for Marine Permian

Faunal divisions for the entire world have been summarized by Waterhouse (1976a) in a text which traces correlative rocks and faunas of eighteen substages, classed in

[1] Manuscript received, December 14, 1976.
[2] Department of Geology and Mineralogy, University of Queensland, Brisbane, Australia.

Article Identification Number: 0149-1377/78/SG06-0022/$03.00/0

eight stages (Table 1) around the globe. This new classification refines subdivisions for the Early and Middle Permian, and substantially changes and subdivides the so-called Late Permian (=Ochoan) Stage still used in works that predate the recent advances in Armenia and China. Stages and substages used here follow the definition in Waterhouse (1976b).

Table 1. Subdivisions of the Marine Permian System.

Series	Stage	Substage	Horizons	P Number	Symbol
Permian or Triassic		Dienerian			
Late Permian	Dorashamian	Griesbachian		19	PDg
		Ogbinan		18	PDo
		Vedian		17	PDv
	Djulfian	Baisalian		16	PJb
		Urushtenian		15	PJu
	Punjabian	Chhidruan		14	Ppc
		Kalabaghian		13	PPk
Middle Permian	Kazanian	Sosnovian		12	PZs
		Kalinovian		11	PZk
	Kungurian		Ufimian	10 c	
		Irenian	Elkin	10 b	PKl
			Nevolin	10 a	
		Filippovian		9	PKf
	Baigendzinian	Krasnoufimian		8	PBk
		Sarginian		7	PBs
	Sakmarian	Aktastinian		6	PSa
Early Permian		Sterlitamakian		5	PSs
		Tastubian		4	PSt
	Asselian	Kurmaian		3	PAk
		Uskalikian		2	PAu
		Surenan		1	PAs
?	Orenburgian			0	
Late Carboniferous	Gshelian			0	
	Kasimovian			0	

Early Permian Series

Asselian Stage

The Asselian Stage was proposed for rocks and faunas of the Ural Mountains near the Kiya, Sintas, and Dombas Rivers, and is subdivided into three substages (Table 1), typified by Fusulinacean zones, including *Pseudofusulina* vulgaris* (Schellwien) etc for the Surenan Substage, *Schwagerina moelleri* Rauser and other species for the Uskaliki-an Substage, and *Schwagerina sphaerica* Rauser with other species for the Kurmaian Substage at the top of the stage. Large brachiopod faunas have been described but need revision and stratigraphic elucidation. Four new ammonoid families, Perrinitidae, Metalegoceratidae, Popanoceratidae, and Paragastrioceratidae, help characterize the stage with *Juresanites* and *Paragastrioceras* appearing in the Uskalikian Substage, and *Sakmarites, Tabantalites* and *Protopopanoceras* appearing first in the Kurmaian Substage.

Ruzencev (1952; 1965) stressed that the ammonoid families allowed discrimination from the supposedly Late Carboniferous Orenburgian "Stage," which lacked these families and was typified by *Uddenites* and *Prouddenites.* But other Soviet authorities maintain (as at the Carboniferous Congress at Moscow, 1975) that the Orenburgian could be basal Asselian, on the basis of Fusulinacea and microfossils. Certainly the basal Asselian or Surenan Substage does not have any typical Asselian ammonoids, and *Uddenites* and *Prouddenites* appear in the upper Gaptank Shale of West Texas with Fusulinacea and Brachiopoda deemed to be basal Permian by Cooper and Grant (1973).

Correlative rocks and faunas are extensive (Table 2), as discussed in detail by Water-house (1976a, p. 53-77). They include at least some faunas referred to the Carbonifer-ous Period, such as the Paren fauna of the Kolyma River region (Zavodowsky et al, 1970) and are widely represented over the fragments of Gondwana in Australia, South America, Africa, India and the Himalayas, and possibly even more northern ranges by glacial tillite overlain by the bivalve *Eurydesma* fauna, of Kurmaian age. The fullest marine sequence in Gondwana lies in Tasmania, and also displays a three-fold subdi-vision (Clarke and Banks, 1975).

Sakmarian Stage

The Sakmarian Stage is also based on sequences in the Urals, especially in the Aktyubin region. Fusulinacea are again critical for recognition of the stage in the Urals, with one suite, including *Rugosofusulina serrata* Rauser, characterizing the basal or Tastubian Substage, and another zone, including *Pseudofusulina devexa* Rauser, characterizing the following Sterlitamakian Substage. According to Ruzencev (1952) the upper Tastubian level witnessed the first appearance of many new genera, includ-ing *Synartinskia, Medlicottia, Metalegoceras,* and *Uraloceras,* but the lower Tastubian yielded only an "advanced" species of *Juresanites* with many Asselian survivors. Nor do ammonoids define the Sterlitamakian Substage clearly in the Urals. Brachiopods require modern study. In spite of these difficulties it has been possible to trace the substages widely elsewhere, as shown by Furnish (1973) on the basis of ammonoids, and Waterhouse (1976a, p. 78-94) who also used Fusulinacea, Brachiopoda, and Bivalvia. Faunas of both substages are distinctive and widely present in both hemi-spheres, except in the Gondwana sequences of Africa, and rarely in India and South America. Sakmarian faunas were especially well developed in Australia, as "Fauna II" of Dickins, but somewhat younger faunas from the Gympie and Sydney basins have also been mistakenly incorporated in Fauna II by Runnegar and Ferguson (1969), and Dickins (1976).

There is some uncertainty whether the Aktastinian Substage should be included with the Sakmarian Stage. This substage is characterized in the Urals by first entries of ammonoids *Neoshumardites* and *Aktubinskia* and various *Pseudofusulina* with the first appearance, in the Urals, of *Parafusulina.* Brachiopods from numerous faunas over the world strongly suggest a closer relationship to Sakmarian than to the overlying Baigendzinian faunas, and early work in Fusulinacea was interpreted in the same way (Rauser-Chernosova, 1949). But now Fusulinacea and Ammonoidea are said to indi-cate closer affinities with the Baigendzinian rather than Sakmarian Stage. Whereas

Table 2. Intercorrelations of the most significant marine sequences in the world for the Permian Period (from Waterhouse, 1976a, Table 2, p. xvi-xvii). Amb Formation of Salt Range here classes with "coaly shales," previously classed at base of Wargal Formation (Waterhouse, 1972). M—Member, F—Formation, H—Horizon, Z—Zone, f—fauna, G—Group, CM—Coal Measures, +—type for Substage.

Stage	Substage	Permian Platform Urals		Salt Range	Armenia
Dorashamian	Griesbachian	TATARIAN	Kutuluk H		
	Ogbinan				<u>Paratirolites</u> Z <u>Comelicania</u> Z + <u>Araxilevis</u> Z
	Vedian		Malokinel H		
Djulfian	Baisalian				<u>Codonofusiella</u> Z = Abadehian)
	Urushtenian		Bolshkinel H	Kathwai M	
Punjabian	Chhidruan			Chhidru F +	Hachik H
	Kalabaghian		Sok H	Kalabagh M +	Gnishik H
Kazanian	Sosnovian	Sosnov H +		Middle Wafgal F	Armik H
	Kalinovian	Kalinov H +		Lower	
Kungurian	Irenian	Ufimian H + Elkin H + Nevolin H +			
	Filippovian	Filippov H +		Amb	Asni H
Baigendzinian	Krasnoufimian	Saranin H		F	
	Sarginian	Sargin H +			
Sakmarian	Aktastinian	Aktastin H +		Sardi F	
	Sterlitamakian	Sterlitamak H +		Warchha F	Davalin H
	Tastubian	Tastub H +		<u>Conularia</u> Beds	
Asselian	Krumaian	Kurmai H +		<u>Eurydesma</u> beds Talchir	
	Uskalikian	Uskalik H +			
	Surenan	Suren H +			

(Continued on next page)

these fossils are relatively rare in so many sequences they cannot be granted the same weight as Brachiopoda but the question is still under debate. The matter is relatively trivial, because it is substages rather than stages that are important for correlation. Certainly a poorly preserved, or ill-studied benthic fauna of actual Aktastinian age can be readily interpreted as "Sakmarian" rather than "Baigendzinian," even though it is impossible to assign the particular substage.

Baigendzinian Stage

The classical "Artinskian" sequences of the Arti region in the Urals are now referred to the Baigendzinian Stage. Current ammonoid usage makes no use of the classic Artinskian Stage (Furnish, 1973). Artinskian has long been a facies term (Dunbar, 1940) that ranged considerably in age, so that its passing may help diminish the confusion that prevailed over its limits. Alternatively, we may prefer to revert to the original understanding of Artinskian, which is to be equated with Baigendzinian. Two levels are recognized, though not especially well defined, on the basis of Fusulinacea

Table 2. Continued

Arctic Canada	USA, Texas	New Zealand	Queensland composite	China
Griesbachian +	Ochoan G	Aperispirifer nelsonensis Z	? Rewan F / Baralaba CM	Changhsing / ? F
		Durvilleoceras Z		Wuchiaping F
	Hegler M	Spinomartinia spinosa Z	Tamaree F	
	Capitan — Cherry Canyon M	Plekonella multicostata Z	Upper Curra Lmst	
		Martiniopsis woodi Z	?	
	Appel Ranch M	Terrakea brachythaerum Z	Pelican Ck f	Maokou F
Cancrinelloides Z	Willis Ranch M	Echinalosia ovalis Notospirifer Z	Scottville f	
	?			
Lissochonetes Z	China Tank M	Terrakea concavum Z	Exmoor f	
Pseudosyrinx Z	Road Canyon M	(Terrakea)	Gebbie IIIc IIIb IIIa	
Sowerbina Z	Cathedral Mt.	Echinalosia prideri Z Martinia adentata Z		
Antiquatonia Z	F			
Jakutoproductus Z	Lenox	Notostrophia zealandicus, N. homeri Zones	Sirius Shale	Chihsia
Tornquistia Z	Hills F		Tiverton F	F
Yakovlevia Z			Lizzie Ck F	
Tomiopsis Z	Neal Ranch ? F	Mourlonia impressa	Burnett F	Maping F
Orthotichia Z				
Kochiproductus Z	U. Gaptank	Atomodesma	Joe Joe F	

(New Zealand column vertical group labels: Maitai G, Productus Creek G, Takitimu G. USA Texas vertical label: Capitan. Arctic Canada vertical labels: Tahkandit F, Jungle Creek F.)

and Brachiopoda, the lower called the Sarginian Substage, followed by the Krasnouf-imian Substage. The latter name is preferred because of the wealth of its brachiopods near Krasnoufimsk to the alternate name Saraninian. Sarginian Fusulinacea include *Pseudofusulina** *makarovi* Rauser, and Saranin beds contain a distinct zone typified in part by *Parafusulina solidissma* (Rauser). Brachiopods also help distinguish the two levels. Characteristic Sarginian ammonoids include *Propinacoceras, Waagenina,* and *Neocrimites* with several other genera appearing for the first time. But as far as can be ascertained, there are no or very few ammonoids from the Krasnoufimian (=Saranin) level. Therefore the view that there is no distinctive Krasnoufimian zone, as expressed by Ruzencev (1956) and ignored by Furnish (1973), has no weight. However the two substages are relatively close to each other, whereas there are strong faunal contrasts between substages of the Asselian and Sakmarian Stages.

The two substages can be traced widely and are well discriminated even in Australia, New Zealand, and Pakistan (Table 2) on the basis of Brachiopoda. Fusulinacea and Ammonoidea were widespread during this stage. Substantial withdrawal of seas led to widespread terrestrial deposits or unconformity over parts of Arctic Canada, the Alps of Europe, and widely in eastern Australia. But marine deposits were more widespread in eastern Australia than realized, and indeed *Neocrimites* is found as a Baigendzinian index in the Bowen basin of Queensland.

Middle Permian Series

Kungurian Stage

The Kungurian Stage lay at the base of the Permian System defined by Murchison (1841) and is thus a very long-established unit, distinguished principally by Brachiopoda, including such species as *Streptorhynchus pelargonatum* (Schlotheim), *Pterospirifer alatus* (Sowerby), *Cleiothyridina pectinifera* (Sowerby), *Dielasma elongatum* (Schlotheim), and others, as well as distinctive bivalves and microfossils. Fusulinacea and Ammonoidea are rare in the sequences of the Russian Platform and Urals. Although Ruzencev (1956) acknowledged that ammonoids were too rare to allow comparison with those of the Baigendzinian Stage, some authorities such as Nassichuk (1970) have treated the Kungurian as part of the Baigendzinian. But this is to ignore other fossil evidence. Indeed, if only one stage is to be recognized, the Kungurian clearly must be granted priority over Baigendzinian. The assertion that the newly proposed Baigendzinian Stage should replace both the classic Artinskian and Kungurian Stages is based on an inflated view of the value of recent ammonoid studies. Miscorrelation is also involved, because these workers also recognize the so-called Roadian Stage, based on the Road Canyon Formation of West Texas, as occurring above the Baigendzinian faunas of the Cathedral Mountain beds. Available evidence indicates that the Roadian Stage is to be correlated with the Filippovian Substage at the base of the Kungurian Stage.

Two substages are recognized in the Kungurian Stage, the Filippovian and Irenian, but the Ufimian level at the top of the Irenian is faunally close to the Kazanian Stage (see Waterhouse, 1976a), and there is some need to improve nomenclature.

Within the Tethyan realm, the Kungurian Stage saw the first entry of primitive *Neoschwagerina,* and primitive members of the ammonoid family Cyclolobidae (*Glassoceras, Mexicoceras, Waagenoceras*). Brachiopod faunas were also widespread, and include those of the lower Zechstein and Magnesian Limestones of western Europe, and the Foldvik Creek beds of eastern Greenland, although these faunas have often been reassigned a later Permian age. Marine incursions occurred widely in eastern Australia and New Zealand.

Kazanian Stage

The Kazanian Stage is the second classic stage of the Permian System, recognized since the days of Murchison over the Russian Platform, principally through its distinctive brachiopods, including *Cancrinelloides* (aff. *Monticulifera*) and members of the *Licharewia* suite, which ranged widely across the Arctic and into eastern Australia and New Zealand. Two benthic zones, not very distinct, are widely present, named the Kalinovian and Sosnovian Substages (Waterhouse, 1976a, p. 127 ff). In more tropical regions the Kazanian faunas are further typified by advanced species of *Neoschwagerina (craticulifera* and *margaritae),* and by advanced *Waagenoceras.*

At the end of the Kalinovian Substage, the sea began to withdraw extensively from continental shelves in the Arctic and in eastern Australia, leading to very substantial emergence, and loss of living space especially in temperate conditions. Andesitic-basaltic island arc vulcanicity largely ceased in the Kazanian in Yunnan, the Himalayas, New Zealand, and probably eastern Australia, but persisted for a little longer in Mexico.

Punjabian Stage

The Punjabian Stage, revived by Stepanov (1973) from the unit proposed for sequences in the Salt Range, Pakistan, contains two substages, the Kalabaghian and Chhidruan, distinguished by brachiopods and apparently sharing the distinctive ammonoid *Cyclolobus* (Grant, 1970). Correlative brachiopod faunas in southern and eastern Asia contain the characteristic *Yabeina-Lepidolina* Fusulinacean assemblage. The fact that these rocks and faunas are typified by sections away from the type sequences of the Soviet Union has caused difficulties for correlation. To Grant (1970) and Waterhouse (1972) *Cyclolobus* was essentially correlative with another ammonoid *Timorites,* found with *Yabeina* in the Capitan formation of Texas, and in the eastern Pacific (Japan, Ussuriland). Indeed the two ammonoids, where found in one sequence, occur together in the Amarassi beds of Timor. But Miller and Furnish (1940) considered *Cyclolobus* to have been derived from *Timorites* and so Furnish (1973) placed the Punjabian Stage (or the Chhidruan Substage) almost at the top of the Permian, above the Djulfian Stage. This view is not followed by Stepanov (1973), Taraz (1973), Grant and Cooper (1973), Kozur (1974), Nakazawa and Kapoor (1977), and others. Rather, faunal evidence accrues yearly to support the basic correlations set out by Waterhouse (1972).

The Punjabian Stage was the last of the extensive marine stages of the Permian, and during and at the end of the stage seas withdrew from northern Siberia, much of the Himalayas, south-east Asia, and western Australia.

Djulfian (Dzhulfian) Stage

The Djulfian Stage is based in part on the Armenian sequences of Djulfa Gorge and Dorasham railway cutting, for the type sequence of the Baisalian Substage (Waterhouse, 1972), with faunas described in Ruzencev and Sarytcheva (1965). A suite of characteristic ammonoids includes species of *Vescotoceras, Djulfoceras, Vedioceras, Araxoceras* (Araxoceratidae), and *Codonofusiella* is a widespread Fusuline, although also found in older beds, so that the absence of *Neoschwagerina, Yabeina,* and *Lepidolina* carries the most weight. Characteristic brachiopods include *Septospirigerella* and *Dorashamia.*

The underlying substage, typified by *Codonofusiella* and the brachiopods *Crurithyris* and *Orbicoelia,* is named the Urushtenian Substage, from beds near Urushten, Greater Caucasus, by Waterhouse (1976a, p. 157), with substantial faunas described in part by Likharev (1936). The substage is also represented by *Codonofusiella* beds in southern Armenia. Taraz (1973) proposed a rival name Abadenian "Stage," here relegated to substage, based on good outcrops in central Iran (Taraz, 1974). Faunas are yet to be fully described from this sequence. The choice of names is of trivial concern, and essentially hinges on whether it is decided to grant custodianship for Permian type sequences to the Soviet Union as much as possible, or whether the good and relatively accessible outcrops in Iran should be granted recognition. There is no debate about their equivalence.

Djulfian faunas are relatively limited in extent. They probably include the late Capitan faunas of New Mexico and Mexico (Waterhouse, 1976a), and the huge Wuchiaping faunas of southern China where diagnostic Fusulinacea and ammonoids accompany large brachiopod faunas that definitely require revision. Waterhouse (1973b) referred the thick Greville beds that contain a Flemingitid ammonoid *Durvilleoceras* to the Baisalian Substage. However, Furnish et al (1976) asserted that the ammonoid was of mid-Scythian, i.e. Smithian age, at the base of the Triassic System. Because the *Durvilleoceras* beds are extensive and are overlain widely by further Late Permian and even Smithian faunas, this would mean we have to hypothesize a huge low-angle thrust, some 400 km long, to have moved Permian-Scythian from an unknown distance over alleged Scythian. The New Zealand sequences are better exposed than any sequences I have seen in the Alps of Europe or the Himalayas, let alone Armenia or the Salt Range, and show no sign of any such postulated thrust.

Another controversial fauna is in the Kathwai beds at the top of the Permian in the Salt Range, Pakistan. The beds contain *Ophiceras connectens* Schindewolf, suggesting an early Scythian (Griesbachian) age, and Permian brachiopod species that indicate a Djulfian age (Waterhouse, 1972), if correctly assessed by Grant (1970), though I do not now feel that a Griesbachian age can be ruled out on sedimentary grounds. Nor can we be too confident about the species similarities proposed by Grant (1970). But the essential point is that as Newell (1973) and Waterhouse (1973b) have shown, whatever the exact age, the fauna belongs to the Permian System and Grant (1970) was basically correct in treating the brachiopods as typically Permian. By contrast, Kummel and Teichert (1964) have repeatedly striven to treat the brachiopods as peculiar and exceptional *survivors*. This is a false and an unwarranted premise based on severe misunderstanding of world *Ophiceras* and *Otoceras* faunas.

Late Permian Series

Dorashamian Stage

The type sections for the Vedian and Ogbinan Substages of the Dorashamian Stage are based also on outcrops in southern Armenia (Waterhouse, 1972, 1976a, p. 168; Rostovstev and Azaryan, 1973). The Vedian Substage is characterized by two brachiopods, *Comelicania* and *Janiceps,* and the ammonoid *Phisonites.* The Ogbinan Substage has some typical brachiopod species and the restricted ammonoid *Paratirolites.* Vedian faunas are well represented in the upper Bellerophonkalk above Baisalian faunas in southeast Europe (Italy, Austria, Yugoslavia, and Hungary), and by the Gujo bivalve fauna in Japan and in the Stephens limestones of New Zealand. The rather poorly known Changhsing Limestone of southeastern China contains *Palaeofusulina,* and probably from stratigraphic position is equivalent to part of the Vedian and Ogbinan Substages. In northwestern Nepal, the faunas of the *"Echinalosia" kalikotei* Zone may belong to the Vedian Substage (see Waterhouse, 1976a, Fig. 30, p. 142).

Permian-Triassic Contact

Traditionally the worldwide transgressive sequence with *Otoceras,* often accompanied by the bivalve *Claraia,* has been regarded as heralding the start of the Triassic Period. Together with overlying *Ophiceras* "zones," these faunas have been incorporated in the Griesbachian Stage, below the Dienerian, Smithian, and Spathian Stages of the former Scythian Stage or Series (Tozer, 1967). But this view follows blindly the concepts expressed by Diener (1909) and others, that typical Permian life, especially brachiopods, perished before *Otoceras, Ophiceras, Claraia,* and other "Triassic genera" appeared. The view is not correct. It has been shown by Kozur (1973), Newell (1973), and Waterhouse (1973b, 1976a, p. 172) that virtually all fossil groups of early and perhaps late Griesbachian age are Permian in appearance, especially the Brachiopoda, on which Diener primarily relied. Only the Fusulinacea are not known in early Griesbachian faunas, suggesting that these perished perhaps during the Ogbinan Substage as the last of a rapidly dwindling remnant. Kozur (1973) has therefore reclassified the units as in Table 3 (reviewed by Waterhouse, 1976c).

It must be agreed that the Gangetian horizon is completely Permian in its attributes, and the Smithian level (which I consider to be a substage) is Triassic in its benthic attributes. The faunal change, which involved the critical loss of significant Permian Productida, occurred at the end of the Gangetian, Ellesmerian, or Gandarian levels.

Because there are reports of Productacea with *Ophiceras* (see Nakazawa et al, 1975) it would seem that the boundary should follow the Ellesmerian level, as Newell (1973) has proposed. But we are concerned now with extremely fine units of time and biochronology, and are hampered primarily by the lack of attention to microfossils and benthic fossils, especially Brachiopoda. The Makarewan fauna of New Zealand is interesting in this regard, because it appears to be of Gangetian or Brahmanian age, and underlies Smithian rocks and faunas. It lacks Productida and other members of the orders and superfamilies that perished during the Permian Period, but the constituent genera are all Permian in nature. Thus the fauna could represent a last true but reduced Permian fauna, or Early Triassic survivors from a catastrophe at the end of the Permian (Waterhouse, 1976a, p. 174; 1976c, p. 380).

Key Fossil Groups

Brachiopoda

Because Brachiopoda are the most common (Fig. 1) and probably most closely studied benthic fossils they must assume the main burden of correlation, and that they can bear this burden is substantiated by the worldwide and detailed correlations amplified in Waterhouse (1976a). Studies are being pursued widely, especially in the Soviet Union, and surely one of the greatest series of monographs of all time is now being issued on the superb brachiopod faunas of the Glass Mountains and West Texas, by Cooper and Grant (1972-1976).

Table 3. Alternate Classification of Early Triassic - Late Permian.

(Tozer, 1967)	(Kozur, 1973)			(Proposed Permian top)
Stage*	Stage*	Substage*	Key Fossils	
Spathian	Olenekian			
Smithian	Jakutian			?Waterhouse
Dienerian	Brahmanian	Gandarian		Newell
Griesbachian		Ellesmerian	Ophiceras commune	Kozur
	Dorashamian (upper)	Gangetian	Otoceras boreale	Tozer
			?Hypo-phiceras triviale	

*Stage and Substage as used by Tozer and Kozur: the rank should probably be scaled down.

Several major problems remain. Of these, the greatest is the need for modern studies of the brachiopods from the Lower Permian in the Urals. If the Soviet Union is to be custodian for the world standard of the Permian System, its scientists have a duty to restudy the brachiopods, because these are the most important for worldwide correlations. It is also imperative that the extensive collections from the Chhidru Formation and Kalabagh Member made by Pakistani and American geologists be monographed, because these world standards have few other important fossils, except the two key ammonoid genera *Xenodiscus* and *Cyclolobus*. A third area of interest lies in the Late Permian brachiopod faunas of southern China, in the Wuchiaping and Changhsing Formations. At present these are dated by stratigraphic succession, and by Fusulinacea and Ammonoidea. The dated descriptions of the brachiopods do not allow ready discrimination from Punjabian faunas of China (Grant, 1970). Presumably, this is because the descriptions need revision, and especially, attention to internal detail, unless there were peculiar, provincial developments.

Another question concerns the age of *Attenuatella* and other genera that widely typify the base of the Permian System in North America, Siberia, and the Arctic. In Verchyoyan perhaps, and certainly in Austria, *Attenuatella* appears to occur in what could be ?Kasimovian faunas as reviewed by Waterhouse (1976a). Faunas like those in Austria have also been described from northwestern Spain by Wagner and Prins (1970). The Austrian occurrence is just above late Moscovian or Kasimovian Fusulinacea described by Pasini (1963) and *Protriticites* and *Pseudotriticites* are found with rich brachiopod faunules just a little higher, indicative of a definite lower Kasimovian age in Prins (*in* Flügel, 1972). This means that *Attenuatella*, like *Tomiopsis*, first appeared during the Carboniferous Period, and that it was reinvigorated and widely dispersed at the start of the Permian Period. Yet Ramovs (1971) has asserted that the Cantabrian (i.e. Kasimovian) faunas of Spain are closest to those of "Orenburgian" not Kasimovian age with *Rugosofusulina alpina antiqua* in the Karavanke Mountains of Yugoslavia. The Orenburgian is, of course, either Permian or Carboniferous, and *Rugosofusulina*, like *Pseudofusulina*, may be limited to the Permian Period.

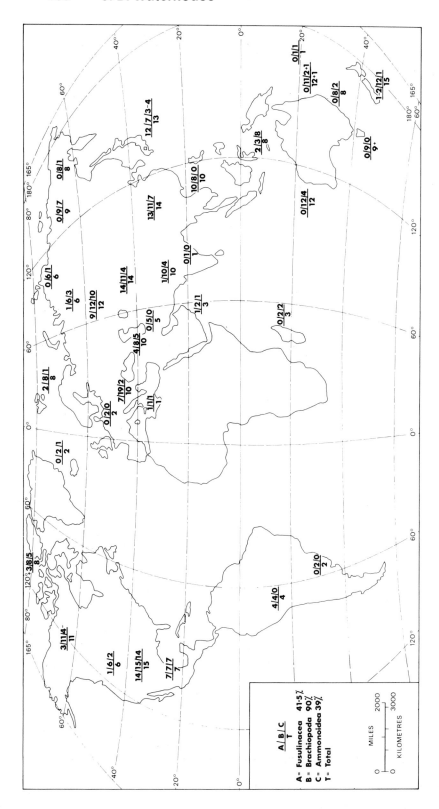

FIG. 1—Provisional estimates of number of substages represented faunally in various regions, with estimates of number of substages bearing Fusulinacea, Brachiopoda clearly were of major significance in Permian sequences as shown by this chart. Ammonoidea were counted even where rare, and gain because the Griesbachian-basal Dienerian is included, when ammonoids for the first time became widespread enough to allow worldwide correlation. Number of substages with Fusulinacea, 41·5%; number of substages with Brachiopoda, 90%; number of substages with Ammonoidea, 39%. Total number of substage stations with marine invertebrate faunas: 265. (from Waterhouse, 1976a, Fig. 7, p. 21).

Fusulinacea

Whether one assigns second place to Fusulinacea or to Ammonoidea may be a matter for argument, but even for the Early Permian standard, Ruzencev had to use fusuline zonation to establish the position of the Ammonoid zones which were much less persistent, and indeed, insignificant or missing from some formations (Barkhatova, 1964). Fusulinacea are indeed limited in their geographic range, but have been closely studied, and are especially useful for subdividing the monotonous carbonate sequences of the paleotropical biome. Agreement with brachiopod subdivisions is remarkably close, as shown by Waterhouse (1973a, 1976a). Recent work by Japanese authorities has added fusuline support for the brachiopod correlations proposed for the Djulfian and Punjabian Stages. Nakazawa and Kapoor (in press) have recently discovered *Neoschwagerina* (perhaps *margaritae*) in the lower Wargal Formation of the Salt Range, Pakistan, which suggest a probable Kazanian age and thus help underpin the validity and correlation of the overlying standard Punjabian Stage.

There are of course some conflicts or inconsistencies between brachiopod and fusuline correlations. There are inconsistencies in the ranges of *Neoschwagerina craticulifer* and *N. margaritae* to cast doubt over whether they were consistently Kazanian, or possibly commenced in the Kungurian Stage (Waterhouse, 1973a).

Toriyama (1973) and Nakazawa and Kapoor (1977) considered that *Lepidolina kumaensis* of southern and eastern Asia belongs to the Djulfian rather than Punjabian Stage, but this does not seem to be very firmly supported as yet by independent evidence, perhaps because fusuline studies have outpaced those on other faunal groups.

For the Glass Mountains, I have followed the conclusions for correlation by Cooper and Grant (1973), but a refreshing challenge is offered to aspects of their work on the basis of Fusulinacea and stratigraphic suggestions by Wilde (1975), involving such problems as the base of the Permian, and the Aktastinian-Baigendzinian equivalents, as discussed previously.

However, the most outstanding problem that aggravates use of key Fusulinacean genera is the appallingly inconsistent use of the genera *Pseudofusulina, Schwagerina*, and *Pseudoschwagerina* by Soviet and American schools, as reviewed by Kahler and Kahler (1966) and Waterhouse (1976a, p. 20). The sooner a reconciliation is achieved, the better, because the present confusion must constitute one of the most glaring examples of a breakdown in Linnean nomenclature to ever bedevil biologic taxonomy.

Ammonoidea

Ammonoid fossils are too rare in most sequences, and indeed are probably absent from far more sections than those in which they are present, to ever be of prime use for local or worldwide correlation. As a rule, they are found too sporadically to allow dating of more than a small part of the column, in only a few parts of a trough, shelf, or basin. Moreover, in spite of what has been written, virtually all species and most if not all genera were strongly provincial in distribution, and certainly no ammonoid genus of Permian age is as widely distributed as a number of brachiopod genera, particularly amongst the Productida and Spiriferida. Nonetheless, ammonoids have always been at the forefront of any scheme to subdivide and correlate Permian sequences, and one of the most outstanding studies in recent years, by Furnish (1973), following work by Ruzencev (1952, 1956), has provided a scheme that (especially for the Early and early Middle Permian) is both in agreement with brachiopod studies, and indeed preceded and guided brachiopod studies. The very rarity of ammonoid shells has encouraged a certain willingness to hazard long-ranging and bold correlations that have frequently been sustained by the more ponderous studies on numerous and diversified benthos.

An example may be taken from the Gympie basin of Queensland where *Neocrimites*, a genus especially typical of the Baigendzinian-Kungurian Stages and indeed found elsewhere in Queensland with Baigendzinian faunas, was described with benthic Fauna II (Runnegar and Ferguson, 1969) usually claimed to be of Sakmarian age. This would imply that the range of the ammonoid has to be extended. But restudy of the benthic fossils may well suggest that the ammonoid is Kungurian and that the Bivalvia, on which prime reliance was placed, and the Brachiopoda are indeed Kungurian as well. The Gympie benthos, in short, was completely miscorrelated by Runnegar and Ferguson (1969).

Most of the present divergence in views between brachiopod and ammonoid workers is concerned solely over one sequence, in the Salt Range of Pakistan. There are now three different views on the Salt Range correlation. On the basis of the fusuline *Colaniella*, Nakazawa and Kapoor (1977) correlated the Chhidru Formation with the Djulfian Stage instead of placing it above (Furnish, 1973), or below (Waterhoust, 1972; Stepanov, 1973; Grant and Cooper, 1973). The Salt Range sequence apart, agreement is very good indeed. I think upper Chhidru *might* be Djulfian.

Conodonts

Conodonts are becoming increasingly useful and potentially may assume an even more critical role although it appears that they failed to penetrate high Permian latitudes to the extent that none have been found in Australia for example, despite intensive search (Nicol, 1975). The various correlations proposed on the basis of conodont studies by W. A. Sweet have been condemned as too deferential by Grant and Cooper (1973), meaning that the work unquestioningly accepted a framework hypothesized from the Miller and Furnish (1940) model of *Cyclolobus* evolution. However, Grant and Cooper (1973) were able to reinterpret the basic ranges successfully. Kozur (1974) has now substantially revised both his own earlier studies and those by Sweet in favor of the brachiopod correlations set out by Waterhouse (1972, 1976a). Agreement is not complete, but is certainly very close.

Standard and Local Biochronologic Nomenclature

Selection and elucidation of type sections and names is now required, and best supervised by the Permian Subcommission of the Commission on Stratigraphic Nomenclature. I anticipate little further use for local stage names. Certainly I proposed some myself some years ago, but this was before the major clarification of the Permian Period, and before it was possible to correlate benthic faunas with faunas of the world standard. Now it is better to integrate local biozones into the world framework.

Yet authors persist in ignoring biozones, and in using a handful of stage names. Perhaps the worst example is that by Carter (1974), who in attacking the straw man offered by the long-outdated New Zealand stage scheme, studiously ignored the current more sophisticated zonal scheme (Waterhouse, 1973c), and even introduced a new biochronologic unit based on one gastropod species with no supporting fauna, no type section, no evidence for age, or any correlation. Such is not acceptable stratigraphy or paleontology. One of the most recent attempts proposed a local stage nomenclature by Clarke and Farmer (1976) for the Permian of Tasmania. This is a well-substantiated scheme, admittedly preceding faunal descriptions, but nonetheless meticulous in its attention to correct procedure over nomenclature and type sections and so on. Alas that the world standard other than for the Late Permian is still not nearly so well known. But I would judge that the biozones elucidated by Clarke and Banks (1975) for Tasmania will prove even more unappropriate for understanding the sequence there.

There will of course be protests from paleontologists in the United States about the failure to preserve the scheme based on the faunas of Texas and New Mexico. However there must be one world scheme, not two, and if the Russian names have priority they presumably may lay claim to the world standard sequences. I suppose we have to endure the platitudes about different standards for different parts of the world, but these are covert excuses for nationalistic or special interests. Instead, the recognition of formal biozones in a proper hierarchy (Waterhouse, 1976b) will meet the needs of biochronology when matched with standard international substages. Further advances would incorporate ecologic hierarchies as well, to recognize provinces and biomes (Waterhouse, 1976b).

Summary

In the preceding section, I have drawn attention to outstanding problems, and the persistence of some discrepancies in marine Permian correlation. But we may conclude that the major fossil groups are in good agreement and that most of the needs lie in the requirement for monographs of certain critical faunas and clarification of nomenclature, both biostratigraphic and systematic. It is doubtful if our understanding has progressed more substantially during the last decade for any period other than the Permian, thanks to the outstanding work by Ruzencev and Sarytcheva (1965), Furnish (1973), Cooper and Grant (1973), and Kozur (1974).

Paleomagnetic Stratigraphy

The Kiaman Magnetic Interval which is now so clumsily rechristened "The Late Paleozoic Interval of Predominantly Reversed Geomagnetic Polarity" provides a robust indication of Permian and Late to middle Carboniferous age (Valencio, 1972). Several short intervals of normal polarity provide prospects for more detailed correlation, including the Oak Creek Magnetic Event, near the Carboniferous-Permian boundary (McElhinney, 1969) and the Quebrada del Pimiento Magnetic Event in the middle Permian (Valencio and Vilas, 1972). An interval of several brief polarity changes called the Illawarra Zone may help define the top of the Permian or beginning of the Triassic Period (Valencio and Mitchell, 1972). These events have been identified in a few places over the globe, but so far the correlations proposed have not been substantiated to any very refined degree, as reviewed by Waterhouse (1974b). The paleomagnetic work is clearly just beginning, and there is some likelihood of discovering further normal events within the Permian segment. But promise is high, and should prove particularly useful for gaining time control of the middle and Late Permian terrestrial sequences.

Radiometric Ages

Recent advances in biostratigraphy for the Permian Period, as set out in Furnish (1973), Kozur (1974), and Waterhouse (1972, 1976a), have demolished the biostratigraphic framework used by Harland et al (1964), Banks (1973), and others for radiometric ages. A number of ages assigned in Harland et al (1964) were either generalized or wrong, and it is necessary to establish a new set of "absolute" values for the Permian Period, based on more recent work, and on reassessment of stratigraphy. The data will be retabulated and discussed in another study, and may be summarized here.

Radiometric Values

Permian-Carboniferous Boundary

Smith, *in* Harland et al (1964), and Banks (1973, p. 6575) accepted an age of about 280 m.y. for the base of the Permian, but Banks (1973) noted this value was minimal. From the revised constants proposed by Beckinsale and Gale (1969) for K-Ar, this age would fall nearer to 290 m.y. B.P. Russian sediments of "Sakmarian" age (now Asselian, Sakmarian), yielded values of 260 m.y., either Surenan or Uskalikian-Kurmaian, no doubt much too low a value, though not impossible to reconcile with younger and older values, provided we ignore values from rocks of similar age.

Asfanas'yev et al (1965, p. 1932) concluded that the base of the Permian could not be older than 270 to 275 m.y., or about 260 m.y. by "western" decay rates. Yet in the very preceding sentence they recorded a result of 310 m.y. for Early Permian or Late Carboniferous trachydacite porphyries.

Data from eastern Australia is of particular interest. In New South Wales, the Patterson toscanite (N1, Fig. 2) lies below the Seaham Varves and has been scraped by ice. It generally is regarded as Late Carboniferous (approximately Stephanian) together with the varves, but the varves could be as young as basal Permian (Kurmaian) because correlative beds contain the bivalve *Eurydesma* (Campbell *in* Packham, 1969; Waterhouse, 1974b). Thus, the toscanite could be latest Carboniferous, at just over 300 m.y. In Queensland, the Nychum Volcanics (Q5 in Fig. 2) with *Glossopteris* and *Cardiopteris* (suggestive of an Early Permian age, Balme, 1973), gives values ranging from 325 to 351 m.y. (Rb-Sr: $\lambda = 1.39 \times 10^{-11} yr^{-1}$) with two "best results" at 294 m.y. B.P. (Th-Pb gave 198 ± 116 m.y. B.P.).

These Australian results would thus suggest a boundary somewhere in the vicinity of 300 m.y. B.P. if not before. Moreover, the results are consistent with rather high values for slightly younger results in Queensland. For instance, the nearby Featherbed

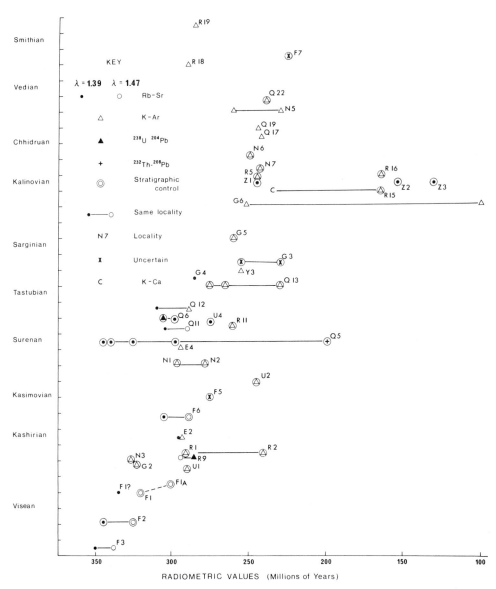

FIG. 2—Distribution of radiometric values for the Permian Period, as well as some Carboniferous and Triassic results. Samples are keyed to study by the author (now in preparation). Although new constants may require some emendation to the chart, the scatter will not be reduced, and the sequence will not be changed.

Volcanics of Queensland (Q6 in Fig. 2), slightly younger than the Nychum Volcanics, gave results of 299 m.y. and the related Elizabeth Creek granite may be of the same age, or older. It may be noted that such ages would conform moderately well with the various European granites that are dated as Stephanian, because the late Stephanian is possibly as young as Early Permian. The Lizzie Creek Volcanics, with Tastubian fauna yield values at 265 or 274 m.y. (K-Ar; now 276 m.y., Green and Webb, 1974), above the Bulgonunna Volcanics of 287 m.y. (Q11, Q13). A slightly younger granodiorite at Ridgelands, Queensland, cuts Sterlitamakian faunas to yield a result of 258 to 263 m.y.

Middle Permian

The Kungurian and Kazanian Stages are represented in New Zealand (Z1-3) and Russia (R5) by results of 235 to 240 m.y. approximately, if we select the better data. On the other hand, somewhat greater ages also are known. For instance the German lower Zechstein (G6) indicates 252 + 19 m.y. B.P. on K-Ar. It is accompanied by many lower values at less than 100 m.y. B.P. which have to be rejected (Kaemmel et al, 1970).

Late Middle, and Upper Permian

The Berkeley Latite in the lower Newcastle Coal Measures of New South Wales (N6 in Fig. 2) is post-Kazanian in age and, in giving a value of 250 m.y. (now ca 260 m.y.) is clearly out of phase with Russian and New Zealand values.

Various intrusives in Queensland have been dated simply as post-Kazanian. In modern terms, they are possibly younger than late middle Permian Flat Top and upper South Curra faunas which are probably Chhidruan in age. The granites presumably postdate the Chhidruan, and were perhaps as early as Urushtenian-Baisalian when uplift became widespread in eastern Australia at the same time as rapid depression of a trough in nearby New Zealand, suggestive of accelerated seafloor spreading. The values range from close to 245 m.y., which appear to fit well with the higher end of the range, to 225 m.y. The K-Ar value of 245 m.y. from the Gyranda Formation in the lower Blackwater Group appears to be well controlled stratigraphically. It supports high age values for the lower middle Permian, and itself is likely to be upper middle Permian.

Late Permian, Early Triassic

The top of the Permian is very poorly controlled. Falke (1972, Fig. 2, p. 298) showed a French result as Late Permian with a value of 225 m.y., but little is available about its stratigraphic control. A Late Permian value from the Stanthorpe granite, Queensland, yields a comparable result at 223 m.y.

The age of 225 m.y. proposed in Harland et al (1964) for the start of the Triassic Period, and upgraded to 230 m.y. by Green and Webb (1974) appears very acceptable, as does the proposed close of the Triassic Period at 200 m.y. B.P. (or 205 m.y.) proposed by Armstrong and Besancon (1970). But these values have no radiometric or stratigraphic support, and are in fact based on stratigraphic interpolation from older and younger ages. As shown later, such procedures are likely to be correct, but they are based on relative, not absolute procedures.

Discussion of Scatter

A somewhat optimized summary of radiometric values for the Permian Period (Fig. 2) shows a disheartening scatter of values that offers little immediate prospect of confidently delimiting the age and duration of the period, let alone delimiting subdivisions within the period. The values show that the base of the Permian is somewhere between 345 m.y. and 198 m.y. B.P. and that the middle Permian is somewhere between 245 m.y. and 165 m.y. In other words, the error is much greater than one period. By leaving out extreme values, the range is reduced to approximately 20 m.y. to 30 m.y., approximately one-half period in duration. Such results are (it must be emphasized) the best and most consistent, leaving aside many values for the sole reason that they look unacceptable. We can only attain any sort of resolution by heavily biasing the data through selection. A single age value, at 240 m.y. B.P. for instance, may be Moscovian (middle Carboniferous), or middle Permian, or, within 10 m.y. of Middle Triassic (Table 4). A single result has a 36% chance of incorrectly indicating the period, and about a 50% chance of falling within the correct third of the period, provided we "optimize" the results.

Alternate Correlation Paths

It is possible to trace alternate paths of absolute ages through the Carboniferous and Permian Periods, as shown in Figure 3. For the higher side of values, using samples that are well controlled stratigraphically, a reasonably straight line is established from

Table 4—Range of Published Radiometric Values (removing extreme deviants)
for Permian and Carboniferous Stages.

Stages	Range of Values (m.y. B.P.)
Late Permian	
Dorashamian Stage	291-300
Middle Permian	
Djulfian Stage	231-240
Punjabian Stage	241-260
Kazanian Stage	241-250
Kungurian Stage	231-260
Early Permian	
Baigendzinian Stage	261-270
Sakmarian Stage	(221-240)* 251-280
Asselian Stage	(200-210)* 251-310 (300-350)+
Late Carboniferous	
Orenburgian Stage	
Gshelian Stage	
Kasimovian Stage	241-300
Middle Carboniferous	
Moscovian Stage	(231-240)* 271-330
Bashkirian Stage	281-330
Early Carboniferous	
Visean Stage	331-350
Tournaisian Stage	341-350

*results too low; must be excluded.
+results too high; must be excluded.

Visean to late middle Permian, in spite of the likelihood that the substages need not be of equal duration, and this line would continue well into the currently accepted results for the Triassic Period. The value $\lambda = 1.39 \times 10^{-11}$ yr^{-1} is preferred for Rb-Sr; and the new constants for K-Ar are preferable to earlier values. The path is most strongly supported for Visean, Early Permian (Queensland), and late middle Permian Australian results. It suggests the following possible ages:

	Million Years B.P.
Start of Triassic (at Smithian Substage)	235-230
Dorashamian (Late Permian) Stage	245
Kungurian (middle Permian) Stage	275 (-5)
Asselian (Early Permian) Stage.	295 (+5)
Moscovian (middle Carboniferous) Stage	325

But these results lie considerably above many values from the middle and Late Carboniferous and middle Permian, so that it is possible to construct a second correlation path at values lower by 20 to 25 m.y. (Fig. 3). Such a large discrepancy, emphatically more than the 5 m.y. variation claimed by Banks (1973), is nearly 10% difference, and would involve, in relative terms, approximately one-half to one-third of the period. The new adjustment for K-Ar constants has considerably reduced the difference between the two correlation paths.

Low path of radiometric values:	Million Years B.P.
middle Permian (Kungurian Stage)	245-250
Early Permian (Asselian Stage).	260-265
middle Carboniferous (Moscovian Stage)	290

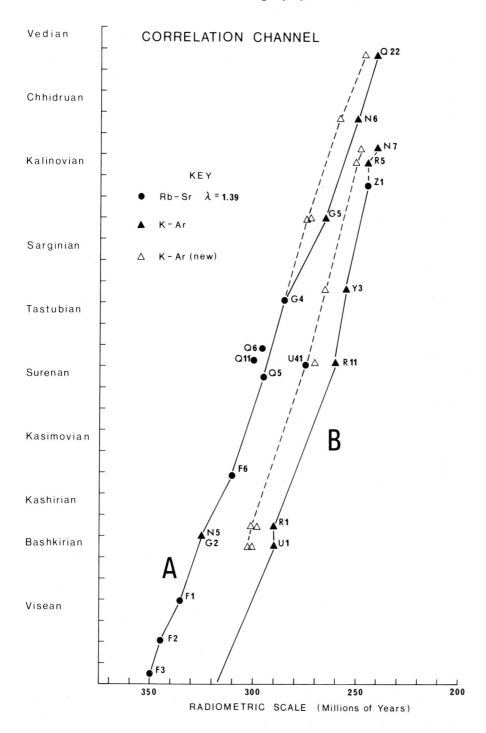

FIG. 3–"Correlation channel" of best or most consistent radiometric values for Permian Period, keyed as in Figure 2 to a study now in preparation. Open triangle of K-Ar uses the Beckingsale and Gale (1969) constant. The higher set of values is preferred herein.

These values agree with those of Harland et al (1964) for the middle Permian, but are otherwise lower.

At present, on radiometric evidence alone it appears difficult to choose between the two paths, which may be conceived as offering a "correlation channel" for the Permian Period. Both paths may be reconciled with underlying and overlying values. Insofar as the path of lower values includes a number of K-Ar results from sedimentary samples, with high possibilities of leakage, the path of higher values is perhaps more feasible even though it differs more from the set of values preferred by Harland et al (1964). It is, of course, all too facile to "correct" various values by explanations of leakage, or initially high concentrates of strontium or argon. These explanations may be correct, but they must first be related to a time line or "cline of values" itself subject to similar adjustments and corrections on a nonstatistical, nonexperimental basis. In that particular aspect, so-called absolute values certainly lack the objectivity presented by paleontologic data, where uncertainty over constants and initial ratios and metamorphic interference form no part of the problem. The problem instead lies more with the paleontologists than with the original materials. The scatter is summarized in Table 5.

Table 5. Summary of Deviant Results from Low or High Correlation Paths.

Interval	Radiometric Value	% deviant results rejecting high or low deviant values	% deviant results retaining high or low deviant values
Ages following high values:			
Middle Permian	240-270 m.y.	22%	22%
Early Permian	270-300 m.y.	45%	68%
Middle, Late Carboniferous	300-330 m.y.	62%	67%
Early Carboniferous	330-350 m.y.	40%	40%
Total		42%	55%
Ages following low values:			
Middle Permian	240-250 m.y.	44%	44%
Early Permian	250-260 m.y.	67%	79%
Middle, Late Carboniferous	260-290 m.y.	65%	67%
Total		59%	67%

Solar Chronology: A New Procedure to Attain Absolute Chronology

There appears to be little prospect of obtaining an absolute time scale for the Paleozoic Era through radiometric procedures unless these can be independently substantiated, because radiometric values do not show sufficient internal consistency. Improved laboratory techniques and improved constants have not reduced the scatter in recent years. Instead the uncertainty grows as more and more data is accumulated to underline the basic fact that original composition, pressure, and temperature were all variables, not constants, against a steady cline for time.

A broader multidisciplinary approach may resolve these difficulties, and enable radiometric values to be treated as metamorphic rather than temporal indices. Towards this aim, it is timely to consider a new and integrated approach towards the problems of both absolute and relative chronology through the use of rhythmic and regular changes in the world's climatic regime. The thesis is based on these assumptions: (1) that biota have responded through migration, death, and evolution to climatic changes, as statistically demonstrated by Waterhouse and Bonham-Carter (1976); (2) that climatic change therefore lies at the root of biochronology—correlation entails the matching of a change (probably in solar energy budget) in one direction all over the globe, more or less simultaneously, although various after-effects and local effects overprint this basic pattern; and (3) radiometric evidence strongly suggests a periodicity of approximately 30 million years for major coolings, and sometimes glaciations, that were also coeval with major changes to life. There are suggestions of changes of lesser duration, and also of very warm intervals and super-cold intervals but these have yet to be analyzed. If we accept the 30 million year periodicity we have a model that enables radiometric values to be tested (Fig. 4).

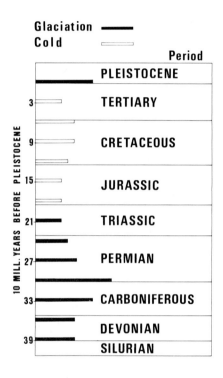

FIG. 4–Distribution through time of major cold intervals that appear to have been spaced at 30 m.y. intervals. Black bars based on major glacial episodes, but poorly based for the Triassic. Dotted bars indicate cold (but not glaciated) intervals, based on oxygen isotope data, poorly established for Early Jurassic.

Various aspects of the geologic record back to the Early Cretaceous strongly suggest a periodicity of close to 30 m.y. for major cold climatic episodes with temperature evaluated from combined techniques of the stratigraphic record, oxygen isotope values, and floral and faunal affinities and diversities, correlation based on stratigraphic, paleontologic and paleomagnetic data, and absolute ages based on radiometric values. An Early Cretaceous cooling within the Neocomian is not so well defined (see Dettmann and Playford, 1969, p. 202-203) although Russian data suggest low temperatures (Teis et al, 1957). The early record is more speculative, but possibly fits the 30 m.y. periodicity. Although early Mesozoic coolings are at present poorly defined, Bowen (1966) recorded indications of a Pliensbachian cooling at 180 m.y. If the 30 m.y. periodicity were maintained, one may also be expected in or near the Oxfordian, assuming that radiometric values correspond with the age assigned to the rocks. There is very poor control for the Triassic Period, where radiometric ages are few, variable, and clearly untrustworthy, and where faunas are little understood in terms of communities and biomes, and effect of climate. Nor is there any marked evidence for severe climatic cooling in the rocks, although the writer has recently discovered a diamictite in the Upper Countess Formation of upper Anisian age in the Countess Range, southern New Zealand. It is believed that New Zealand then lay close to the South Pole, from paleomagnetic and faunal evidence (see Waterhouse, 1974a), and the diamictite might have been caused by a short-lived glaciation, apparently of post-Smithian age and apparently before the upper Anisian.

By comparison, Permian stratigraphy, climates, and faunas and radiometry are under much better control (Fig. 5). A number of glaciations have been established, but it is clear that there have been ripple effects, with a major glaciation followed fairly shortly by one or two further glacial episodes so that very probably the first of each of the group may well have been at the 30 m.y. mark or at least at some regular spacing.

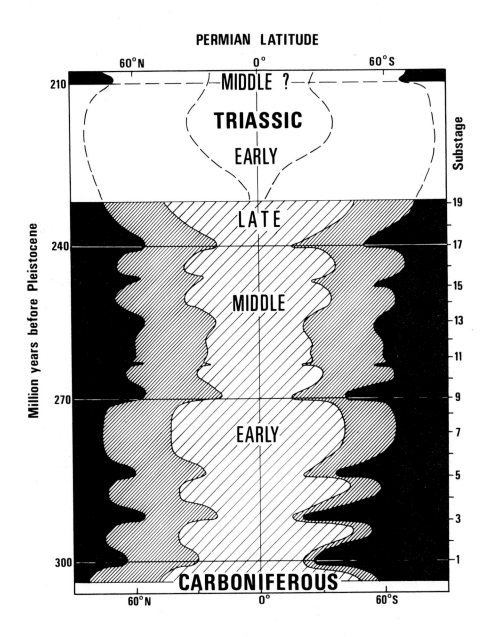

FIG. 5—Climatic change through the Permian Period, as indicated by movement of faunal biomes (major faunal assemblages) traced by Waterhouse and Bonham-Carter (1975). Expansion of the cold (black) biome at the expense of the temperate (close strips) and tropical (wide strips) biome indicates a cold interval. Thus geographic movements of life reinforce sedimentary and geochemical evidence for solar chronology.

If we assume a 30 m.y. spacing for the first of each of the three clusters, we have a first cluster at the start of the Permian, a second in the middle Permian, and a third in the late Permian. This would suggest a longevity for the period of nearly 70 m.y. rather than the estimate of 55 m.y. in Harland et al (1964) and other works. Since it is now shown that the Permian of the post-Kazanian age is much longer than envisaged by Harland et al (1964) the difference in values is well founded, and the period must clearly have lasted longer than previously envisaged (see Table 6).

Table 6—Proposed Episodicity for Major Dramatic Coolings.

Millions Years B.P.	Stage (Substage)	Period
180	Pliensbachian	Jurassic
210	Early Anisian	Triassic
240	Vedian	Permian
270	Filippovian	Permian
300	Surenan	Permian
330	?Vereiian (?Bashkirian)	Pennsylvanian

These values agree well with the higher side of the correlation channel for the Permian Period set out previously, allow a modest duration for the Triassic Period, and fit well with Mississippian (Lower Carboniferous) values from the Visean Stage.

The earlier glacial record is moderately well defined for the rest of the late Paleozoic Era, with two extensive glacial episodes and corresponding faunal changes in the Early and Late Devonian Period (Waterhouse, 1974c). At a 30 m.y. periodicity this would involve only slight corrections to the age assigned in Harland et al (1964) to the Famennian (i.e. 360 m.y. instead of 353 m.y.), and the start of the Gedinnian Stage (at 390 instead of 395 m.y.). Of course, this good agreement is contingent on the correctness of the higher values of my Permian correlation channel.

The Late Ordovician glaciation would almost fall on the same trend, if the Silurian Period were allotted a duration close to 32 m.y. as proposed by Boucot (1975, p. 65).

Leaving aside the implications for planetary motion and solar budgets for the earth, it may be seen that the 30 m.y. periodicity may indeed provide a framework against which to test radiometric values. In this framework, stratigraphic and paleontologic procedures are critical for assessing the geologic positions of the radiometric samples, and stable isotopes also assume major importance because they can be used to calibrate temperature changes. Whether finer subdivisions within the 30 m.y. cycle are possible remains to be examined. For the Permian Period (where studies are relatively advanced and climatic cycles are unusually well defined), the initial glacial episode was followed by two further glaciations before the start of the next cycle. Thus minor cycles were superimposed on the major curves, and these must offer good prospects for age determinations by means of the criteria set out previously. But the duration of the intervals requires much more sophisticated work by radiometric study with support from oxygen isotope studies.

References Cited

Armstrong, R. L., and J. Besancon, 1970, A Triassic time scale dilemma—K-Ar dating of Upper Triassic mafic igneous rocks, eastern U.S.A. and Canada and post-Upper Triassic plutons, western Idaho, U.S.A.: Eclogae Geol. Helvetiae, v. 63, no. 1, p. 15-28.

Afanas'yev, G. D., et al, 1965, Basic data on the age of boundaries between certain geological systems and epochs: Internat. Geology Rev., v. 7, p. 1928-1948.

Balme, B. E., 1973, Age of a mixed Cardiopteris-Glossopteris flora: Geol. Soc. Australia Jour., v. 20, no. 1, p. 103-104.

Banks, P. O., 1973, Permian-Triassic radiometric time scale in the Permian and Triassic Systems and their mutual boundary: Canadian Soc. Petroleum Geologists Mem. 2, p. 669-677.

Barkhatova, V. P., 1964, The status of the Schwagerina beds in connection with the problem of the boundary between the Carboniferous and the Permian: 5th Cong. Int. Strat. and Geol. Carbonifere C. R., v. 1, p. 273-282.

Beckingsale, R. D., and N. H. Gale, 1969, A reappraisal of the decay constants and branching ratio of K^{40}: Earth and Planetary Sci. Letters, v. 6, p. 289-294.

Boucot, A., 1975, Evolution and extinction rate controls: New York, Elsevier, p. 1-427.

Bowen, R., 1966, Paleotemperature analysis: Amsterdam, Elsevier, 265 p.

Carter, R. M., 1974, A New Zealand case study of the need for local time scales: Lethaia, v. 7, no. 3, p. 181-202.

Clarke, M. J., and M. R. Banks, 1975, The stratigraphy of the lower (Permo-Carboniferous) parts of the Parmeener Super-Group, Tasmania: 3rd Gondwana Sym. (Canberra) Papers; Canberra, Australian Nat. Univ. Press, p. 453-467.

—— and N. Farmer, 1976, Chronostratigraphic nomenclature for late Palaeozoic rocks in Tasmania: Royal Soc. Tasmania, Pap. Proc. 110, p. 91-109.

Cooper, G. A., and R. E. Grant, 1972-6, Permian brachiopods of West Texas; Parts I-IV: Smithsonian Contr. Paleobiology, no. 14 (1972); no. 15 (1974); no. 19 (1975); no. 21 (1976).

—— —— 1973, Dating and correlating the Permian of the Glass Mountains in Texas: Canadian Soc. Petroleum Geologists Mem. 2, p. 363-377.

Dettmann, M. E., and G. Playford, 1969, Palynology of the Australian Cretaceous, in K. S. W. Campbell, ed., Stratigraphy and paleontology: Canberra, Australian Nat. Press, p. 174-201.

Dickins, J. M., 1976, Correlation chart for the Permian System in Australia: Australian Bur. Mineral Resources Geology and Geophysics Bull., v. 156B, p. 1-26.

Diener, C., 1909, Lower Triassic Cephalopoda from Spiti, Malla Johar, and Byans: India Geol. Survey Mem. Palaeon. Indica, ser. 15, v. 6, no. 1, p. 1-186.

Dunbar, C. O., 1940, The type Permian—its classification and correlation: AAPG Bull., v. 24, no. 2, p. 237-281.

Falke, H., 1972, The palaeogeography of the continental Permian in central, west, and in part of south Europe, in H. Falke, ed., Rotliegend, Essays on European lower Permian: Leiden, E. J. Brill, p. 281-299.

Flügel, H. W., 1972, Fuhrer zu den excursionen der 42 Jahresversammlung der Palaontogischen Gesellschaft in Graz.

Furnish, W. M., 1973, Permian stage names: Canadian Soc. Petroleum Geologists Mem. 2, p. 522-548.

—— et al, 1976, Reinterpretation of ceratitic ammonoids from the Greville Formation, New Zealand: Geol. Mag., v. 113, no. 1, p. 39-46.

Grant, R. E., 1970, Brachiopods from Permian-Triassic boundary beds and age of Chhidru formation, West Pakistan, in B. Kummel and C. Teichert, eds., Stratigraphic boundary problems: Lawrence, Kansas, Kansas Press, p. 117-153.

—— and G. A. Cooper, 1973, Brachiopods and Permian correlations: Canadian Soc. Petroleum Geologists Mem. 2, p. 572-595.

Green, D. C., and A. W. Webb, 1974, Geochronology of the northern part of the Tasman Geosyncline, in The Tasman geosyncline—a symposium: Geol. Soc. Australia, Queensland Division, p. 275-293.

Harland, A. B., A. G. Smith, and B. Wilcock, eds., 1964, The Phanerozoic time scale, a symposium: Geol. Soc. London Quart. Jour., v. 120, supp., 458 p.

Kaemmel, T., et al, 1970, Radiogeochronologische daten vom Perm der DDR zur Gewinnung von Eichpunkten fur die internationale geochronologische Skala: Zeitschr. Angew. Geologie, v. 16, no. 2, p. 57-63.

Kahler, F., and G. Kahler, 1966, Fusulinida (Foraminiferida), in F. Westphal, ed., Fossilum catalogue; 1 Animalia: The Hague, Parts 111-114, 974 p.

Kozur, H., 1973, Beitrage sur stratigraphie und Palaontologie der Trias: Geol.-Paleontol. Mitt. Innsbruck, v. 3, no. 1, p. 1-30.

—— 1974, Zur Altersstellung des Zechsteinkalkes (ca1) innerhalb der "tethyalen" Permgliederung: Freiberger Forshungshefte Reihe C., no. C298, p. 45-50.

Kummel, B., and C. Teichert, 1964, The Permian-Triassic boundary in the Salt Range of West Pakistan: 22nd Int. Geol. Cong. (India) Rept. Vol. Abs., p. 120.

Likharev, B. K., 1936, Brachiopoda of the Permian System of the USSR: Central Geol. and Prosp. Inst. Paleon. USSR Mon., v. 39, no. 1, p. 1-152.

Marcour, J., 1859, Dyas et Trias ou le Nouveau gres rouge en Europe, dans l'Amerique du nord et dan l'Inde: Archives Sci., v. 5, p. 5-37, 116-146.

McElhinny, M. W., 1969, The palaeomagnetism of the Permian of southeast Australia and its significance regarding the problem of intercontinental correlation: Geol. Soc. Australia Spec. Pub., v. 2, p. 61-67.

Miller, A. K., and W. M. Furnish, 1940, Permian ammonoids of the Guadalupe Mountain Region and adjacent areas: Geol. Soc. America Spec. Paper, v. 26, p. 1-244.

Murchison, R. I., 1841, Letter to Dr. Fisher v. Waldheim: Philos. Mag., N.S., v. 19, p. 418.

Nakazawa, K., and H. M. Kapoor, (1977), Correlation of the marine Permian in Gondwana and the Tethys: 4th Int. Gondwana Symposium (Calcutta), India Geol. Survey, p. 1-18.

—— et al, 1975, The Upper Permian and Lower Triassic in Kashmir, India: Kyoto Univ. Coll. Sci. Mem. (Ser. Mineral.), v. 42, no. 1, p. 1-106.

Nassichuk, W. W., 1970, Permian ammonoids from Devon and Melville Islands, Canadian Arctic Archipelago: Jour. Paleontology, v. 44, p. 77-97.

Newell, N. D., 1973, The very last moment of the Palaeozoic era: Canadian Soc. Petroleum Geologists Mem. 2, p. 1-10.

Nicol, R. S., 1975, The effect of Late Carboniferous-Early Permian glaciation on the distribution of Permian conodonts in Australasia: Geol. Soc. America Abs. with Programs, v. 7, no. 6, p. 828-829.

Packham, G. H. (ed.), 1969, The geology of New South Wales: Geol. Soc. Australia Jour., v. 16, no. 1, p. 1-654.

Pasini, M., 1963, Alcuni Fusulinida della serie del monte a Auernig (Alpi Cammieche) e loro significato stratigrafico: Riv. Italiana Paleontologia e Stratigrafia, v. 69, no. 3, p. 337-371.

Ramovs, A., 1971, Connections of the Upper Carboniferous brachiopod faunas from the Carnic Alps and Karavanke Mountains with those of the Cantabrian Mountains (Spain), in The Carboniferous of northwest Spain, part II: Oviedo, Univ. Fac. Cienc., no. 4, p. 373-377.

Rausser-Chernosova, D. M., 1949, The stratigraphy of Upper Carboniferous and Artinskian deposits of Bashkirian Cisuralia: Akad. Nauk USSR Inst. Geol. Sci. Tr., v. 105, p. 3-21.

Rostovtsev, K. O., and N. R. Azaryan, 1973, The Permian-Triassic boundary in Transcaucasia: Canadia Soc. Petroleum Geologists Mem. 2, p. 89-99.

Runnegar, B., and J. A. Ferguson, 1969, Stratigraphy of the Permian and Lower Triassic marine sediments of the Gympie District, Queensland: Queensland Univ. Dept. Geology Papers, v. 6, no. 9, p. 247-281.

Ruzencev, V. E., 1952, Biostratigraphy of the Sakmarian Stage in the province of Aktyubinsk, Kazakh USSR: Akad. Nauk SSR Lab. Trudy Paleon. Inst., v. 42, p. 22-87.

—— 1956, Lower Permian ammonites of the southern Urals II, Ammonites of the Artinskian stage: Akad. Nauk SSSR Lab. Trans Paleon. Inst., v. 60, p. 1-275.

—— 1965, Principal ammonoid assemblages of the Carboniferous period: Paleon. Zh., v. 27, p. 3-17.

—— and T. G. Sarytcheva, 1965, The development and succession of marine organisms at the Paleozoic Mesozoic boundary: Akad. Nauk SSSR Paleon. Inst. Tr., v. 108, p. 1-431.

Stepanov, D. L., 1973, The Permian System in the USSR: Canadian Soc. Petroleum Geologists Mem. 2, p. 120-136.

Taraz, H., 1973, Correlation of the uppermost Permian of Iran, central Asia and south China: AAPG Bull., v. 57, no. 6, p. 1117-1133.

—— 1974, Geology of the Surmaq-Deh Bid Area, Abadeh Region, central Iran: Geol. Surv. Iran Rept. 37.

Teis, R. V., M. A. Chupakhin, and D. P. Naidin, 1957, Determination of paleotemperatures from isotopic composition of oxygen in calcite of certain Cretaceous fossil shells from Crimea: Geokhim. (Akad. Nauk SSSR), v. 9, p. 323-329.

Toriyama, R., 1973, Upper Permian fusulinacean zones: Canadian Soc. Petroleum Geologists Mem. 2, p. 498-512.

Tozer, E. T., 1967, A standard for Triassic time: Canada Geol. Survey Bull., v. 156, p. 1-103.

Valencio, D. A., 1972, Intercontinental correlation of Late Palaeozoic South American rocks on the basis of their magnetic remanences: Acad. Brasiliera Cienc. Anais, v. 44, Supp., p. 357-364.

—— and J. Mitchell, 1972, Palaeomagnetism and K-Ar ages of Permo-Triassic igneous rocks from Argentina and the international correlation of Upper Palaeozoic-Lower Mesozoic formations: 22nd Int. Geol. Cong. (Montreal) Sec. 3, Tectonics, p. 189-195.

—— and J. F. Vilas, 1972, Palaeomagnetism of Late Palaeozoic and Early Mesozoic rocks of South America: Earth and Planetary Sci. Letters, v. 15, p. 75-85.

Wagner, R. H., and C. F. Winkler Prins, 1970, The stratigraphic succession, flora and fauna of Cantabrian and Stephanian A rocks at Barruelo, (prov. Palencia), N.W. Spain, in Coloquie Strat. Carbonifer: Univ. Liege, v. 55, p. 487-551.

Waterhouse, J. B., 1967, Proposal of series and stages for the Permian in New Zealand: Royal Soc. New Zealand Geol. Trans., v. 5, no. 6, p. 161-180.

—— 1972, The evolution, correlation and paleogeographic significance of the Permian ammonoid family Cyclolobidae Zittel: Lethaia, v. 5, p. 230-252.

—— 1973a, Permian brachiopod correlations for south-east Asia: Geol. Soc. Malaysia Bull., v. 6, p. 187-212.

—— 1973b, An Ophiceratid ammonoid from the New Zealand Permian and its implications for the Permian-Triassic boundary: Geol. Mag., v. 110, no. 4, p. 305-329.

—— 1973c, Communal hierarchy and significance of environmental parameters for brachiopods, the New Zealand Permian model: Royal Ontario Mus. Life Sci. Contributions, v. 92, p. 1-49.

—— 1974a, The significance of Permian faunal biomes in reconstructing Gondwana: Acad. Brasiliera Cienc. Anais, v. 44, Supp., p. 365-381.

—— 1974b, The Permian Carboniferous boundary for Gondwana: Acad. Brasiliera Cience. Anais, v. 44, Supp., p. 383-391.

—— 1974c, Upper Palaeozoic era: Encyclopedia Britannica, v. 15, p. 921-930.

—— 1976a, World correlations for marine Permian faunas: Queensland Univ. Dept. Geology Papers, v. 7, no. 2, p. i-xvii; 1-232.

—— 1976b, The significance of ecostratigraphy and need for biostratigraphic hierarchy in stratigraphic nomenclature: Lethaia, v. 9, p. 317-325.

—— 1976c, Permian-Triassic boundary in New Zealand: New Zealand Jour. Geology Geophysics, v. 19, no. 3, p. 373-384.

—— and G. Bonham-Carter, 1975, Global distribution and character of Permian biomes based on brachiopod assemblages: Canadian Jour. Earth Sci., v. 12, no. 7, p. 1085-1146.

—— —— 1976, Range, proportionate representation and demise of brachiopod families through Permian Period: Geol. Mag., v. 113, no. 5, p. 401-428.

Wilde, G. L., 1975, Fusulinid-defined Permian stages, *in* Permian exploration, boundaries, and stratigraphy: West Texas Geol. Soc. and Permian Basin Sec. SEPM Pub., p. 67-83.

Zavodowsky, V. M., et al, 1970, Polyroy atlas permskoy fauny: flory Severo-Vostoka, SSSR, *in* M. V. Kulikov, ed., Atlas of Permian faunas and floras of northeastern USSR: 407 p.

Report on Isotopic Dating of Rocks in the Carboniferous System[1]

A. BOUROZ[2]

Abstract Three main stratigraphic scales are used regionally for the Carboniferous System: U.S.A., Western Europe, and U.S.S.R. Correlation problems among these three classifications are far from having precise solutions. It is necessary for each absolute age measurement, to point out the regional classifications to which the analyzed sample belongs, because in any case the measured age cannot be considered as having a world value. The partial results acquired by now are as follows:

    (1) Devonian-Carboniferous boundary, 330 m.y. (Western Europe, Armorican Massif); (2) Mississippian-Pennsylvanian boundary, less than 320 m.y. (U.S.A., under the gap between the two subsystems); (3) Bashkirian-Moscovian boundary, 308 m.y. (U.S.S.R., Donetz basin); (4) Middle-Upper Carboniferous, (Westphalian-Stephanian) boundary, 300 m.y. (Western Europe, Cevennes basin); and (5) Carboniferous-Permian boundary, 290 m.y. (U.S.S.R., Akchatan Granite).

Introduction

If we want to measure absolute ages of successive boundaries in a sedimentary sequence, these measurements are subject to two restrictions: these measurements must be made on samples which have not suffered subsequent metamorphism; and these samples should come from levels as close as possible to the stratigraphic level to be dated.

This is self-explanatory but for the Carboniferous System, a third problem arises because we are using three different stratigraphic scales in three regions of the equatorial belt (on a pre-drift model): USA, Western Europe, and the USSR. It is important to note that correlations among these three stratigraphic scales are far from having precise solutions up to now. They are still studied by all the working groups of the Subcommission on Carboniferous Stratigraphy (SCCS) and concerning the series boundaries, by the joint working-groups Subcommission on Devonian Stratigraphy - SCCS and SCCS - Subcommission on Permian Stratigraphy which are beginning work at present.

Consequently (concerning all absolute age measurements obtained to date), it is absolutely necessary to indicate what regional scale the analyzed sample belongs to, because any one age measurement cannot be considered as having a worldwide value.

We use some data on isotopic dating in the Carboniferous System especially from *The Phanerozoic Time Scale* (Harland et al, 1964) and *Certain Key Data for the Phanerozoic Time Scale* (Afanas'yev, 1970) but for the Devonian-Carboniferous and the Westphalian-Stephanian boundaries, we have new data obtained from the publications concerning those symposia.

©Copyright 1978 by The American Association of Petroleum Geologists. All rights reserved.

AAPG grants permission for a *single* photocopy of this article for research purposes. Other photocopying not allowed by the 1978 Copyright Law is prohibited. For more than one photocopy of this article, users should send request, article identification number (see below), and US $3.00 per copy to Copyright Clearance Center, Inc., One Park Ave., New York, NY 10006.

[1] Manuscript received, October 5, 1976.

[2] Charbonnages de France, 75008 Paris, France (Past chairman, Geological Survey).

Article Identification Number: 0149-1377/78/SG06-0023/$03.00/0

Devonian-Carboniferous Boundary

The dating concerning this boundary agrees near 330 m.y.; a new measurement (see Table 1) was done last year on a welded-tuff (ignimbrite) occurring in the well-defined base of the Tournaisian located in the Laval basin, Armorican Massif, France. The time determined was 330 ± 6 m.y.

It is in agreement with the data admitted on that limit before, if we keep the base of the *Gattendorfia* Zone as a limit. But the discussions occurring in the IUGS Devonian-Carboniferous Joint Working Group might change this result if the working group chooses the base of the Wocklumeria Zone, or another one, as a new boundary (in Alaska for instance, between the Wocklumeria and Gattendorfia Zones, there are 800 or 1,000 m of sediments).

Table 1. Radiometric Dating for the Devonian-Carboniferous Boundary

Speakers: C. Boyer,[1] A. Pelhate,[2] P. Vidal[3]
Reporting: Devonian-Carboniferous Boundary
Sample One:

> *Name, Stratigraphic Unit* – lower part, Huisserie Formation
> *Geographic Location* – Laval basin, Amorican massif (France)
> *Stratigraphic Position* – lower Tournaisian
> *Petrologic Designation* – ignimbrite (welded tuff)
> *Analytic Details* –
>> (1) *method used:* rubidium-strontium
>> (2a) *age and uncertainty (1 sigma or 2 sigma):* 2 sigma, 330 ± 6 m.y.
>> (2b) *if isochron method applied, which regression program used:* Brooks et al, 1972
>> (3) *decay constant used:* $\lambda Rb^{87} = 1.47 \cdot 10^{-11} y^{-1}$
>> (4) *evaluation of reliability:*
>> (5) *additional comments:* $Sr^{87}/Sr^{86} = 0.7095 \pm 0.005$; Wendt Index = 1.1; isochron established with six whole-rock samples.
>
> *Analyst* – P. Vidal
> *Institution* – Centre Amorician d'Etude Structurale des Socles
> *Published Reference* – unpublished

Sample Two: (Fill in Details as Above)
Sample Three: (Fill in Details as Above)

---

[1] Laboratoire de Pétrographie Volcanologie, Université de Paris, Orsay, France.
[2] Laboratoire de Geologie, Université de Bretagne Occidentale, Brest, France.
[3] Centre Amorician d'Etude Structurale des Socles, Rennes, France.

Mississippian-Pennsylvanian Boundary

During the last Carboniferous Congress held at Moscow (September 1975) a proposition was made for unifying the Lower-Middle Carboniferous boundary in the whole of the Equatorial Belt (Bouroz et al, in press). If the proposition is accepted, the boundary will be approximately at the base of the *Reticuloceras* Zone which characterizes (but with a gap) the Mississippian-Pennsylvanian boundary in North America, and the Namurian A-Namurian B of the Heerlen Classification (1935). It would be desirable that new measurements be made in eastern Europe and Siberia at the Serpukhovian-Bashkirian boundary which has the same definition (base of the *Reticuloceras* Zone).

Mamet made a measurement on this boundary in North America, at the base of the gap between the two subsystems (in the *Eumorphoceras* Zone). It must be considered as slightly too old compared with the base of the *Reticuloceras* Zone.

Bashkirian-Moscovian Boundary

This boundary, well defined in the Donetz basin and in the southern Ural area (especially by marine microfauna) is equivalent approximately to the Westphalian B-Westphalian C limit but a precise correlation has not been done until now because of a lack of significant marine microfauna in northwestern Europe where the stratotype of the Westphalian is located.

A measurement was done on samples of a glauconite level located in the Vereyan Substage, at the base of the Moscovian of the Donetz basin (Afanas'yev, 1970). The time found was 308 ± 10 m.y.

That boundary is nearly equivalent to the Morrowan-Atokan boundary in the USA but this correlation needs to be confirmed by more precise work.

Middle and Upper Carboniferous (Westphalian-Stephanian?) Boundary

Historical definition of the Stephanian (Munier-Chalmas and de Lapparent, 1893) was based essentially on a floral variation. Later the same variation was observed first in Donetz basin and afterwards, in northwestern Spain (Wagner et al, 1973). In the USSR, it coincides with the appearance of the fusulinid genus *Protriticites*. In Spain, occurrence of several levels of limestone with fusulinids among an abundant flora will allow satisfactory correlations with the Donetz basin on the one hand, and with the French Stephanian (stratotype) on the other hand, but the lack of marine fauna in the French Stephanian will lead to the choice of a new stratotype in the Russian or perhaps in the Spanish area.

An absolute age measurement was done on a cinerite of the Cevennes basin (Bouroz et al, 1972) located in a sequence equivalent to the upper Cantabrian (Bouroz, 1970). The age found was 300 ± 10 m.y. (Table 2).

Table 2. Radiometric Dating for the Upper Cantabrian.

Speakers: A. Bouroz,[1] M. Roques,[2] and Y. Vialette[2]
Reporting: upper Cantabrian (Stephanian)
Sample One:

     Name, Stratigraphic Unit—Faisceau de St. Jean—zone 2
     Geographic Location—Cevennes basin (France)
     Stratigraphic Position—upper Cantabrian
     Petrologic Designation—cinerite
     Analytic Details—
          (1) *method used:* rubidium-strontium
          (2a) *age and uncertainty (1 sigma or 2 sigma):* 2 sigma, 300 ± 10 m.y.
          (2b) *if isochron method applied, which regression program used:* McIntyre et al, 1966
          (3) *decay constant used:* $\lambda = 1.47 \cdot 10^{-11} \cdot y^{-1}$
          (4) *evaluation of reliability:*
          (5) *additional comments:* $Sr^{87}/Sr^{86} = 0.7036 \pm 0.0014$, the isochron was established
              with three samples—whole-rock, 2.63<d<2.89, and d<2.63
     Analyst—Y. Vialette
     Institution—Geological Institute, Clermont-Ferrand University (France)
     Published Reference—Colloque sur les Methodes et les Tendances de la Stratigraphie, Fac.
              des Sci., Orsay, 1970.
Sample Two: (Fill in Details as Above)
Sample Three: (Fill in Details as Above)

[1] Past-chairman, Geological Survey, Charbonnages de France, Paris, France.
[2] Laboratoire de Geologie, Faculté des Sciences, Clermont-Ferrand, France.

At the Krefeld Congress (September 1971), a proposal was made identifying the base of the Cantabrian with the base of the Stephanian in the broader sense (Bouroz et al, 1972) but the definition of that base was founded on the evolution of floral species only. New proposals made during the Moscow Carboniferous Congress (September, 1975; Bouroz et al, in press) could change that definition, more particularly if the marine fauna will be considered as more prevalent for remote correlations. In that case, and seeing the comparison between Spanish-French and the Russian sequences, it would be possible (and even desirable) that a new definition of the Middle-Upper Carboniferous boundary should be put higher in the present Cantabrian. So, we can state in first approximation, the age of 300 m.y. for the Westphalian-Stephanian boundary.

In the United States, correlations are not precise enough at present to accept the equivalence with the Desmoinesian-Missourian boundary.

Carboniferous-Permian Boundary

Measurements done on Akchatan granites gave the following results (Afanas'yev, 1970): (1) 39 analysis (U, Th-Pb), 293 ± m.y.; (2) 10 measurement, 287 ± m.y.; and (3) Rb-Sr isochron, 291 ± m.y.

Those measurements are important because the analyzed samples come from an area neighboring the stratotype area where the Permian was defined; moreover, several analyses made in France and Germany (Leutwein, 1970) indicated about 295 ± 10, 295 ± 5, 297 ± 5 m.y. for Stephanian intrusive rocks, and 280 ± 5 and 280 ± 10 m.y. for Permian intrusive rocks. The stratigraphic positions of these samples are not exact enough to give a more precise result. Finally, it seems possible, on the base of the Akchatan Granite, to assign the following age: Carboniferous-Permian boundary, 290 m.y. (subject to a new definition of that boundary by the IUGS Carboniferous-Permian Joint Working Group which must discuss the question as soon as possible).

Summary

The geochronologic scale data for the Carboniferous can be summarized as follows: Carboniferous-Permian, 290 m.y.; Westphalian-Stephanian (in the broader sense), 300 m.y.; Bashkirian-Moscovian, 308 m.y.; Mississippian-Pennsylvanian, slightly less than 320 m.y.; Devonian-Carboniferous (base of the Gattendorfia Zone), 330 m.y.

References Cited

Afanas'yev, G. D., 1970, Certain key data for the Phanerozoic Time Scale: Eclogae Geol. Helvetiae, v. 63, no. 1, p. 1-7.

Bouroz, A., 1970, Synthese des correlations par les cinerites des bassins stephaniens francais (Massif Central, Alpes externes, Jura): Acad. Sci. (Paris) C.R. Ser. D., v. 271, no. 14, p. 1171-1175.

——— M. Roques, and Y. Vialette, 1970, Etude de la cinerite au sommet de la zone 2 du Bassin des Cevennes: (France) Bur. Recherches Geol. et Minieres Bull., v. 77, no. 1, p. 503-507.

——— et al, 1972, Sur la limite Westphalien-Stéphanien et sur les subdivisions du Stéphanien inferieur sensu lato: 7th Int. Cong. Carboniferous Stratigraphy and Geology, Proc. (Krefeld).

——— et al, (in press) Proposals for an international chronostratigraphic classification of the Carboniferous: 8th Cong. Int. Stratigr. Geol. Carbonifere, (Moscow) Proc.

Brooks, C., S. R. Hart, I. Wendt, 1972, Realistic use of two error treatments as applied to Rb-Sr data: Geophysics and Space Physics, v. 10, no. 2, p. 551-577.

Harland, W. B., A. G. Smith, and B. Wilcock, eds., 1964, The Phanerozoic time scale—a symposium: Geol. Soc. London Quart. Jour., v. 120, Supp., p. 260-2.

Leutwein, F., 1970, Etude geochronologique de l'evolution du Permien continental en France: (France) Bur. Recherches Geol. et Minieres Bull., v. 77, no. 2, p. 973-977.

McIntyre, G. A., et al, 1966, The statistical assessment of Rb-Sr isochron: Jour. Geophys. Research, v. 71-72, p. 5459-5468.

Munier-Chalmas, E., and A. de Lapparent, 1893, Note sur la nomenclature des terrains sedimentaires: Soc. Geol. Fr. Bull., v. 21, no. 3, p. 438-493.

Wagner, R. H., et al, 1973, The post-Leonian basin in Palencia, A report on the stratotype of the Cantabrian Stage: Int. Union Geol. Sci., Subcom. Carb. Strat. (Prague) Proc.

The Mississippian-Pennsylvanian Boundary[1]

MACKENZIE GORDON, JR.[2] and B. L. MAMET[3]

Abstract In the type regions for the Mississippian and Pennsylvanian Systems, the boundary between the two systems occurs at a hiatus that varies laterally in age. The section across the boundary seems to be complete in the central Appalachian region, where a reference section has been established for the Pennsylvanian in West Virginia. However, the scarcity of marine beds in the Lower Pennsylvanian in that area makes it extremely difficult to establish detailed correlations with areas of predominantly marine deposition. Search has been underway by a working group studying this boundary for a section of continuous (and principally marine) deposition from Late Mississippian into Early Pennsylvanian, which could serve as a boundary stratotype for the transition between the two. Areas where deposition seems to have been continuous across the boundary are known in Tennessee, Alabama, Arkansas, Oklahoma, Utah, Nevada, Idaho, Wyoming, and Alaska.

The Lower-Middle Carboniferous boundary of the official classification[4] in the Soviet Union approximates the Mississippian-Pennsylvanian boundary. Field trips of the Eighth Carboniferous Congress, in 1975, provided the opportunity for geologists to study sections in the Moscow basin, southern Ural Mountains, Donets basin, northern Caucasus, Kuznetsk basin, and parts of Middle Asia. Large hiatuses are present at this boundary in the Moscow and Kuznetsk basins. A hiatus is also present in the southern Urals at the boundary between the Lower and Middle Carboniferous (base of the Bashkirian Stage as revised in November, 1974). In other sections, particularly one near Chimkent in south-central Asia, deposition appears to have been continuous across this boundary.

At present, a radiometric date of roughly 320 m.y. for the Mississippian-Pennsylvanian boundary is reached by extrapolation of rather meager data. Before the age of the boundary itself can be determined accurately, a boundary stratotype must be designated.

Introduction

In the United States, rocks classified in most other parts of the world as the Carboniferous System are regarded as constituting two systems. The older of these is the Mississippian System, which consists, in its type region, of predominantly carbonate rocks associated in its lower and upper parts with fine to moderately coarse clastic rocks. The younger is the Pennsylvanian System and consists in the American Midcontinent and East largely of coal-bearing clastic rocks and relatively minor carbonate rocks; in the American West, carbonate and clastic rocks are both abundant in the Pennsylvanian and very little coal is present. The boundary between the two systems, over much of the North American craton, is marked by a regional unconformity.

The hiatus between the Mississippian and Pennsylvanian proved a convenient stratigraphic "break" between the two systems for early investigators of Carboniferous strata in the United States. However, modern stratigraphy demands that the systemic

[1] Manuscript received, March 7, 1977.
[2] U.S. Geological Survey, U.S. National Museum of Natural History, Washington, D.C. 20244.
[3] Département de Géologie, Université de Montréal, Montréal, Quebec, Canada.
[4] Lower and Middle Carboniferous are used in the formal sense throughout this paper.

Article Identification Number: 0149-1377/78/SG06-0024/$03.00/0

boundary be established in sections of continuous deposition—or if a hiatus necessarily is present, it must be minimal with respect to geologic time.

In 1971, a working group was formed within the Subcommission on Carboniferous Stratigraphy of the IUGS Commission on Stratigraphy (with M. Gordon, chairman) to study the Mississippian-Pennsylvanian boundary. The problem required three lines of study. First, biostratigraphic investigations were necessary in North America in sequences of continuous deposition across the boundary to determine the faunal characteristics of the beds missing in the type regions. Second, comparisons were to be made with the boundary between the Lower and Middle Carboniferous of Western Europe and the Soviet Union, with which the Mississippian-Pennsylvanian boundary should coincide very closely. Third, radiometric ages were to be based upon samples taken at or near the boundary, if possible.

Type Mississippian Region

What began as the Mississippi Group (Winchell, 1869, p. 79) eventually became the Mississippian System. The type region for the Mississippian is in the central part of the drainage basin of the Mississippi River, in southeastern Iowa, northeastern Missouri, and western Illinois. Sections along the bluffs of the Mississippi River have been studied by many geologists for a long time. The type Mississippian section is a composite one and contains gaps in the section caused by wide shifts in the shorelines of the shallow seas that covered the North American craton during much of the Carboniferous. The uppermost division of the Mississippian, the Chesterian Series, is a cyclic sequence typically exposed in southwestern Illinois. The uppermost Chesterian formation, the Grove Church Limestone, contains foraminifers of Mamet Zone 18, as does a considerable thickness of the rocks beneath. It is correlated with the upper part of the Upper *Eumorphoceras* (E_2) ammonoid zone of the northwestern European section, which contains similar foraminifers. Evidence from northern Arkansas, where a rather complete ammonoid sequence has been found, indicates that the seas completed their withdrawal from the shelves following deposition of ammonoid beds that can be correlated with Subzone E_2b of the northwestern European section. Ammonoid faunas of the E_2c Subzone characterized by the genus *Nuculoceras*, or of the H_1 and H_2 Zones characterized by *Homoceras* and related genera, are not known in the United States. In the few places where rocks of this age are known, a brachiopod-rich facies generally is present.

Type Pennsylvanian Region

The name Pennsylvanian was introduced to replace the descriptive term "Coal Measures" (Williams *in* Branner, 1891, p. 135). The type region is in the northern Appalachian coal region and in the state of Pennsylvania, where the Pennsylvanian rocks are almost entirely continental in origin. As a more complete section of the lower half of this system has been found in southern West Virginia and adjacent parts of western Virginia, the subdivisions of the Pennsylvanian System have been defined mainly in terms of the stratigraphy of this more southern area. The Mississippian-Pennsylvanian boundary is placed at the base of the Pocahontas Formation, which, in the vicinity of its type locality at the town of Pocahontas, Virginia, rests on the Bramwell Member at the top of the Bluestone Formation. The Bramwell Member contains a marine fauna predominantly of bryozoans, brachiopods, and pelecypods of Late Mississippian (late Chesterian) age; it has not yielded a single foraminifer.

The Pocahontas Formation is a large deltaic deposit containing several coal beds and a flora typified by *Neuropteris pocahontas* White. Englund and Delaney (1966) and Englund (1969) showed, by sections derived from a series of seven drill holes (the farthest one a little more than 80 km west of Pocahontas), that the Pocahontas Formation fingers westward into the Bluestone Formation beneath overlying deposits of the Lee Formation and disappears near the Kentucky state line, where it is truncated by the overlying deposits. Therefore, in the subsurface, a considerable part of the Bluestone Formation may be Pennsylvanian in age.

Type Morrowan Series

The difficulty of correlating the predominantly continental deposits of the Appalachian region with other regions of predominantly marine deposition farther west in the United States gave rise to a third Carboniferous type region. A composite section of predominantly marine beds was established in Arkansas, Oklahoma, Missouri, and Iowa. The oldest Pennsylvanian beds are those of the Morrowan Series in Washington County, Arkansas (Sutherland and Henry, 1977). These are shelf deposits resting unconformably on eroded rocks of late Chesterian (Late Mississippian) age. The earliest Pennsylvanian fossils include ammonoids of the lower *Reticuloceras* (R_1) Zone and foraminifers of Mamet Zone 20. Most American biostratigraphers accept the base of the Morrowan as the base of the Pennsylvanian.

Precise identification of the base of the Morrowan is difficult; because of an onlap relationship, the lowest bed in any one section is probably not the same age as the lowest bed in another section. Basinward, where the hiatus seems minimal, the presumed Pennsylvanian beds are devoid of marine fossils.

Paleontologic Control

Five groups of fossils have been useful in delimiting the Mississippian-Pennsylvanian boundary and other groups have helped locally. However, each group has its own drawbacks.

Ammonoids

Ammonoids have been the classical guide fossils upon which much biostratigraphy of the Carboniferous is based. However, they are limited to marine rocks of several facies and are absent in many other facies. The Mississippian-Pennsylvanian boundary, and the boundary between the Lower and Middle Carboniferous of the Soviet Union, is placed at the base of the lower *Reticuloceras* (R_1) Zone. *Reticuloceras* in the United States is found in the Hale Formation of northwestern Arkansas. Because the formation occurs in an onlap relationship and because ammonoids of the underlying *Homoceras* Zone are not found in the United States, one cannot determine whether the lowest *Reticuloceras* occurrences in the section are actually from the base of the zone.

Brachiopods

The productoids are particularly helpful in recognizing the boundary locally. To mention a few, *Inflatia, Diaphragmus, Carlinia,* and *Ovatia* are common genera in the very Late Mississippian, but are superseded by *Rugoclostus, Tesuquea, Sandia,* and *Linoproductus* in the Early Pennsylvanian. However, the species tend to be restricted to depositional basins and do not range widely geographically; even some of the genera are restricted to eastern and western seas.

Smaller Foraminifera

Smaller Foraminifera have the advantage of occurring in fairly large assemblages in limestones, however they are generally absent in clastic and siliceous rocks. The Mississippian-Pennsylvanian boundary has been placed between Mamet Zone 19 (characterized by the acmes of *Eosigmoilina* and *Hemiarchaediscus*) and Mamet Zone 20 (characterized by the first appearance of *Globivalvulina* and *Millerella sensu stricto*), which Mamet and his associates equated with the *Homoceras* and *Reticuloceras* ammonoid zones, respectively. Recently, some workers in the Donets basin in the southwestern USSR have claimed that Mamet Zone 19 is restricted there to rocks below *Homoceras* and that both the *Homoceras* and *Reticuloceras* Zones belong in Mamet Zone 20 (Aisenverg, 1975, oral commun.). This problem requires immediate attention.

Conodonts

Platform conodonts have been of limited use in determining the Mississippian-Pennsylvanian boundary, as both the zonation and taxonomy of the conodonts have been in a state of flux. Lane and Straka (1974), placed the boundary above their

Rachistognathus muricatus Zone, which is equivalent to the top of the *Gnathodus girtyi simplex* Zone of Webster (1969), and below their *Rachistognathus primus* Zone. *R. primus* occupies a well-defined zone at the base of the Pennsylvanian, both in the American West and Midcontinent, and within that zone is intergradational with *R. muricatus* (Lane and Straka, 1974, p. 99-100). Although *R. muricatus* is common in the highest Mississippian beds, it also occurs in Lower Morrowan equivalents in Nevada (Lane and Straka, 1974, p. 98).

Plant Megafossils

In general terms, Late Mississippian (Chesterian) beds are characterized by plants of Read and Mamay's (1964) floral Zone 3, which shows an affinity to the Namurian A flora of Europe. Early Pennsylvanian plants of Zone 4, found in the Pocahontas Formation of Virginia and West Virginia, bear more resemblance to the Namurian B floras of Europe. These similarities to European floras were pointed out by Jongmans (1937). Read and Mamay (1964) placed the base of the Pennsylvanian at the first appearance of *Neuropteris pocahontas*, by far the most common plant in the Pocahontas Formation. *N. pocahontas* is found in the base of the formation but also occurs locally in sandy layers in the Bluestone Formation, mentioned previously in this article.

Jongmans (1937) also recorded the presence of plants resembling the Namurian A flora in two collections from the basal part of the Pocahontas Formation, below Pocahontas coal no. 1. This suggests that the early appearance of *Neuropteris pocahontas* is with plants that normally would be considered Mississippian in age. Flora of this region are under study by Gillespie and Pfeffercorn (1976) who have published a preliminary list.

Structural Control of Carboniferous Deposition

Figure 1 shows, in simplified form, the major structural features of the North American craton that controlled the distribution of seas during the Mississippian; the features in western Canada are approximately as depicted by Ziegler (1969, p. 16). Land areas are marked by a stippled pattern and deep troughs are shown in dark gray. The seas are shown at their maximum extent, so far as known; therefore the sketch map does not represent any single moment in geologic time. Major features include an eastern and a western interior sea, divided by a Transcontinental arch that extended southwestward from the Canadian Shield. The southwestern part of this peninsula was relatively unstable and changed considerably with time, allowing the seas to break through locally.

The eastern interior sea was bordered at the east by a large land mass and (at the south) contained the Ouachita trough which subsided rapidly and was quite deep during the Late Mississippian and Early Pennsylvanian. The western interior sea was bordered at the continental edge by the Antler orogenic belt, shown here as a chain of islands, the waters between which permitted some contact with world seas. West of this belt lay a deep eugeosynclinal trough and an outer island arc. The miogeosynclinal and eugeosynclinal belts extended northwestward through western Canada to Alaska. The sea reached their greatest extent in the late Early Mississippian. During the Late Mississippian, shorelines of both inland seas fluctuated wildly, culminating in a general withdrawal of the seas from the North American craton at the close of the Mississippian.

Foraminiferal Zone 19

One foraminiferal zone seems to be missing in most parts of North America in the unconformity that is present at the Mississippian-Pennsylvanian contact. This is Zone 19 which Mamet regards as equivalent to the *Homoceras* ammonoid zone of Europe. A search for Zone 19 foraminifers in North America is likewise a search for areas where the seas persisted and deposition was continuous across the systemic boundary. So far, Zone 19 has been recognized at 10 localities in six states of the United States and one locality in northwesternmost Canada. These are places where beds younger than the uppermost Mississippian of the Mississippi River Valley and older than the oldest

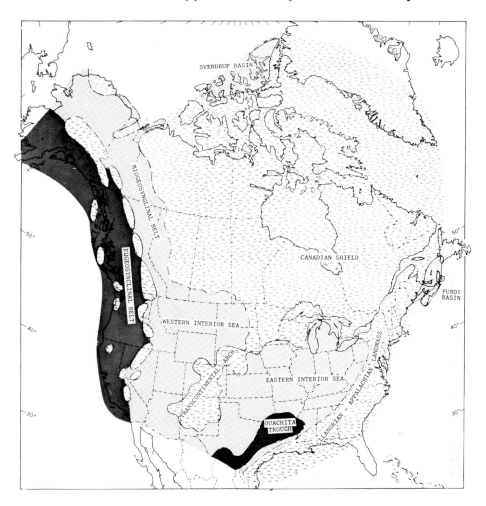

FIG. 1—Map showing main structural features controlling deposition during
the Mississippian.

marine Pennsylvanian of the American Midcontinent have been found. Their localities
are shown in Figure 2.

Localities 7 through 11 are along a belt 400 km long in the Brooks Range, Alaska,
and extending 40 km into Canada. This is the only area of significant size in North
America where beds containing Zone 19 foraminifers are known. In much of southeastern Alaska and western Canada, foraminifers of Zone 20 are found in beds that
rest unconformably on beds containing foraminifers of Zone 18. The few Zone 19
occurrences within the conterminous United States are widely scattered. However, our
data are not complete enough to rule out extensive development of beds of this age
within the Cordilleran miogeosyncline.

Comparison With Soviet Sections

The second line of investigation, comparison with what is considered to be the same
boundary in the Soviet Union—the one between the Lower and Middle Carboniferous
Series—was also pursued. Field trips connected with the Eighth International Congress
of Carboniferous Stratigraphy and Geology provided the opportunity for foreign
geologists to examine at first hand Carboniferous sections in several regions within the
Soviet Union.

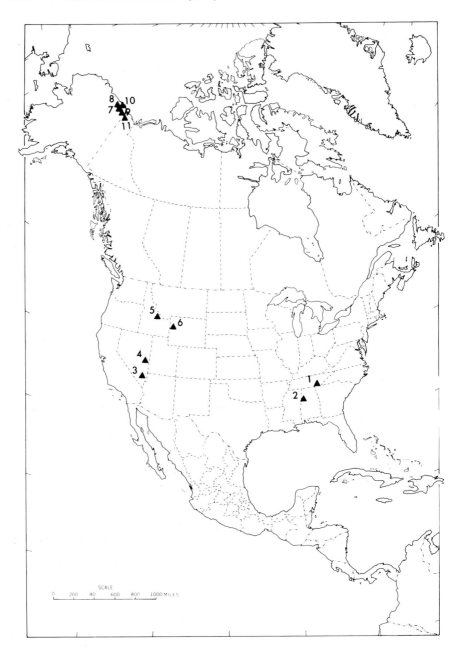

FIG. 2—Sites (indicated by black triangles) where latest Mississippian Zone 19 Foraminifera have been found in North America.

1. Tennessee: Daisy, Hamilton County, Pennington Formation (so-called pelletoidal limestone extremely rich in Archaediscidae—Typical Pennington is devoid of foraminifers; unpublished).

2. Alabama: Pickens County, unknown formation, Shell Robinson well, cuttings from depth 5,960 to 5,970 ft, sec. 14, T20S, R16W (unpublished).

3. Nevada: Arrow Canyon Range, Clark County, top of Indian Springs Formation. (Mamet and Skipp, 1970, p. 343 and 348; Brenckle, 1973, p. 17).

4. Nevada: East slope of Schell Creek Range, near Major's Place, White Pine County, lower part of Ely Limestone (Mamet *cited by* Gordon, 1971, p. 51).

5. Idaho: Hawley Mountains, Lost River Range, top of Surrett Canyon Formation, localities 118 and 125 (Mamet et al, 1971, p. 22, 23).

6. Wyoming: Berry Creek Section, Teton Range, Teton County, collection 153 (Mamet, 1975, p. B6 and B43).

7. Alaska: Ikiakpuk Creek, 68 A-1, Franklin Mountains, Brooks Range, top of Alapah Limestone (Armstrong et al, 1970, p. 694-695).

8. Alaska: Sadlerochit Mountains 68 A-3, top of Alapah Limestone (Armstrong et al, 1970, p. 694-695).

9. Alaska: Clarence River section 71 A-1-2, top of Alapah Limestone (Armstrong and Mamet, 1975, p. 20).

10. Alaska: Eastern Sadlerochit Mountains 68 A-4A and 68 A-4B, top of Alapah Limestone (Armstrong and Mamet, 1975, p. 20).

11. Canada: Joe Creek section 117 B-20, British and Buckland Mountains, Yukon Territory, top of Alapah Limestone (Mamet *in* Bamber and Waterhouse, 1971, p. 199).

Visitors to the region around Moscow were able to see that the Moscow syneclise is not a region where continuous deposition took place across the boundary between the two series. Several hiatuses are present in the late Lower Carboniferous and early Middle Carboniferous sections, by far the greatest occurring at the series boundary. The Serpukhovian, which has its type locality in this region, was suggested recently as a stage name to replace Namurian A, to which Soviet geologists previously have restricted their Namurian. The Serpukhovian follows the Visean and is succeeded by the Bashkirian Stage, the lowest division of the Middle Carboniferous. The type section of the Serpukhovian is incomplete; the top is deeply eroded. Its base is poorly defined in a platform-type condensed sequence. It contains a platform fauna with low diversity. Many macrofacies are dolomitized and recrystallized. The equivalents of the Bashkirian Stage are only locally present. Over considerable areas, rocks of the Moscovian Stage (which normally overlies the Bashkirian) rest directly upon Lower Carboniferous beds.

In the southern Ural Mountains, where the type Bashkirian rocks are exposed, the section is more complete. As of November 1974, the base of the Bashkirian was officially lowered to coincide with the base of the Middle Carboniferous. This has posed some problems. The type Bashkirian section at Mount Uklykaya along the Zilim River near the village of Tashasty cannot now be used internationally as a stratotype because the base, as revised, is covered. We have examined an auxiliary section, proposed by O. L. Einor (see Sinitsyna et al, 1975) along the Askyn River near the village of Solontsky. The proposed boundary for the Lower and Middle Carboniferous (base of the Bashkirian Stage) is apparently at a sharp lithologic break. It marks the boundary between two formations. The underlying limestone of the Ustarbaisky horizon[5] is crowded with shells of the brachiopod *Striatifera* and contains foraminifers of Mamet Zone 18. The contact is uneven and depressions are filled with fine-grained gray limestone of the overlying Syuransky horizon, which contains Zone 20 foraminifers. Therefore, a hiatus is present, representing at least all of Zone 19. It is notable that the same boundary problem that is so widespread in the American Midcontinent exists in this part of the Ural Mountains.

We have not seen evidence of the *Homoceras* Zone in the south Urals, although these beds are known to crop out in the south-central Urals. Perhaps a complete section occurs in that area. If not, perhaps one of the carbonate sections we saw on the Askyn and Zilim Rivers could be combined with a boundary stratotype section from elsewhere in the Ural Mountains as a standard for the Bashkirian Stage.

The section in the Donets basin in the southern Ukraine is regarded as exceptionally complete; it combines interfingering sequences of both continental and marine beds. Faunal evidence indicates that the Lower-Middle Carboniferous boundary lies near the

---

[5] Soviet geologists use the term "horizon" to denote a stage-stratigraphic unit of complex substantiation, that is, a rock unit based upon a combination of biostratigraphic, lithostratigraphic, and climatic stratigraphic criteria, any one of which may predominate in defining the unit (Khalfin, 1969).

top of the Protvinsky horizon. The highest gigantoproductid fauna occurs in zone $C_1{}^n d_1$, near the middle of that formation and the earliest *Choristites* at the base of zone $C_1{}^n e_1$ in the overlying Krasnopolyansky horizon. These two faunal occurrences bracket the series boundary. Intervening zone $C_1{}^n d_2$ at the top of the Protvinsky horizon has no characteristic brachiopod fauna, but in the upper limestones of this zone, both homoceratid ammonoids and *Reticuloceras* are known. Presumably the series boundary lies within this zone. Zone 19 foraminifers have been described in the upper part of the Upper *Eumorphoceras* (E_2) and in the lower and middle *Homoceras* Zone. During the Congress, Aisenverg stated that all of Zone 19 lies below the *Homoceras* Zone, which now belongs in Zone 20. The stratigraphic relationships between the foraminiferal, ammonoid, and conodont zones seem to require more study in this basin, as they do in other parts of the world.

Those geologists participating in the field excursion to middle Asia had the privelege of examining an unbroken sequence across the Lower-Middle Carboniferous boundary. A continuous succession of marine carbonate rocks is exposed in the Koikebiltau Mountain section near Chimkent, Uzhbekhistan. The Seslavinsky horizon of the Upper Namurian Uzunbulaskaya suite contains a continuous sequence of forambearing strata. Within this formation, foraminiferal zones 19, pre-20, and early 20 can be recognized.

Radiometric Dating

Our preliminary studies have shown that in two rather remote parts of the world—middle Asia and the Brooks Range, Alaska—sections across the Mississippian-Pennsylvanian boundary and its equivalent, the Lower-Middle Carboniferous boundary can be studied. Although we do not have unequivocal foraminiferal evidence, similar areas of continuous deposition may possibly be proved in the Great Basin of Utah and Nevada.

The Brooks Range sections have glauconite pellets scattered from the highest Mississippian to the lowermost Pennsylvanian, but because the rocks have been tectonically deformed, no radiometric ages should be expected. Moreover access to these sections is difficult and they are far remote from the original Mississippian-Pennsylvanian region. An easier solution to the problem would lie in Tennessee, where correlations with the type Pennsylvanian are relatively easy. However, the sections are quite condensed and diastems are probable; in particular, the top of the Pennington is certainly eroded.

How can radiometric dating of the Mississippian-Pennsylvanian boundary be approximated? If we use Sando's extrapolation (Sando, 1975) of the radiometric data of Francis and Woodland (1964) and Lambert (1971), we find that the boundary is in the vicinity of 320 m.y., with a probable error of 10 m.y. Slightly older ages are presented by Armstrong (this volume), but the base of the Upper Carboniferous is drawn at the Visean-Namurian boundary (that is, nearly a stage older than the Mississippian-Pennsylvanian).

We wish to conclude by quoting Hedberg's statement, used to open this symposium: "It is urged that the boundary-stratotype principle be used throughout the whole Standard Geochronological Scale. . . so that there will be a uniform and comparable basis for all units of this scale." Until a precise boundary-stratotype for the Mississippian-Pennsylvanian is designated, geochronologic definition of that boundary will remain a *will o' the wisp*.

References Cited

Armstrong, A. K., and B. L. Mamet, 1975, Carboniferous biostratigraphy, northeastern Brooks Range, Arctic Alaska: U.S. Geol. Survey Prof. Paper 884, 29 p.

—— B. L. Mamet, and J. T. Dutro, Jr., 1970, Foraminiferal zonation and carbonate facies of Carboniferous (Mississippian and Pennsylvanian) Lisburne Group, central and eastern Brooks Range, Arctic Alaska: AAPG Bull., v. 54, no. 5, p. 687-698.

Bamber, E. W., and J. B. Waterhouse, 1971, Carboniferous and Permian stratigraphy and paleontology, northern Yukon Territory, Canada: Canadian Petroleum Geology Bull., v. 19, no. 1, p. 29-250.

Branner, J. C., 1891, Introduction to the geology of Washington County: Arkansas Geol. Survey, Ann. Rept. 1888, v. 4, p. xi-xiv.

Brenckle, P. L., 1973, Smaller Mississippian and Lower Pennsylvanian calcareous foraminifers from Nevada: Cushman Found. Foram. Research Spec. Pub., v. 11, 82 p.

Englund, K. J., 1969, Relation of the Pocahontas Formation to the Mississippian-Pennsylvanian systemic boundary in southwestern Virginia and southern West Virginia (abs.): Geol. Soc. America Abs. with Programs, v. 1, pt. 4, p. 21.

—— and A. O. Delaney, 1966, Intertonguing relations of the Lee Formation in southwestern Virginia: U.S. Geol. Survey Prof. Paper, 550-D, p. D47-D52.

Francis, E. H., and A. W. Woodland, 1964, The Carboniferous period, in W. B. Harland, A. G. Smith, and B. Wilcock, eds., The Phanerozoic time scale; a symposium: Geol. Soc. London Quart. Jour., v. 120, supp., p. 221-232.

Gillespie, W. H., and H. W. Pfefferkorn, 1976, Plant fossils in early and middle parts of the proposed Pennsylvanian System stratotype in West Virginia: Geol. Soc. America, Field Trip Guidebook no. 3 Northeast and and Southeast Sections Joint mtg., p. 6-10.

Gordon, M., Jr., 1971, Biostratigraphy and age of the Carboniferous formations, in D. A. Brew, ed., Mississippian stratigraphy of the Diamond Peak area, Eureka County, Nevada: U.S. Geol. Survey Prof. Paper 661, p. 34-55.

Jongmans, W. J., 1937, Comparison of the floral succession in the Carboniferous of West Virginia with Europe: 2e Congres pour l'Avancement de Stratigraphie Carbonifere (Heerlen), v. 1, p. 393-412.

Khalfin, L. L., ed., 1969, Problemy stratigrafii (Classification in Stratigraphy): Sib. Nauchno-Issled. Inst. Geologii, Geofiziki Mineral'nogo Syr'ya Trudy, no. 94, 168 p.

Lambert, R. St. J., 1971, The pre-Pleistocene Phanerozoic time-scale; a review, in W. B. Harland, and E. H. Francis, eds., The Phanerozoic time-scale; a supplement: Geol. Soc. London Spec. Pub. 5, p. 9-31.

Lane, H. R., and J. J. Straka, II, 1974, Late Mississippian and Early Pennsylvanian conodonts, Arkansas and Oklahoma: Geol. Soc. America Spec. Paper 152, 144 p.

Mamet, B. L., 1975, Carboniferous Foraminifera and algae of the Amsden Formation (Mississippian and Pennsylvanian) of Wyoming: U.S. Geol. Survey Prof. Paper 848-B, p. B1-B18.

—— and B. A. Skipp, 1970, Preliminary foraminiferal correlations of early Carboniferous strata in the North American Cordillera: Univ. Liege Congres et colloques, Colloque sur la stratigraphie du Carbonifere, v. 55, p. 327-348.

—— et al, 1971, Biostratigraphy of Upper Mississippian and associated Carboniferous rocks in south-central Idaho: AAPG Bull., v. 55, no. 1, p. 20-33.

Read, C. B., and S. H. Mamay, 1964, Upper Paleozoic floral zones and floral provinces of the United States: U. S. Geol. Survey Prof. Paper 454K, K1-K35.

Sando, W. J., 1975, Diastem factor in Mississippian rocks of the northern Rocky Mountains: Geology, v. 3, no. 11, p. 657-660.

Sinitsyna, Z. A., et al, 1975, Carboniferous sections of south Urals: 8th Int. Cong. Carboniferous Stratigraphy and Geology (Moscow), field trip guidebook, 183 p.

Sutherland, P. K., and T. W. Henry, 1977, Carbonate platform facies and new stratigraphic nomenclature of the Morrowan Series (Lower and Middle Pennsylvanian), northeastern Oklahoma: Geol. Soc. America Bull., v. 88, p. 425-450.

Webster, G. D., 1969, Chester through Derry conodonts and stratigraphy of northern Clark and southern Lincoln Counties, Nevada: California Univ. Pubs. Geol. Sci., v. 79, 105 p.

Winchell, A., 1869-1870, On the geological age and equivalents of the Marshall Group: Am. Philos. Soc. Proc., v. 11, p. 57-82 (1869), p. 385-418 (1870).

Ziegler, P. A., 1969, The development of sedimentary basins in western and arctic Canada: Alberta Soc. Petroleum Geol., 89 p.

Devonian[1]

WILLI ZIEGLER[2]

Abstract Evaluation of time and correlation work within the Devonian suffer greatly from the lack of internationally agreed upon boundaries. The only boundary of the whole system which has been internationally agreed upon is its lower boundary—an excellent example of a precisely defined boundary based on biochronologic evidence. The IUGS Subcommission on the Silurian-Devonian Boundary which defined it suggested this boundary (the stratotype of which is in Czechslovakia) to be within a time span of between 100,000 to 300,000 years, as estimated from biologic evidence.

There is no absolute dating in the Devonian that could compete with such a highly precise biochronologic date. The range of error of any published absolute data is many times larger than the time span represented by this lower boundary of the Devonian.

Radiometric Data

Radiometric ages are few. The older ones date back to Kulp (1961) and Harland et al (1964) and give the duration of the system as about 50 to 60 m.y. Only Starik (*in* Nalivkin, 1973) has placed a duration of 80 m.y. for the system. The most recent data are all from the new world and there are at present no absolute age data available from the classic areas from which studies on Devonian stratigraphy originated (i.e. in Germany and Czechoslovakia).

Boucot (1975) discussed absolute data from the Appalachians and one date from Australia and reached the reasonable conclusion that the data are not adequate enough to calculate the duration or boundaries of the Devonian standard stages (as there are Gedinnian, Siegenian . . .). However, these data suggested to him that the Devonian may have had its lower boundary at about 405 m.y. and its upper boundary at 350 m.y. It may have lasted approximately 55 m.y. which is in agreement with Harland's and Kulp's 50 m.y. but in contrast to the Russian 80 m.y. The data demonstrate that they do not aid much in the assumption about relative duration of time of the Devonian standard divisions (Table 1).

Other Data

Therefore it is obvious that other criteria must be used in an approach to estimate relative duration of Devonian units. Boucot (1975), starting from the fact that relative time duration of standard units is based on the rate of evolution of fossils, concluded that paleontologists or biostratigraphers after some time with a fossil group may be able to get an impression of the relative amounts of evolution. From these, the time involved in fossil-based stratigraphic units may be reasoned. However, these impres-

[1] Manuscript received, January 17, 1977.
[2] Fachbereich Geowissenschaften, Lahnberge, 3550 Marburg/Lahn, Federal Republic of Germany.

Article Identification Number: 0149-1377/78/SG06-0025/$03.00/0

Table 1. Absolute Data from Eastern North America
and Australia (discussed in Boucot, 1975).

Frasnian-Famennian Boundary	362 ± 6 m.y.	Cerberean (Australia) Volcanics
Eifelian-Givetian Boundary	355 to 380 m.y.	Acadian granites in fossiliferous Eifelian beds
Siegenian	385 ± 15 m.y.	Shiphead Bentonite
Gedinnian	408 ± 3 m.y.	Eastport Volcanics

sions of the respective paleontologists are difficult to assess quantitatively.

A poll among Anglo-American paleontologists analyzed by Boucot (1975) resulted in an estimation of the duration of Devonian standard stages (Fig. 1), expressed by indices. An assumption on the duration of Devonian Stages taken from German literature (Fig. 1) differs strongly in the Lower Devonian but is otherwise rather congruent with those of Boucot (1975) as compared in Table 2.

Although it is obvious that this approach is vague, it seems that it is more reliable than estimates based on maximum thickness of sediments. The method of estimating time by evolution of organisms may become more reliable as more paleontologists think about it and as more radiometric ages become available.

Table 2. Relative length of time units. Estimations in right hand column are based on Solle (1972, p. 64) for the Lower Devonian units. Basis for the relative paleontologic unit duration of present paper is the concept that Lower Devonian as a whole is of the same duration as in Boucot's column.

	Relative Paleontologic Unit Duration Index	
	(Boucot, 1975)	(Present Paper)
Famennian	1.4 = 18.9%	1.5 = 20.3%
Frasnian	1.2 = 16.2%	1.3 = 17.6%
Givetian	1.2 = 16.2%	1.1 = 14.9%
Eifelian	1.0 = 13.5%	0.9 = 12.2%
Emsian	1.0 = 13.5%	1.3 = 17.6%
Siegenian	0.8 = 10.8%	0.6 = 8.1%
Gedinnian	0.8 = 10.8%	0.7 = 9.5%
	7.4 = 100%	7.4 = 100%

Summary

Radiometric ages are too few in the Devonian and have too wide a range of error to be especially helpful in a system in which the wealth of biochronologic data is overwhelming (ammonoids, conodonts, ostracods). And yet there is a need for more radiometric data. One example sheds light on the problem.

In the Upper Devonian, conodont zonation comprises 30 zonal subdivisions (Ziegler, 1971). It would be important to learn whether they represent 10, 15, or more million years of time, as this relation would clarify whether a conodont zone lasted 0.6 or 0.3 (or less) million years of time. The worldwide recognizable conodont zonation would provide an even greater help for the geologist than it does today as such a refined biochronologic standard.

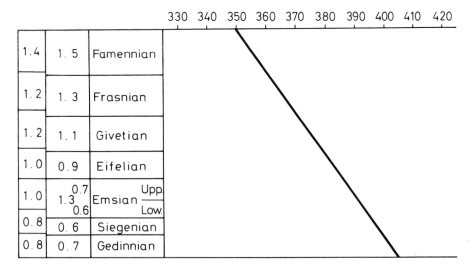

FIG. 1–Plot of radiometric age determination against relative time units scaled off through estimate of the amount of organic evolution within each relative time unit. Radiometric age line from Boucot (1975, Fig. 19) who determined it as an average of several radiometric ages considering the range of error. This average absolute time line pretends to be sharper than it is. Any radiometric dates arrived at through drawing a horizontal line from the boundaries of standard units to cut the time graph certainly have an error of several million years (±). The left two columns from Table 2 (this paper) show estimates of relative paleontologic unit duration.

The example also demonstrates that radiometric age dating methods are far from producing data so precise as to compete with the highly accurate biochronologic divisions. Present problems in Devonian chronology lie in the correlation of series and stage boundaries. However, concerning time, these problems fall far below the resolution of presently available radiometric ages.

References Cited

Boucot, A. J., 1975, Evolution and extinction rate controls, *in* Developments in paleontology and stratigraphy: Amsterdam, Elsevier, v. 1, 427 p.

Harland, W. B., A. G. Smith, and B. Wilcock, eds., 1964, The Phanerozoic time-scale; a symposium: Geol. Soc. London Quart. Jour., v. 120, supp., 458 p.

Kulp, J. L., 1961, The geological time scale, *in* F. E. Wickman and E. Welin, eds., Pre-Quaternary absolute age determinations: 21st Int. Geol. Cong. (Norden) Proc. Sect. 3, p. 18-27.

Nalivkin, D. V., 1973, Geology of the USSR: Edinburgh, Oliver and Boyd, 855 p.

Solle, G., 1972, Abgrenzung und Untergliederung der Oberems-Stufe, mit Bemerkungen zur Unterdevon-Mitteldevon-Grenze: Hesse Landesamt Bodenforschung Notizbl., v. 100, p. 60-91.

Ziegler, W., 1971, Conodont stratigraphy of the European Devonian, *in* W. C. Sweet, and S. M. Bergstrom, eds., Symposium on conodont biostratigraphy: Geol. Soc. America Mem., v. 127, p. 227-284.

The Silurian System[1]

NILS SPJELDNAES[2]

Abstract The newly agreed-upon Silurian-Devonian boundary is an excellent example of a modern high precision definition of a stratotype boundary, based on biostratigraphic evidence. The Ordovician-Silurian boundary is being studied by a working group under two subcommittees (Ordovician and Silurian). The biostratigraphic factual material forms an excellent base for relative-time stratigraphy in the Silurian (both for the boundaries to other systems, and internally). Stratigraphically relevant isotope datings are few. Although many datings have been published covering the general time interval of the Silurian, only a few have been adequately tied to the fossiliferous sequence, and there is not much good evidence for the precise absolute dates of either the upper or the lower boundary. The result is that estimates of the duration of the Silurian varies with as much as 100%. The available evidence is critically examined and a dating program suggested.

The Silurian System

Relative (paleontologic) dating within the Silurian System is probably one of the best of the pre-Cenozoic systems. There are about 20 well-defined zones which can be applied internationally. They have been used for about 75 years and are based mostly on graptolites. The system has been continuously refined and improved. Where graptolites are rare or absent, it has in most cases been possible to correlate the shelly sequences with the standard graptolite scheme with considerable reliability and precision.

The Silurian-Devonian boundary has been defined by using the graptolite zones (the base of the *Monograptus uniformis* zone), and also other important groups which have been correlated into the standard sequence to give an easily identified and very precise boundary. A working group is now elaborating the Ordovician-Silurian boundary according to the same principles.

Precision attained by paleontologic datings is very high. Standard zones give a precision in the range of 1 to 2 m.y.; values which are unattainable with the present isotope-dating techniques. By using evolutionary series in single species, the precision may be increased to 200,000 to 400,000 years. These extremely low values are due to the fact that the precision of the paleontologic dating methods are almost independent of absolute age. Similar precisions may be attained locally also in the Cambrian, Ordovician, and Devonian, but the Silurian appears to be the only system where such levels of precision can be obtained regularly.

[1] Manuscript received, October 5, 1976.
[2] Paleoecology Department, Aarhus University, DK-8000, Aarhus, Denmark.

The author is indebted to a number of colleagues for kind advice and information. Gratitude is expressed to J. M. Berdan for clearing up the biostratigraphic problems of the Pembroke and Eastport Formations, and to M. A. Lanphere, M. Churkin, Jr., and G. D. Eberlein, for allowing me to use their date from the *Monograptus cyphus* zone in Alaska (in press).

Article Identification Number: 0149-1377/78/SG06-0026/$03.00/0

Isotope Dating

In contrast to the excellent paleontologic datings, isotope values given for the Silurian reveal a rather confused picture. The numerous estimates for the duration, both the beginning and the end of the system, vary considerably (Fig. 1). The only other systems where the uncertainties in duration are of the same size are the Precambrian and the Quaternary. This is partly due to the low absolute precision of isotope dating methods in this time interval, and to the difficulties in finding good dating material at the critical levels in the Silurian. However, mostly it is because only a few geochronologists have concentrated on Silurian problems.

Most isotope datings in this time interval have been on metamorphic rocks (Dewey et al, 1970; Naylor, 1971; Reynolds et al, 1973) and mineral veins (Mitchell and Ineson, 1975). These studies are of considerable actual and potential value for petrogenetic and tectonic work, but it is difficult to tie them to the fossiliferous sequence with sufficient precision; therefore they are not very relevant for the time scale discussion.

A critical examination of the available evidence for the proposed boundaries shows that the widely discrepant age for the Silurian-Devonian boundary is partly due to the fact that both the Geological Society of London's *Phanerozoic Time Scale* (Harland et al, 1964), and the *Elsevier Time Scale* (Eysinga, 1975) rely strongly on the *estab-*

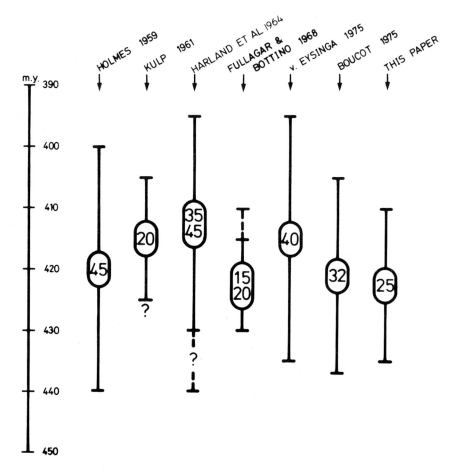

FIG. 1–Range of some recent estimates of the duration of the Silurian in millions of years. The Geological Society of London scale is taken from Harland et al (1964).

lished age of the Shap Granite (Kulp, 1961; Lambert and Mills, 1961). This granite intrudes into a sedimentary complex where the youngest paleontological dating is in the lower Ludlow, and pebbles of it are found in overlying unaltered sediments of Early Carboniferous age. In such situations (Fig. 2A), sound geologic thinking suggests that the date is closer to the higher than to the Lower paleontologic date, and that there is also an Upper Devonian bed concordantly below the lower Carboniferous rocks indicating that the granite is not younger than the early Late Devonian. Lambert and Mills (1961) showed that the critical point is whether the unfossiliferous Mell Fell Conglomerate, found far north of the Shap Granite, is: (1) deposited before or after the intrusion of the granite; and (2) of Early Devonian age or younger. None of these questions have been answered with sufficient accuracy to be of value for time scale work. The gap between the definitely fixed paleontologic dates is quite considerable, and even if the estimate seems reasonable from this individual case, it has probably caused an underestimation in the absolute age of the Silurian-Devonian boundary.

The much higher ages suggested by Bottino and Fullagar (1966) and Fullagar and Bottino (1968) are based on isotope dating of extrusive volcanics dated by lenses of fossiliferous sediment, (Fig. 2B). This situation may also cause some errors and uncertainties because there may be hidden time gaps at easily overlooked disconformities in such a sequence. The discrepancies are likely to be small, especially considering the fact that the isotope datings are much less precise than the paleontologic dating in this interval.

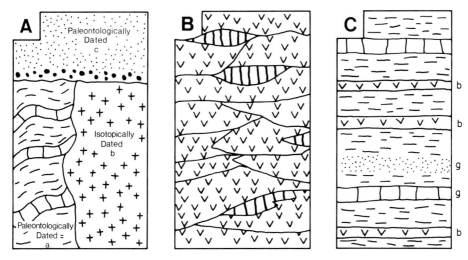

FIG. 2—Diagrammatic sketches of the common situations in correlating isotope dates with the fossiliferous sequence.

A—An intrusive, with isotope age *b*, cutting fossiliferous sediments of age *a*, and overlain by unaltered fossiliferous sediments of age *c*. There *a>b>c*, and normally *a—b<b—c*. Precision depends on *a—c*, which for geologic reasons normally is quite sizable.

B—A situation where fossiliferous lenses of sediments are found between lavas or tuffs. It may be difficult to get the exact tie-in between dated volcanics and fossiliferous lenses, but uncertainties will normally be considerably smaller than with A.

C—Bentonites and glauconites in an undisturbed, fossiliferous sequence. Here the sources of error are limited to the isotope methods, and to changes in dating materials due to halmyrolysis and diagenesis.

Paleontologic dating of the sediments (the Pembroke and Eastport Formations) are now rather good (Berdan, 1966, 1971a, 1971b). Previously there was some confusion, both as to the real paleontologic age, and as to the position of the Silurian-Devonian boundary and other changes in stratigraphic terminology. Even if the paleontologic date is fixed very close to the presently accepted Silurian-Devonian boundary, there is still some slight uncertainty, because it is not known exactly where the dated samples were taken in relation to the fossiliferous sedimentary lenses. However. the discrepancy is very small and the figure given by Bottino and Fullagar (1966) and Fullager and Bottino (1968) of 412 ± 5 m.y. is a realistic value for the age of the boundary. Whereas the samples are possibly taken just below the boundary the actual age may be slightly lower, even if Boucot's (1975) figure of 405 m.y. must be regarded as too low. In Figure 1, 410 m.y. is chosen as a conservative estimate, and the available material points to an age in the range 410 to 415 m.y. This estimate is supported by other estimates and measurements by Harper (1968), Naylor (1971) and Reynolds et al, (1973), even if it must be admitted that their dating points are far from the boundary and the tie-in with the fossiliferous sequence is rather indirect. Before this age problem can be settled, many more dates are needed on suitable rocks from the proximity of the paleontologically defined boundary.

The Ordovician-Silurian boundary is even more uncertain than the Silurian-Devonian boundary. The spread in estimates is somewhat lesser, but the data on which they are based is weak indeed. Most of them appear to be extrapolations from Middle Ordovician dates or based on metamorphic events of a rather uncertain time.

Recently Lanphere et al (in press) obtained a date from a volcanic breccia in the *Monograptus cyphus* zone (Llandovery, Early Silurian) in Alaska. On the basis of this age (433 ± 3 m.y.) they estimated the boundary to be at about 435 to 437 m.y. Fullagar and Bottino (1968) dated the supposedly basal Silurian Arisaig Volcanics at 430 ± 15 m.y. In this case the paleontologic dating is not perfect, but certainly close. Dewey et al (1970) suggested, from an analysis of mostly metamorphic material, that the Ordovician-Silurian division occured about 440 m.y. ago.

Problems connected with the dating of this time frame are aggravated by the fact that it occurs very close in time to Late Ordovician glaciation. This event has caused considerable transgressions and regressions, which complicated the stratigraphy. It is also calculated to have resulted in abnormal rates both of sedimentation and evolution of organisms in this interval. Therefore it is necessary to use even more stringent controls on the stratigraphic and paleontologic evidence than usual.

Very tentatively, it may be estimated that the available observations indicate an age of about 435 m.y. for the Ordovician-Silurian boundary. Since both isotope dates from Alaska and from the Arisaig Volcanics are minimum ages, based on rocks on the Silurian side of the boundary, the age could not be much lower unless there are errors in these measurements. An upper limit is given by the consistent Middle Ordovician ages based on bentonites, both from Europe and North America. These dates indicate that the upper Caradocian has an age of about 445 m.y. (Adams et al, 1958; Edwards et al, 1959; Kulp, 1961; Faul, 1961; Adams and Rogers, 1961). With the present estimate, this leaves only 10 m.y. for the uppermost Caradocian and Ashgillian. This may be a realistic figure, but it can hardly be much lower.

There is very little reliable isotope dating relevant to the subdivisions of the Silurian. There are a few glauconite dates (Sanford and Mosher, 1975), but those are sometimes difficult to place in relation to the Llandovery-Wenlock and the Wenlock-Ludlow boundaries.

The available rocks are there for solving this problem. Numerous bentonites are found especially close to the Llandovery-Wenlock boundary, in Scandinavia, the Baltic region, and a number of other regions close to the Caledonian orogen. Bentonites are also found in the lower Ludlow, in Scandinavia, Poland, and Podolia. As shown by Adams and Rogers (1961) bentonites form the best available link between the fossiliferous sequence and the easily dated volcanic materials. They can only be rivaled by the comparatively rare cases where extrusive volcanics and pyroclastics are found in undisturbed fossiliferous sediments and where sediments can be dated directly. A detailed isotope dating program on these bentonites may yield not only better dating on the Silurian but also an estimate on the duration of its subdivisions (Fig. 2c).

This is of special interest because the biologic evolution is well known in the Silurian, and important questions on the duration of species and fluctuation in the rate of evolution with time depends on a more precise dating of the Silurian and its subdivisions. This rather fundamental problem in stratigraphy and paleontology was recently discussed in detail by Boucot (1975).

References Cited

Adams, J. A. S., and J. J. W. Rogers, 1961, Bentonites as absolute time-stratigraphic calibration Points: New York Acad. Sci. Ann., v. 91, p. 390-396.
—— et al, 1958, Absolute dating of bentonites by strontium-rubidium isotopes: Geol. Soc. America Bull., v. 69, p. 1527.
Berdan, J. M., 1966, Baltic ostracodes from Maine: U.S. Geol. Survey Prof. Paper 55.
—— 1971a, Late Silurian-Early Devonian ostracode assemblages from Maine: U.S. Geol. Survey Prof., Paper 750-A, p. A132.
—— 1971b, Silurian to Early Devonian ostracodes of European aspect from the Eastport Quadrangle Maine (Abs.): Geol. Soc. America, v. 3, p. 18.
Bottino, M. L., and P. D. Fullagar, 1966, Wholerock rubidium-strontium age of the Silurian-Devonian boundary in northeastern North America: Geol. Soc. America Bull., v. 77, p. 1167-1176.
Boucot, A. J., 1975, Evolution and extinction rate controls: New York, Elsevier, 427 p.
Dewey, J. F., W. S. McKerrow, and S. Moorbath, 1970, The relationship between isotopic ages, uplift and sedimentation during Ordovician times in western Ireland: Scottish Jour. Geol., v. 6, p. 133-145.
Edwards, G., et al, 1959, Further progress in absolute dating of the Middle Ordovician: Geol. Soc. America Bull., v. 70, p. 1596.
Eysinga, F. W. B., 1975, Geological time scale: New York, Elsevier, 3rd ed.
Faul, H., 1961, Some Paleozoic dates in Maine, western Europe and southern United States: New York Acad. Sci. Ann., v. 91, p. 369-371.
Fullagar, P. D., and M. L. Bottino, 1968, Rb-Sr whole-rock ages of Silurian-Devonian volcanics from eastern Maine: EOS, v. 49, p. 346.
Harland, W. B., A. G. Smith, and B. Wilcock, 1964, The Phanerozoic time-scale: Geol. Soc. London Quart. Jour., v. 120, supp., 458 p.
Harper, C. T., 1968, Isotope ages from the Appalachians and their tectonic significance: Canadian Jour. Earth Sci., v. 5, p. 49-59.
Holmes, A., 1959, A revised geological time-scale: Edinburgh Geol. Soc. Trans., p. 183-216.
Kulp, J. L., 1961, Geologic time scale: Science, v. 133, p. 1105-1114.
Lambert, R. St. J., and A. A. Mills, 1961, Some critical points for the Paleozoic scale from the British Isles: New York Acad. Sci Ann., v. 91, p. 378-389.
Lanphere, M. A., M. Churkin, Jr., and G. D. Eberlein, in press, Radiometric age of the *Monograptus cyphus* graptolite zone in southeastern Alaska—an estimate of the age of the Ordovician-Silurian boundary: Geol. Mag.
Mitchell, J. G., and P. R. Ineson, 1975, Potassium-argon ages from the graphite deposits and related rocks of Seathwaite, Cumbria: Yorkshire Geol. Soc. Proc., v. 40, p. 413-418.
Naylor, R. S., 1971, Acadian Orogeny, An abrupt and brief event: Science, v. 172, p. 558-560.
Reynolds, P. H., E. E. Kublick, and G. K. Muecke, 1973, Potassium-argon dating of slates from the Meguma Group, Nova Scotia: Canadian Jour., Earth Sci., v. 10, p. 1059-1067.
Sanford, J. T., and R. E. Mosher, 1975, Radiometric age determination of Silurian glauconite: AAPG Bull., v. 59, p. 1201-1203.

Ordovician Geochronology[1]

REUBEN J. ROSS, JR.,[2] CHARLES W. NAESER,[2] and RICHARD S. LAMBERT[3]

Abstract Fission track analysis of pyrogenic apatite from a bentonite in the Acton Scott beds of the type section of Murchison's Caradoc Series has produced minimum ages of 443 ± 27 m.y. (external detector method) and 459 ± 27 m.y. (population method). That the sample was never heated as high as 80°C after deposition of the ash was determined by the conodont "color thermometer."

No other new Ordovician radiometric information has become available since the thorough and demanding review presented at the Ordovician System Symposium at Birmingham in 1974.

The most reliable published old ages for Ordovician rocks are by the potassium-argon method on micas from the post-Middle Ordovician Thetford Mines Gabbro (Canada; 480 m.y.) and from biotite from an intrusive that cuts the Girvan Ballantrae Complex (475 m.y.). All other age determinations related to the Ordovician are significantly younger than these.

Attempts have been made recently to place limits on Ordovician time by considering tectono-magmatic events in their stratigraphic frameworks. On the assumption that the Highland Border Group and the MacDuff slates in Scotland are both Dalradian and Arenig, the metamorphic peak in the Dalradian (490 to 500 m.y.) must be Llanvirnian or younger, depending on the Rb^{87} half-life and exact interpretation of isochrons and chrontours. However, setting the metamorphic peak post-Llanvirnian creates many difficulties and 495 ± 5 m.y. is therefore preferred for the Llanvirnian.

Radiometric calibration of provincial stratotype sections will prove essential to understanding Ordovician biostratigraphy because of the wide separation of tectonic plates into differing latitudinal environments in Ordovician time.

Discussion

The newest information on Ordovician geochronology comes from Caradocian bentonites both in England and in the eastern United States. Fission track analysis of pyrogenic apatite from a volcanic ash in the Acton Scott beds of the type section of Murchison's Caradoc Series gave ages of 443 ± 27 m.y. (external detector method), and 459 ± 27 m.y. (population method), averaging 451 ± 21 m.y. (Ross et al, 1976). We consider this to be a *minimum age* for the late Caradocian. Use of the conodont "color thermometer" devised by Epstein et al (1975) has demonstrated that these samples from the stratigraphic section along the Onny River, Shropshire, had never been heated higher than 80°C after deposition of the ash. Because fission tracks in apatite are annealed at higher temperatures, this information is critical in aiding interpretation. In this case the conodonts in the bracketing limestone beds and in the bentonite itself were pale yellow to very pale brown in color.

[1] Manuscript received, November 22, 1976.

[2] U.S. Geological Survey, Denver, Colorado, 80225.

[3] Department of Geology, University of Alberta, Edmonton, Alberta, Canada.

The authors acknowledge, with many thanks, critical review of the manuscript by Z. E. Peterman and J. D. Obradovich.

Article Identification Number: 0149-1377/78/SG06-0027/$03.00/0

Two other analyses on Caradocian strata deal with bentonites at the top of the Tyrone Limestone in well cores from northern Kentucky. On biostratigraphic grounds the Tyrone is considered to be older than the Acton Scott beds of the type Caradoc, and the same age as the Carters Limestone of Alabama and Tennessee.

From core of a well, drilled at Falmouth, Kentucky, bentonite logged 1 ft below the top of the Tyrone was collected by E. R. Cressman (USGS) and biotite from it was analyzed by J. D. Obradovich (USGS, written communication, 1968; USGS Isotope Lab, Sample DKA-1538). The resulting potassium-argon (K-Ar) age was 435 ± 15 m.y.

Ages of Caradocian samples, one from Shropshire and one from Kentucky, are within analytical error of one another, although the youngest biostratigraphically seems to be the oldest isotopically. Somewhat disturbing is the dating of Lower Silurian volcanic breccia (zone of *Monograptus cyphus*) from Esquibel Island, southeastern Alaska, by Lanphere et al, (in press). The age was found to be 433 ± 3 m.y. Because the sample is not oldest Silurian, the age of the Ordovician boundary was estimated as 435 to 437 m.y. seeming as old as two of the three Caradocian samples. Although only the Silurian age was calculated using new decay constants the conflict will still exist when the Caradocian potassium-argon ages are similarly corrected.

Not all of these ages can be correct if the biostratigraphic succession is correct. But biostratigraphy cannot be responsible for the very different ages given to the uppermost Tyrone Limestone. To the conflicting evidence can be added that from bentonite in the correlative Carters Limestone (and equivalent beds) of Tennessee and Alabama (Harland et al, 1964, p. 357; Lambert, 1971) which produced ages ranging from 419 ± 5 m.y. (K-Ar) and 440 ± 5 m.y. (U-Pb on zircon) using new decay constants for uranium to 436, 460, 462, and 478 m.y. (Rb-Sr on biotite, with assumed initial ratios: $T\frac{1}{2} Rb^{87} = 4.85 \cdot 10^{10}$ y).

Biotite and sanidine from a bentonite in the *Chasmops* Limestone at Kinnekulle, Vestergotland, Sweden, has a mean K-Ar age of 444 ± 4 m.y. (Bystrom-Aklund et al, 1961; Harland et al, 1964, item 157, p. 360). Graptolites indicated that the dated strata are approximately equivalent biostratigraphically to the Acton Scott beds in Shropshire. This age, particularly if allowance is made for the small change in decay constants, is well within the spread of the fission track age of the Acton Scott strata.

In the preceding discussion, it was shown that the Caradocian has provided a wide range of ages. Earlier parts of the Ordovician are equally as problematic. Data are available from Thetford Mines, Quebec; the Bay of Islands and Nipper's Cove ophiolites, Newfoundland; the Girvan-Ballantrae complex, southwest Scotland; and the Dalradian of the Grampians, Scotland. The first three areas are all now believed to be Ordovician ocean-floor complexes, all of which are extensively disturbed tectonically. The fourth is a high-grade metamorphic complex, polydeformed and multiply intruded, so it is not surprising that interpretation of radiometric ages from these areas is difficult.

The Thetford Mines ophiolite (Laurent, 1975) is intruded by adamellites which have yielded muscovite K-Ar ages of 477 and 481 m.y. (Harland et al, 1964, item 170). Laurent believed the adamellites are related to subduction processes associated with the emplacement of the Thetford ophiolite and uses these ages as corroboration. However, these argon determinations are suspect, belonging as they do to a series of published results in which yields of 99% and 100% radiogenic argon were reported. Although such yields are possible, several redeterminations of other samples in this series have shown the earlier reported ages to be too high. We urge caution in the use of these figures for defining any portion of Ordovician time.

The ophiolites of Newfoundland have received extensive isotopic studies, reported by Mattinson (1975), Dallmeyer and Williams (1975), and Archibald and Farrar (1976). Mattinson determined zircon U-Pb ages (Jaffey constants) of 508 ± 5 m.y. for a trondhjemite in Trout River ophiolite, correlative with the Bay of Islands ophiolite, and 463 ± 6 m.y. for a trondhjemite cutting the Nipper's Harbor ophiolite (believed to be part of the Bett's Cove ophiolite). Each age is defined by the upper intercept with concordia of a line drawn through points for two magnetically-separated zircon fractions. Mattinson did not examine the lower intercepts of these same lines. Discussion of the 508 ± 5 m.y. age will be included in the next section of this paper, but the other age, 463 m.y., is difficult to reconcile with other facts unless the Ordovician is very much shorter than at present believed. The Bett's Cove ophiolite is now defined as being the lower part of the Snooks Arm Group (Upadhyay et al, 1971) which includes, in its upper portion, supposedly "Early Ordovician" sediments. However, the paleontologic evidence as published (Snelgrove, 1931, p. 488) is slight: ". . .graptolites occur in the black graphitic slates of the Snooks Arm Series. These fossils have been determined by R. Ruedemann as *Loganograptus logani* and *Didymograptus gracilis*. . .(of) the Middle Arenig of Great Britain." It is also possible, but not very likely, that the Nipper's Cove trondhjemite dike is substantially younger than the Bett's Cove complex, so the figure of 463 ± 6 m.y. is a minimum for the Early Ordovician.

The Bay of Islands ophiolite is part of the Humber Arm allochthon, which contains several ophiolites and overlying clastic sediments. The general interpretation is that it is Early Ordovician ocean floor, sheared and thrust onto continental marginal crust, and the whole then obducted as the Humber Arm allochthon. This multistage history complicates interpretation of ages. Mattinson's (1975) figure of 508 ± 5 m.y. is most reasonably regarded as the age of formation of the Trout River ophiolite. Dallmeyer and Williams (1975) reported three total fusion Ar^{40}/Ar^{39} ages of 458, 466, and 477 (± 10) m.y., but incremental heating age patterns showed "excess" argon at low temperatures and a main plateau at 460 ± 5 m.y. They interpreted this latter age as being the synchronous metamorphic age of all three samples. The samples came from the metamorphic aureole below the Bay of Islands ophiolite, which is apparently a fairly normal contact aureole of 300 m width (Williams and Smyth, 1973), with hornblende-hornfels facies rocks grading structurally downwards into greenschist-type hornfels. Williams and Smyth (1973) believed that this aureole was created dynamothermally during obduction, because the ultramafics at the contact grade upward into a normal ophiolite complex with gabbros, sheeted dikes, pillow lavas etc, and the aureole is remarkably constant over large distances. However, there are numerous problems with such an interpretation, notably lack of a cause of the thermal energy for the metamorphism. Dallmeyer and Williams (1975), following the dynamothermal aureole hypothesis, concluded that 460 ± 5 m.y. dates the Arenig phase of obduction.

Archibald and Farrar (1976) also applied conventional K-Ar and the Ar^{40}/Ar^{39} incremental heating technique to hornblendes from the Bay of Islands ophiolite. The aureole yielded results similar to those of Dallmeyer and Williams (1975), ranging from 458 to 548 m.y. (conventional-argon), with a clear correlation of high age and low potassium. An isochron plot showed that these seven amphiboles could have been formed at 454 ± 9 m.y. if each had an equal quantity of excess argon at 2.10^{-7} cm^3 STP/g. A single Ar^{40}/Ar^{39} incremental heating experiment gave a plateau at 460 m.y. These data corroborate those of Dallmeyer and Williams in detail, and the interpretation of formation at 460 m.y. in an environment rich in Ar^{40} seems logical. Archibald and Farrar (1976) also reported six conventional K-Ar ages on amphibole from the gabbros and sheeted dike complex of the Bay of Islands ophiolite, ranging from 442 to 461 m.y. with an average of 452 ± 12 m.y. The ages agree within the analytical error at the 2σ level and the average is preferred. It is considered to be the age of cooling after obduction. They also reported five K-Ar ages on amphibole from a foliated amphibolite and two ages on hornblende from the Little Port complex (which includes the Trout River ophiolite of Mattinson, 1975), which range from 457 ± 13 to 477 ± 13 m.y. with an average of 469 ± 14 m.y. They concluded that this age is a minimum for removal of the complex from a region of later deformation, brecciation, dike intrusion, and prehnite-grade alteration, prior to or during the assembly of the Humber Arm allochthon.

As noted previously, this occurred in two stages, in Arenig time (Stevens, 1970) and between the middle Llanvirn (the youngest fossiliferous unit in the allochthon) and the late Llandeilo (the oldest overlying formation)—see Bergstrom et al (1974) for the paleontologic evidence. Archibald and Farrar (1976) related their ages of 460 m.y. and 469 m.y. to the Arenig event, utilizing the data of Harris et al (1965) which is discussed below as supporting evidence. Also involved in this conclusion is Stevens' observation (1970) that ophiolite detritus first arrives as one of the components in arkose units of the unfossiliferous Blow Me Down-Brook Formation, which gradationally overlies the late Arenig-middle Arm Point Formation. Archibald and Farrar then related the 452 m.y. ages to the second sequence of tectonic events in late Llanvirn-early Llandeilo time.

In Scotland, a similar degree of complexity exists. The Girvan-Ballantrae ophiolite complex, which is certainly pre-Caradocian and which contains associated middle Arenig lavas, was dated by Harris et al (1965), who determined an argon age of 475 m.y. for the Colmonell Gabbro, which intrudes the serpentinite. They also dated biotite from a mica-andesite lava of lower Caradocian age from Bail Hill (Ayrshire) at 445 m.y. The 475 m.y. figure is the average of five analyses at two laboratories, ranging from 467 to 483 m.y., while the 445 m.y. result came from a range of 437 to 455 m.y. from four runs at three different laboratories. Harris et al (1965) concluded that their averages gave a minimum age for each unit.

Within the Grampian Highlands, the MacDuff Slates are now regarded as extending up into the Arenig (Downie et al, 1971), after which time they were folded and cut by the Insch Gabbro at 486 ± 17 m.y. (Pankhurst, 1970). The slates themselves have given a whole-rock isochron of 448 ± 7 m.y., (Pankhurst, 1974) related to either further metamorphism or cooling. Both of these Rb-Sr figures ($T\frac{1}{2}$ Rb^{87} = 4.85 - 10^{10} y) are minimal for the Arenig, if the micropaleontologic evidence is correct.

Somewhat more indirect conclusions concerning Arenig time are found in the Dalradian metamorphic pattern as a whole. The major metamorphism of the Iltay Nappe was complete in the vicinity of the Highland Boundary fault by 503 ± 6 m.y. ago, from whole-rock, K-Ar datings on slate (Harper, 1967). On the opposite side of the fault, part of the Highland Border Series is regarded as stratigraphically and structurally similar to part of the Dalradian of the Iltay Nappe, being both Dalradian and Arenig (Johnson and Harris, 1967; Downie et al, 1971). Lambert and McKerrow (in press) have adopted this view in a geotectonic synthesis of Cambro-Ordovician events affecting the Scottish and Irish Dalradian, but there are many difficulties attached to placing an age of 503 m.y. within the Arenig or later stratigraphic time.

Therefore, there appear to be two possible time scales for the Ordovician, one of which places Arenig events in the range 475 to 465 m.y., Llanvirnian events around 450 m.y., and Caradocian events around 440 m.y. On this scale, there are numerous difficulties with Scottish and Irish tectonics. The other time scale leans heavily on the Scottish Grampian data, and on Mattinson's 508 m.y. zircon age, and assigns the 460 m.y. or so ages from Newfoundland to Llanvirnian events rather than Arenig events. On this scale, Arenig time would be about 500 to 480 m.y., Llanvirn-Llandeilo 480 to 455 m.y. and Caradoc time around 450 m.y. Most of the K-Ar ages obtained on biotite and hornblende would be minimal. There does not seem to be any rigorous way of checking which scale is preferable, or whether either is right. Much seems to hinge on the origin of the aureole of the Bay of Islands ophiolite: whether it is a true thermal aureole or was formed during transport, and whether or not it has been reheated during either the earlier or later stages of obduction. The latest data on the Caradocian tend to support the older scale, but high errors on the relevant ages do not permit a decisive answer.

Summary

In the preceding discussion, all K-Ar ages have been cited using author's constants and have not been corrected. The maximum variation in any quoted age will be about 1.5% and will not affect the discussion.

Finally, we would remark that biostratigraphers working on lower Paleozoic strata are becoming more acutely aware of the influence of ancient ecologies on composition of fossil assemblages. The realization that almost all the continental plates moved through different latitudinal (and presumably climatic) zones during the early Paleozoic is less than a decade old. The discrimination of fossils that are reliable as indicators of relative geologic time from those that are more faithful indices of environment than of time becomes increasingly complex. In making the distinction, biostratigraphers could use the assistance of isotopic geochronologists. It may be that an uncomfortably large number of supposedly correlative fossil assemblages did not actually live coevally. However, we must be convinced that isotopic methods are at least as accurate and definitive as those based on the comparison of fossils, and from this review, it looks as if we are some distance from that ideal.

The senior author notes that where the most recently derived radiometric ages of Ordovician rocks are compared with biostratigraphically controlled correlations, a chaotic chart results (Table 1). Part of the confusion may be attributed to uncertainty concerning decay constants or to the reliability of fossil identifications. But the most critical factor in any assessment is the basic geologic interpretation. For instance, until the geologic origins of the various components of so-called ophiolite complexes are fully understood one can hardly have complete confidence in isotopic ages derived therefrom to determine the age of the rock and to establish the time scale on which the age of the rock is dated. Or we may wonder what geologic factor may have upset the isotopic ratios in bentonites of the Carters and Tyrone limestones, particularly when we suspect that more than argon may have been affected.

References Cited

Archibald, D. A., and E. Farrar, 1976, K-Ar ages from the Bay of Islands ophiolite and the Little Port complex, western Newfoundland, and their geological implications: Canadian Jour. Earth Sci., v. 13, p. 520-529.

Bergström, S. M., J. Riva, and M. Kay, 1974, Significance of conodonts, graptolites and shelly faunas from the Ordovician of western and north-central Newfoundland: Canadian Jour. Earth Sci., v. 11, p. 1625-1660.

Byström-Aklund, A. M., H. Baadsgaard, and R. E. Folinsbee, 1961, K-Ar age of biotite, sanidine and illite from Middle Ordovician bentonites at Kinnekulle, Sweden: Geol. Fören. Stockholm Förh., v. 8, p. 92-96.

Dallmeyer, R. D., and H. Williams, 1975, Ar^{40}/Ar^{39} ages from the Bay of Islands metamorphic aureole; their bearing on the timing of Ordovician ophiolite obduction: Canadian Jour. Earth Sci., v. 12, p. 1685-1690.

Downie, C., et al, 1971, A palynological investigation of the Dalradian rocks of Scotland: Great Britain Inst. Geol. Sci. Rep., no. 71/9, p. 30.

Epstein, A. G., J. B. Epstein, and L. D. Harris, 1975, Conodont color alteration—an index to organic metamorphism: U.S. Geol. Survey Open File Rept. No. 75-379, 54 p.

Faul, H., 1960, Geologic time scale: Geol. Soc. America Bull., v. 71, p. 637-644.

Harland, W. B., A. G. Smith, and B. Wilcock, 1964, The Phanerozoic time-scale; a symposium: Geol. Soc. London Quart. Jour., v. 120, supp., 458 p.

Harper, C. T., 1967, The geological interpretation of potassium-argon ages of metamorphic rocks from the Scottish Caledonides: Scottish Jour. Geology, v. 3, p. 46-66.

Harris, P. M., et al, 1965, Potassium-argon age measurements on two igneous rocks from the Ordovician system of Scotland: Nature, v. 205, p. 352-353.

Johnson, M. R. W., and A. L. Harris, 1967, Dalradian-?Arenig relations in parts of the Highland Border, Scotland, and their significance in the chronology of the Caledonian orogeny: Scottish Jour. Geology, v. 3, p. 1-16.

Lambert, R. St. J., 1971, The pre-Pleistocene Phanerozoic time-scale—a review: Geol. Soc. London Spec. Pub. 5, p. 9-27.

—— and W. S. McKerrow, in press, The Grampian orogeny: Scottish Jour. Geology.

Lanphere, M. A., M. Churkin, Jr., and G. D. Eberlein, in press, Radiometric age of the *Monograptus cyphus* graptolite zone in southeastern Alaska—an estimate of the age of the Ordovician-Silurian boundary: Geol. Mag.

Laurent, R., 1975, Occurrences and origin of the ophiolites of southern Quebec, northern Appalachians: Canadian Jour. Earth Sci., v. 12, p. 433-455.

Mattinson, J. M., 1975, Early Paleozoic ophiolite complexes of Newfoundland, isotopic ages of zircons: Geology, v. 3, p. 181-183.

Pankhurst, R. J., 1970, The geochronology of the basic igneous complexes: Scottish Jour. Geology, v. 6, p. 83-107.

—— 1974, Rb-Sr whole rock chronology of Caledonia events in northeast Scotland: Geol. Soc. America Bull., v. 85, no. 3, p. 345-350.

Ross, R. J. Jr., C. W. Naeser, and G. A. Izett, 1976, Apatite fission track dating of a sample from the type Caradoc (Ordovician) series in England: Geology, v. 4, no. 8, p. 505-506.

Snelgrove, A. K., 1931, Geology and ore deposits of Betts Cove - Tilt Cove area, Notre Dame Bay, Newfoundland: Canadian Mining and Metall. Bull., v. 24, p. 477-519.

Stevens, R. K., 1970, Cambro-Ordovician flysch sedimentation and tectonics in west Newfoundland and their possible bearing on a proto-Atlantic ocean: Geol. Assoc. Canada Spec. Paper 7, p. 165-178.

Table 1. Ordovician Geochronology—A Case of Chaos Contracted by Ross (August, 1976).

MILLIONS OF YEARS	STABLE SHELF			CONTINENTAL BORDERS			
	SHROPSHIRE TYPE CARADOC	SWEDEN	APPALACHIANS ALA., KY.	QUEBEC	NEWFOUNDLAND	SCOTLAND	ALASKA
420			419±5 K-AR CARTERS LS C.BICORNIS ZONE (FAUL, 1960)				425±3 K-AR (OLD CONST) ESQUIBEL I. MONOGRAPTUS CYPHUS; 433± K-AR (NEW CONST) (LANPHERE ET AL, in press) [SILURIAN]
430			435±15 K-AR TYRONE LS (OBRADOVICH, written commun.)				
440		444±4 K-AR CHASMOPS LS PRE D.CLINGANI (BYSTROM-ASKLUND ET AL, 1961)	440±5 U-PB ZR CARTERS LS (NEW CONST)		POSSIBLY OVERLAIN BY TOP LLANDEILO OR LOW CARADOC	445 K-AR BAIL HILL ANDESITE LOW CARADOC GRAPTS (HARRIS ET AL, 1965); 448±7 WHOLE ROCK MACDUFF SLATE (PANKHURST, 1974) MINIMUM FOR ARENIG	
450	451±21 FT APATITE ACTON SCOTT BEDS (ROSS ET AL, 1976)				452±12 K-AR BAY OF I. OPHIO. COOL POST OBDUCT (ARCHIBALD & FARRAR, 1976)		
460					460±5 ^{40}AR/^{39}AR BAY OF I. OPHIO. METAM AUREOLE (DALLMEYER & WILLIAMS, 1975)		

CARADOCIAN

Table 1. (continued)

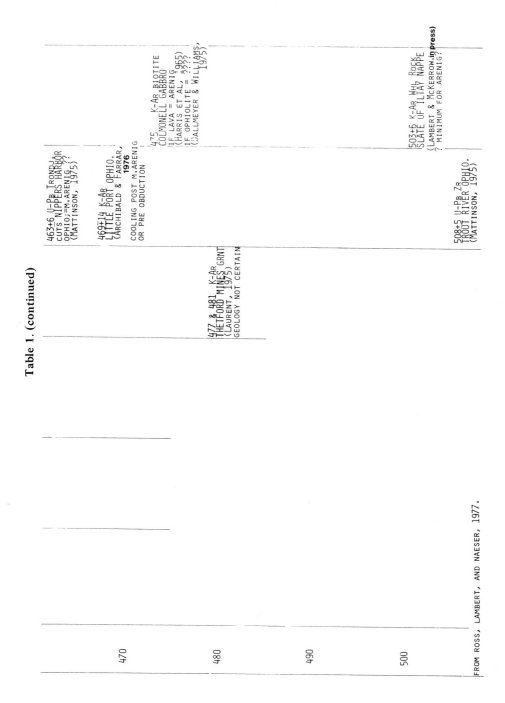

463±6 U-Pb TRONDJ.
CUTS NIPPERS HARBOR
OPHIO;=M.ARENIG ??
(MATTINSON, 1975)

469±14 K-AR
LITTLE PORT OPHIO.
(ARCHIBALD & FARRAR,
1976
COOLING POST M.ARENIG
OR PRE OBDUCTION

475 K-AR BIOTITE
COLMONELL GABBRO
IF LAVA = ARENIG,
(HARRIS ET AL, 1965)
IF OPHIOLITE = ????
(DALLMEYER & WILLIAMS,
1975)

477 & 481 K-AR
THETFORD MINES GRNT
(LAURENT, 1975)
GEOLOGY NOT CERTAIN

503±6 K-AR WH. ROCK
SLATE OF TILTAY NAPPE
(LAMBERT & MCKERROW.**in press**)
? MINIMUM FOR ARENIG?

508±5 U-Pb ZR.
TROUT RIVER OPHIO.
(MATTINSON, 1975)

470

480

490

500

FROM ROSS, LAMBERT, AND NAESER, 1977.

Sweet, W. C., H. Harper, Jr., and D. Zlatkin, 1974, The American Upper Ordovician standard, XIX, a Middle and Upper Ordovician reference standard for the eastern Cincinnati region: Ohio Jour. Sci., v. 74, no. 1, p. 47-54.

Upadhyay, H. D., J. F. Dewey, and E. R. W. Neale, 1971, The Betts Cove ophiolite complex, Newfoundland, Appalachian oceanic crust and mantle: Geol. Assoc. Canada Proc., v. 24, p. 27-34.

Williams, H., and W. R. Smyth, 1973, Metamorphic aureoles beneath ophiolite suites and Alpine peridotites; tectonic implications with West Newfoundland examples: Am. Jour. Sci., v. 273, p. 594-621.

The Cambrian System[1]

J. W. COWIE[2] and S. J. CRIBB[2]

Abstract Forty selected radiometric ages are plotted to illustrate a time scale for the Cambrian System and strata immediately above and below. Uncertainty is unavoidable at present, not only because the boundaries of the system are the subject of international working group investigations which are not complete, but also because the radiometric age determinations are too few and their association with stratigraphical data gives rise to manifest imprecision. Tentative ages of 560 m.y. and 485 m.y. are suggested for the base of widely accepted minimum limits of the system. If certain fossiliferous beds were to be included in the Cambrian System and excluded from pre-Cambrian sequences in an agreed international stratigraphic decision, then the base could be lowered as far as 590 m.y. Similarly Tremadoc sequences currently under discussion may be retained in the Cambrian changing the age of the top to 465 m.y. Working groups are also active on intra-systemic boundaries; dates of 530 m.y. and 505 m.y. are suggested for the traditional Lower - Middle and Middle - Upper Cambrian boundaries respectively. All the dates, except those taken from Glaessner's manuscript (Ⓔ and △, used in Figure 1) have been recalculated using the new decay constants decided upon at the Symposium on Geochronology held at the 25th International Geological Congress.

Introduction

Further review has been carried out along the general lines of the one previously undertaken by Cowie which was published in 1964 in *The Phanerozoic Time Scale*, a symposium organized on behalf of the Geological Society of London (Harland et al, 1964). New results are included which were summarized by Lambert in 1971 in *The Phanerozoic Time Scale; A Supplement* or have been sent to the author in response to a circular to the Project Working Group on the Precambrian-Cambrian Boundary, IUGS/IGCP and the Subcommission on Cambrian Stratigraphy of the IUGS. Vendian (pre-Cambrian) dates were more fully summarized in Cowie and Glaessner (1975) and have not been further quoted here.

In Table 1, selected items are given with some data. These have been recalculated except for those taken from Glaessner's manuscript, Ⓔ and △ (see Fig. 1). In most original works, the decay constants used in the age calculations are stated and all these dates have been recalculated using the new decay constants agreed upon at the Symposium on Geochronology held at the 25th International Geological Congress in Sydney, Australia, during August, 1976 (see notes on Fig. 1). In cases where the constants are not given or are unknown, the dates have been recalculated assuming the most likely constants. The information on the 40 points in this table is amplified, where necessary, later in this paper.

[1] Manuscript received, January 17, 1977.

[2] Department of Geology, University of Bristol, Bristol, England.

A. P. Gotto, research assistant, is thanked for help at all stages of preparation. M. F. Glaessner, S. Orlowski, L. N. Repins, and V. V. Khomentovskiy sent information and(or) drew attention to publications.

Article Identification Number: 0149-1377/78/SG06-0028/$03.00/0

355

Table 1. Radiometric Dates for the Cambrian and Contiguous Strata.

Serial No.	Mineral or Rock	Younger bracketing of stratigraphic age	Recalculated Age (m.y.)	Older bracketing of stratigraphic age	Location
31	?	Tommotian	553±17†	Tommotian	River Aldan USSR (135°E)
30	?	Tommotian	587±18†	?Tommotian	River Yudoma USSR (137°E)
29	?	Early Cambrian	582±17†	Early Cambrian	Olenek, USSR (123°E)
28	Illite	Lower Ordovician	457±14†	Upper Cambrian	Oklahoma, USA
27	Shale	Medial Upper Cambrian	495±15†	Medial Upper Cambrian	Southern Sweden
26	Glauconite	Lower Tremadoc	473±14†	Lower Tremadoc	Poland
25	Biotite	Lower Ordovician	484±15† 467±14† 476±14† 477±14† 478±14†	Lower Ordovician	Scotland
24	Whole rock	Late Precambrian	557±4 578±27	Late Precambrian	Massachusetts, U.S.A.
23	Glauconite	Early Lower Cambrian	511±15† 516±15† 518±16† 522±16†	Early Lower Cambrian	Leningrad & Byelorussia, U.S.S.R.
22	Biotite	-	518±16†	Early Middle Cambrian	Nova Scotia, Canada
21	Illite	Middle Cambrian	530±16†	Middle Cambrian	Wyoming, U.S.A.
20	Glauconite	Tommotian	504±15†	Tommotian	River Aldan, USSR (135°E)
19	Glauconite	Late Precambrian	548±16†	Late Precambrian	Southern Urals, U.S.S.R.
18	Glauconite	Early Cambrian	522±16†	Early Cambrian	Olenek, USSR (123°E)
17	Glauconite	Early Cambrian	522±16†	Early Cambrian	River Yenisei, USSR (85°E)
16	Glauconite	Lower Cambrian	572±29	Lower Cambrian	Tennessee, USA.
15	Biotite	Medial Cambrian	553±10	Late Precambrian	Normandy, France
14	Glauconite	Early Cambrian	526±16†	Early Cambrian	Olenek, USSR (123°E)
13	Glauconite	Early Cambrian	527-556	Early Cambrian	Olenek, USSR (123°E)
12	Glauconite	Early Cambrian	558±17†	Early Cambrian	Olenek, USSR (123°E)
11	Glauconite	Early Cambrian	545±16†	Early Cambrian	River Chaya, USSR (108°E)
10	Glauconite	Early Cambrian	569±17†	Early Cambrian	River Chikanda, USSR (118°E)
9	Glauconite	Late Precambrian	564±17† 573±17† 581±17†	Late Precambrian	Ukraine, USSR
8	Glauconite	Early Cambrian or Precambrian	549±16†	Early Cambrian or Precambrian	River Asha, Urals, USSR
7	Whole rock	Lower Cambrian	594±11†	Precambrian	Newfoundland, Canada
6	Glauconite	-	550±17†	-	Ukraine, USSR
5	Glauconite	?Early Cambrian (Tommotian)	578±17†	Early Cambrian	River Aldan, USSR (135°E)

TABLE I (continued)

Serial No.	Mineral or Rock	Younger bracketing of stratigraphic age	Recalculated Age (m.y.)	Older bracketing of stratigraphic age	Location
4	Glauconite	?Early Cambrian or Precambrian (pre-Tommotian)	579±17†	?Early Cambrian or Precambrian (pre-Tommotian)	River Kolaykhan, USSR (103°E)
3	Glauconite	Late Precambrian	589±18† 592±18†	Late Precambrian	S. Urals, USSR
2	Uraninite	–	~614±20	Late Precambrian	Katanga, Congo, Africa.
1	Glauconite	?Early Cambrian (Tommotian)	587±10	?Early Cambrian (Tommotian)	River Yudoma, USSR (137°E)

In Figure 1, the dates are plotted as error bars with age in millions of years as one ordinate against geographic location approximately related to longitude (east and west) on the other ordinate. When errors have been quoted with a date in the original works these values have been retained but otherwise (in order to produce a realistic picture of the determinations) a *suggested* error bar of ±3% has been constructed. These values are marked by (+) in Table 1 and are plotted as dashed lines in Figure 1. Longitude is a convenient factor to utilize because it gives some impression of geographic location around the world, separates the points for clarity, and suggests provenance in North America, Europe and Africa, Asia (east of the Urals in the USSR), and Australia.

General Stratigraphical Comments

The divisions in Figure 1 are a provisional interpretation which could well be changed on stratigraphical grounds due to the activities of working groups of the International Union of Geological Sciences and projects of the International Geological Correlation Program: the Precambrian-Cambrian Boundary Working Group (IUGS - IGCP - UNESCO), the Cambrian-Ordovician Working Group, the Commission on Stratigraphy (IUGS), and the working group belonging to the Subcommission on Cambrian Stratigraphy (IUGS)—the Cambrian Correlation Working Group. Subdivisions of pre-Cambrian strata and the position of the Precambrian-Cambrian Boundary, wherever it may be selected to lie, are also of concern to the Subcommissions on Precambrian and Cambrian Stratigraphy.

In the interpretation of Figure 1, it has therefore been decided to put the Tommotian (type area in eastern Siberia) with a question mark between the Vendian (type area in the Russian platform), which is usually accepted as Precambrian and contains the Ediacaran fauna (see following discussion) in its upper part, and the Cambrian. Recommendations of the Project Working Group on the Precambrian-Cambrian Boundary, adopted at the Paris conference in 1974 (Cowie and Rozanov, 1975) included: (1) that the "Ediacara" type fauna should be considered Precambrian; and (2) that the Olenellid-Fallotaspid trilobite faunas should be considered as Cambrian.

The Tremadoc sequences in Britain are included by the official British Institute of Geological Sciences in the Cambrian System whereas different views are taken by others. Because this subject is also under international discussion the Tremadoc is therefore placed with a question mark between Cambrian and Ordovician.

The provisional ages in millions of years for the boundaries in Figure 1 are a new estimate by the joint authors, utilizing, as far as possible from the available facts, decay constants agreed upon at the International Geological Congress at Sydney, August, 1976 (see Fig. 1).

Items by Serial Number

The ages given below are those from the original paper or personal communication before recalculation for this contribution, but in some cases the decay constants originally used were not given. In these cases, assumptions have been made by the authors under three general cases:

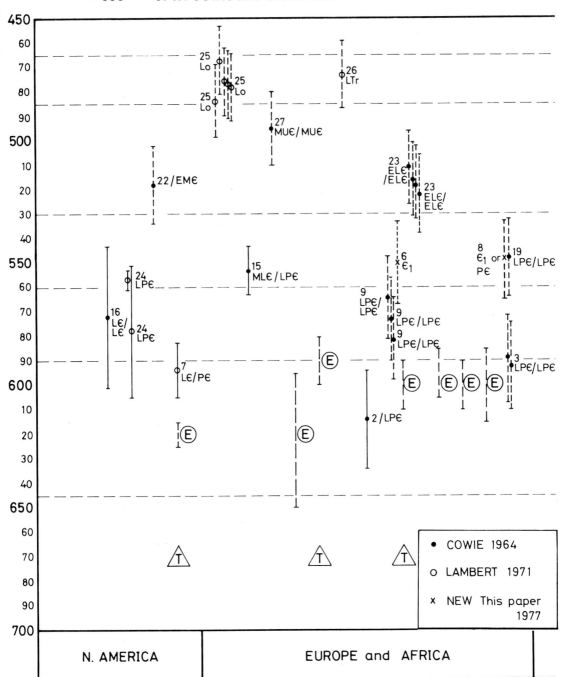

FIG. 1—Forty selected radiometric dates plotted with ages in millions of years against approximate geographical longitude. The points are categorized as: ●, referred to in Cowie, 1964; 0, referred to in Lambert, 1971; X, newly added in this paper in 1976.

Each point has a serial number (see Table 1) and a coded stratigraphic age or two bracketing ages: Pε, Precambrian; LPε, Late Precambrian; To, Tommotian; Eε, Early Cambrian, ELε, Early Lower Cambrian; Lε, Lower Cambrian; MLε, Medial Lower Cambrian; EMε, Early Middle Cambrian; Mε, Middle Cambrian; MUε, Medial Upper Cambrian; Uε, Upper Cambrian; L Tr, Lower Tremadoc; and LO, Lower Ordovician.

The symbol ⓔ refers to occurrences of Ediacaran fossils and △T̲ refers to the youngest approximately dated tillites found below such occurrences (data provided to the author from

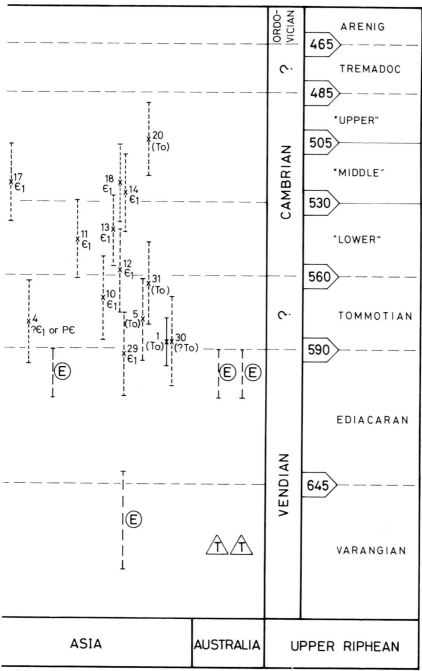

manuscript by Glaessner, in press). The dates represented by the symbols Ⓔ and △ were presented in 1975 by Glaessner to the International Geological Correlation Program's Symposium in Moscow on "Correlation of the Precambrian" (Glaessner, in press). Data from items 21 and 28 have not been plotted because of the generally accepted unreliability of potassium-argon determinations on illite. All data corrected as far as possible to the following decay constants: λU^{238} = 1.55125 x $10^{-10} y^{-1}$; λU^{235} = 9.8485 x $10^{-10} y^{-1}$; λRb^{87} = 0.142 x $10^{-10} y^{-1}$; $\lambda_\beta K^{40}$ = 4.963 x $10^{-10} y^{-1}$; $\lambda_e K^{40}$ = 0.5811 x $10^{-10} y^{-1}$.

The right-hand stratigraphic scale is a provisional interpretation only.

(A) $K^{40}\lambda_\beta = 4.72$ x $10^{-10} y^{-1}$, $\lambda_e = 0.585$ x $10^{-10} y^{-1}$;
(B) $K^{40}\lambda_\beta = 4.72$ x $10^{-10} y^{-1}$, $\lambda_e = 0.557$ x $10^{-10} y^{-1}$; and
(C) $K^{40}\lambda_\beta = 4.68$ x $10^{-10} y^{-1}$, $\lambda_e = 0.585$ x $10^{-10} y^{-1}$.

1. The information which is the basis for this item (and items 4, 5, 6, 8-14, and 9-20) was pointed out by L. N. Repina and V. V. Khomentovskiy, Novosibirsk, USSR. Translation from the Russian was informally done by C. D. Offord of the Department of Russian Studies, University of Bristol.
Region, River: Yudoma-Maya region, River Yudoma on the right bank above River Tnallakh, USSR.
Suite: Lower part of the Pestrotsvet Limestone, Tommotian.
Age: 587 ± 10 m.y.
Constant Used: Assumption (A).
Reference: Khomentovskiy et al, 1972.
  2. Item 55 *in* Harland et al, 1964; Cowie, 1964 (U-Pb).
  3. Item 118 *in* Harland et al, 1964; Cowie, 1964, (K-Ar).
  4. Region, River: Anabar uplift, River Kotuykan, USSR.
Suite: Nemakit-Daldyn horizon, Precambrian-Cambrian.
Age: 580 m.y.
Constant Used: Assumption (A).
Reference: Kazakov and Knorre, 1973.
  5. Region, River: Aldan Shield, River Aldan, USSR.
Suite: Pestrotsvet, Sunnaginicus zone, Tommotian, ?Early Cambrian.
Age: 578 m.y.
Constant Used: Assumption (A).
Reference: Khomentovskiy et al, 1972.
  6. Region, River: Volyno-Podolian Platform.
Suite: Supra-Laminarites horizon.
Age: 575 m.y.
Constant Used: Assumption (B).
Reference: Semenenko et al, 1960.
  7. Item 352 *in* Lambert, 1971 (Rb-Sr).
  8. Region, River: Urals, River Asha, USSR.
Suite: Ashinsk, lower parts, Early Cambrian.
Age: 573 m.y.
Constant Used: Assumption (B).
Reference: Polevaya et al, 1962.
  9. Item 117 *in* Harland et al, 1964; Cowie, 1964 (K-Ar).
  10. Region, River: Uchura-Maya, mountain range Udokan, River Chikanda, USSR.
Suite: Lower part of Aldan stratum, ϵ_1, Lower Cambrian.
Age: 570 m.y.
Constant Used: Assumption (A).
Reference: Kazakov and Knorre, 1973.
  11. Region, River: Potom Highlands, River Chaya by the stream Lempey, USSR.
Suite: Usatov ϵ_1, Lower Cambrian
Age: 570 m.y.
Constant Used: Assumption (B).
Reference: Zharkov and Chechel, 1964.
  12-13. Region, River: North Siberia, Olenek uplift, USSR.
Suite: Kessyussa, ϵ_1, Lower Cambrian.
Age: lower 558 m.y., upper 527 to 556 m.y.
Constant Used: Assumption (A).
Reference· Salop, 1973.
  14. Region, River: North Siberia, Olenek uplift, River Olenek, USSR.
Suite: Kessyussa, ϵ_1, Lower Cambrian.
Age: 550 m.y.
Constant Used: Assumption (B).
Reference: Kazakov et al, 1965.

15. Item 42 *in* Harland et al, 1964; Cowie, 1964 (K-Ar).
16. Item 183 *in* Harland et al, 1964; Cowie, 1964 (Rb-Sr).
17. Region, River: Yenisei, Island of Plakhin, USSR.
Suite: Aldan ϵ_1, Lower Cambrian.
Age: 545 m.y.
Constant Used: Assumption (B).
Reference: Kazakov et al, 1966.
18. Region, River: North Siberia, Karaulakh, USSR.
Suite: Tyusser ϵ_1, Lower Cambrian.
Age: 545 m.y.
Constant Used: Assumption (B).
Reference: Kazakov et al, 1966.
19. Item 116 *in* Harland et al, 1964; Cowie, 1964 (K-Ar).
20. Region, River: Aldan Shield, River Aldan.
Suite: Pestrotsvet ?ϵ_1 (Tommotian), ?Lower Cambrian-Precambrian.
Age: 527 m.y.
Constant Used: Assumption (B).
Reference: Polevaya et al, 1962.
21. Item 48 *in* Harland et al, 1964; Cowie, 1964 (K-Ar).
22. Item 70 *in* Harland et al, 1964; Cowie, 1964 (K-Ar).
23. Item 100 *in* Harland et al, 1964; Cowie, 1964 (K-Ar).
24. Item 353 *in* Lambert, 1971 (Rb-Sr).
25. Item 350 *in* Lambert, 1971 (K-Ar).
26. Item 348 *in* Lambert, 1971 (K-Ar).
27. Item 34 *in* Harland et al, 1964; Cowie, 1964 (U-Pb).
28. Item 47 *in* Harland et al, 1964; Cowie, 1964 (K-Ar).
29. Region, River: Olenek, USSR.
Suite: lower horizon ϵ_1 (the upper part of Kessyussa formation).
Age: 583 m.y.
Constant Used: Assumption (C).
Reference: Shishkin, 1975.
30. Region, River: Middle reaches of the Yudoma River, USSR.
Suite: Sunnaginian (?) horizon, ?Lower Cambrian (Tommotian)
Age: 588 m.y.
Constant Used: Assumption (C).
Reference: Khomentovskiy et al, 1972.
31. Region, River: River Aldan, USSR.
Suite: Sunnaginian horizon, ?Lower Cambrian (Tommotian)
Age: 578 m.y.
Constant Used: Assumption (B).
Reference: Khomentovskiy et al, 1972.

In items 1, 4, 5, 6, 8, 10, 11, 12, 13, 14, 17, 18, and 20 where constants are not quoted in the originals it is assumed that the constants used would have been those current in the country at the time of publication.

References Cited

Burchart, J., 1971, Absolute ages of rocks from Poland—a review of geochronological data: Pol. Tow. Geol. Rocz., v. 41, no. 1, p. 241-255.
Cowie, J. W., 1964, The Cambrian period, *in* W. B. Harland, A. G. Smith, and B. Wilcock, eds., The Phanerozoic time-scale: Geol. Soc. London Quart. Jour., v. 120 supp., p. 255-258.
—— and M. F. Glaessner, 1975, The Precambrian-Cambrian Boundary—a symposium: Earth Sci. Rev., v. 11, p. 209-251.
—— and A. Y. Rozanov, 1975, Precambrian-Cambrian boundary working group in Normandy and Paris, IUGS/IGCP: Geol. Mag., v. 112, no. 2, p. 197-198.
—— A. W. A. Rushton, and C. J. Stubblefield, 1972, Special Cambrian report of the Geological Society: Geol. Soc. London, Rept. No. 2, p. 1-42.
Glaessner, M. F., in press, The Ediacara fauna and its place in the evolution of the Metazoa: IGCP (Moscow) Sym. on Correlation of the Precambrian.

Harland, W. B., A. G. Smith, and B. Wilcock, 1964, The Phanerozoic time-scale; a symposium: Geol. Soc. London, v. 120, supp., 458 p.

Kazakov, G. A., and K. G. Knorre, 1973, The geochronology of the upper Precambrian Uchur-Maya region of the Siberian platform: Akad. Nauk SSSR Kom. Opred. Absol. Vozrasta Geol. Form. Tr., no. 16, p. 192-205.

—— —— and L. N. Prokofieva, 1965, The absolute age of Precambrian sedimentary rocks, "Olenek" uplift, eastern Siberia: Geokhimiya No. 11, p. 1313-1317.

—— —— and V. P. Strizhov, 1966, The absolute age of Precambrian sediments of the western margin of the Siberian platform of the Yenisei ridge, Turukhan and Chadobetsk uplift: Akad. Nauk SSSR, Kom. Opred. Absol. Vozrasta Geol. Form., Tr., no. 13, p. 312-316.

Khomentovskiy, V. V. et al, 1972, Standard sections of the deposits of the Upper Precambrian and Lower Cambrian of the Siberian platform: Akad. Nauk SSSR, Sib. Otd. Inst. Geol. Geofiz., Tr., no. 141, 356 p.

Lambert, R. St. J., 1971, The pre-Pleistocene Phanerozoic time-scale—a review, *in* The Phanerozoic Time-Scale; a supplement: Geol. Soc. London, Spec. Pub. 5, p. 9-31.

Polevaya, N. I., G. A. Murina, and G. A. Kazakov, 1962, New data for the more precise definition of the post-Cambrian scale of absolute geochronology: 10th Ses. Comm. on determination of absolute age of geological formations Trans.

Salop, L. I., 1973, The general stratigraphic scale of the Precambrian: Leningrad, Nedra, 310 p.

Semenenko, N. P., E. S. Burkser, and M. N. Ivantishin, 1960, Age groups of rock mineralization of the strata of the Ukraine in absolute chronology: 21st Int. Geol. Cong. Doklady Sovet. Geol. Prob. 3, p. 112-131.

Shishkin, V. V., 1975, Stratigraphy of the Cambrian-Precambrian boundary in the Igarka region, *in* Analogues of the Vendian Complex in Siberia: "Nauka" Moscow, USSR Acad. Sci. Siberian Div. Trans. Inst. Geol. Geophys., v. 232, p. 192-195.

Zharkov, M. A., and E. I. Chechel, 1964, Late Precambrian Sediments of the River Chaya basin, western slope of the North Baikal Mountains: Izv. Acad. Sci. USSR, v. 159, no. 1.

Numerical Correlation of Middle and Upper Precambrian Sediments[1]

MICHEL G. BONHOMME[2]

Abstract Many age determinations on sedimentary Precambrian sequences have been published. These measurements were made on whole-rock samples, clay fractions, or interbedded volcanics. A new approach leads to new possibilities about numerical stratigraphy of old nonfossiliferous sediments. This paper is a first attempt to build a nearly worldwide correlation of middle and upper Precambrian sediments based on age determinations.

The main problem with these results is in their interpretation. The age may be the age of the source area of the sediment, the time of deposition or early diagenesis, the time of epimetamorphism, or the time of late diagenesis. Some criteria useful in the interpretation of age determinations in sediments are defined.

Discussion

Many age determinations on sedimentary Precambrian sequences have been published. These age measurements were obtained either with whole-rock samples, clay fractions, or interbedded volcanics. A new approach, defined by Bonhomme and Clauer (1972) and more precisely by Clauer (1976), leads to new possibilities about numerical stratigraphy of old nonfossiliferous sediments. A project devoted to that matter has been presented at the IGCP program and accepted. This paper is a first attempt to use results, already published and in press, to build a nearly worldwide correlation of middle and upper Precambrian sediments based on age determinations.

The main problem with these results is in their interpretation. Interpretation may correspond either to the age of the source area of the sediment, to the time of deposition or early diagenesis, to the time of epimetamorphism, and even to the time of late diagenesis. Bonhomme and Clauer (1972) defined some criteria useful in the interpretation of age determinations on sediments. Clay minerals represent a unique fraction of silicate minerals in sediments, the geochemistry of which allows them to equilibrate with surface environments under the thermodynamic conditions of the hydrosphere (Millot, 1964). The mineralogy of clays can be used to trace the geochemical history of the sediment because that history can be correlated with the isotopic results—that is, with the development of significant age measurements.

These criteria are: (1) kaolinite and 2 Md-illite in siltstones mainly indicate a detrital origin, which commonly leads to a scatter of points on a Rb-Sr diagram or to an inherited isochron from the source area; (2) occurrence of smectite, interlayered clay minerals 1 Md-illite and associated chlorite show generally the age of early diagenesis and represent the best opportunity to approach as nearly as possible the age of sedimentation; and (3) evidence of epimetamorphic illites (2 Md-illite), as defined by Kubler (1966) and Dunoyer de Segonzac (1969), means that the age obtained is that

[1] Manuscript received, October 7, 1976.

[2] Centre National, de la Recherche Scientifique, 67 Strasbourg, France.

Article Identification Number: 0149-1377/78/SG06-0029/$03.00/0

of the thermo-tectonic event and is likely to be much younger than the time of deposition.

As a control, it is suggested that a whole stratigraphic sequence be analyzed as Russian authors have done on Russian and Siberian platforms by the K-Ar method (Keller et al, 1975). Clauer (1975) published ages on the upper Precambrian Atar series of Mauritania which were obtained by the Rb-Sr method. Another criterion is the initial Sr^{87}/Sr^{86} ratio; if the ranges between 0.700 and 0.709, respectively for very old and uppermost Precambrian (see references in Clauer, 1976), the age obtained more likely corresponds to the time of deposition or early diagenesis. If it is greater than 0.710, there is a strong probability that the age is related to a postdepositional transformation which has to be defined. Locally, as in Atar (Clauer, in press), some inversions appear in the regularly decreasing age succession, in contradiction with the stratigraphic sequence. Here lies a real difficulty, because it is very difficult to see such a phenomenon if the study is restricted to a unique level. It could appear also if the age is uniform throughout a large stratigraphic sequence. Even in this case, if there are many carbonaceous beds the large amount of common strontium might lead to an undisturbed initial ratio. The clay mineralogy could also remain more or less undisturbed. Such an interpretation might be proposed for both Little Belt and Big Belt Mountains, and the Sun River series (Obradovich and Peterman, 1968).

Published results have been recalculated using constant: $\lambda Rb^{87} = 1.39 \ 10^{-11} \ yr^{-1}$. There are three main remarks about numerical correlations:

1. Two middle Precambrian series are defined in central Africa and northern South America. Both lie on 2 billion-year old cratons. The Franceville series remains, until the present, the oldest typically sedimentary sequence, even if it has undergone some strong diagenesis under a rock cover of 4,000 to 5,000 m.

2. After these sediments, which postdate the cratons and limit the orogenic episodes, there was a long interval without sedimentation in Africa and South America. On the contrary in eastern Europe and northern Asia, there was a more or less continuous sedimentation after 1,650 m.y. In North America, the oldest sedimentary rock date is defined in the Belt series. In fact, the possibility for a late diagenesis to occur in that unit means that the date of 1,325 m.y. might be a youngest limit for the time of deposition, thus unknown between 1,325 and 1,700 m.y. In central Africa, the Roan Formation (Cahen, 1973) is considered to be deposited at the same time—that is, between 888 ± 44 m.y. and the Kibarian orogeny, which ends at 1,300 m.y., or middle or upper Riphean (Keller et al, 1975).

3. In all other countries of western and central Africa, South America, and northern Europe, sedimentation began between 1,000 and 1,100 m.y. or less. This means that, in West Africa, sediments of lower and middle Riphean are not present; they were eroded or not deposited.

In all formations under reference which are dated as upper Precambrian, tillites have been described. In northern Europe, western Africa, and the USSR, a tillite system has been dated as Eocambrian. Pringle (1973) obtained an age of 668 ± 23 m.y. for an argillaceous level situated between two tillite horizons of that age. It was dated in Mauritania as older than 613 ± 51 m.y. (Clauer, 1975), in Ghana as older than 620 m.y. (Boshko et al, 1971), and as older than 645 ± 30 m.y. in Zaire (Cahen, 1973). Moreover, it is considered to be younger than 680 ± 20 m.y. in the USSR (Keller et al, 1975). Thus, the age of that worldwide extended tillite episode must be the age measured by Pringle (1973). As a consequence, the good agreement of all determinations, from Australia (Dunn et al, 1971) to northern Europe, implies that this tillite is a very good stratigraphic correlation horizon.

A tillite is also well known in South America beneath the Bambui Formation where it has been dated as older than 1,000 to 1,050 m.y.; the age is not well defined for interpretative reasons. In southern Zaire, the Great Conglomerate is considered older than 888 ± 44 m.y. These two occurrences might be of the same age. Such a tillite is unknown in the USSR succession.

Stratigraphy of the Riphean in the USSR is mainly based on stromatolitic assemblages (Keller et al, 1975). A comparison between Africa and the USSR was made by Bertrand-Sarfati and Raaben (1970). As a result, the West African assemblage is correlated to the upper Riphean. All dates determined on the sediments containing the first African stromatolitic association with essentially *Conophyton* show an age of 900 to 1,000 m.y. This confirms the belief that stromatolite populations seem to be (until the present) a correlation item at least as good as isotopic age determinations. The slight differences in time between exposures generated in countries situated far away from each other is probably lower than physical uncertainties in isotopic dating.

Conclusion

There is a fairly good agreement between isotopic dating and tillitic or fossil occurrences where the clay mineralogy is precise enough to allow well-defined interpretation of radiometric data along with the stratigraphic succession. For that reason, it is recommended that only regularly decreasing age successions be studied.

Today, it is the way to avoid the risk of late diagenesis and subsequent age modification. Age dating, tillites, and fossils can be used separately or together for the last 500 m.y. of the Precambrian. These items could probably be used successfully between 1,100 and 1,800 m.y., but for rocks older than 1,800 m.y. the present lack of information by isotope dating in sedimentary sequences does not provide such correlations.

References Cited

Bassot, J. P., et al, 1963, Mesures d'ages absolus sur les series precambriennes et paleozoiques au Senegal oriental: Soc. Geol. France Bull., ser. 7, v. 5, no. 3, p. 401-405.

Bertrand-Sarfati, J., and M. E. Raaben, 1970, Comparaison des ensembles stromatolitiques du Precambrien superieur du Sahara occidental et de l'Ougnat: Soc. Geol. France Bull., ser. 7, v. 12, no. 2, p. 364-371.

Bonhomme, M., and N. Clauer, 1972, Possiblites d'utilisation stratigraphique des datations directes rubidium-strontium sur les mineraux et les roches sedimentaires, *in* Colloque sur les methodes et tendances de la stratigraphie: (France) Bur. Recherches Geol. et Minieres Mem. 77, v. 2, p. 943-950.

—— and U. G. Cordani, 1976, Rb-Sr dating of Upper Precambrian sediments from north-eastern Brazil: 4th European Coll. Geochronology, Cosmochronology, Isotope Geol. (Amsterdam), Als, 10 p.

—— F. Weber, and R. Faure-Mercuret, 1965, Age par la methode rubidium-strontium des sediments du bassin de Franceville (Republique Gabonaise): Alsace-Lorraine Service Carte Geol. Bull., p. 243-252.

Boshko, N. A., et al, 1971, Nouvelles determinations de l'age absolu de glauconies d'Afrique Occidentale: Akad. Nauk SSSR Doklady, v. 198, p. 1401-1402, (in russian).

Cahen, L., 1973, Correlation de certaines series du Precambrien superieur du Zaire a la lumiere de l'etude des stromatolites et des donnees de geochronologie radiometrique: Mus. Royal Afrique Centrale Annales, Dept. Geol. Mineral, Rapp. Ann. (1972), p. 38-51.

Cahen, L., and N. J. Snelling, 1974, Potassium-Argon ages and additions to the stratigraphy of the Malagarasian (Bukoban system of Tanzania) of SE Burundi: Geol. Soc. London Jour., v. 130, p. 461-470.

Clauer, N., 1975, Dating of sedimentary minerals and rocks, possibilities of the Rb/Sr method and application to the Upper Precambrian from the West African craton: Corr. Symp. of the Precambrian (Moscow), Als., p. 67-68.

—— 1976, Geochimie isotopique du strontium des milieux sedimentaires. Application a la geochronologie de la couverture du craton ouest-africain: Sci. Geol. Mem., Strasbourg, 256 p.

Dunn, P. R., B. P. Thomson, and K. Rankama, 1971, Late Precambrian glaciation in Australia as a stratigraphic boundary: Nature, v. 231, no. 5304, p. 498-502.

Dunoyer de Segonzac, G., 1969, Les mineraux argileux dans la diagenese; passage au metamorphisme: Alsace-Lorraine Service Carte Geol. Mem. 29, 317 p.

Hagemann, R., et al, 1974, Chemical and isotopic studies of the Oklo fossil reactor: Int. Mtg. Geochronology, Cosmochronology, Isotope Geol. (Paris), Als., 1 p.

Keller, B. M., et al, 1975, The Phiphean of the USSR and the problems of the general Time-Stratigraphic Scale of the Upper Precambrian: Corr. Symp. of the Precambrian (Moscow), Als, p. 63-66.

Kubler, B., 1966, La cristallinite de l'illite et les zones tout-a-fait superieures du metamorphisme, *in* Colloque sur les etages tectoniques: Neuchatal Univ. Inst. Geol., p. 105-122.

Loubet, M., et al, 1974, Oklo natural reactor. Age, nuclear characteristics, chemical history: Int. Mtg. Geochronology, Cosmochronology, Isotope Geol. (Paris), Als., 1 p.

Millot, G., 1964, Geologie des argiles; alterations, sedimentologie, geochimie: Paris, Masson et C°, 499 p.

Obradovich, J. D., and Z. E. Peterman, 1968, Geochronology of the Belt Series, Montana: Canadian Jour. Earth Sci., v. 5, p. 737-747.

Priem, H. N. A., et al, 1973, Age of the Precambrian Roraima Formation in northeastern South America; evidence from isotopic dating of Roraima pyroclastic volcanic rocks in Suriname: Geol. Soc. America Bull., v. 84, no. 5, p. 1677-1684.

Pringle, I. R., 1973, Rb-Sr age determination on shales associated with the Varanger ice age: Geol. Mag., v. 109, no. 6, p. 465-472.

Weber, F., and M. Bonhomme, 1975, Donnees geochronologiques nouvelles sur le Francevillien et son environment, *in* The Oklo Phenomenon: New York, Int. Atomic Energy Agency, ed., p. 17-35.

Aspects of the Revised South African Stratigraphic Classification and a Proposal for the Chronostratigraphic Subdivision of the Precambrian[1]

L. E. KENT[2] and P. J. HUGO[2]

Abstract In 1970, the National Committee for Geological Sciences accepted in principle the ISSC recommendations for stratigraphic classification and a South African Committee for Stratigraphy (SACS) was appointed. A local code was published in 1971 and, through working groups, SACS has revised and completely reclassified the South African stratigraphic column. The former and the revised principal stratigraphic subdivisions are presented in tables.

The principal lithostratigraphic units in the main Precambrian basins of deposition are being termed supergroups, groups, and sequences; the former Swaziland, Witwatersrand, and Ventersdorp "Systems" are now Supergroups; the Pongola, Dominion Reef, and Waterberg "Systems" are Groups; and the Transvaal "System" is now the Transvaal and Griqualand-West sequences with component groups. The term Bushveld Complex replaces Bushveld Igneous Complex.

The Gariep and Nosib Groups (formerly "Systems") have been placed in the Damara Supergroup. The Nama "System" which spans the Precambrian-Phanerozoic boundary has become a Group; its counterpart in the southwestern Cape Province, which is intruded by the Cape Granite, is now termed the Malmesbury Group.

For the Phanerozoic, SACS decided that the Cape "System" becomes the Cape Supergroup and the Karoo "System" a sequence. In the Cape Supergroup the major units are now termed the Table Mountain, Bokkeveld, and Witteberg Groups. In the Karoo sequence, the Dwyka Tillite Formation forms the base followed by the Ecca, Beaufort, and Drakensberg Groups. The widespread dolerites intrusive into the Karoo are still collectively referred to as Karoo Dolerites.

In the Jurassic-Cretaceous, the term *Beds* has been abandoned and Suurberg, Uitenhage, and Zululand Groups are defined.

The problem of subdividing the Precambrian chronostratigraphically is discussed. In South Africa there are unrivalled, relatively undisturbed, and only slightly metamorphosed Precambrian successions that cover, with only minor hiatuses, the time span from 3,750 m.y. to the Phanerozoic. A wide span of reliable radiometric age determinations, and absence of major facies changes, has led to their formal subdivision into the following erathems: Swazian (3,750+ to 2,870 m.y.), Randian (2,870 to 2,630 m.y.), Vaalian (2,630 to 2,070 m.y.), Mogolian (2,070 to 1,080 m.y.) and Namibian (1,080 to 570 m.y.).

Introduction

Geological Survey organizations and geologists working in southern Africa have long followed the so-called dual stratigraphic nomenclature adopted by the 2nd International Geological Congress (Bologna, 1881) with time units in the hierarchy: era, period, epoch, and age; and parallel units for the rock sequences formed during the respective divisions of time: group, system, series, and stage.

In the Phanerozoic, the two major successions were termed the Cape System and Karroo System. They were established in 1888 by A. Schenck as "Kapformation" and

[1] Manuscript received, September 30, 1976.

[2] Geological Survey of the Republic of South Africa, Pretoria, South Africa.

The authors wish to record their indebtedness to present and past fellow members of the South African Committee for Stratigraphy and the conveners and members of the working groups whose reports have been used in preparing this paper.

Article Identification Number: 0149-1377/78/SG06-0030/$03.00/0

in 1857 by A. G. Bain as "Karroo or Reptiliferous (Lacustrine) Series", respectively, at the same time as some of the classic systems of the Northern Hemisphere were being named. The designations Cape System and Karroo System were officially adopted by the Geological Commission of the Cape of Good Hope (the precursor of the Geological Survey of South Africa) in 1902.

The use of the terms *system* and *series* was extended to major subdivisions of the pre-Phanerozoic. G.A.F. Molengraaff, State Geologist of the Zuid-Afrikaansche Republiek, for example, defined the Transvaal System in 1901. The thick, predominantly sedimentary succession which hosts the auriferous conglomerates of the Transvaal was named the Witwatersrand Series by W. H. Penning in 1888; it was changed to Witwatersrand System by E. T. Mellor in 1917.

Over the years the number of named systems increased, no less than 15 (12 Precambrian) being recognized on the Geological Map of the Republic of South Africa and the Kingdoms of Lesotho and Swaziland published by the Geological Survey in 1970 (Fig. 1). Most were divided into series and, in the Transvaal System, even stages and substages were defined.

Successions that did not justify "system" rank were named *formations*. In the *Information Brochure for Professional and Technical Officers* (Geological Survey of South Africa, 1947) this term is stated to have ". . .no stratigraphic significance and, like beds, should be used as a collective term for a succession of sediments, including effusives and pyroclastics, which shows no stratigraphic hiatus and whose stratigraphic position or taxonomic rank is unascertained." Examples are the Matsap Formation and the Koras Formation in northwestern Cape Province. Formations were divided into unnamed units based on lithology.

The term *Beds* was used loosely for some units in the "series" of the Witwatersrand System and for successions of rather limited extent ranging in age up to the Alexandria and Bredasdorp "Beds" of the Tertiary. The terms formation and beds were thus used in essentially the same way as the lithostratigraphic units of the International Subcommission on Stratigraphic Classification (Hedberg, 1976).

The South African representative on the ISSC did not agree with the Statement of Principles presented at its meeting during the 21st Congress (Copenhagen, 1960). He considered the chronostratigraphic units used, such as series, prefixed when necessary by the qualifiers litho-, bio-, or chronostratigraphic, were preferable to a separate series of terms for each system of classification. However, in 1970 the National Committee for Geological Sciences accepted the ISSC recommendations in principle and in 1971, conjointly with the South African Scientific Committee for the International Union of Geological Sciences, appointed the South African Committee for Stratigraphy (SACS). Its first task was the preparation of a stratigraphic code (1971). Subsequently through nine working groups in which, over the years, many geologists have served, a complete reclassification of the stratigraphic column for South Africa has been made (Fig. 2). A general account is being prepared for publication by the Geological Survey and this will be followed by detailed reports (complete with stratotype descriptions) on each unit.

Following the acceptance in principle of the ISSC recommendations by the National Committee for Geological Sciences, many authors have coined stratigraphic terms. Some of them have been accepted by SACS; others are still informal.

Precambrian

In devising a lithostratigraphic subdivision of the former Swaziland System and equivalent rocks, it was decided that the Barberton Mountain Land be taken as the type locality and the tract surrounding the Barberton greenstone belt be used as a reference area for the classification of the granites of the Kaapvaal Craton.

The whole volcano-sedimentary accumulation comprising the Barberton greenstone belt constitutes a distinctive and well-defined supergroup—the Swaziland Supergroup. The rocks can be divided into a dominantly volcanic lower assemblage and a dominantly sedimentary upper assemblage. The volcanic assemblage has been termed the Onverwacht Group and the sedimentary one has been divided into the Fig Tree and Moodies Groups, the former being largely argillaceous and the latter arenaceous.

Approximate age in million years	Era	CHRONOSTRATIGRAPHIC SYSTEM	FORMAL SUBDIVISION: SERIES/FORMATION	FORMAL SUBDIVISION: SYSTEM
		TERTIARY TO QUATERNARY		TERTIARY TO QUATERNARY
	MESOZOIC	CRETACEOUS		CRETACEOUS
	MESOZOIC	TRIASSIC TO JURASSIC	Stormberg Beaufort	KAROO
200	PALEOZOIC	UPPER CARBONIFEROUS TO UPPER PERMIAN	Ecca Dwyka	KAROO
	PALEOZOIC	ORDOVICIAN TO LOWER CARBONIFEROUS	Witteberg Bokkeveld Table Mountain	CAPE
	PALEOZOIC	CAMBRIAN	Klipheuwel Formation	
	PRECAMBRIAN	UNDIFFERENTIATED		NAMA/DAMARA NOSIB
	PRECAMBRIAN	UNDIFFERENTIATED	Koras Formation	
1080 1950	PRECAMBRIAN	UNDIFFERENTIATED		WATERBERG BUSHVELD IGNEOUS COMPLEX
	PRECAMBRIAN	UNDIFFERENTIATED	Pretoria Dolomite Black Reef	TRANSVAAL
	PRECAMBRIAN	UNDIFFERENTIATED		VENTERSDORP
2640	PRECAMBRIAN	UNDIFFERENTIATED	Kimberley–Elsburg Main–Bird Jeppestown Government Reef Hospital Hill	WITWATERSRAND
2800	PRECAMBRIAN	UNDIFFERENTIATED		DOMINION REEF
	PRECAMBRIAN	ARCHAEAN COMPLEX	Mozaan Insuzi	PONGOLA
3100– 3300	PRECAMBRIAN	ARCHAEAN COMPLEX		MOODIES KHEIS
	PRECAMBRIAN	ARCHAEAN COMPLEX	Fig Tree Onverwacht	SWAZILAND

FIG. 1–Former chronolithostratigraphic subdivision of the South African column (after Van Eeden, 1972).

Approx. age in million years	CHRONOSTRATIGRAPHIC UNIT	MAJOR LITHOSTRATIGRAPHIC UNIT			
		Formation	Subgroup	Group	Supergroup/Sequence
2	QUATERNARY			Various formations and groups	
65	TERTIARY				
140	CRETACEOUS	Various		Uitenhage Suurberg / Zululand	
195	JURASSIC	Several Several Clarens Elliot Molteno		Lebombo Drakensberg	KAROO
230	TRIASSIC	Various Various Various Dwyka	Tarkastad Adelaide	Beaufort	
280	UPPER CARBONIFEROUS			Ecca	
325	DEVONIAN	Various		Natal Witteberg Bokkeveld Table Mountain	CAPE
395	ORDOVICIAN, SILURIAN			Nama/Malmesbury Gariep, Nosib	DAMARA
1080	NAMIBIAN	Various formations and subgroups		Koras Waterberg/Soutpansberg	
2070	MOGOLIAN	Various	Kransberg Matlabas Nylstroom	Rooiberg Olifantshoek Pretoria/Postmasburg Chuniespoort/Campbell/Griquatown Wolkberg	TRANSVAAL/ GRIQUALAND WEST
2630	VAALIAN	Various Black Reef	Malmani	Pniel Platberg Klipriviersberg	VENTERSDORP
2800	RANDIAN	Various	Turffontein Johannesburg Jeppestown Government Reef Hospital Hill Messina Shanzi	Central Rand West Rand Dominion Limpopo (Beit Bridge)	WITWATERSRAND
2900	SWAZIAN	Various	Mozaan Nsuze Geluk Tjakastad	Pongola Moodies Figtree Onverwacht	SWAZILAND
3750					

FIG. 2 – Lithostratigraphic subdivision of the South African column.

The majority of layered ultramafic bodies of the Barberton area are classed as intrusive complexes within the lower ultramafic Tjakastad Subgroup of the Onverwacht Group. Granitoid rocks have intruded and in part deformed the rocks of the Swaziland Supergroup. Tonalitic diapir domelike plutons (3,310 to 3,170 m.y.)[3] which were the result of reactivation of an ancient basement are confined to the western and southern flanks of the Barberton area. This event was succeeded by a period of migmatization and injection of sheetlike bodies, derived in part from the earlier granitoids, as evidenced by the Nelspruit and Lochiel granitoids having isotopic ages of 3,160 to 2,990 m.y. The final nonorogenic granitoid event was the emplacement of transgressive plutons (3,130 to 2,880 m.y.).

Strata assigned to the Pongola Group are exposed in three separate but extensive areas extending from southern Swaziland, across the southeastern Transvaal into northern Natal. It is divided into the Nsuze Subgroup (mainly volcanic rocks) which is overlain disconformably by the Mozaan Subgroup, comprising mainly shale and quartzite. The Usushwana Complex is intrusive into the succession and its date of 2,874 ± 30 m.y. records a minimum age for the Pongola Group (Davies et al, 1969, p. 583). A lava in the Nsuze succession dated at 3,090 m.y. is accepted as the minimum age for this subgroup.

In earlier literature all granitoid gneisses in the Messina area were included in the basement and only the remaining rocks were considered to have been Messina Formation. However, recent work has shown that most of the granitic gneisses form part of a conformable succession. These gneisses have therefore been reallocated to the cover rocks as the Shanzi Subgroup which, with the Messina Subgroup, constitutes the Beit Bridge Group[4]. The Bulai and Singelele granitoid gneisses, which are appreciably younger than the basement, represent intrusive phases possibly originating from remobilized basement and stratigraphically lower cover rocks. The granitoids have been dated at 2,690 ± 60 m.y.

The term Witwatersrand Triad is used informally to tie together the Dominion Reef, Witwatersrand, and Ventersdorp "Systems" which are closely associated geographically.

The predominant lithology of the renamed Dominion Group is andesite-rhyolite, with associated tuffs and basal quartzite within which the auriferous uranium-bearing conglomerates, the Dominion reefs, occur.

The Witwatersrand Supergroup is divided into a lower West Rand Group and an upper Central Rand Group. In further subdivision, well-entrenched names have been retained as far as possible. In the West Rand Group the unifying lithologies are the alternating quartzites and shales of the Hospital Hill, the Government Reef, and the Jeppestown Subgroups. The Central Rand Group which replaces the former Upper Witwatersrand Division, is a predominantly quartzose succession comprising the Johannesburg Subgroup, Booysens Formation, and the Turffontein Subgroup.

In the Ventersdorp Supergroup, two groups have been distinguished: a lower Klipriviersberg and a middle Platberg (Winter, 1976). The remaining formations have been assigned provisionally to a Pniel group.

The Transvaal sequence[5] not only incorporates all the series of the old Transvaal System but also the Wolkberg System (Truter, 1949). The Black Reef Quartzite is not assigned to any group and stands alone as a formation overlying the basal Wolkberg Group.

The Chuniespoort Group is equivalent to the old Dolomite Series and comprises principally chemical sediments. It is overlain by the Pretoria Group (equivalent to the old Pretoria Series) which is composed mainly of arenaceous and argillaceous sediments, with some lavas, including the Hekpoort Andesite Formation.

The succession of rocks in Griqualand West previously also grouped under the Transvaal System has now been renamed the Griqualand West Sequence. The Campbell

---

[3] For sources of radiometric age determinations and limits of accuracy, see Burger and Coertze (1973) and the supplements published in the *Annals of the Geological Survey of South Africa*.

[4] Beit Bridge Group was provisionally styled "Limpopo Group," and is shown by that name on Figure 2.

[5] The term "sequence" is discussed under the heading, *Phanerozoic*.

Group at the base embraces the old Black Reef Series (now Vryburg Formation) and the chemical sediments of the old Dolomite Series (Campbell Rand Series). The ironstone and jaspilites of the former Upper Dolomite Series are now included in the Griquatown Group, while the Pretoria Series has become the Postmasburg Group. The mainly sedimentary Olifantshoek Group which follows includes the Hartley Andesite and consists of the old Matsap Formation and the new Groblershoop Formation, the latter succession being previously regarded as part of the Kheis System. The remaining rock successions of the former Kheis System constitute the Marydale and Wilgenhoutdrif Formations. Towards the west, these rocks became involved in the Namaqua tectonic event of 1,100 m.y. ago, the products of which are grouped under the general term Namaqualand Metamorphic Complex.
of which are grouped under the general term Namaqualand Metamorphic Complex.

The Rooiberg (Felsite) Series which disconformably overlies the Transvaal sequence has been renamed the Rooiberg Group.

The Waterberg Group embraces a succession of rocks previously grouped under the Loskop and Waterberg Systems. Three subgroups are distinguished: a lower Nylstroom comprising mainly sandstone and graywacke, a middle Matlabas consisting of grit and siltstone, and an upper Kransberg composed largely of sandstone.

Recent fieldwork indicated that the Soutpansberg succession represents one sedimentary and volcanic cycle, with the beds probably laid down in a large deep trough or aulacogen. The middle part of the succession overlaps sediment of the late Waterberg basin (Jansen, 1975). A Soutpansberg Group comprising several formations is now recognized.

Although the formal stratigraphic code is not particularly applicable to the rocks of the Bushveld Complex, (in keeping with the nongenetic terminology the term *igneous* has been dropped) a classification has nevertheless been attempted using many of the guidelines in the code. In subdividing the complex, *zone* is informally used as the fundamental unit of lithostratigraphic classification in the layered sequence of basic rocks. Although at variance with the stratigraphic code, it conforms to international usage and does not entail drastic changes in terminology.

At group level, the Bushveld Complex (which partly overlaps in age the Waterberg Group) is subdivided into three lithostratigraphic units. In chronological order they are the Rustenburg Layered Suite, the Rashoop Granophyre Suite, and the Lebowa Granite Suite. The Rustenburg Layered Suite corresponds to the basic intrusive phase and is made up of five zones; these conform to the zonal subdivision that has been used for a long time.

The remaining suites comprise the acid rocks of the complex. Both the Rashoop and Lebowa Suites contain a number of informally named rock units (such as the Beestkraal granophyre and the Klipkloof granite). Thus far, only two formal units have been recognized, the Nebo Granite and the Makhutso Granite, both forming part of the Lebowa Granite Suite.

Three formations are recognized in the Marico Diabase Suite which corresponds to what was known as the "Sill Phase" of the complex (Hall, 1932).

The former Koras Formation has been given group status (1,090 ± 20 m.y.).

Extending from southern South West Africa[6] into northwestern Cape Province the former Nosib and Damara Systems are now the Nosib and Gariep Groups of the Damara Supergroup. The Nama System which spans the boundary between the Precambrian and the Phanerozoic has been renamed the Nama Group. It comprises in ascending order the Kuibis, Schwarzrand, and Fish River Subgroups. Its counterpart in southwestern Cape Province, the Malmesbury Group which embraces the Swartland and Boland Subgroups, is intruded by the well-known Cape Granite (550 m.y.). The inliers of pre-Cape successions in the southern Cape Province have been placed in a Cango Group.

Phanerozoic

For the Phanerozoic, SACS has decided that the Karroo System should be termed a *sequence*. This use of the term is from an ISSC suggestion that the Karroo might be

---

[6] Groups that occur only in South West Africa are not dealt with in this paper.

considered an informal system or supersystem, or be termed informally a "sequence" by analogy with North American usage where the term is used for ". . .great aggregations of strata lying between major regional unconformities and frequently comprising the rocks of several systems, e.g. Sauk Sequence" (Hedberg, 1971).

For the Cape Supergroup two separate stratigraphic columns are distinguished, one for western and one for eastern Cape Province, respectively, reflecting both facies changes and correlation problems. The succession comprises three groups, embracing 24 formations in western and 23 in eastern Cape Province. The Table Mountain, Bokkeveld, and Witteberg Groups replace the former Table Mountain, Bokkeveld, and Witteberg Series.

As recent paleontologic evidence suggests that the strata formerly grouped under the term Table Mountain Series in Natal may be partly or entirely the lateral equivalent of the Witteberg Group (Lock, 1973), it has been decided that they be placed in a Natal Group rather than the Table Mountain Group.

The Karoo successions (note: new spelling with one "r" only) are partly Paleozoic and partly Mesozoic and in South Africa include strata of no less than four internationally accepted periods—Carboniferous, Permian, Triassic, and Jurassic. Collectively they make up the Karoo sequence.

In the main basin, the Karoo rocks have been placed provisionally in a supergroup subdivided into three groups, each containing several formations, and four ungrouped formations. The succession commences with the Dwyka Tillite Formation which, unlike the former Dwyka Series, is restricted to distinctly glacial material (diamictite, varved shale with dropped pebbles, fluvioglacial gravel and conglomerate). The former Upper Dwyka Shales are henceforth to be grouped with the overlying Ecca Group which consists essentially of shale, with sandstone prominently developed only at the basin margins.

The Beaufort Group, which is essentially the former Beaufort Series, is composed of mudstone and sandstone which indicates a change in environment, from deposition in a large body of water (possibly marine in the case of the Ecca) to generally continental (mainly fluviatile) conditions in the Beaufort. It is recognized that this change is unlikely to have taken place at the same time everywhere and that the Ecca-Beaufort boundary is a diachronous one.

As there are no unifying lithologic features in the various units constituting the old Stormberg Series the use of "Stormberg" as a formal group designation cannot be justified. Its three units are now named Molteno, Elliot, and Clarens Formations, respectively, corresponding with the former Molteno Beds, Red Beds, and Cave Sandstone (Du Toit, 1954).

The name, basic concept, and boundaries of the Drakensberg Beds consisting of up to 1,200 m of basalt (Du Toit, 1954), remain basically the same in the new Drakensberg Group. However, in view of important facies and age differences between the Karoo lavas in the Lesotho and Lebombo areas it was thought inadvisable to attempt any correlation between the two. Accordingly, a Lebombo Group has now been recognized and a set of independent formation names adopted for the three main lithologic units embraced by it. The stratigraphic terminology of the isolated outliers of Karoo strata occurring north of the main basin in the Transvaal has not yet been finalized.

Dolerites intrusive into the Karoo sequence are still collectively referred to as Karoo Dolerites. They are not treated as formations, but merely added to the stratigraphic column (or map) with distinguishing symbol, i.e. Jd (Jurassic dolerite).

The best continuous succession of rocks of Jurassic-Cretaceous age is exposed in the Algoa Basin (Cape Province) where the Suurberg Group containing volcanics of Jurassic age is overlain by the sedimentary Upper Jurassic - Lower Cretaceous Uitenhage Group, the latter term superseding the Uitenhage Series of earlier workers. In northern Natal, the Jurassic Jozini and Bumbeni volcanics of the Lebombo Group are overlain by a Zululand Group extending over the whole of the Cretaceous.

Tertiary deposits are scattered. However, it was possible to assign certain units with comparative assurance to a particular series and sometimes to a particular stage. Chronostratigraphic subdivision of the Quaternary System is much more difficult, because subdivisions are shorter and are partly based on climatic fluctuations of the Northern Hemisphere, which are difficult to identify in South Africa and may indeed not have affected it or have been out of phase. The system has therefore been conventionally divided into Pleistocene and Holocene Series with the boundary at 10,000 B.P., and radiometric dates established in other countries have been adopted for the subdivision of the Pleistocene.

Proposed Chronostratigraphic Subdivision of the Precambrian

It has long been an objective of stratigraphers to achieve a systematic hierarchy of chronostratigraphic units in the Precambrian. Many local systems were established during the early attempts to unravel its stratigraphy. Recently several have been redefined as lithostratigraphic groups, e.g. the Cuddapah, Dharwar, and Delhi Groups replace the former systems with the same names in India, but with the retention, at least in the interim, of the component "series" and "stages" (Pichamuthu, 1967). In a general account of the Precambrian of the Congo, Rwanda, and Burundi, Cahen and Lepersonne (1967) noted that the former systems are referred to without a suffix as, for example, Katangan and Roan; the former series are sequences, and stages and beds are groups and formations. As a final example, in Scotland such units as the Moinian, Lewisian, and Dalradian that were formerly termed series, are now referred to as assemblages with components of group and complex status (Anderson, 1965).

However, in Australia local systems for the largely unmetamorphosed upper Precambrian have recently been formally established, e.g. in 1950 the Adelaidean which comprises lithostratigraphic units of group and supergroup status (Dunn et al, 1966). This is in keeping with the intention of Ashley et al (1933) when they drew up the rules for stratigraphic classification and nomenclature that were the progenitor of the American and the international stratigraphic codes. They regarded a group as ". . .a local or provincial subdivision of a system, based on lithological features. It is usually less than a standard series and contains two or more formations."

The problem of the status of major stratigraphic units in the Precambrian was stressed by Rankama (1963): "While the magnitude of a system is not standardized, the Precambrian, whether an erathem, or a supersystem, or a group, is still not covered by internationally accepted rules for its stratigraphic classification. . . ." The ISSC (Hedberg, 1976) also remarked that although the Precambrian represents some 85% of geologic time, it has not been divided systematically into globally recognized chronostratigraphic units. The view was expressed that sound *local* chronostratigraphy should be built up in appropriate areas and then by proceeding to regional and continental scale, worldwide scope should ultimately be attained. In amplifying this, Hedberg (1974) recommended ". . .concentration of chronostratigraphically oriented studies on the rock strata of those local areas in the world where thick, relatively continuous, and relatively unmetamorphosed sequences of Precambrian strata are known with minimum geologic complications and minimum possibilities of having had their radiometric clocks reset." Cloud (1971) seemed to have been thinking on similar lines when he wrote, "Meanwhile it seems desirable to follow the lead of the South Africans in defining and formally naming a succession of local rock systems within the Precambrian and within a geochronological framework, pending the day when it may be possible to integrate them into a more comprehensive general nomenclature."

Hedberg (1974), in discussing chronostratigraphic classification of the Precambrian, noted that, "Stratotype definition of Precambrian units also has been called impracticable because of the lack of long unbroken sections for types. . . ." This objection, which he considers invalid, applies to only a small fraction of the South African Precambrian successions. As Cloud (1971) pointed out, in this part of the world there are unrivalled successions of only slightly metamorphosed rocks with ages of up to more than 3,500 m.y. having a wide span of well-determined radiometric dates. Furthermore these successions are almost continuous and thus cover the time span from 3,800 m.y. to 550 m.y. with no major hiatuses. The ill-defined Precambrian-Cambrian boundary is spanned by the fossil-bearing Nama Group.

As regards the duration of major chronostratigraphic units in the Precambrian, it is considered that standards other than those internationally accepted for the Phanerozoic should apply. Neglecting the Quaternary, the Phanerozoic systems range in duration from 35 to 70 m.y. The major units for the Precambrian Eon should cover time spans of the order of the Phanerozoic (560 m.y.) and should thus be of at least erathem rank.

Thus establishment of a local but formally named chronostratigraphic subdivision of the South African Precambrian was decided on by SACS in 1975 and the outline on which Figure 3 is based was subsequently adopted. The time spans of the proposed erathems range from 240 to 990 m.y.

It was decided that the very well known and established system names—Swaziland, Witwatersrand, and Transvaal should be used as a basis for naming the older erathems—Swazian, Randian, and Vaalian. Mogolian is from the Mogol River in the type area of the Waterberg Group and Namibian is from the coastal Namib Desert of South West Africa.

It may be noted that Truswell (1967) in a critical review of South African stratigraphic terminology, although advocating the adoption of lithostratigraphy, strongly favored rentention of the Witwatersrand, Ventersdorp, Transvaal, and Nama as local Precambrian systems. However, the Swaziland and Waterberg systems were regarded by him as lithologic groups.

In adopting the "*ian*" terminations the international usage for the Paleozoic has been followed. Except for the Swazian (symbol Z), the initial letters of the erathems which will be used in map legends do not duplicate those of the internationally accepted system/period names in the Phanerozoic.

Due to absence of major facies changes in the pre-Namibian, it is considered that the boundaries proposed for the erathems are isochrons within the limits that are reasonable for the time spans involved.

Within the erathems, sequences, supergroups, and groups have been established (Fig. 3). The Swazian erathem, for example, occupies a longer period of time than that involved in the accumulation of the Swaziland Supergroup.

For further chronostratigraphic subdivision there are, for example in the Swaziland and Witwatersrand successions, relatively thin but widespread deposits of volcanic ash and also tillites which could serve as reliable chronohorizons. The Witwatersrand Supergroup (and to a lesser extent the Swaziland Supergroup, the Ventersdorp Supergroup, and the Transvaal sequence) has been intensely studied and described in minute detail. This study was possible due to the remarkably low degree of metamorphism and will facilitate future chronostratigraphic subdivision. The relative freedom from metamorphism has also permitted the preservation of fossil algae (Barghoorn and Schopf, 1966) in rocks of the Fig Tree Group (3,100 m.y.) and of widespread stromatolites in the Vaalian dolomite (2,250 m.y.). Similar fossils have been used in Russia and elsewhere (Cloud and Semikhatov, 1969; Gebelin, 1974; Walter and Preiss, 1972) for biostratigraphic subdivision of the Precambrian and for intercontinental correlations. It is hoped that more intensive studies of the South African stromatolites, which also occur in the Swaziland Supergroup and the Ventersdorp Supergroup, will be made with this end in view.

Boundaries of the proposed erathems are at ± 1,080, 2,070, 2,630, 2,870, and 3,750 m.y. with an order of accuracy of the underlying radiometric dating of 30 to 70 m.y. Using the tectonic cycle approach, Semikhatov (1972) distinguished the following boundaries in the Precambrian which he considered to be of planetary extent—1,700 and 2,600 ± 100 m.y. The last coincides with the boundary between the Randian and the Vaalian. The periods of major tectonism and magmatism in South Africa are the 3,100 m.y. Northern Cape - Transvaal, 2,000 m.y. Bushveld and Limpopo, and the 1,100 m.y. Namaqualand events.

Some of the more recent proposals for the chronostratigraphic subdivision of the Precambrian are given in Figure 4.

As regards the Australian Precambrian systems, they rely for their standing on their use in publications of the Bureau of Mineral Resources, Geology and Geophysics, and state geological surveys. They have not been considered by the Stratigraphic Nomenclature Committee of the Geological Society of Australia. Although there is some doubt as to the age of the base of the Adelaidean, it and the Carpentarian are used for

CHRONOSTRATIGRAPHIC UNIT	Approx. time span in million years	Key radiometric date in million years	MARKER UNIT	LITHOSTRATIGRAPHIC UNIT
CAMBRIAN (€)	70	550 ± 20 — 570	Cape Granite/Salem Granite (S.W.A.)	NAMA GROUP
NAMIBIAN (N)	510	800 ± 50 — 1080 ± 70	Naauwpoort lava in Nosib Group (S.W.A.) — Quartz porphyry in Koras Group	DAMARA SUPERGROUP
MOGOLIAN (M)	990	1670 ± 25, 1920 ± 30, 2050 ± 30, 2070 ± 30, 2098 ± 12	Makhutso Granite, Bushveld granite, Bushveld granophyre, Hartley Andesite, Bushveld ferrogabbro	WATERBERG GROUP (BUSHVELD COMPLEX)
VAALIAN (V)	560	2224 ± 50 — 2630 ± 60	Hekpoort Andesite — Quartz porphyry at top of Ventersdorp	TRANSVAAL SEQUENCE
RANDIAN (R)	240	2750, 2800	Contorted Bed, Lava in Dominion Group	VENTERSDORP SUPERGROUP, WITWATERSRAND SUPERGROUP, DOMINION GROUP
SWAZIAN (Z)	880+	2874 ± 30, 3090 ± 90, 3360 ± 100, 3750 ± 60	Usushwana Complex, Nsuze lava, Lava in Onverwacht Group, Granite boulder in Moodies conglomerate	PONGOLA GROUP, SWAZILAND SUPERGROUP

FIG. 3.—Chronostratigraphic subdivision of the Precambrian in South Africa.

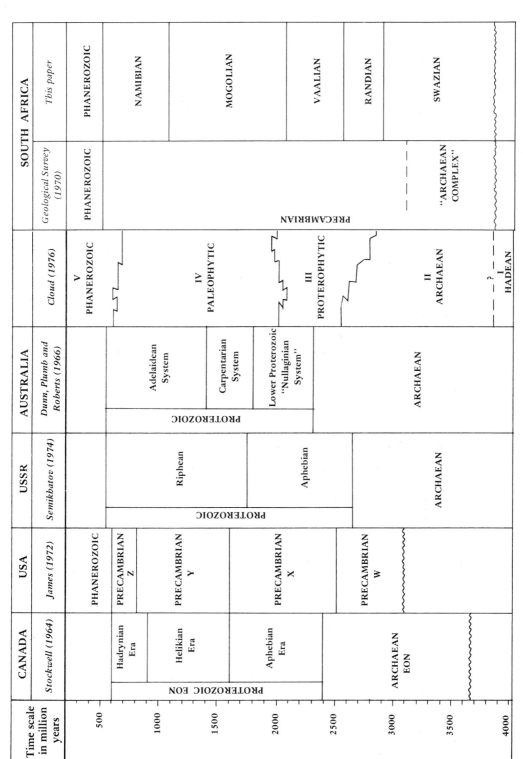

FIG. 4—Selected current Precambrian time and time-rock classifications.

the Proterozoic rocks of northern, central, and eastern Australia on the 1:2,500,000 geologic map of Australia, 1976 (I. R. McLeod, personal communication). However, the Nullaginian has not received wide acceptance in Australia. Below 2,600 m.y., the Australian record is blurred by metamorphism (Dunn et al, 1966).

The Nullaginian-Carpentarian (1,800 m.y.) and the Carpentarian-Adelaidean (1,400 m.y.) boundaries are not markedly in evidence in the South African successions, neither are those of the Vendian-Karatan (670 m.y.), Karatan-Urmatin (950 m.y.), and the Urmatin-Burzyan (1,400 m.y.) of the Riphean of Russia (Harland, 1974).

The major time boundaries proposed by Cloud (1972, 1976), i.e. Hadean-Archean (3,800 m.y.), Archaean-Proterophytic (diachronous 2,450 to 2,750 m.y.), Proterophytic-Paleophytic (2,000 m.y.), and Paleophytic-Phanerozoic (700 m.y.) which are, respectively: the postulated date for the appearance of a solid crust, the time span of major cratonization, the beginning of an oxygenous atmosphere, and the first appearance of Metazoa are represented on the proposed subdivision of the South African Precambrian.

Finally, we agree with Hedberg (1974) in not favoring the numerical schemes, such as the I-VI that has been used in Russia and the W-Z of the United States Geological Survey, for subdividing the Precambrian on the basis of time and not of rock units.

On the 1955 edition of the 1:1,000,000 geological map of the country, the Geological Survey of South Africa divided the Precambrian into Archaeozoic and Proterozoic, the boundary being the base of the "Dominion Reef System" (i.e. 2,800 m.y.). On the 1970 edition, this was abandoned; the pre-"Pongola System" rocks were, however, grouped as Archaean Complex with an upper time boundary of 3,100 m.y. Due to the upper boundary of the Archaean being in such dispute, the term is not being used by SACS.

References Cited

Anderson, J. G. C., 1965, The Precambrian of the British Isles, *in* The Precambrian, Vol. 2; New York, Wiley-Interscience, p. 25-111.

Ashley, G. H., et al, 1933, Classification and nomenclature of rock units: Geol. Soc. America Bull., v. 44, p. 423-459.

Barghoorn, E. S., and J. W. Schopf, 1966, Micro-organisms three billion years old from the Precambrian of South Africa: Science, v. 152, no. 3723, p. 758-763.

Burger, A. J., and F. J. Coertze, 1973, Radiometric age measurements on rocks from southern Africa to the end of 1971: South Africa Dept. Mines Geol. Survey Bull., v. 58, 45 p.

Cahen, L., and J. Lepersonne, 1967, The Precambrian of the Congo, Rwanda and Burundi, *in* The Precambrian, Vol. 3; New York, Wiley-Interscience, p. 143-290.

Cloud, P., 1971, Precambrian of North America: Geotimes, v. 16, no. 3, p. 13-19.

―― 1972, A working model of the primitive earth: Am. Jour. Sci., v. 272, p. 537-548.

―― 1976, Major features of crustal evolution: 13th A. L. du Toit Memorial Lecture, Geol. Soc. South Africa.

―― and M. A. Semikhatov, 1969, Proterozoic stromatolite zonation: Am. Jour. Sci., v. 267, p. 1017-1061.

Davies, R. D., et al, 1969, Sr-Isotopic studies on various layered intrusions in Southern Africa, *in* Symposium on the Bushveld Igneous Complex and other layered intrusions: South Africa Geol. Soc. Spec. Pub. 2, p. 576-593.

Dunn, P. R., K. A. Plumb, and H. G. Roberts, 1966, A proposal for time-stratigraphic subdivision of the Australian Precambrian: Geol. Soc. Australia Jour., v. 13, no. 2, p. 593-608.

Du Toit, A. L., 1954, The geology of South Africa: Edinburgh, Oliver and Boyd, 3rd ed., 611 p.

Gebelin, G. D., 1974, Biologic control of stromatolite microstructure; implication for Precambrian time stratigraphy: Am. Jour. Sci., v. 274, p. 575-598.

Geological Survey of South Africa, 1947, Information brochure for professional and technical officers: South Africa Geol. Survey, 69 p.

―― 1970, Geological map of the Republic of South Africa and the Kingdoms of Lesotho and Swaziland.

Hall, A. L., 1932, The Bushveld Igneous Complex of the Central Transvaal: South Africa Dept. Mines Geol. Survey Mem., v. 28, 560 p.

Harland, W. B., 1974, The Precambrian-Cambrian boundary, *in* Lower Palaeozoic rocks of the world, Vol. 2: London, John Wiley.

Hedberg, H. D., 1974, Basis for chronostratigraphic classification of the Precambrian: Precambr. Res., v. 1, p. 165-177.

—— ed., 1971, Preliminary report on chronostratigraphic units: 24th Int. Geol. Cong. Proc., Rep. 6, 39 p.

—— ed., 1976, International stratigraphic guide—a guide to stratigraphic classification, terminology, and procedure: New York, John Wiley and Sons, 200 p.

James, H. L., 1972, Subdivision of Precambrian: an interim scheme to be used by U.S. Geological Survey: AAPG Bull., v. 56, p. 1026-1030.

Jansen, H., 1975, Precambrian basins on the Transvaal Craton and their sedimentological and structural features: South Africa Geol. Soc. Trans. and Proc., v. 78, p. 25-33.

Lock, B. E., 1973, The Cape Supergroup in Natal and the Northern Transkei: Geol. Mag., v. 110, p. 485-486.

Mellor, E. T., 1917, The geology of the Witwatersrand. An explanation of the geological map of the Witwatersrand Goldfield: South Africa Geol. Survey, 46 p.

Molengraaff, G. A. F., 1901, Géologie de la République Sud-Africaine du Transvaal: Soc. Geol. France Bull., 4 ser., p. 13-92.

Penning, W. H., 1888, The South African goldfields: Jour. Soc. Arts. (London), v. 36, p. 433-443.

Pichamuthu, C. S., 1967, The Precambrian of India, *in* The Precambrian,Vol. 3: New York, Wiley-Interscience, p. 1-96.

Rankama, K., ed, 1963, The geologic systems, The Precambrian, Vol. 7: New York, Wiley-Interscience.

Semikhatov, M. A., 1972, On the general Precambrian stratigraphic scale: 24th Int. Geol. Cong. Proc., Sec. 1, p. 273-277.

—— 1974, Proterozoic stratigraphy and geochronology (in Russian): Akad. Nauk SSSR Izv. Ser. Geol., v. 256, 302 p.

South African Committee for Stratigraphy, 1971, South African code of stratigraphic terminology and nomenclature: South Africa Geol. Soc. Trans. and Proc., v. 74, p. 111-131.

Stockwell, C. H., 1964, Principles of time stratigraphic classification in the Precambrian: Royal Soc. Canada Spec. Pub. 8, p. 53-60.

Truswell, J. F., 1967, A critical review of stratigraphic terminology as applied in South Africa: South Africa Geol. Soc. Trans. and Proc., v. 70, p. 81-116.

Truter, F. C., 1949, A review of volcanism in the geological history of South Africa: South Africa Geol. Soc. Trans. and Proc., v. 52, p. xxix-xxxix.

Van Eeden, O. R., 1972, The geology of the Republic of South Africa—an explanation of the 1:1,000,000 map (1970 ed.): South Africa Dept. Mines Geol. Survey Spec. Pub. 18, 85 p.

Van Niekerk, C. B., and A. J. Burger, in press, The Moodies conglomerate boulders, *in* Mineralization in metamorphic terranes: South Africa Geol. Soc. Spec. Pub.

Walter, M. R., and W. V. Preiss, 1972, Distribution of stromatolites in the Precambrian and Cambrian of Australia: 24th Int. Geol. Cong. Proc., Sec. 1, p. 85-93.

Winter, H. d. l. R., 1976, A lithostratigraphic classification of the Ventersdorp Succession: South Africa Geol. Soc. Trans. and Proc., v. 79, p. 31-48.

Explanation of Indexing

References are indexed according to the important, or "key" words. Authors and titles are also represented here; where more than one author has contributed to a paper, each person is cited, alphabetically, according to his last name.

In the column to the left of the keyword entry, the first letter represents the AAPG book series from which the reference originated (In this case, S stands for Studies series. Every five years, AAPG merges all its indexes together, and the letter S will differentiate this reference from those of Memoir Series, or from the AAPG Bulletin). The following number is the Series number (in this case, Studies Number 6), and the last entry is the page number in this volume.

A small dagger symbol (†) is used to highlight a manuscript title.

Note: This index is set up for a single-line entry. Where entries exceed one line of type (this is especially evident in title indexing), the line is terminated. The reader sometimes must be able to realize the keywords, although out of context.